Structured C for Engineering and Technology

Third Edition

TOM ADAMSON

JAMES L. ANTONAKOS
Broome Community College

KENNETH C. MANSFIELD JR.
Broome Community College

Prentice Hall

Upper Saddle River, New Jersey   Columbus, Ohio

Library of Congress Cataloging-in-Publication Data

Adamson, Thomas A.
Structured C for engineering and technology / Tom Adamson,
James L. Antonakos, Kenneth C. Mansfield, Jr. — 3rd ed.
p. cm.
Includes index.
ISBN 0-13-625229-X
1. C (Computer program language) 2. Structured programming.
I. Antonakos, James L. II. Mansfield, Kenneth C. III. Title.
QA76.73.C15A33 1998
005.265—dc21 97-20691
CIP

Editor: Charles E. Stewart, Jr.
Production Editor: Mary Harlan
Design Coordinator: Karrie M. Converse
Text Designer and Production Coordinator: Custom Editorial Productions, Inc.
Cover Designer: Russ Maselli
Cover Photo: Alfred Pasieka/Science Photo Library
Production Manager: Pamela D. Bennett
Illustrations: Custom Editorial Productions, Inc.
Marketing Manager: Debbie Yarnell

This book was set in Times Roman by Custom Editorial Productions, Inc., and was printed and bound by R. R. Donnelley & Sons Company. The cover was printed by Phoenix Cover Corp.

Printed in the United States of America

10 9 8

ISBN: 0-13-625229-X

Prentice-Hall International (UK) Limited, *London*
Prentice-Hall of Australia Pty. Limited, *Sydney*
Prentice-Hall Canada Inc., *Toronto*
Prentice-Hall Hispanoamericana, S. A., *Mexico*
Prentice-Hall of India Private Limited, *New Delhi*
Prentice-Hall of Japan, Inc., *Tokyo*
Simon & Schuster Asia Pte. Ltd., *Singapore*
Editora Prentice-Hall do Brasil, Ltda., *Rio de Janeiro*

To my wife Susan, my son Kenny,
and my twin daughters Kristen and Kimberly,
I'll make up for those missed nights and weekends.

Kenneth C. Mansfield Jr.

To my four cats Anselle, Tasha, Louis, and Connor,
who kept my wife Michele company
while I was writing.

James L. Antonakos

Preface

Introduction

The C programming language is finding its way into more and more curriculums and schools every day. The need for a textbook that teaches the fundamental programming concepts required by today's students is great. All technology students who use C discover that it is powerful and yet simple to learn. The ability to control a computer at the hardware level with a high-level language such as C is important and necessary—especially in today's society, where more and more tasks are being delegated to the computer.

The purpose of this textbook is to describe the C programming language, by example, to a nonprogrammer. The basic things every programmer does (loops, calculations, formatting, I/O) are covered in detail. Over 400 *tested and working programs* are used to illustrate the necessary C concepts. This book is suitable for students in any technological field, particularly students studying electrical and computer technology.

Chapter Topics

Chapter 1 discusses the fundamental concepts of the C programming language. Structured programming, variables, operators, functions, and I/O are all described.

Chapter 2 emphasizes the correct way to write a structured program in C. The operating details of functions are expanded.

Chapter 3 covers the use of various control statements, such as `if`, `if...else`, and `switch`. Logical operations are also covered.

Chapter 4 shows how loops are written and used in C. Examples include the use of `for`, `while`, and `do while` loops, as well as nested loops.

Chapter 5 discusses the relationships among variables, pointers to variables, and the memory space required for variables of certain types. The local scope and global scope of variables are also covered.

Chapter 6 illustrates how character strings are handled in C. String initialization, passing a string as a function parameter, and built-in string handling functions are explained.

Chapter 7 contains a collection of typical array applications, such as sorting (many different methods), and recursive problem solving.

Chapter 8 covers the basic techniques of structuring data using `enum` and `typedef`. Advanced data structures, such as linked-lists and binary trees, are also covered.

Chapter 9 explains the details involved in file operations. Numerous examples are provided to show how a file is accessed and utilized by a C program.

Chapter 10 shows how simple graphing operations can be performed. Examples include drawing of color lines, rectangles, and circles. Coordinate transformations and scaling are also discussed. A 3D virtual reality program is included.

Chapter 11 provides the essential details needed to understand the connection between the statements in a C program and the machine language instructions necessary to perform the work of the C statements. The interface between C and 80x86 assembly language is covered.

Chapter 12 introduces the object oriented language C++. Many examples are provided to demonstrate the many new features of C++.

Changes and Additions to the Third Edition

The following major changes and additions have been made to create the third edition of this textbook:

- **The old Chapter 6 has been split into Chapters 6 and 7.** Strings and arrays are now covered in separate chapters. A new Case Study has been developed for Chapter 7 (Arrays), solving the Eight-Queens puzzle via recursion.
- **The old Chapter 7 has been split into Chapters 8 and 9.** Data structures and disk I/O are now covered in separate chapters. A new Case Study has been developed for Chapter 8 (Data Structures) called MiniMicro, where the operation of a simple microprocessor is emulated.
- **C++ is introduced in a brand new final chapter.** The elements of C++ are introduced, including classes and objects, inheritance, constructors and destructors, operator overloading, and much more.
- **A Troubleshooting Techniques section has been added to each chapter.** The authors' experience with writing and troubleshooting software has been added to each chapter. Many tips and techniques are provided to help the student develop good troubleshooting habits.
- **Executables are now included on the companion disk.** For students who do not have a C compiler of their own, the executables for each Case Study (plus the virtual reality game) are provided to allow them to run the programs from the textbook.

In addition, many new figures, questions, example programs (such as the Virtual Maze program from Chapter 10), and programming problems have been added. Overall, *Structured C for Engineering and Technology,* Third Edition, is better organized and has much more material than before.

Acknowledgments

We would like to thank all of the students and instructors who used the second edition, and who contributed many helpful comments.

We would also like to thank our editor, Charles Stewart, and his assistant, Kate Linsner, for their guidance and patience. Two other individuals were especially helpful during the production phase. Laura Bofinger, our production supervisor at Custom Editorial Productions, Inc., and Cindy Lanning, our copyeditor, did outstanding jobs and deserve our thanks as well.

James L. Antonakos
antonakos_j@sunybroome.edu

Kenneth C. Mansfield Jr.
mansfield_k@sunybroome.edu

Contents

1 C Fundamentals

Objectives

This chapter gives you the opportunity to learn:

1. Why C is used as a programming language and what you need to use C in a computer.
2. Some beginning C commands and how to use them to develop a **structured program.**
3. How to recognize block structure and use it in program development.
4. What a programmer's block is and why it is useful.
5. The basic elements needed to write a program in C.
6. The purpose of include files.
7. How the `printf()` function operates.
8. What identifiers are and how to use them.
9. The keywords used in C.
10. The importance of declaring variables and how to do this.
11. The methods C uses to perform arithmetic operations.
12. Ways of formatting your output using the `printf()` function.
13. The development of the `scanf()` function for getting user input.
14. Some common reasons for bugs in a C program.
15. The steps used to develop a technology program using the C language.

Key Terms

C Environment
Editor
Compiler
Include Files
Library Files
Linker

Source Code
Program Structure
Programmer's Block
Remarks
Top-down Design
Computer Memory
Token
Keyword
Character
String
Integer
Float
Data Type
Statement
Expression
Single Statement
Compound Statement
Standard Output
Format Specifier
Argument
Field
Function
Declaration
Identifier
Type Specifier
Assignment Operator
Arithmetic Operator
Precedence
Compound Assignment Operator
Escape Sequence
Field Width Specifier
E Notation

Outline

Introduction

This chapter introduces you to some of the fundamentals of the C language. You will learn how to display values on the monitor screen and how to get values from the program user. This chapter also shows you how to do basic arithmetic operations with C.

When you complete this chapter, you will be able to write some of your first technology programs in C. The chapter concludes with the design of an actual technology program, and some example C programs to illustrate technology applications of C.

1.1 The C Environment

Overview

This section presents what is called the **C environment**. An environment in programming means all the different programming tools you will need in order to work with a particular programming language.

When you program in the programming language called BASIC, the environment is usually built into the microcomputer (in its ROM). In this case, all that is usually necessary is to turn the computer on, start typing in BASIC commands, and type in a command to instruct the computer to execute the program. This is not the case with other programming languages, especially C.

The C Environment

The C environment contains an **editor, compiler, include files, library files, linker,** and much more. Each of these does the following:

Editor Allows you to enter and modify your C source code.

Compiler A program that converts the C program you have developed into a code understood by the computer.

Include Files Files that consist of many separate definitions and instructions that may be useful to a programmer in certain instances.

Library Files These are previously compiled programs that perform specific functions. These programs can be used by you to help you develop your C programs. For example, the C function that allows you to display text on your monitor screen (the `printf()` function) is not in the C language. Instead, its code is in a library file. The same is true of many other functions, such as graphics, sound, and working with the disk and printer, just to name a few. You can also create your own library files of routines developed by you that you use over and over again in different C programs. By doing this, you can save hours of programming time and prevent programming errors.

Linker Essentially, the linker combines all of the necessary parts (such as library files) of your C program to produce the final executable code. Linkers play an important and necessary role in all of your C programs. In larger C programs, it is general practice to break the program down into smaller parts, each of which is developed and tested separately. The linker will then combine all of these parts together to form your final executable program code.

You will also need some kind of a disk operating system to assist you in saving your programs. The main parts of the C environment are shown in Figure 1.1.

Conclusion

In this section, you were introduced to the C environment. Here, you got an idea of what it takes to enter a C program into your computer and what else is needed in order to get the program to do what you want it to do (executing the program).

In the next section, you will learn some of the reasons why C is such an important programming language and how you can use it in the field of technology.

1.1 Section Review

1. State the purpose of an editor. Why is it needed?
2. Give the reason for using a compiler.
3. Explain the action of a compiler.

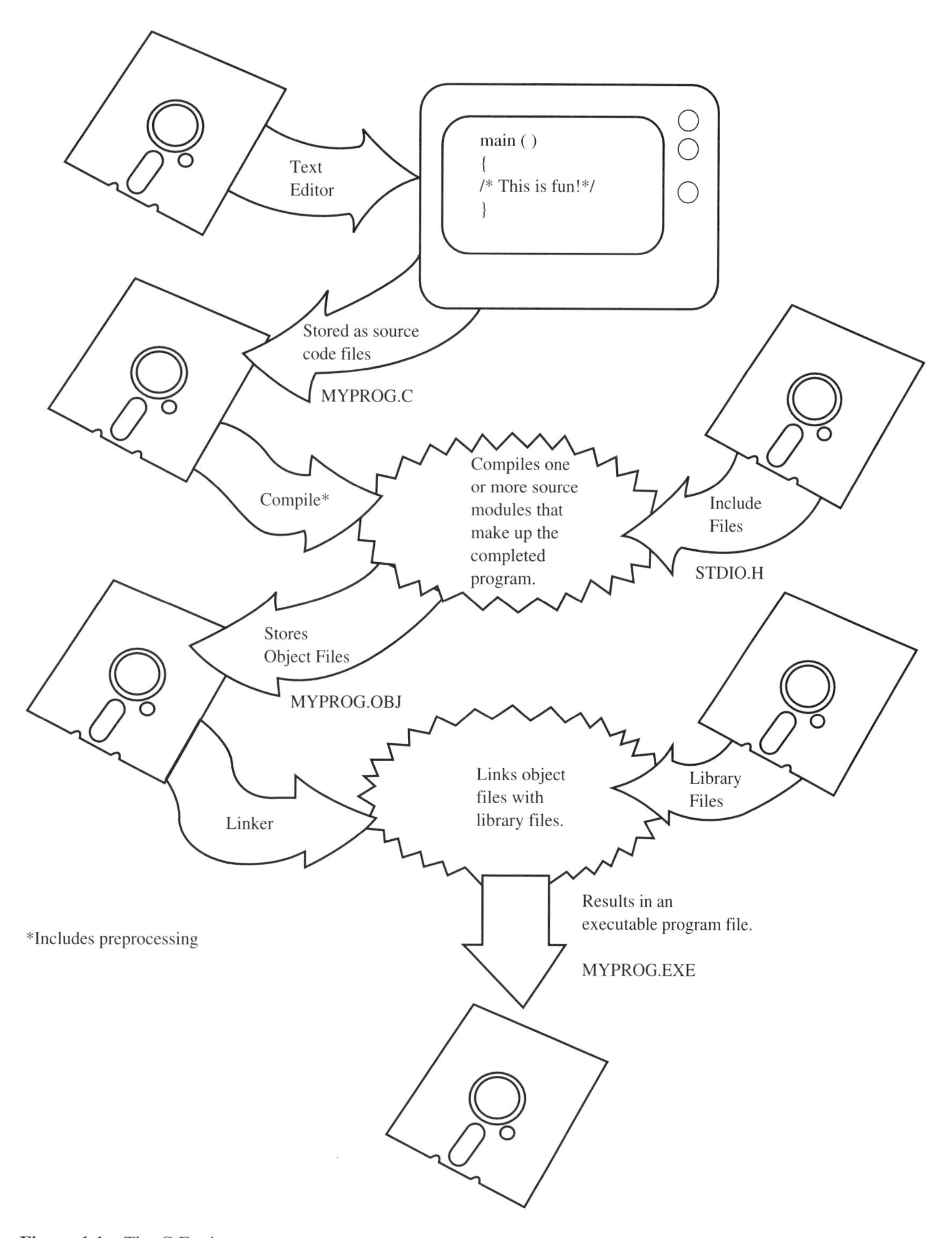

Figure 1.1 The C Environment

1.2 Why C?

Background

C is an increasingly popular programming language in industry and schools and for personal use. The many reasons for this are outlined in Table 1.1.

Disadvantages of C

Since C is such a versatile language, its code can be written with such brevity that it becomes almost unreadable. This style of programming is not encouraged and is not presented in this book. Because of its flexibility, C will allow you to write programs that could wind up containing very difficult-to-find "bugs." Programming in C is like having a sports car with a 500 mph top speed and very few highway regulations—you have to be very careful how you handle such unrestricted power.

What C Looks Like

Here is a C program:

Program 1.1

```
#include <stdio.h>

/* This is a C program. It will print a message on the */
/* computer screen. */

main()
{
     printf("Send me a 10 ohm resistor.");
}
```

Table 1.1 The Power of C

Advantage	What This Means to You
Designed for top-down programming	Your programs will be easier to design.
Designed for structure	Your programs will be easier to read and understand.
Allows modular design	Enhances the program's appearance so others can easily follow and modify. Makes it easier for you to debug.
An efficient language	More compact, quicker-running programs.
Portability	A program you write on one computer will operate on another computer system with few if any changes.
Computer control	You exercise almost absolute control over your computer.
Flexibility	You can easily create other languages and operating systems.

Table 1.2 Major Parts of a C Program

Item	Purpose
`#include <stdio.h>`	Tells the compiler to *include* the standard input/output file.
`main`	This marks the point where the C program begins execution. Required for all programs.
`()`	Must appear immediately after `main`. Usually information that will be used by the program is contained within these parentheses.
`/*    */`	These symbols are optional and are used to enclose comments. Comments are remarks used by you to help clarify the program for people. They are ignored by the compiler.
`;`	Each C statement ends with a semicolon. For now, think of a C statement as consisting of a C command.
`{}`	The braces are required in all C programs. They indicate the beginning and the end of the program instructions.

Note several things about Program 1.1. It is easy to read. No one doubts what the program will do. The program has a structure that makes it easy to read. There are some "strange" symbols (such as the /*, which will be explained shortly). What you see in this program is the **source code**. This is what you, the programmer, would type in.

In Program 1.1, many of the items required by a C program are present. These items are the keywords `include`, `main`, and `printf`.

When the program is executed, it displays

```
Send me a 10 ohm resistor.
```

on the monitor screen.

The Different Parts

As outlined in Table 1.2, all C programs must have certain parts. Don't worry about the other items (such as the `printf`) used in Program 1.1. They were used to assist C in displaying the sentence on the monitor. You will learn the meanings of these and where to use them shortly.

Conclusion

Congratulations! You have just seen your first C program. It just displayed a message on the screen, something any computer can easily do without using C. It was shown here just to keep things as simple as possible so you could concentrate on the main items of a C program.

In the next section, you will learn some important points about structure and why you should use it.

1.2 Section Review

1. State three advantages of the C programming language.
2. Explain what is meant by portability in a programming language.

3. What C command must all C programs start with?
4. Give an example of a comment in a C program.
5. What is the purpose of the braces { } in a C program?

1.3 Program Structure

Discussion

In this text, all C programs will conform to a specific structure. You can think of a **program structure** as the format you use when entering the program. Program 1.1 had a structure:

```
#include <stdio.h>
/* This is a C program. It will print a message on the */
/* computer screen. */
main()
{
     printf("Send me a 10 ohm resistor.");
}
```

When compiled and executed, this program would cause the monitor to display (on the first line starting at the left of the screen)

```
Send me a 10 ohm resistor.
```

Program 1.1 could just as well have been written with a different structure:

Program 1.2

```
#include <stdio.h>

/* This is a C program. It will print a message on the */
/* computer screen. */
main () {printf("Send me a 10 ohm resistor."); }
```

When compiled and executed, Program 1.2 would still do exactly the same thing. The difference is that the structure has been changed. The program is a little harder to read. You could have written the same program as

Program 1.3

```
#include <stdio.h>

main() {printf("Send me a 10 ohm resistor."); }
```

Again, when compiled and executed, Program 1.3 would have done the same thing. The point is that program structure is considered good structure when the structure is easy for people to read and understand.

For the simple program introduced here, program structure may seem to make very little difference. However, in larger programs, program structure becomes very important.

Structured vs. Unstructured Programming

One of the measures of a "good" computer program is that it can be read and understood by anyone—even if that person does not know how to program.

For example, the first C program presented in this book was easy to understand. Admittedly it didn't do much, but the point is that if good structure is observed, the program will be easier to understand, modify, and debug when necessary. An unstructured program is one in which little or no effort has been made to help people read and understand it. Remember, the structure of a program makes no difference to the computer, only to the people who have to work with it.

Program Blocks

You can think of a structured program as having certain parts of the program located at particular positions in the program document. These positions can be thought of as *blocks* of information. A letter contains blocks of information:

John Student
123 Page Mill Road
Programville, USA
01234

The Resistor Company
321 Mill Page Road
Sourcecode, USA
43210

Dear Sir:

Send me a 10 ohm resistor.

Sincerely,

John Student

P.S. Thanks for the prompt delivery of my last order.

The letter above could just as well have been written as follows:

John Student 123 Page Mill Road Programville, USA 01234
The Resistor Company 321 Mill Page Road Sourcecode, USA 43210
Dear Sir: Send me a 10 ohm resistor. Sincerely, John Student
P.S. Thanks for the prompt delivery of my last order.

The difference between the two letters is that one is structured and the other isn't. The unstructured letter would take less time to write, less time to print, and use less paper. Over a period of time, if all your letters were done this way, you might even save some postage. But they would be difficult to read, not because of the writing style but because of their structure.

The same letter is shown again below, but this time its block structure is emphasized:

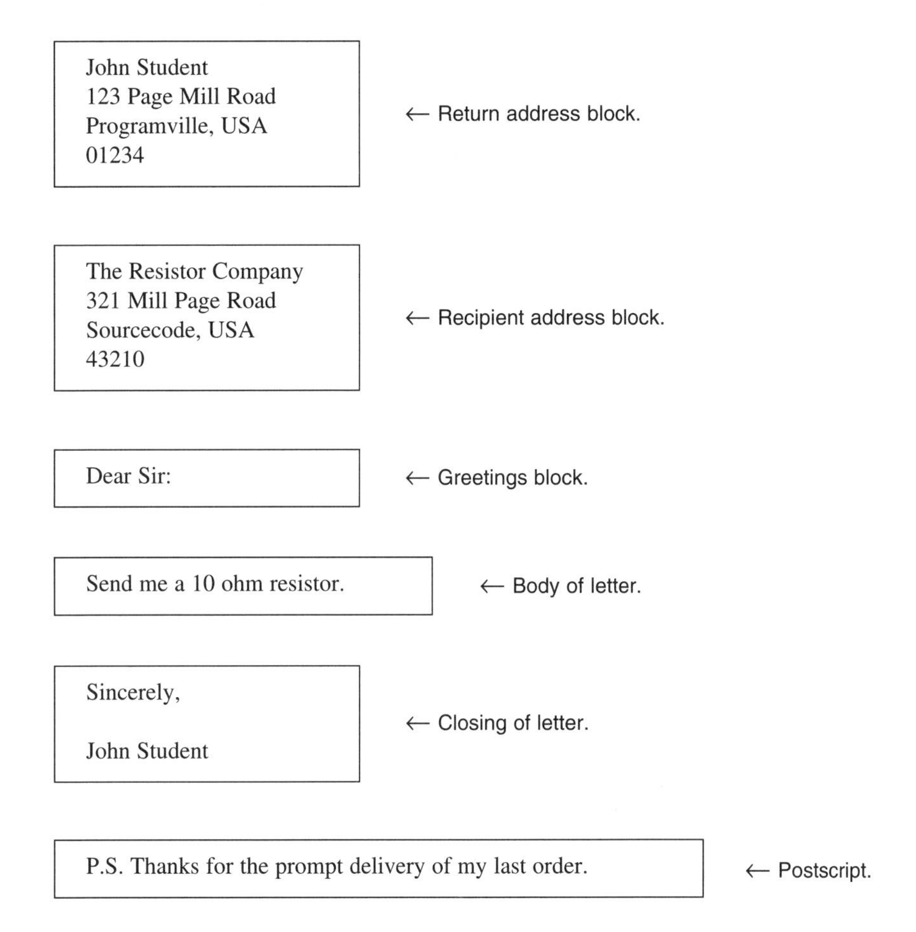

The block structure of the letter makes the letter easy to read. In a similar fashion, when block structure is used in programs, they too are easier to read. Just as there is an agreed-upon block structure for letters, in this text, there will be an agreed-upon block structure for writing your C programs.

The Programmer's Block

All well-documented C programs will start with a block called the **programmer's block**. It consists of **remarks** that contain the following information:

1. Program name
2. Developer
3. Description of the program
4. Explanation of all variables
5. Explanation of all constants

You already know enough about C to do this now. Suppose you need to develop a C program that would solve for the voltage drop across a 10 ohm resistor with a specified amount of current. The formula for this relationship is

$$V = I \times R$$

Where

V = The voltage across the resistor measured in volts.
I = The current in the resistor measured in amps.
R = The value of the resistor measured in ohms.

The C program would start with the programmer's block. This is illustrated in Program 1.4.

Program 1.4

```
#include <stdio.h>

/*****************************************************/
/*                  Voltage Solver                   */
/*****************************************************/
/*          Developed by : A. Good Structure         */
/*                                                   */
/*****************************************************/
/*      This program will solve for the voltage drop  */
/* across a 10 ohm resistor. The program user must   */
/* enter the value of current.                       */
/*****************************************************/
/*                   Variables used:                 */
/*---------------------------------------------------*/
/*            V= Voltage across resistor.            */
/*            I= Current in resistor.                */
/*****************************************************/
/*                  Constants used:                  */
/*---------------------------------------------------*/
/*            R= 10 (A 10 ohm resistor)              */
/*****************************************************/

main()

{
    /* Body of the program to do the above. */
}
```

Program 1.4 will compile and execute, but nothing will happen because no commands were put into the program. The program consists only of remarks and the essential elements of a C program. But the programmer's block is complete. It tells you exactly what the program will do, and, just as important, it defines all of the variables (V and I) and the value of the constant R that will be used in the program. When doing **top-down design** (presented later in this chapter), stating the programming problem in words is the essential first step in programming design. Selecting the variables and defining them is another important step. For now, just know what must be included in a programmer's block and know the mechanics of how to write such a block using C.

Advantages and Disadvantages of Structured Programming

The main disadvantage of structured programming affects primarily those who have been programming in an unstructured way. Old habits die hard. If you've never programmed before, then structured programming will not have any disadvantages for you. The only other disadvantage of structured programming is that it makes short, unstructured programs longer.

In the past, **computer memory** was relatively expensive and limited. Thus, at one time it did pay to conserve computer memory space by making programs as brief as possible. Today, this is no longer the case. Even pocket computers have more memory than many of the older large machines. Thus, there is no longer any necessity to be brief in programming. There is, however, the necessity to be clear in programming and to develop good programming habits that result in completed programs that are easy to understand, easy to modify, and easy to correct. That's the purpose of this book.

What you learn in this text can be applied to any structured language such as Pascal and even BASIC. When you complete this text you will be proficient in creating programs using structured programming techniques.

Figure 1.2 illustrates the structure of a complete C program using ANSI (American National Standards Institute) prototyping standards. The programming details will be covered in the following chapters. For now, look through the program to get the general idea.

Conclusion

This section presented a general idea of the difference between structured and unstructured programming. You will learn much more about these areas, including the power of C as a programming language. Test your new skills in the following section review.

1.3 Section Review

1. Define the term "structure" as used in programming.
2. Is it necessary to give a structure to C in order for it to compile without errors?
3. Describe a block structure and give an example.
4. Explain the reasons for structured programming.
5. Describe a programmer's block and state what it must contain.

1.4 Elements of C

Discussion

This section lays the ground rules for the fundamental elements of all C programs. In this section you will see many new definitions. These definitions will set the stage for the remainder of this chapter as well as the remainder of this text.

What C Needs

To write a program in C, you use a set of characters. This set includes the uppercase and lowercase letters of the English alphabet, the ten decimal digits of the Arabic number system, and the underscore (_) character. Whitespace characters (such as the spaces between words) are used to separate the items in a C program, much the same

```
#include <stdio.h>
#include <math.h>
#define PI 3.14159
#define SQUARE(x) ((x) * (x))
/***************************************************************/
/*                        Circle Area                          */
/***************************************************************/
/*              Developed by: A. G. Programmer                 */
/*                                                             */
/***************************************************************/
/*        This program will solve for the area of a            */
/*   circle.  The programmer need only enter the radius.       */
/*   Value returned is in square units.                        */
/***************************************************************/
/*                     Function Prototypes                     */
/*-------------------------------------------------------------*/
void explain_program(void);
/* This function explains the purpose of the program to user.  */
/*-------------------------------------------------------------*/
float get_value(void);
/* This function prompts the user for the value of the circle  */
/* radius.  It returns the value entered by the program user.  */
/*-------------------------------------------------------------*/
float circle_area(float radius);
/*            radius = the radius of the circle                */
/* This function calculates the area of a circle.              */
void display_answer(float area);
/*         area = area of the circle in square units           */
/* This function displays the area of the circle.              */
/***************************************************************/
main()
{
     float radius;        /* Radius of the circle. */
     float area;          /* Area of the circle.   */

     explain_program();             /* Explains program to user. */
     radius = get_value();          /* Get radius from user.     */
     area=circle_area(radius);      /* Compute the circle area.  */
     display_answer(area);          /* Display the answer.       */
}
void explain_program()              /* Explains the program.     */
{
     printf("This program calculates the area of a circle.\n");
     printf("Just enter the value of the radius and press -RETURN-\n");
     printf("\n");                  /* Put in a blank line.      */
}
float get_value()                   /* Gets radius from user.    */
{
     float input_value;             /* Value entered by the user.*/
     printf("Value of the radius ==> ");
     scanf("%f",&input_value);
     return(input_value);
}
float circle_area(float radius)     /* Compute the circle area.  */
{
     float area;                    /* Area of the circle.       */
     area = PI * SQUARE(radius);
     return(area);
}
void display_answer(float area)     /* Display the answer.       */
{
     printf("\n\n");                /* Print two blank lines.    */
     printf("The area of the circle is %f units.",area);
}
/***************************************************************/
```

Annotations: Preprocessor directives; Programmer's block; Purpose of each function; Explain all variables; Main function; Other functions

Figure 1.2 Structure of a C Program

as they are used to separate words in this book. These whitespace characters also include the tab and carriage return, as well as other control characters that produce white spaces.

Tokens. In every C source program, the most basic element recognized by the compiler is a single character or group of characters known as a **token**. Essentially, a token is source program text that the compiler will not break down any further—it is treated as a fundamental unit. As an example, in C `main` is a token; so is the required opening brace (`{`) as well as the plus sign (+).

ANSI C Keywords. **Keywords** are predefined tokens that have special meanings to the C compiler. Their definitions cannot be changed; thus they cannot be used for anything else except the intended action they have on the program in which they are used. These keywords are as follows:

```
auto      double    int       struct    break     else
long      switch    case      enum      register  typedef
char      extern    return    union     const     float
short     unsigned  continue  for       signed    void
default   goto      sizeof    volatile  do        if
static    while
```

Types of Data. The C language allows three major types of data: numbers, characters, and strings. A **character** is any item from the set of characters used by C. A **string** is a combination of these characters.

Numbers Used by C. C uses a wide range of numbers. Numbers used by C fall into two general categories: **integer** (whole numbers) and **float** (numbers with decimal points). These two main categories can be further divided as shown in Table 1.3.

Table 1.3 Subdivisions of C Data Types

Type Identifier	Meaning	Range of Values (IBM PC)
`char`	character	–128 to 127
`int`	integer	–32,768 to 32,767
`short`	short integer	–32,768 to 32,767
`long`	long integer	–2,147,483,648 to 2,147,483,647
`unsigned char`	unsigned character	0 to 255
`unsigned`	unsigned integer	0 to 65,535
`unsigned short`	unsigned short int	0 to 65,535
`unsigned long`	unsigned long int	0 to 4,294,967,295
`enum`	enumerated	0 to 65,535
`float`	floating point	3.4E +/– 38 (7 digits)
`double`	double floating point	1.7E +/– 308 (15 digits)
`long double`	long double floating point	1.7E +/– 4932 (15 digits)

As you can see in Table 1.3, C offers a rich variety of **data types**. Generally speaking, the larger the value range of the data type, the more computer memory it takes to store it. As a general rule, you want to use the data type that conserves memory and still accomplishes the desired purpose. As an example, if you needed a data type for counting objects—such as the number of resistors for a parts order—the type `int` would probably do the job. However, if you were writing a program for as much precision as possible you might consider the type `double`, which can give you 15-digit accuracy.

As you may note in Table 1.3, some of the data types produce the same range of values (such as `int` and `short`). On other computer systems, `short` and `int` may actually have different ranges.

C Statements

A C **statement** controls the flow of the execution of a program. It consists of keywords, **expressions**, and other statements. In C, an expression is a combination of operands and operators that expresses a single value (such as `answer = 3 + 5;`).

There are two types of statements in C. These are **single statements** and **compound statements**. A *compound statement* is delimited by braces (`{}`) while a single statement ends with a semicolon (`;`). As you progress through this book, reference will be made to C statements.

Conclusion

This section presented the elements of a C program. The rest of this chapter will show you how to use them. Check your understanding of this section by trying the following section review.

1.4 Section Review

1. State what is used to write a C program.
2. Explain what is meant by a token in C.
3. What are the major data types used by C?
4. Explain what is meant by a keyword. Give an example.
5. What data type handles the largest number?

1.5 The printf() Function

This section introduces you to the powerful `printf()` function used in C. This is actually a separate C function (as `main()` is a function) and is contained in the standard library that comes with your C system.

What printf() Does

The `printf()` function is used to write information to **standard output** (normally your monitor screen). You saw the `printf()` function used earlier in this chapter. The structure of this function is

```
printf(character strings with format specifiers, variables or values);
```

Characters are set off by single quotes (such as `'a'`), and strings are set off by double quotes (such as `"This is a string."`). A **format specifier** instructs the `printf()` function how to convert, format, and print its **arguments**. For now, think of an argument as the actual values that are within the parentheses of the function. A format specifier begins with a percent (`%`) character. As an example,

```
printf("This is a C statement.");
```

when executed produces

```
This is a C statement.
```

With a format specifier and an argument,

```
printf("The number 92 in decimal is %d.",92);
```

when executed produces

```
The number 92 in decimal is 92.
```

Another way of producing the same output is

```
printf("The number %d in decimal is %d.",92,92);
```

Table 1.4 lists the various type fields used by the `printf()` function in format specifiers.

Program 1.5 illustrates the use of the different **field** type format specifiers.

Program 1.5

```
#include <stdio.h>

main()
{
    printf("The value 92 using field type d is %d. \n", 92);
    printf("The value 92 using field type i is %i. \n", 92);
    printf("The value 92 using field type u is %u. \n", 92);
    printf("The value 92 using field type o is %o. \n", 92);
    printf("The value 92 using field type x is %x. \n", 92);
    printf("The value 92 using field type X is %X. \n", 92);
    printf("The value 92.0 using field type f is %f. \n", 92.0);
    printf("The value 92.0 using field type e is %e. \n", 92.0);
    printf("The value 92.0 using field type E is %E. \n", 92.0);
    printf("The value 92.0 using field type g is %g. \n", 92.0);
    printf("The value 92.0 using field type G is %G. \n", 92.0);
    printf("The value 92 using field type c is %c. \n", 92);
    printf("The character '9' using field type c is %c. \n", '9');
    printf("The string 92 using field type s is %s. \n"," 92");
}
```

Execution of Program 1.5 results in the following output:

```
The value 92 using field type d is 92.
The value 92 using field type i is 92.
The value 92 using field type u is 92.
```

```
The value 92 using field type o is 134.
The value 92 using field type x is 5c.
The value 92 using field type X is 5C.
The value 92.0 using field type f is 92.000000.
The value 92.0 using field type e is 9.20000e+01.
The value 92.0 using field type E is 9.20000E+01.
The value 92.0 using field type g is 92.
The value 92.0 using field type G is 92.
The value 92 using field type c is \.
The character '9' using field type c is 9.
The string 92 using field type s is 92.
```

For now, don't worry about the \ symbol you see in each `printf()` function. You will learn more about these symbols in this chapter. What they do here is to cause a carriage return and a new line of output. Without them in Program 1.4, all of the output would run together.

Note that the character argument used the single quote while the string argument used a double quote. Also note that the numerical value of the backslash (\) is 92. (Refer also to Appendix C, which contains the ASCII Table.)

More Than One

You can use more than one format specifier in a `printf()` function. However, you must have at least as many arguments as format specifiers; if not, the results will be unpredictable. You can have more arguments than format specifiers; however, the extra arguments will just be ignored.

Table 1.4 Field Type Format Specifiers Used by `printf()`

Character	Argument	Resulting Output
`d`	integer	Signed decimal integer.
`i`	integer	Signed decimal integer.
`o`	integer	Unsigned octal integer.
`u`	integer	Unsigned decimal integer.
`x`	integer	Unsigned hexadecimal integer using lowercase letters.
`X`	integer	Unsigned hexadecimal integer using uppercase letters.
`f`	floating point	Signed floating point number.
`e`	floating point	Signed floating point number using e notation.
`E`	floating point	Signed floating point number using E notation.
`g`	floating point	Signed decimal number in either e form or f form, whichever is shorter.
`G`	floating point	Signed decimal number in either E form or f form, whichever is shorter.
`c`	character	A single character.
`s`	string	Prints character strings.
`%`	none	Prints the % sign.

Note: Use a `l` prefix with `%d`, `%u`, `%x`, `%o`, to specify long integer (for example `%ld`).

An example of using more than one format specifier is

```
printf("A character is %c and a number is %d.",'a',53);
```

Note that the arguments are separated by commas.

Conclusion

This section introduced you to the `printf()` function in C. You saw how to use this function to display strings, characters, and numbers. You also saw what a format specifier is and how to use various field type format specifiers. Check your understanding of this section by trying the following section review.

1.5 Section Review

1. State the purpose of the `printf()` function.
2. How are characters distinguished from strings?
3. What is the purpose of a format specifier in the `printf()` function?
4. What is an argument?
5. State the rule concerning the number of format specifiers and arguments.

1.6 Identifying Things

Discussion

In this section you will learn how to use words to identify specific parts of your C program. Here you will also get an introduction to functions. In this section you will see how to give names to strings, values, and parts of your program. Knowing how to do this will make your programming tasks in any programming language much easier.

What Is a Function?

A **function** is an independent collection of **declarations** and statements. A declaration states the relationship between the name and type of a variable or other functions. You will see more about this later in this chapter. What is important for now is that you realize that a function is usually designed to perform one task. Every C program must have at least one function called `main()`. Dividing tasks into separate parts in a program makes programs easier to design, correct, understand, and modify.

What Needs Identification?

When you create a function, it's best to give it a descriptive name. As an example, a function that would calculate the total power dissipated in a resistor could be named

```
resistor_power()
```

This is more descriptive than naming it

```
function_1()
```

These two examples both use **identifiers** to distinguish one function from the other. An identifier is nothing more than the name you give to a part of your C program. Identifiers can be used to name parts of a formula, such as

```
total = resistor_1 + resistor_2;
```

An identifier can be used to assign a constant value that describes the value and can then be used in your program:

```
PI = 3.14159;
circle_area = PI * radius * radius;
```

As you can see from the examples above, identifiers play an important role in a C program.

Creating Your Own Identifiers

There are some rules to follow when making your own identifiers. First, every identifier must start with a letter of the alphabet (uppercase or lowercase) or the underscore (_). The remainder of the identifier may use any arrangement of letters (uppercase or lowercase), digits (0 through 9), and the underscore—and that's it—no other characters are allowed. This means that spaces are not allowed in identifiers. Most C compilers will distinguish among at least the first 31 characters of an identifier. Example 1.1 illustrates.

EXAMPLE 1.1

Which of the following are legal C identifiers?

A. `This_1`
B. `_This1`
C. `1_for_the_road`
D. `One_for_the_road`
E. `One for the road`
F. `The_Following:`
G. `E_=_IR`

Solution

Keep in mind the rules for identifiers explained prior to this example.

A. `This_1`	A legal C identifier.
B. `_This1`	A legal C identifier.
C. `1_for_the_road`	Not a legal C identifier. It does not start with a letter of the alphabet or underscore.
D. `One_for_the_road`	A legal C identifier.
E. `One for the road`	Not a legal C identifier. Spaces are not allowed.
F. `The_Following:`	Not a legal C identifier. The : symbol is not allowed.
G. `E_=_IR`	Not a legal C identifier. The = sign is not allowed.

Case Sensitivity

Identifiers in C are case sensitive. This means that C makes a distinction between uppercase and lowercase letters in an identifier. Thus, as far as C is concerned, the following identifiers are not equal:

```
pi PI Pi pI
```

Neither are the following:

`This_One THIS_ONE this_one`

All of the identifiers above are legal in C; it is just that they are not equal. What this means is that you must be careful with your use of identifiers in C. As an example, if you define the identifier `pi` to be equal to 3.14159, you must use the lowercase letters `pi` anywhere in the program you expect the identifier to equal 3.14159. If you use `PI` or `Pi`, neither of which you assigned a value to, then the program will contain errors.

Keywords

An identifier cannot have the same spelling and case as a C keyword. Example 1.2 illustrates.

EXAMPLE 1.2

State which of the following identifiers are legal and which are reserved, and state if any are equal.

A. `Ohms_Law` D. `reconstruct`
B. `continue` E. `Reconstruct`
C. `Ohms Law` F. `_continue`

Solution

Keep in mind the discussions presented prior to this example.

A. `Ohms_law`	Legal C identifier, not reserved.
B. `continue`	Legal C reserved word.
C. `Ohms law`	Not a legal C identifier. Spaces are not allowed.
D. `reconstruct`	Legal C identifier, not reserved.
E. `Reconstruct`	Legal C identifier, not reserved.
F. `_continue`	Legal C identifier, not reserved. (However, it is so close to looking like a reserved word that this practice is discouraged.)

Conclusion

This section presented the important concepts for naming parts of your C program. You learned what an identifier is and how it can be used to name (identify) parts of your C program. Check your understanding of this section by trying the following section review.

1.6 Section Review

1. What is a C function?
2. What is an identifier?
3. What are the rules for creating your own identifiers?
4. How many characters of an identifier are recognized by C?

1.7 Declaring Things

In this section you will learn that all variables must be declared. This means that you must let the compiler know ahead of time, before you use the variables, the identifier that will be used for each variable as well as the type of variable you will be using. At first, this may seem like a lot of extra work. But you will find that by doing this you will actually reduce the chances for program errors.

What Is a Variable?

You can think of a variable as a specific memory location set aside for a specific kind of data and given a name for easy reference. Essentially, you use variables so that the same memory space can hold different values of the same type at different times. As an example, if you were calculating the voltage across a fixed value of resistance as the current changed, the voltage variable would have a different value each time the current changed.

Declaring Variables

In C, you must declare all variables before using them. To declare a variable, you must declare its type and identifier. Table 1.5 presents the fundamental **type specifiers** that will be used for variables.

Program 1.6 illustrates how to declare variables.

Program 1.6

```
#include <stdio.h>

main()
{
     char a_character;         /* This declares a character. */
     int an_integer;           /* This declares an integer. */
     float floating_point;     /* This declares a floating point.*/

     a_character = 'a';
     an_integer = 15;
     floating_point = 27.62;

     printf("%c is the character.\n",a_character);
     printf("%d is the integer.\n",an_integer);
     printf("%f is the floating point.\n",floating_point);
}
```

Table 1.5 Fundamental C Type Specifiers

Integers	Floating Point	Other Types
char	double	const
enum	float	void
int	long double	volatile
long		
short		
signed		
unsigned		

When executed, Program 1.6 produces

```
a is the character.
15 is the integer.
27.620000 is the floating point.
```

In the program, the variable declarations are

```
char a_character;     /* This declares a character. */
int an_integer;       /* This declares an integer. */
float floating_point;/* This declares a floating point. */
```

Observe that declaring a variable consists of stating its type and its identifier. In the body of the program, each variable was assigned a value:

```
a_character = 'a';
an_integer = 15;
floating_point = 27.62;
```

This was done with the **assignment operator** (=). You will learn more about this operator in the next section.

Initializing Variables

You can combine a variable declaration with the assignment operator, thus giving the variable a value at the same time that it is declared. This is shown in Program 1.7.

Program 1.7

```
#include <stdio.h>

main()
{
    char a_character = 'a';         /* This declares/assigns a character.  */
    int an_integer = 15;            /* This declares/assigns an integer.   */
    float floating_point = 27.62;  /* This declares/assigns a float point.*/

    printf("%c is the character.\n",a_character);
    printf("%d is the integer.\n",an_integer);
    printf("%f is the floating point.\n",floating_point);
}
```

Note that in each case, the variable declaration is followed by a comment that states the purpose of each variable. It's good to get in the habit of doing this kind of program documentation.

More of the Same Type

In C, if you have variables of the same data type, you can declare them as follows:

```
int number_1, number_2, number_3;
```

Even though this is legal in C, this practice will not be used in this text because it discourages you from commenting on each variable used in the program.

Why Declare?

When you declare variables, you gather all the information about them at one place in the program. This allows everyone reading your source code to quickly identify the data that will be used (assuming that you add comments that give good explanations). Doing this also forces you to do some planning before leaping into the program code. Another important reason for doing this is that it prevents you from misspelling a variable within the program. If there were no requirements to declare, then you could create a new identifier without knowing it. This could cause disastrous problems that are very difficult to find and correct. The more you program, and the more complex and practical your programs are, the more thankful you will be that variables must be declared before you can use them.

1.7 Section Review

1. How can you think of a variable in terms of a C program?
2. Explain what it means to declare a variable.
3. Name three fundamental C type specifiers.
4. Explain what is meant by initializing a variable.
5. What is one good reason for declaring variables?

1.8 Introduction to C Operators

Discussion

In this section you will discover how to cause a C program to act on variables. This means you will learn about arithmetic operations as well as other operations that are unique to C.

What Are C Operators?

A C operator causes the program to do something to variables. Specifically, an **arithmetic operator** allows an arithmetic operation (such as addition, +) to be performed on variables. This section will introduce you to the most commonly used C arithmetic operators.

Arithmetic Operators

The common arithmetic operators used by C are listed in Table 1.6.

Table 1.6 Common Arithmetic Operators

Symbol	Meaning	Example
+	Addition	`answer = 3 + 5 (answer → 8)`
-	Subtraction	`answer = 5 - 3 (answer → 2)`
*	Multiplication	`answer = 5 * 3 (answer → 15)`
/	Division	`answer = 10 / 2 (answer → 5)`
%	Remainder	`answer = 3 % 2 (answer → 1)`

Program 1.8 illustrates the use of the first four arithmetic operators in C.

Program 1.8

```
#include <stdio.h>

main()
{
     float number_1 = 15.0;     /* First arithmetic operator.          */
     float number_2 = 3.0;      /* Second arithmetic operator.         */
     float addition_answer;     /* Answer to addition problem.         */
     float subtraction_answer;  /* Answer to subtraction problem.      */
     float multi_answer;        /* Answer to multiplication problem.   */
     float division_answer;     /* Answer to division problem.         */

     addition_answer = number_1 + number_2;
     subtraction_answer = number_1 - number_2;
     multi_answer = number_1 * number_2;
     division_answer = number_1 / number_2;

     printf("15 + 3 = %f\n",addition_answer);
     printf("15 - 3 = %f\n",subtraction_answer);
     printf("15 * 3 = %f\n",multi_answer);
     printf("15 / 3 = %f\n",division_answer);
}
```

Execution of Program 1.8 produces

```
15 + 3 = 18.000000
15 - 3 = 12.000000
15 * 3 = 45.000000
15 / 3 = 5.000000
```

Note that all of the numbers used in Program 1.8 were of type `float`. It's important to note that in C, division of type `int` variables will truncate the answer. This means that 5/2 = 2 and that 2/3 = 0. The remainder operator (`%`) requires the use of whole numbers. This operator is also called the *modulo* operator.

Important Considerations

C does not allow a reverse type of assignment structure. This means that you cannot have an assignment to an expression:

```
6 + 3 = answer;   ←Not allowed!
```

Nor can you have an assignment to a constant:

```
3 = answer;   ←Not allowed!
```

Both of the above attempts will produce compile time errors.

Table 1.7 Precedence of Operations

Priority	**Operation**
First	( )
Second	Negation (assigning a negative number)
Third	Multiplication *, Division /
Fourth	Addition +, Subtraction –

Precedence of Operations

Precedence of operations is simply the order in which arithmetic operations are performed. As an example, consider the expression

```
X = 5 + 4/2;
```

Precedence of operations requires that division be done before addition. If this were not the case, then the operation above could be interpreted in two different ways. If the division were performed first it would yield (5 + 2 = 7). If addition were performed first, 5 would be added to 4 and then the sum would be divided by 2 (9/2 = 4.5). The interpretation of any expression must be consistent for reliable and predictable program results. For example, if you want to indicate that 5 is to be added to 4 first and the result then divided by 2, parentheses must be used:

```
X = (5 + 4)/2;
```

Table 1.7 indicates the precedence of operations for C.

In all cases, operations proceed from left to right. Example 1.3 illustrates.

EXAMPLE 1.3

Determine the results of the following operations:

A. $Y = 6 + 12/6 * 2 - 1$
B. $Y = (6 + 12)/6 * 2 - 1$
C. $Y = 6 + 12/6 * (2 - 1)$
D. $Y = 6 + 12/(6 * 2) - 1$

Solution

Using precedence of operations, the results are:

A. $Y = 6 + 12/6 * 2 - 1$
   $Y = 6 + 2 * 2 - 1$ (Operations from left to right. Division precedence.)
   $Y = 6 + 4 - 1$ (Multiplication precedence.)
   $Y = 9$ (Operations from left to right.)

B. $Y = (6 + 12)/6 * 2 - 1$
   $Y = 18/6 * 2 - 1$ (Work inside parentheses done first.)
   $Y = 3 * 2 - 1$ (Operations from left to right.)
   $Y = 6 - 1$ (Multiplication precedence.)
   $Y = 5$

C. $Y = 6 + 12/6 * (2 - 1)$
   $Y = 6 + 12/6 * 1$ (Work inside parentheses done first.)
   $Y = 6 + 2 * 1$ (Operations from left to right.)
   $Y = 6 + 2$ (Multiplication precedence.)
   $Y = 8$

D. $Y = 6 + 12/(6 * 2) - 1$
$Y = 6 + 12/12 - 1$ (Work inside parentheses done first.)
$Y = 6 + 1 - 1$ (Division precedence.)
$Y = 6$ (Operations from left to right.)

Compound Assignment Operators

In C, **compound assignment operators** combine the simple assignment operator with another operator. For example, consider the statement:

```
answer = answer + 5;
```

What this statement means is that the memory location called "answer" will be assigned the new value of its old value plus 5. It does *not* mean "answer equals answer plus 5." The = sign does *not* mean equal in C; it means *assigned to* and is called the assignment operator. Thus, in the example above, if `answer` had the value of 10, then

```
answer = answer + 5;
```

would cause the new value of `answer` to be 15.

The expression above can be shortened in C by using the compound assignment

```
answer += 5;
```

Table 1.8 lists the compound assignments of the arithmetic operators presented at the beginning of this section.

Use of the compound assignment operators is illustrated in Program 1.9.

Program 1.9

```
#include <stdio.h>

main()
{
      int number = 10;    /* Value of number for example. */

      number += 5;
      printf("Value of number += 5 is %d\n",number);

      number -= 3;
      printf("Value of number -= 3 is %d\n",number);

      number *= 3;
      printf("Value of number *= 3 is %d\n",number);

      number /= 5;
      printf("Value of number /= 5 is %d\n",number);

      number %= 3;
      printf("Value of number %%= 3 is %d\n",number);
}
```

Table 1.8 Common Compound Assignments in C

Symbol	Example	Meaning
+=	X += Y;	X = X + Y;
–=	X –= Y;	X = X – Y;
*=	X *= Y;	X = X * Y;
/=	X /= Y;	X = X / Y;
%=	X %= Y;	X = X % Y;

Execution of Program 1.9 produces

```
Value of number += 5 is 15
Value of number -= 3 is 12
Value of number *= 3 is 36
Value of number /= 5 is 7
Value of number %= 3 is 1
```

Observe the last `printf()` function in the program. In order to get a printout of the `%` sign, a double `%%` was used.

Conclusion

This section presented the basic arithmetic operations in C. Here you learned the meaning of an assignment operator as well as compound assignments and order of precedence. Check your understanding of this section by trying the following section review.

1.8 Section Review

1. List the common arithmetic operators used in C.
2. In C, what does integer division do to the remainder? What is the significance of this?
3. In C, is the following allowed: `3 - 2 = result;`? Explain.
4. What is meant by precedence of operation?
5. Give an example of a compound assignment used in C.

1.9 More printf()

Discussion

This section presents some more details about the `printf()` function. You have already been using the `\n` as a part of this function and were briefly told that this causes a carriage return and a new line. In this section you will learn more about the power of the `printf()` function.

Escape Sequences

The `\n` is an example of an **escape sequence** that can be used by the `printf()` function. The backslash symbol (\) is referred to as the escape character. You can think of an escape sequence used in the `printf()` function as an escape from the normal

Table 1.9 Escape Sequences

Sequence	Meaning
`\n`	New line
`\t`	Tab
`\b`	Backspace
`\r`	Carriage return
`\f`	Form feed
`\'`	Single quote
`\"`	Double quote
`\\`	Backslash
`\xdd`	ASCII code in hexadecimal (ex: \x41 equals 'A')
`\ddd`	ASCII code in octal (ex: \101 equals 'A')

Note: The double quote and the backslash can be printed by preceding them with the backslash.

interpretation of a string. This means that the next character used after the \ will have a special meaning, as listed in Table 1.9.

The Interactive Exercises for this chapter will give you some practice in using the `printf()` escape sequences in your system.

Field Width Specifiers

The `printf()` function allows you to format your output. Recall from the previous programs that when you printed the output of a type `float` it would appear as: `16.000000`; even though you didn't need the six trailing zeros, they were still printed. The `printf()` function provides **field width specifiers** so that you can control how printed values will appear on the monitor. The syntax is:

```
%<width>.<digits>F
```

Where

`%` = The format specification.
`<width>` = The width of the field.
`<digits>` = Number of digits to the right of the decimal.
`F` = The format specifier.

As an example, the statement

```
printf("The number %5.2f uses them.",6.0);
```

would display

```
The number   6.00 uses them.
```

Note that the width first indicates how the number is justified (five spaces), and then the number of digits following the decimal point is specified. For additional examples, review Program 1.5.

Conclusion

This section presented some more important details about the `printf()` function. Here you saw the other escape sequences that could be used with the `printf()` function as well as the format specifiers. Test your understanding of this section by trying the following section review.

1.9 Section Review

1. Explain the use of an escape sequence in the `printf()` function.
2. What is the backslash character (\) sometimes called when used in a `printf()` function?
3. Give three escape sequences that are used in the `printf()` function.
4. What is a field width specifier as used by the `printf()` function?

1.10 Getting User Input

Discussion

The real power of a technical C program is its ability to interact with the program user. This means that the program user gets to input values of variables. As you might guess, there is a built-in C function that lets you make this happen.

The scanf() Function

The `scanf()` function is a built-in C function that allows your program to get user input from the keyboard. You can think of it as doing the opposite of the `printf()` function. Its use is illustrated in Program 1.10.

Program 1.10

```
#include <stdio.h>

/*                    Getting user input.                     */
main()
{
     float value;     /* A number inputted by the program user. */

     printf("Input a number => ");
     scanf("%f", &value);

     printf("The value is => %f", value);
}
```

When Program 1.10 is executed, the output will appear as follows (assuming the program user inputs the value of 23):

```
Input a number => 23
The value is => 23.000000
```

Note that the `scanf()` function has a similar format to that of the `printf()` function. First, it contains the `%f`, enclosed in quotation marks. This tells the program that a value will be entered that will be a floating point type. Next, it indicates the variable identifier where this value will be stored. It indicates this by using a comma outside the quotes and then an `&` (ampersand sign) immediately followed by the name of the variable identifier (`&value`). Now the value that the user inputs will be the value of the variable `value`. Also note that the result of using the `scanf()` function is a carriage return.

Format Specifiers

The format specifiers for the `scanf()` function are similar to those for the `printf()` function. This is illustrated in Table 1.10.

It should be noted that either the `%f` or the `%e` format specifier may be used for accepting either exponential or decimal notation.

The `scanf()` function can accept more than one input with just one statement, as shown below:

```
scanf("%f%d%c",&number1, &number2, &character);
```

In the preceding case, the variable `number1` will accept a type `float`, the variable `number2` a type `int`, and `character` a type `char`. In this case, the program user would have to type in three separate values separated by a space. As an example:

```
52.7 18 t
```

Since it is easy for the program user to make errors entering data in this manner, multiple inputs with the `scanf()` function will not be used in this text.

A Real Technology Program

You now know how to input data, do a basic computation, and display the results. Program 1.11 solves for the voltage across a resistor when the value of the current and resistance are known. The mathematical relationship is:

$$\text{Voltage} = \text{Current} \times \text{Resistance}$$

Table 1.10 `scanf()` Format Specifiers

Specifier	Meaning
`%c`	A single character.
`%d`	Signed decimal integer.
`%e`	Exponential notation.
`%f`	Floating point notation.
`%o`	Unsigned octal integer.
`%u`	Unsigned decimal integer.
`%x`	Unsigned hexadecimal integer.

Program 1.11

```
#include <stdio.h>

/*                          Ohms Law                              */

main()
{
     float voltage;         /* Value of the voltage.     */
     float current;         /* Value of the current.     */
     float resistance;      /* Value of the resistance. */

     printf("Input the current in amps => ");
     scanf("%f", &current);

     printf("Input the resistance in ohms => ");
     scanf("%f", &resistance);

     voltage = current * resistance;     /* Compute the voltage. */

     printf("The value of the voltage is %f volts", voltage);

}
```

Assume that the program user will enter the value of 3 for the current and 4 for the resistor value. Then execution of Program 1.11 would yield

```
Input the current in amps => 3
Input the resistance in ohms => 4
The value of the voltage is 12.000000 volts
```

There are several key points to note concerning this program:

- All variables were declared, and a comment was made about each one.
- Each `scanf()` function used `%f` to indicate that the input would be a floating point and the `&variable` to indicate which variable would store the user input value.
- The actual calculation was commented.
- The `printf()` function used the `%f` to indicate that the numeric output would be floating point and the variable identifier whose value was to be displayed was indicated at the end of the quotes.

What you just observed was a fundamental problem in technology solved by the C language. This was a very simple problem that you could have easily solved on your pocket calculator. But, for now, the point is to keep the problems simple so that they don't get in the way of understanding the C language. As you progress through the text, your programs will become much more powerful.

Using the E Notation

Some mention should be made of using **E (exponential) notation** with C. As pointed out earlier, C will accept E notation numbers for floating point types. You can do this for input as well as output. For example, in the last program, the program user could have inputted the more practical values of 0.003 amps for the current and 2000 ohms for the resistance. This could have been done using E notation:

```
Input the current in amps => 3E-3
Input the resistance in ohms => 2E3
The value of the voltage is 6.000000 volts
```

It should be pointed out that C will accept either an uppercase E or a lowercase e for this kind of data representation. There are other C functions, such as `gets()`, `atoi()`, and `atof()`, that can be used to perform conversions. We will examine these functions when we learn about string functions.

Conclusion

This section brought you to the point where you can now develop some very basic technology programs using the C language. In the next section, you will learn how to perform more arithmetic functions using the C language. This will be an important step enabling you to handle almost any type of technology formula. Check your understanding of this section by trying the following section review.

1.10 Section Review

1. State the purpose of the `scanf()` function.
2. How does the `scanf()` function know what variable identifier to use for inputting the user data?
3. Does use of the `scanf()` function produce a carriage return to a new line?
4. How may floating point values be entered by the program user in a C program?
5. State what you must do in order to have values outputted to the screen in E notation.

1.11 Troubleshooting Techniques

Discussion

The material in this section is designed to help you minimize some of the most common errors encountered by beginning C programmers, by the use of a number of simple troubleshooting techniques. Troubleshooting is required whenever an error message shows up during compilation.

Types of Error Messages

Depending upon the type of compiler you are using, you will receive different kinds of error messages. Most compilers produce three types of error messages:

1. Fatal error messages
2. Compilation error messages
3. Warning messages

A *fatal error* message will immediately terminate the compiling process. It will not check for any further errors and stops with a message display of what may be the most likely problem. Most C systems will move the cursor to the line of source code where it "thinks" the error is located. You may find, depending on the nature of the error, that where the cursor is placed is not the actual location of the error. You must remember that no error location scheme is perfect, and not all possible errors are predictable. Your compiler is giving you its best guess of what is causing the error. It's up to you to determine exactly where the error is located and to correct it accordingly.

A *compilation error message* is a result of a less severe error. In these cases, the compiler will try to keep compiling and will produce other compilation error messages if it encounters more errors of this type. In some cases, it may not be able to continue the compilation process and a fatal error will result. In any case, no object code is produced, and you will wind up with a list of error messages. You can then use this list to help determine where the problems are in your source code.

Warning messages are the results of programming errors that will allow your program to compile and link, but will not allow your program to be executed. These warning messages are listed for you as a reference to use in determining the problems with your source code.

Case Sensitivity

Recall that an important aspect of C programming is that its identifiers are case sensitive. This means that C does make a distinction between uppercase and lowercase letters. Program 1.12 will not compile because of the uppercase M used in the function `main()`.

Program 1.12

```
/*      Case sensitivity example        */

{
     Main()

     /* This program will not compile! */
}
```

Case sensitivity applies to all identifiers. As an example, Program 1.13 will not execute because the programmer capitalized the first letters of the identifiers when they were declared but forgot the capitalization when they were used in the program.

Program 1.13

```
/*      Another case sensitivity example        */

main()
{
     float This_One = 2; /* Program constant. */
     float That_One;          /* Program variable. */

     That_one = 2 * This_one;

     /* This program will not compile! */
}
```

There are two bugs in Program 1.13, both of the same type. The declared variables `This_One` and `That_One` use two capital letters each. However, in the body of the program, the programmer forgot to use a capital O:

```
That_one = 2 * This_one;
```

Therefore, the compiler will think these are new identifiers which have not been declared! A good troubleshooter always checks the case of each identifier.

The Semicolon

Another error common to beginning C programmers is the omission of the semicolon. In C, the semicolon identifies the end of a statement or instruction. The semicolon is actually a part of the C statement and must be included. A missing semicolon will always prevent your program from executing. However, a missing semicolon can really confuse the compiler as to what is causing the problem. You may get error messages from missing semicolons that make no mention of them. A good rule to remember is that if the warning message doesn't make sense, check first for missing semicolons.

As examples, Programs 1.14 and 1.15 both have missing semicolons; however, each produces different error messages (the kind you get will depend on your system). In both cases, the error messages will fail to recognize that the problem is a missing semicolon.

Program 1.14

```
/*     Missing semicolon.        */

main()
{
     float This_One = 2         /* Program constant. */
     float That_One;            /* Program variable. */

     That_One = 2 * This_One;

     /* This program will not compile! */
}
```

Program 1.15

```
/*     Another missing semicolon.       */

main()
{
     float This_One = 2;       /* Program constant. */
     float That_One       /* Program variable. */

     That_One = 2 * This_One;

     /* This program will not compile! */
}
```

Note that Program 1.14 has a semicolon missing from the constant declaration, while Program 1.15 has a semicolon missing from the variable declaration.

The best rule to follow is always to make a good visual inspection of your C program before attempting to compile it. Always look for missing semicolons.

Incomplete or Nested Comments

Another common source of program bugs is in the use of comments. These usually are of one of two types. In the first, the user produces the starting comment delimiter /* but forgets to include the ending one. This error is shown in Program 1.16.

Program 1.16

```
/*      Omitting comment delimiters.        */

main()
{
     float constant_1 = 2;           /* Program constant. */
     float variable_1;               /* Program variable.

     variable_1 = constant_1 + constant_1;

     /* This program will not compile! */
}
```

Program 1.16 will not compile because of the missing ending delimiter */ for the comment: /* Program variable. However, this so confuses the compiler that when a warning message is displayed, it does not indicate that the problem is with the comment. This brings up another good troubleshooting technique. When you get an error message you don't understand, after checking to ensure all statements have their required semicolons, then check to make sure each ending comment delimiter is present.

Program 1.17 gives an example of nested comments, the second potential source of comment bugs.

Program 1.17

```
/*        Nested comments        */

main()
{
     float constant_1 = 2;           /* Program constant. */
     float variable_1;               /* Program variable. */

     /* This is a comment...

     variable_1 = constant_1 + constant_1; /* Another comment. */

        end of nested comment */
}
```

Normally, Program 1.17 will not execute because one comment is contained inside the other. Some compilers do allow you to set the operating environment so that you can nest comments. However, this practice is not recommended and will not be used in this book.

Example 1.4 gives you practice in using your visual inspection skills to spot some common program bugs.

EXAMPLE 1.4

Determine if there is a bug in any of the following three programs that will prevent execution of the program. If so, state what you would do to correct it (them).

Program 1.18

```
/*      Is there a bug here?      */

main()
{
     float This_Value = 15;          /* Program constant. */
     float That_Value;               /* Program variable. */

     That_Value = this_Value + this_value;
}
```

Program 1.19

```
/*      Any bugs here?      */

main()
{
     float this_value = 15;          /* Program constant. */
     float that_value;               /* Program variable. */

     that_value = this_value + this_value
}
```

Program 1.20

```
/*      Any bugs here?      */

main()
{ float this_value = 15;        /* Program constant. */
  float that_value;             /* Program variable. */

     that_value = this_value + this_value; }
```

Solution

Always perform a good visual observation of your source code before attempting to compile it. Doing this will train you to catch your own mistakes quickly. The ability to find your errors comes with a great deal of practice.

Program 1.18

The case of the declared constant identifier was changed.

```
That_Value = this_Value + this_value;
```

Program 1.19
Missing semicolon at end of a C statement.

```
that_value = this_value + this_value
```

Program 1.20
There are no bugs in this program. However, the style of having the opening and closing braces on the same line with program code is discouraged because they are not as easy to see. Good programming practice in C devotes a whole line to each one.

Conclusion

This section presented some of the most common program bugs encountered by beginning C programmers. Here you saw that good programming practice requires that you carefully check your programs for required semicolons, case sensitivity, and comment delimiters. Check your understanding of this section by trying the following section review.

1.11 Section Review

1. State the three types of error messages found in C.
2. Which type of error message will terminate the compilation process?
3. Explain how case sensitivity will cause your C program not to execute.
4. Why is it helpful to look for missing semicolons if you do not understand how an error message applies to your program?
5. What is meant by a nested comment? Is this legal?

1.12 Example Programs

We will now look at four programs designed to show how practical computations are performed in C. There are many times when we need a simple program to perform a task for us. For example, we are not very good at performing conversions or complex math problems (such as "How many hours are there in 25 years?") in our heads.

Programs 1.21 and 1.22 are two examples of how standard conversions or calculations are implemented in C. Program 1.21 converts speed in miles per hour to speed in feet per second. Program 1.22 computes the equivalent resistance for two resistors in parallel. Let us look at Program 1.21.

Program 1.21

```
#include <stdio.h>
main()
{
	float mph;               /* Speed in miles per hour. */
	float fpm = 5280.0;      /* Feet per mile.           */
	float sph = 3600.0;      /* Seconds per hour.        */
	float fps;               /* Feet per second.         */

	printf("Enter speed in miles/hour => ");
	scanf("%f",&mph);
```

```
    printf("\n");
    fps = mph * fpm / sph;
    printf("The speed in feet/second is %f\n",fps);
}
```

A sample execution of Program 1.21 is as follows:

```
Enter speed in miles/hour => 60

The speed in feet/second is 88.000000
```

Here the user entered 60 and the program computed 88 as the equivalent speed.

Program 1.22 also implements a simple formula.

Program 1.22

```
#include <stdio.h>
main()
{
    float R1;       /* Resistor 1.             */
    float R2;       /* Resistor 2.             */
    float REQ;      /* Equivalent resistance.  */

    printf("Enter resistor 1 value => ");
    scanf("%f",&R1);
    printf("Enter resistor 2 value => ");
    scanf("%f",&R2);
    printf("\n");
    REQ = (R1 * R2) / (R1 + R2);
    printf("The equivalent resistance is %f\n",REQ);
}
```

Program 1.22 is executed with R1 and R2 equal to 1000 and 4000 ohms, respectively.

```
Enter resistor 1 value => 1000
Enter resistor 2 value => 4000

The equivalent resistance is 800.000000
```

Can you think of a way to modify these two programs to solve similar types of problems, such as converting square inches to square meters, or finding the equivalent resistance of three resistors in parallel?

Other types of computations may require the use of special mathematical functions beyond addition, subtraction, multiplication, and division. For example, the question "What is 200 raised to the 7th power?" requires math skills few people can easily perform without a pencil and paper. Fortunately, C comes equipped with many additional math functions. Program 1.23 utilizes one of these functions to compute the total value of an investment over a period of years by using the following formula:

$$\text{Accumulated Value} = \text{Amount Invested} * \text{Interest}^{\text{Years}}$$

Since we need to raise a number to a power, we need to use a function designed for that purpose. The function required is `pow()` and is found in the include file `math.h`. Examine Program 1.23 to see how `pow()` is used in the calculation.

Program 1.23

```
#include <stdio.h>
#include <math.h>
main()
{
        float amount;        /* Dollars to invest.     */
        float interest;      /* Yearly interest rate.  */
        float years;         /* Number of years.       */
        float total;         /* Total accumulation.    */

        printf("Enter amount to invest => ");
        scanf("%f",&amount);
        printf("Enter yearly interest rate => ");
        scanf("%f",&interest);
        printf("Enter the number of years => ");
        scanf("%f",&years);
        printf("\n");
        interest /= 100.0;
        total = amount * pow((1.0 + interest),years);
        printf("The total accumulation is %f\n",total);
}
```

Program 1.24 use the `sqrt()` function from `math.h` to compute the length of the hypotenuse of a right triangle.

Program 1.24

```
#include <stdio.h>
#include <math.h>

main()
{
        float sidea, sideb, sidec;

        print("Enter side A => ");
        scanf("%f",&sidea);
        printf("Enter side B => ");
        scanf("%f",&sideb);
        sidec = sqrt(sidea*sidea + sideb*sideb);
        printf("\nThe length of the hypotenuse is %5.1f",sidec);
}
```

It might be worthwhile to modify Program 1.24 so that all of the the interior angles of the triangle are also computed. This would require the use of the `sin()`, `cos()`, or `tan()` functions from `math.h`. Note that these functions require the input angle to be in radian form.

Some of the other useful functions found in `math.h` are:

```
abs()   atof() cos()  exp()  log()
log10() pow()  sin()  sqrt() tan()
```

You are encouraged to experiment with these functions by modifying Program 1.23. Refer to Appendix B for more information on `math.h`.

Conclusion

As you can see, it does not take very many statements to make a useful C program. With the special functions available through include files, practically every task can be solved through the use of a carefully constructed C program. The next section will show you how to create a new C program. For now, test your understanding with the following section review.

1.12 Section Review

1. What is the formula to convert centimeters to inches? *Note:* There are 2.54 centimeters in one inch.
2. How can Program 1.22 be modified to solve for three resistors in parallel?
3. Why are `float` data types used in Programs 1.21 through 1.24?

1.13 Case Study: Temperature Conversion

Discussion

The topic of this case study is the design of a program that converts a temperature reading from degrees Fahrenheit to degrees centigrade. This is an easy conversion to perform on most pocket calculators. It is used here to illustrate some of the fundamental elements of program design. The problem is intentionally kept simple so as not to dominate the programming development; yet the problem is rigorous enough to require most of the new material presented in this chapter.

First Step—Stating the Problem

The first, most important step in program design is to state the problem in writing. This problem statement must include some specifics about the requirements of the computer program itself. The specifics are:

- Purpose of the Program
- Required Input (source)
- Process on Input
- Required Output (destination)

The problem can be stated as follows:

- Purpose of the Program: Convert a temperature reading from degrees Fahrenheit to degrees centigrade.
- Required Input: Temperature in degrees Fahrenheit. (Source: Keyboard)

- Process on Input: Temperature centigrade = 5/9 × (temperature Fahrenheit – 32).
- Required Output: Temperature in degrees centigrade.

After reading this statement, you (and others who may be working with you) clearly know the purpose of the program. It is evident that the information for the program will be received from the keyboard and that it will consist of a single temperature reading in degrees Fahrenheit. The program will then process this input according to a very specific formula. The output will simply be displayed on the screen. It will be the temperature expressed in degrees centigrade.

For this simple problem, this information may seem obvious. However, for more complex programs where some information will be received from the keyboard while other information may be received from the disk, the processes may not be so obvious. Stating them in writing in this manner ensures that you and the person you are designing the program for are in agreement as to what is to be done and—just as important—what is not to be done.

Blocking Out the Program

Once you have completed the problem statement, the next step is to block out ("comment out") your C program. This is illustrated in Program 1.25.

Program 1.25

```
/**************************************************************/
/*                 Fahrenheit to centigrade                   */
/**************************************************************/
/*              Developed by: Able Programmer                 */
/*                                                            */
/**************************************************************/
/*         This program converts a temperature reading        */
/*    in degrees Fahrenheit to its equivalent in degrees      */
/*    centigrade.                                             */
/**************************************************************/
main()
{
     /* Explain program to user. */

     /* Get Fahrenheit value from user. */

     /* Do computations. */

     /* Display the answer. */
}
```

Program 1.25 is a program outline that gives the program a title and provides the programmer's name and a description of what the program does. The rest is an outline of the major sections of the program. This will act as your guide as you develop the required code. Most important, your outline will actually become a part of your program— the comments for each of the major parts of your program. Once the outline is completed, it is saved to the disk, and a printed copy is made of it. This printed copy can verify that the outline fits the original intent of the program, and can serve as a source of documentation.

Note that the essential `main()` and `{` and `}` are included so that the program will compile. The compiling should be done at this point to ensure that the commented sections do not contain any bugs (such as nested comments, as explained in Section 1.11).

The Next Step

Once you have the program outlined, the next step is to develop each section, one section at a time. This process reduces the amount of program debugging time. For this case study, the first step is to code the program explanation, compile it, and observe the output. This is shown in Program 1.26. A new C function is introduced in this program called `puts()`. This function produces an automatic carriage return to a new line. It is a convenient function to use for just displaying text.

Program 1.26

```
#include <stdio.h>
/****************************************************************/
/*                Fahrenheit to centigrade                      */
/****************************************************************/
/*             Developed by: Able Programmer                    */
/*                                                              */
/****************************************************************/
/*         This program converts a temperature reading          */
/*    in degrees Fahrenheit to its equivalent in degrees        */
/*    centigrade.                                               */
/****************************************************************/
main()
{
     /* Explain program to user. */

     puts("");
     puts("This program will convert a temperature reading in");
     puts("degrees Fahrenheit to its equivalent temperature in");
     puts("degrees centigrade.");

     puts("");
     puts("You only need to enter the temperature in Fahrenheit");
     puts("and the program will do the rest.");

     /* Get Fahrenheit value from user. */

     /* Do computations. */

     /* Display the answer. */
}
```

When executed, Program 1.26 will display

```
This program will convert a temperature reading in
degrees Fahrenheit to its equivalent temperature in
degrees centigrade.

You only need to enter the temperature in Fahrenheit
and the program will do the rest.
```

The `puts()` function requires the `#include <stdio.h>` at the beginning of the program.

The point here is that the program is being developed in a step-by-step manner. This reduces the potential frustration of trying to discover program bugs when developing a large program. If there were a bug at this early stage, the programmer would know that it was confined to this part of the program. Once the bugs are worked out here, then the next section is developed, compiled, and executed. Again, if there are bugs, they must be confined to the most recent code entered (the exception to this is if the most recent code interacts with prior code—still, this method is the preferred method).

Completing the Study

The next step, of course, is to code and compile the next part of the program outline. This is done in Program 1.27.

Program 1.27

```
#include <stdio.h>
/*************************************************************/
/*                Fahrenheit to centigrade                   */
/*************************************************************/
/*             Developed by: Able Programmer                 */
/*                                                           */
/*************************************************************/
/*          This program converts a temperature reading      */
/*   in degrees Fahrenheit to its equivalent in degrees      */
/*   centigrade.                                             */
/*************************************************************/
main()
{
     /* Explain program to user. */

     puts("");
     puts("This program will convert a temperature reading in");
     puts("degrees Fahrenheit to its equivalent temperature in");
     puts("degrees centigrade.");

     puts("");
     puts("You only need to enter the temperature in Fahrenheit");
     puts("and the program will do the rest.");

     /* Get Fahrenheit value from user. */

     puts("");
```

```
    printf("Enter temperature value in Fahrenheit => ");
    scanf("%f", &fahrenheit_temp);

    /* Do computations. */

    /* Display the answer. */
}
```

Execution of Program 1.27 yields

```
This program will convert a temperature reading in
degrees Fahrenheit to its equivalent temperature in
degrees centigrade.

You only need to enter the temperature in Fahrenheit
and the program will do the rest.
Enter temperature value in Fahrenheit =>
```

If you enter a value at this point in the development of the program, no further processing will take place. But again, you have ensured that there are no compile time bugs up to this point.

Next, in Program 1.28, the processing part of the program is developed. Note the declaration of the two temperature variables.

Program 1.28

```
#include <stdio.h>
/****************************************************************/
/*                 Fahrenheit to centigrade                     */
/****************************************************************/
/*              Developed by: Able Programmer                   */
/*                                                              */
/****************************************************************/
/*       This program converts a temperature reading            */
/*   in degrees Fahrenheit to its equivalent in degrees         */
/*   centigrade.                                                */
/****************************************************************/
main()
{
    float fahrenheit_temp;
    float centigrade_temp;

    /* Explain program to user. */

    puts("");
    puts("This program will convert a temperature reading in");
    puts("degrees Fahrenheit to its equivalent temperature in");
    puts("degrees centigrade.");

    puts("");
    puts("You only need to enter the temperature in Fahrenheit");
    puts("and the program will do the rest.");
```

```
    /* Get Fahrenheit value from user. */

    puts("");
    printf("Enter temperature value in Fahrenheit => ");
    scanf("%f", &fahrenheit_temp);

    /* Do computations. */

    centigrade_temp = 5/9 * (fahrenheit_temp - 32);

    /* Display the answer. */
}
```

Note an important addition in Program 1.28. Since variables had to be used, they first had to be declared. Thus, the declaration block had to be added to the program so that the type of each variable could be stated. As before, these variables will be of type `float`. Each variable is commented, as required by good programming practice. Note that the `*` sign is used in the formula to indicate that multiplication is taking place.

Note: An error has been introduced into the program. You will be shown where it is shortly. Can you spot the error now?

The last programming step is to cause the results to be displayed, as shown in Program 1.29.

Program 1.29

```
#include <stdio.h>
/*************************************************************/
/*              Fahrenheit to centigrade                     */
/*************************************************************/
/*           Developed by: Able Programmer                   */
/*                                                           */
/*************************************************************/
/*      This program converts a temperature reading          */
/*   in degrees Fahrenheit to its equivalent in degrees      */
/*   centigrade.                                             */
/*************************************************************/
main()
{
    float fahrenheit_temp;
    float centigrade_temp;

    /* Explain program to user. */

    puts("");
    puts("This program will convert a temperature reading in");
    puts("degrees Fahrenheit to its equivalent temperature in");
    puts("degrees centigrade.");

    puts("");
    puts("You only need to enter the temperature in Fahrenheit");
```

```
    puts("and the program will do the rest.");

    /* Get Fahrenheit value from user. */

    puts("");
    printf("Enter temperature value in Fahrenheit => ");
    scanf("%f", &fahrenheit_temp);

    /* Do computations. */

    centigrade_temp = 5/9 * (fahrenheit_temp - 32);

    /* Display the answer. */

    puts("");
    printf("A temperature of %f degrees Fahrenheit \n", fahrenheit_temp);
    printf("is equal to %f degrees centigrade. \n", centigrade_temp);
}
```

Good programming practice would now require the variable definitions to be placed in the programmer's block.

The last part of the program, the "display the answer" section, is coded, and the program is compiled again to test for any compile time errors. Note that the `%f` is used so that a float type may be displayed and that the `\n` is used to indicate a new line of display. All this being done, there is one final and very important step left.

Checking the Output

Up to this point in the case study, you have confirmed that there are no compile time errors. However, to ensure that there are no run time errors, the program is checked for several different input values. The output results are then checked for accuracy. Several temperatures are entered (negative values, as well as positive values and zero), and the results are checked by a calculator by more than one person. Doing this ensures that the program process was correctly coded to produce what was intended by the program designer.

A sample execution of the program yields:

```
This program will convert a temperature reading in
degrees Fahrenheit to its equivalent temperature in
degrees centigrade.

You only need to enter the temperature in Fahrenheit
and the program will do the rest.

Enter temperature value in Fahrenheit => 212

A temperature of 212.000000 degrees Fahrenheit
is equal to 0.000000 degrees centigrade.
```

As demonstrated in this execution of the program, there is a run time error! This is why it is so important to test the program with different values and check the answers. Just because there are no compile time errors does not mean the program is doing what you intended it to do. No compile time errors simply means that your coding is passable; it says nothing about your design of the program. The reason why you are getting 0.00000 for the centigrade temperature is because you are using integers (5/9) in a division problem. Recall from the discussion on integers in this chapter that division by integers does not leave a remainder (it can't, they are integers). Thus 5/9 will return a value of 0. This 0 value is then multiplied by the term `(fahrenheit_temp - 32)`, which again results in zero (any number multiplied by 0 is 0). Thus, no matter what value you put into the program for the Fahrenheit temperature, the result will always be zero. In order to correct this, the conversion constant 5/9 needs to be changed from an integer. This is done by simply adding a decimal point followed by a zero:

```
centigrade_temp = 5.0/9.0 * (fahrenheit_temp - 32);
```

Conclusion

This section presented your first real case study for the development of a technical C program. You were introduced to the concept of first describing the program requirements in writing. You then saw how to develop an outline of the actual program using comments. From this, each comment section was developed and tested separately. Once the program was completely coded, its accuracy was tested for several different values.

Check your understanding of this section by trying the following section review.

1.13 Section Review

1. What is the first step in the development of a program?
2. State the items that should be included in the program problem statement.
3. Give the first step in the actual coding of the program.
4. Explain the process used to develop the final program.

Interactive Exercises

DIRECTIONS

These exercises require that you have access to a computer and software that support C. They are provided here to give you valuable experience and, most important, immediate feedback on what the concepts and commands introduced in this chapter will do.

Exercises

1. Predict what the output of Program 1.30 will be and then try it.

Program 1.30

```
#include <stdio.h>

main()
{
```

```
    printf("The number is %d\n", 15);
    printf("The number is %i\n", 15);
    printf("The number is %x\n", 15);
    printf("The number is %o\n", 15);
}
```

2. Program 1.31 now uses e and E. What is different about the outputs?

Program 1.31

```
#include <stdio.h>

main()
{
    printf("The number is %e\n", 15.0);
    printf("The number is %E\n", 15.0);
}
```

3. Program 1.32 is a fun one. See if you can predict what the output will be before you try it.

Program 1.32

```
#include <stdio.h>

main()
{
    char letter;

    letter = 'b';

    printf("This is %d or %c or %x.", letter, letter, letter);
}
```

4. Figure out Program 1.33 with pencil and paper. Then try it. Do your results agree with the computer?

Program 1.33

```
#include <stdio.h>

main()
{
    float result = 10;

    result = 2 * (3 + 5)/8 - 3;

    printf("The result is %f.\n", result);

}
```

5. Program 1.34 is a good test question. Make sure you try it!

Program 1.34

```
#include <stdio.h>

main()
{

     printf("What is this ==> \\\n", result);
     printf("and this ==> \"");

}
```

6. For Program 1.35, see if you can predict what the output will be in each case. Then enter and execute the program. Be sure to write what your system gave you in your notes.

Program 1.35

```
#include <stdio.h>

     float number_1 = 125.738;

main()
{

     printf("In decimal notation 125.738 = %d\n", number_1);
     printf("In float notation 125.738 = %f\n", number_1);
     printf("In E notation 125.738 = %e\n", number_1);

}
```

Self-Test

DIRECTIONS

Program 1.36 was developed by a beginning C student. It may contain some errors. Answer the questions that follow by referring to this program.

Program 1.36

```
#include <stdio.h>
/****************************************************************/
/*                     Ohms Law Program                         */
/****************************************************************/
/*              Developed by: A. Good Programmer                */
/*                                                              */
/****************************************************************/
/*         This program will solve for the circuit current.     */
/*    The program user must input the values of the circuit     */
/*    voltage and the circuit resistance.                       */
/****************************************************************/
```

```
/*                       Variables used:                         */
/*---------------------------------------------------------------*/
/*           current = Circuit current in amps.                  */
/*           voltage = Circuit voltage in volts.                 */
/*        resistance = Circuit resistance in ohms.               */
/*****************************************************************/
main()
{
     /* Declarations block */

     float voltage;        /* Value of the voltage.    */
     float current;        /* Value of the current.    */
     float resistance;     /* Value of the resistance. */

     /* Explain the program to the user. */

     puts("This program will compute the value of the ");
     puts("circuit current in amps.");
     puts("");
     puts("You must input the value of the circuit");
     puts("voltage in volts and the circuit");
     puts("resistance in ohms.");
     puts("");
     puts("");

     /* Get input from user. */

     printf("Value of the circuit voltage = ");
     scanf("%f", &voltage);

     printf("Value of the circuit resistance =  ");
     scanf("%f", &resistance);

     /* Do the calculations. */

     current = voltage / resistance;    /* Compute the voltage. */

     /* Display the answer. */

     puts("");
     puts("The current in a circuit with a total resistance");
     printf("of %e ohms and a total voltage of %e\n", resistance, voltage);
     printf("volts is %e amps.\n", current);
}
```

Questions

1. Will Program 1.36 compile and execute on your system? If not, why not?
2. Explain what the program does. How did you determine this?
3. How many variables are there in the program? What type are they? How did you determine this?
4. State why the `puts()` function was used to explain the program to the user.
5. State how the user may input the values of the voltage and the resistance. How did you determine this?
6. In what manner will the output values be displayed? How did you determine this?

End-of-Chapter Problems

General Concepts

Section 1.1

1. What type of a program is used in order to enter C source code?
2. State the program that converts your source code into something the computer understands.

Section 1.2

3. What must all C programs start with?
4. State the purpose of the `/*    */` used in C.
5. What indicates the beginning and the end of program instructions in the C language?

Section 1.3

6. What in a program makes it easy to understand, modify, and debug?
7. State the purpose of the programmer's block.
8. What is contained in the programmer's block?
9. Is it necessary to have a programmer's block in order for a C program to compile?

Section 1.4

10. What is a token in C?
11. Name the symbol other than the letters of the English alphabet and the ten decimal digits of the Arabic number system that can be used to develop a C program.
12. State what is meant by a predefined token.

Section 1.5

13. What is the main difference between how strings and characters are represented in a `printf()` function?
14. State what in a `printf()` function specifies how it is to convert, print, and format its arguments.
15. What is the name given to the actual values within the parentheses of a function?
16. How many arguments must a `printf()` function contain?

Section 1.6

17. What is the name given to an independent collection of declarations and statements in C?
18. State what all identifiers must start with in C.
19. How many characters of an identifier are recognized by C?
20. What is an identifier?

Section 1.7

21. Name the three fundamental C type specifiers.
22. How do you set aside a specific memory location to later receive a value in C?
23. In C, what prevents a new variable from coming into your program as a result of a typing error?
24. What is it called when you combine a variable declaration with an assignment operator?

Section 1.8

25. State the unique characteristic of integer division in C.
26. Is there an order in which arithmetic operations must be performed? What is this called?
27. What does the C statement `result *= 5;` mean?
28. What kind of statement is the statement of problem 27 called?

Section 1.9

29. What in a `printf()` function causes a vast departure from the normal interpretation of a string?
30. Name the character in the `printf()` function that is referred to as the escape character.
31. State what determines the number of digits to the right of the decimal point in a displayed value when the `printf()` function is used.

Section 1.10

32. Give an example of the use of the `scanf()` function to enter the value of a declared constant `constant_1` in floating point.
33. Explain the purpose of the `scanf()` function as described in this chapter.
34. Give an example of outputting the value of a declared constant `constant_1` in E notation.
35. For the C statement

```
scanf("%f",&value);
```

in what numerical format may the program user enter values?

Section 1.11

36. What are the three types of compile time error messages used in C?
37. State the action of a fatal error message.
38. What are the three most common errors made by beginning C programmers as presented in this chapter?
39. Do compiler error messages always identify the problem in your program? Explain.

Section 1.12

40. How are specialized functions brought into a C program?
41. If `int` data types are used in Programs 1.21 through 1.24, how does the execution change?

Section 1.13

42. List the four specific areas, as presented in this chapter, that should be stated in writing when designing a program.
43. State the first step in the coding of a program.

Program Design

You now have enough information to actually develop all of the source code for the following programs. As with all programs you design, you are expected to use the steps presented in the Case Study section of this chapter.

44. Write a C program that determines how many binary bits are required to represent an unsigned integer. For example, 3 bits are needed to represent the value 7, four bits are needed for numbers 8 through 15, and so on. Use the `log()` or `log10()` functions in your program. The user must enter the unsigned integer.
45. Develop a C program that evaluates the following expressions:

    N^2 N^3 2^N 3^N

    for values of N equal to 1, 2, 5, 10 and 20.
46. Develop a C program that solves for the power dissipation of a resistor when the voltage across the resistor and the current in the resistor are known. The relationship for resistor power dissipation is

$$P = I \times E$$

Where

P = Power dissipated in watts.
I = Resistor current in amps.
E = Resistor voltage in volts.

47. Create a C program that would solve for the current in a series circuit consisting of three resistors and a voltage source. The program user must input the value of each resistor and the value of the voltage source. The relationship for total current is

$$I_t = V_t/(R_1 + R_2 + R_3)$$

Where

I_t = Total circuit current in amps.
V_t = Voltage of voltage source in volts.
R_1, R_2, R_3 = Value of each resistor in ohms.

48. Develop a C program that will compute the inductive reactance for a particular frequency. The program user must input the value of the inductor and the frequency. The relationship for inductive reactance is

$$X_L = 2\pi fL$$

Where

X_L = Inductive reactance in ohms.
f = Frequency in hertz.
L = Value of inductor in henrys.

49. Create a C program that will solve for the capacitive reactance of a capacitor. The program user must input the value of the capacitor and the frequency. The relationship for capacitive reactance is

$$X_c = 1/(2\pi fC)$$

Where

X_c = Capacitive reactance in ohms.
f = Frequency in hertz.
C = Value of capacitor in farads.

50. Write a C program that will find the total impedance of a series circuit consisting of a capacitor and an inductor. Program user is to input the value of the inductor, capacitor, and applied frequency. The relationship for series circuit impedance consisting of a capacitor and inductor is

$$Z = X_L - X_c$$

Where

Z = Circuit impedance in ohms.
X_L = Inductive reactance in ohms.
X_c = Capacitive reactance in ohms.

51. Create a C program that will compute the number of items made in an eight-hour day assuming that the same number of items are made each hour. User input is the number of pieces manufactured in one hour.
52. Develop a C program that will compute the area of a circle. User input is the radius of the circle.
53. Write a C program that will allow the program user to determine the maximum and minimum sizes of a `float` for his or her computer system.
54. Write a C program that will compute the volume of a room. User inputs are the height, length, and width of the room.
55. Create a C program that will compute the number of acres of land. The program user is to input the width and length (assume a perfect rectangle) of the land in feet. The program is to display the answer in acres, without E notation.

Formula: 1 acre = $43{,}560\ ft^2$

56. Develop a C program that will convert from degrees centigrade to degrees Fahrenheit. User input is the temperature in centigrade. The relationship is:

$$F = \frac{9}{5}C + 32$$

Where

C = Temperature in centigrade.
F = Temperature in Fahrenheit.

57. Write a C program that will compute a 6% sales tax on a purchase. The program user is to input the total amount of purchase. The program is to return the original amount of purchase, the tax, and the total of the two.
58. Write a C program that computes the length of the opposite and adjacent sides of a right triangle, given the length of the hypotenuse and an angle, as shown in Figure 1.3. The user must enter the length of the hypotenuse and the angle (in degrees).

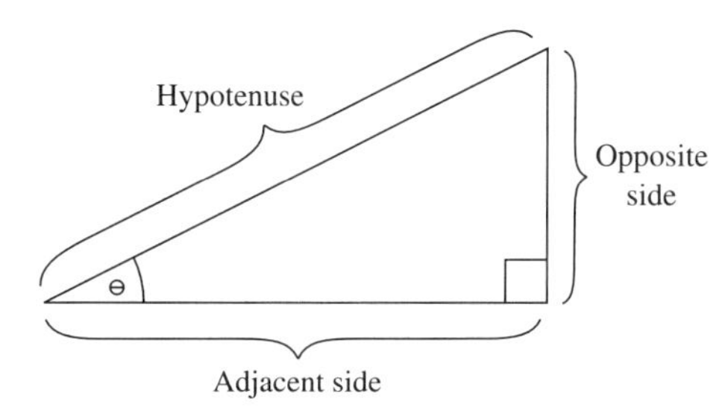

Figure 1.3 For Question 59

2 Structured Programming

Objectives

In this chapter, you will have the opportunity to learn:

1. How to recognize a block structured program.
2. How to develop block structure using C.
3. A very important theorem about programming.
4. The use of C functions in the development of a block structured program.
5. How to pass values between functions.
6. The meaning and use of a formal parameter.
7. The meaning and use of an actual parameter.
8. The meaning of preprocessing and preprocessor commands.
9. The use of the `#define` directive.
10. How to develop and save your own header file.
11. How to use top-down design in the design of a technology problem-solving program.

Key Terms

Program Block
Block Structure
Block Separator
Sequential Block
Loop Block
Branch Block
Function
Function Prototype
Formal Parameter List
Value Passing
Actual Parameter
Formal Parameter
Multiple Arguments
Calling Functions
Recursion
Macro

Header File
Token Pasting
String-izing
Conditional Compilation
Prologue

Outline

Introduction

Chapter 1 got you started developing your own C programs. You learned how to use C to get values from the program user, perform operations on those values, and then display the results.

From this point on, you will learn information that will help you develop technical programs in C that are not so easily solved with the simple scientific pocket calculator. This means that your programs will become longer and perform many kinds of useful operations. Because of this, it is very important that you be introduced to the concept of structured programming. Doing this now will help you develop good programming skills early in your learning of the C language. These good programming habits will then be used in the remaining chapters of this book.

As you will see, it is the purpose of this chapter to help you design programs that are easy to read, understand, debug, and modify. This is a very important chapter.

2.1 Concepts of a Program Block

Discussion

You were introduced to the use of block structure in programming in Chapter 1. There you saw that dividing the program into distinct blocks makes the program easier to read and modify. Remember the analogy to the structure of a business letter. The structure makes the letter easier to read.

This section will present more detailed information on how to develop block structured programs using C.

A Blocking Example

Consider Program 2.1. It does little more than put information out to the screen. However, it does illustrate one of the features of a block structure.

Program 2.1

```
/**************************************************************/
/*                    Typical C Program                       */
/**************************************************************/
/*           Developed by: A. Good Programmer                 */
/*                                                            */
/**************************************************************/
/*       This program will illustrate nothing more than a     */
/*   typical unstructured C program.                          */
/**************************************************************/
main()
{
        /* Explain program to user. */

        puts("This is a C program that illustrates the typical");
        puts("unstructured approach to writing a program in C.");
        puts("When the program is executed, the program user");
        puts("can't tell that the program is unstructured, only");
        puts("the programmer can.");
        puts("Hence a structured program in C is useful only to the");
        puts("programmer, the programmer's boss, the programmer's");
        puts("teacher, those who need to modify the program, and those");
        puts("who are living with the programmer while the programmer");
        puts("is trying to find bugs in the program.");
}
```

Blocking the Structure

You probably had no problem in understanding what Program 2.1 will do. It simply prints a bunch of strings out to the monitor. So why structure it?

You should consider several points. First, the program looks boring. Every program line starts at the same place—at the same column on the left, one `puts` function after another. The structure could be made more readable and a little more interesting if paragraphs could be used. This would help distinguish one part of the program from another. Each paragraph can be thought of as a **program block**. Each program block can be thought of as having one main idea needed in the program. Recall that the case study in Chapter 1 presented this same concept. It shall be used again here.

Program 2.2 shows a simple way of making Program 2.1 a little more interesting.

Program 2.2

```
/**************************************************************/
/*                    Typical C Program                       */
/**************************************************************/
/*           Developed by: A. Good Programmer                 */
/*                                                            */
/**************************************************************/
/*       This program will illustrate the most simple form    */
```

```
/*   of block structured program.                              */
/****************************************************************/

main()
{
        /* First paragraph of explanation. */

        puts("This is a C program that illustrates the typical");
        puts("unstructured approach to writing a program in C.");

        /* End first paragraph of explanation. */

/*-------------------------------------------------------------*/

        /* Second paragraph of explanation. */

        puts("When the program is executed, the program user");
        puts("can't tell that the program is unstructured, only");
        puts("the programmer can.");

        /* End second paragraph of explanation. */

/*-------------------------------------------------------------*/

        /* Third paragraph of explanation. */

        puts("Hence a structured program in C is useful only to the");
        puts("programmer, the programmer's boss, the programmer's");
        puts("teacher, those who need to modify the program, and those");
        puts("who are living with the programmer while the programmer");
        puts("is trying to find bugs in the program.");

        /* End third paragraph of explanation. */
}
```

Note the use of the dashes in Program 2.2 to emphasize each block of the program. Also note the comments at the beginning and end of each block. These tell you the purpose of the block you are entering and state it again when you are leaving it. For the simple program here, the added structure did nothing to help you understand it, but it does serve as a model for what is to follow.

Defining a Block Structure

Block structure means that the program will be constructed so that it consists of a few groups of instructions rather than one continuous listing of instructions. This is the way the longer programs in the previous chapters were presented.

Each group or block of instructions starts with a remark statement telling what the block is to do. An example from Program 2.2 is

```
/* Second paragraph of explanation. */
```

Each program block is defined by separating spaces called **block separators.*** An example of a block separator is

```
/*-----------------------------------------------------------*/
```

although a few blank lines will also separate blocks adequately.

The body of every program block is highlighted by being indented from the left column. The exact number of spaces from the left is not critical. What is important is that you make a distinction between the body of the block and its beginning and ending comments. An example from Program 2.2 is

```
    /* Second paragraph of explanation. */

    puts("When the program is executed, the program user");
    puts("can't tell that the program is unstructured, only");
    puts("the programmer can.");

    /* End second paragraph of explanation. */
```

Some Important Rules

What is important in a structured program is that there is no jumping around. If you don't know how to do this (if you have never used a GOTO) then consider yourself lucky! People can understand things a lot easier if they can follow them logically from one step to another. Hence, you should always follow these important rules when doing block structured programming:

1. All blocks are entered from the top.
2. All blocks are exited from the bottom.
3. When the computer finishes one block, it goes on to another or ends.

Can you design every program this way? Yes, you can—with absolutely no exceptions. There is no excuse for writing a program in C that doesn't contain block structure.

Types of Blocks

No matter what programming language you use, there are only three necessary types of blocks:

1. Sequential block
2. Loop (or iteration) block
3. Branch (or selection) block

What does each of these blocks do? A **sequential block** is the simplest kind of programming block. It is nothing more than a straight sequence of statements, one following the other. A **loop block** can cause the program to go back and repeat a part of the program over again, while a **branch block** gives the option of performing a different sequence of instructions. The concepts of these three different kinds of blocks are illustrated in Figure 2.1.

---

*Use of block separators is optional. To conserve space, they are not always used in the sample programs.

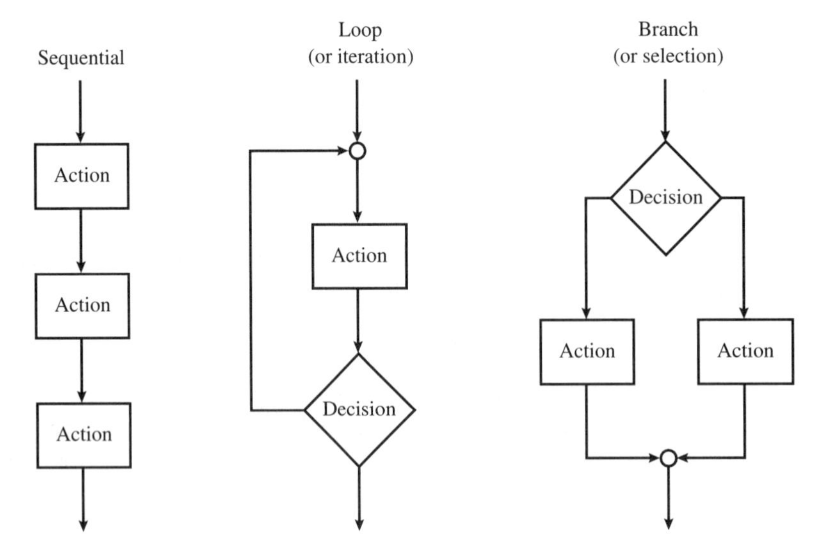

Figure 2.1 Concepts of the Three Kinds of Programming Blocks

Conclusion

This section emphasized the importance of block structure in programming and presented a specific example. You were also introduced to a method of breaking a C program into blocks. For now, test your understanding of this section by trying the following section review.

2.1 Section Review

1. Does a C compiler require that programs be structured? Explain.
2. Define the term "block structure."
3. State one way of beginning a program block.
4. Explain how program blocks may be separated.
5. Explain how the body of a program block is highlighted.
6. Name the three types of blocks.

2.2 Using Functions, Part 1

Discussion

In the last section you were reintroduced to the concept of block structured programming. In this section, you will see how a C program is designed for this kind of structure. This section will serve as an introduction to this important concept of C. You will have an opportunity to increase your understanding of it as you progress through the text.

What Is a C Function?

A C function is an independent collection of source code designed to perform a specific task. All C programs have at least one function called `main()`.

You have already used other functions that are built into your C library, such as `puts()`, `printf()`, and `scanf()`. All of these built-in functions actually do something.

Making Your Own Functions

You can also make your own functions in C. Doing this allows you to create a function and tell C what the function is to do. Then you can use it over and over again just like the built-in functions of C. This means that you could create a function to solve for an electrical series circuit, one to solve for a parallel circuit, or one to solve for the electrical characteristics of a transistor amplifier—to name just a few. Define it only once, give it a name, and then call on it any time you want—just like you call on `puts()` or `printf()` any time you want. As you can see from this, to call a function, you simply use its name. You could even create your own library of functions, save them on your disk, and then invoke them into your program the same as you do `puts()`, `printf()`, and `scanf()`.

This concept is illustrated in Figure 2.2.

What Makes a Function?

In C, when you create a **function** other than `main()`, you first declare it and then define it. When you declare a function, you code in what is called the **function prototype**. A function prototype gives the function name plus other important information concerning the function. This appears at the beginning of the program before `main()`.

When you define a function, you again give the name of the function and other information about it (just as you did for the prototype), and you also produce the body of the

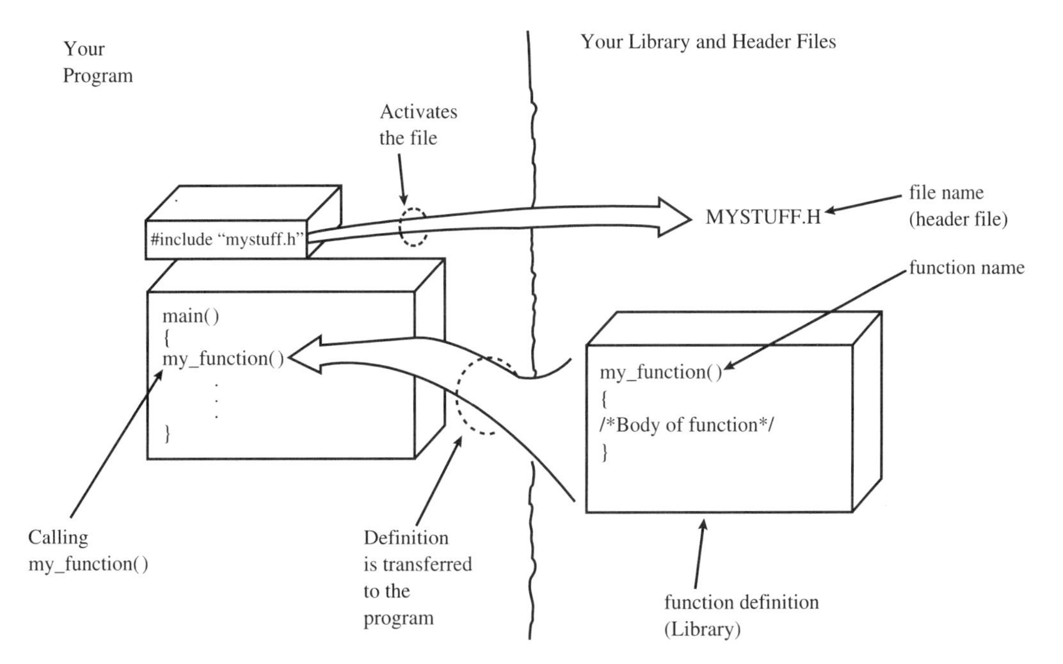

Figure 2.2 Concept of Using Your Own C Functions

function, which contains all of the source code to be used by that function. Everything said here is shown in the following example.

An Example. Program 2.3 is the program from the last section, but restructured so that it can use C functions. Its output to the screen will be exactly the same as it was before. Since you know exactly what the program will do, you can now concentrate on how it is going to do it.

Program 2.3

```
/**************************************************************/
/*                    Typical C Program                       */
/**************************************************************/
/*            Developed by: A. Good Programmer                */
/*                                                            */
/**************************************************************/
/*       This program will illustrate the most simple form    */
/*   of block structured program.                             */
/**************************************************************/

/*   Function prototypes                                      */
/*----------------------------------------------------------*/

        void first_paragraph(void);
/*   This function performs the first paragraph for this      */
/*   program.                                                 */
/*----------------------------------------------------------*/

        void second_paragraph(void);
/*   This function performs the second paragraph for this     */
/*   program.                                                 */
/*----------------------------------------------------------*/

        void third_paragraph(void);
/*   This function performs the third paragraph for this      */
/*   program.                                                 */
/*----------------------------------------------------------*/

main()
{
        first_paragraph();       /* First paragraph of explanation.  */
        second_paragraph();      /* Second paragraph of explanation. */
        third_paragraph();       /* Third paragraph of explanation.  */
        exit(0);
}

void first_paragraph()
{
        /* First paragraph of explanation. */

        puts("This is a C program that illustrates the typical");
        puts("unstructured approach to writing a program in C.");

        /* End first paragraph of explanation. */
```

```
}

void second_paragraph()
{
        /* Second paragraph of explanation. */

        puts("When the program is executed, the program user");
        puts("can't tell that the program is unstructured, only");
        puts("the programmer can.");

        /* End second paragraph of explanation. */
}

void third_paragraph()
{
        /* Third paragraph of explanation. */

        puts("Hence a structured program in C is useful only to the");
        puts("programmer, the programmer's boss, the programmer's");
        puts("teacher, those who need to modify the program, and those");
        puts("who are living with the programmer while the programmer");
        puts("is trying to find bugs in the program.");

        /* End third paragraph of explanation. */
}
```

Program Analysis

The key to analyzing Program 2.3 is to note what the function `main()` now contains:

```
main()
{
        first_paragraph();  /* First paragraph of explanation. */
        second_paragraph(); /* Second paragraph of explanation.*/
        third_paragraph();  /* Third paragraph of explanation. */
        exit(0);
}
```

The `main()` function now consists of nothing more than a series of calls to other functions. The first function is called `first_paragraph()`. What it does is go to the definition of the function by the same name:

```
void first_paragraph()
{
        /* First paragraph of explanation. */

        puts("This is a C program that illustrates the typical");
        puts("unstructured approach to writing a program in C.");

        /* End first paragraph of explanation. */
}
```

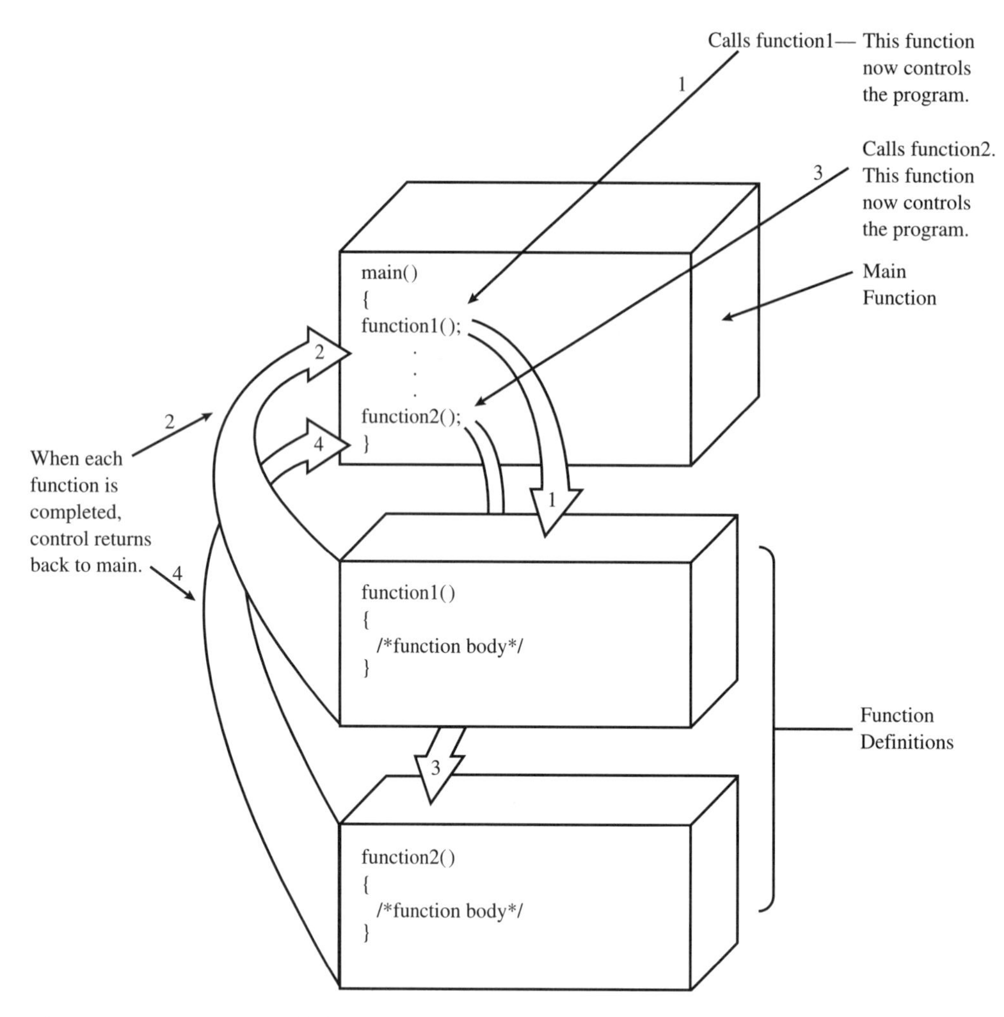

Figure 2.3 The Process of Calling Functions from `main()`

It is here that the function `first_paragraph()` is defined. The definition consists of the program code put between the `{` and the `}`. This is exactly the same thing you have been doing with the C reserved function `main()`!

This process of calling each function, in the order needed, is done by `main()`. This is illustrated in Figure 2.3.

There are some other important things to point out. First note the statement of what each function will do under the heading of `Function prototype`:

```
/* Function prototypes                                    */
/*-------------------------------------------------------*/
      void first_paragraph(void);
/* This function performs the first paragraph for this   */
/* program.                                               */
```

```
/*------------------------------------------------------*/
      void second_paragraph(void);
/* This function performs the second paragraph for this */
/* program.                                             */
/*------------------------------------------------------*/
      void third_paragraph(void);
/* This function performs the third paragraph for this  */
/* program                      .                       */
/*******************************************************/
```

Note that all three of the functions defined in the program (`first_paragraph`, `second_paragraph`, and `third_paragraph`) are not commented out (do not have `/*` and `*/` around them). This means that they represent some kind of instruction to C. The instruction actually tells the compiler that there will be a function by a certain name to be defined later in the program. This is the function prototype. The function has a type (just like variables and constants have types), and for this program, the type is `void`. `void` means that the function will not return a value. This is so because the definition of the function only causes some words to appear on the monitor; it isn't computing anything or doing anything else with values. The second thing it states is that there will not be anything contained in its **formal parameter list** (between the `(` and `)`)*. This is important because doing this instructs the compiler how much and what kind of memory to save for these functions. This is also declared as `void` because no values will be placed here.

The concept of a function prototype is illustrated in Figure 2.4.

Note that each function prototype ends with a semicolon. Also note that when the function was called in `main()` it also ended with a semicolon. Now note—and this is important—that when the function is defined, it does not end with a semicolon. This is no different from what you have been doing with `main()`—when defining it, you don't end it with a semicolon.

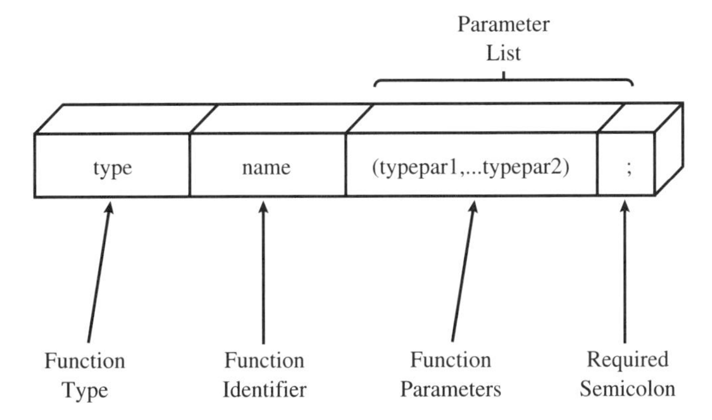

Figure 2.4 Concept of a Function Prototype

*Information contained within the parentheses are parameters. The actual values used for this information are called *arguments*.

Note the structure of the function definition:

```
void first_paragraph()
{
        /*  First paragraph of explanation. */
        puts("This is a C program that illustrates the typical");
        puts("unstructured approach to writing a program in C.");
        /* End first paragraph of explanation. */
}
```

The definition states the function type (`void`—because it isn't returning any value to the calling function). It states its name, and it is immediately followed by the required `()`—which don't contain anything between them because it was already stated in the prototype declarations that they wouldn't (they were `(void)`). There is no semicolon (as stated before). The body of the function is defined by the opening brace `{` and then the closing brace `}`.

The structure of a function definition is illustrated in Figure 2.5.

There is one last important point. Look again at the function `main()`:

```
main()
{
        first_paragraph();  /* First paragraph of explanation.  */
        second_paragraph(); /* Second paragraph of explanation. */
        third_paragraph();  /* Third paragraph of explanation.  */
        exit(0);
}
```

Its last function call is `exit(0);`. `exit` is a C function that causes normal termination of a C program.

Conclusion

This section contained a lot of new information. The importance of this section was not so much the exact details (although they will be important when it comes to programming) but more the concepts—the concept of how a program may be divided into discrete parts

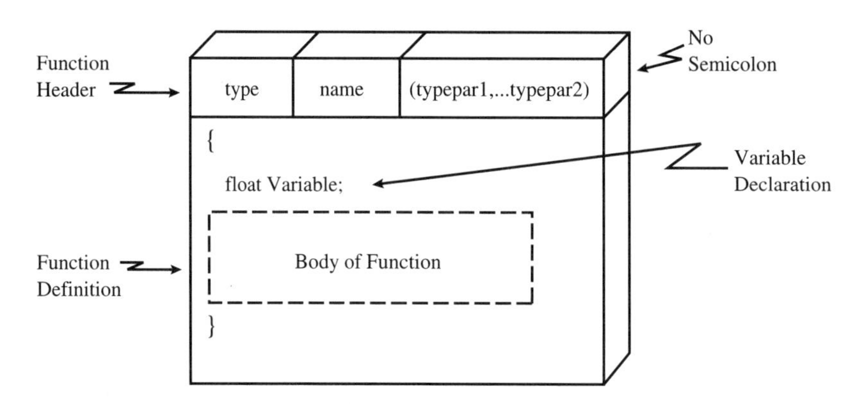

Figure 2.5 Structure of a Function Definition

called functions, and how one function, the `main()` function, can then call these other functions in an order that determines how the program is executed. Check your understanding of this section by trying the following section review.

2.2 Section Review

1. What is a C function?
2. Explain what is meant by a type `void`. Give an example.
3. State the meaning of a function prototype.
4. What is the purpose of using a function prototype?
5. State what is meant by calling a function.
6. Why is the `exit()` used as the last function call in `main()`?

2.3 Inside a C Function

Discussion

In this section, you will discover how functions may be used to actually return a value. This means you can now use functions to perform complex calculations for you and then return the results of these computations to the calling function.

Basic Idea

Program 2.4 gets a value from the program user and squares it.

Program 2.4

```
#include <stdio.h>

main()
{
        float number;    /* Number to be squared.     */
        float answer;    /* The square of the number. */

        printf("Give me a number and I'll square it => ");
        scanf("%f", &number);

        answer = number * number;

        printf("The square of %f is %f", number, answer);

        exit(0);
}
```

When Program 2.4 is executed and the user enters the value 3, the results are

```
Give me a number and I'll square it => 3
The square of 3.000000 is 9.000000
```

To illustrate how a C function can be used in Program 2.4, consider Program 2.5.

Program 2.5

```
#include <stdio.h>

/* Function prototype. */

float square_it(float number);
/* This is the function that will square the number. */

main()
{
        float value;    /* Number to be squared.      */
        float answer;   /* The square of the number.  */

        printf("Give me a number and I'll square it => ");
        scanf("%f", &value);

        answer = square_it(value);        /* Call the function. */

        printf("The square of %f is %f", value, answer);

        exit(0);
}

float square_it(float number)
{
        float answer;             /* The square of the number. */

        answer = number * number;

        return(answer);
}
```

Program Analysis

Program 2.5 has created a separate function called `square_it`. This new function computes the square of a number. Where does it get this number? It gets it from the main function `main()`. How does it get it? It is passed to it from `main()`. How is it passed to it? It is passed to it through its parameter argument (`float number`). Figure 2.6 shows how **value passing** is done.

Note from Figure 2.6 that the formal parameter list contains declarations of the function parameters. In this case, the function has only one, called `parameter_1`. This formal parameter is present in the function prototype as well as in the head of the function definition. However, when the function is called (as called from `main()`), its actual parameter need not have the same name as its formal parameter (it must still have the same type). For example, in Program 2.5, the **actual parameter** was `value` and the formal parameter was `number`. Thus the **formal parameters** define the types and numbers of function parameters, whereas the actual parameters are used when calling the function.

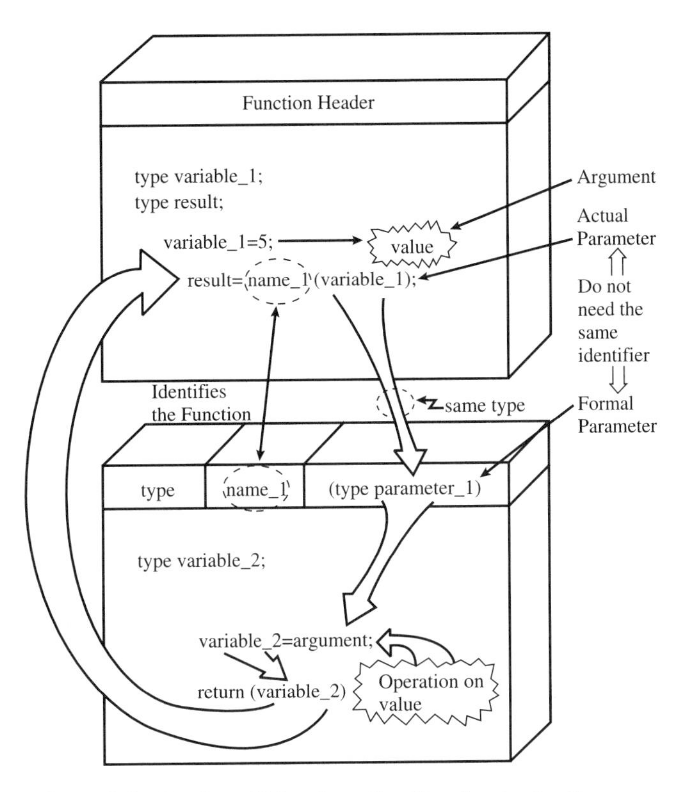

Figure 2.6 Passing a Value from One Function to Another

As you can see from the figure, the parameter of the function `square_it` is no longer void (it is not `square_it(void)` but is instead `square_it(float number)`). This was first done in the prototype part of Program 2.5:

```
/* Function prototype.  */

float square_it(float number);
/* This is the function that will square the number.  */
```

Note that since this function will return a value, it is no longer of type `void`. Hence the compiler must be told what type it is to be. Keeping with what you have been doing up to this point, it is given the type `float`. This means that the answer it gets after squaring the number will be of type `float`. The number that it will square must also have a type. The compiler is also told this in the prototype declaration:

```
(float number);
```

Thus, the prototype declaration for this function tells the compiler to expect a function definition in the program that is identified as `square_it` and that will return a value of type `float` and will do its operation (squaring in this case) on another number identified as `number`, which is also of type `float`. All of this necessary information is given in the prototype declaration:

```
float square_it(float number);
```

Note that the declaration ends with the necessary semicolon.

Next look at the function definition:

```
float square_it(float number)
{
        float answer;    /* The square of the number.    */
        answer = number * number;
        return(answer);
}
```

The function definition starts out exactly like the prototype; the only important difference is that *it does not end with a semicolon*! This is important because it tells the compiler that what is to follow is the definition of this function. It also defines its type and the type of its parameter. This is exactly what was done in the prototype. Remember, the prototype tells the compiler what kind of memory to reserve; the function definition describes exactly what action is to take place.

There is a new C term used in this function definition. It is

```
return();
```

This tells the computer which value is to be returned to the calling function. In this case, the value of the variable `answer` is to be returned. Thus, within the argument of `return`, the variable identifier `answer` is placed:

```
return(answer);
```

And this ends with the required semicolon. Thus, this function will get a value from the calling function. This value will be of type `float` and will be placed in the variable identifier `number`. A computation will be performed:

```
answer = number * number;
```

the results of which will be placed in the variable identifier called `answer`. The value contained in `answer` will then be passed back to the calling function by the command

```
return(answer);
```

Note: on may C compilers the parentheses are optional in the `return` statement.

Next, look at the function `main()`:

```
main()
{
        float value;     /* Number to be squared.     */
        float answer;    /* The square of the number. */

        printf("Give me a number and I'll square it => ");
        scanf("%f", &value);

        answer = square_it(value);       /* Call the function. */

        printf("The square of %f is %f", value, answer);

        exit(0);
}
```

Table 2.1 Important Function Definitions

Term	Example	Comments
Formal parameters [used when defining the function].	`square_it(float number)`	`number` is a formal parameter. It states that a quantity of type `float` will be passed to the function called `square_it(quantity)`.
Actual parameters [used when calling the defined function].	`square_it(value)`	`value` is an actual parameter. The identifier used as the actual parameter may be different from that of the formal parameter, but it must be of the same type.
Arguments [actual values passed to the called function].	`value = 5;` `square_it(value);`	The number 5 is the argument because it is the actual value passed to the function `square_it(value)`.

The function `main()` uses two variables. One variable is the value to be squared (`value`), and the other is needed to contain the answer of the result (`answer`). The program simply asks the program user for a number to be squared and then gets the number from the keyboard. This is done with the `printf` and the `scanf` functions. But now look at what is different: to do the computation, the function that defines squaring need only be called. The value `value` is placed as a function parameter, and this function is made equal to the variable `answer`. The `exit(0);` terminates the program.

Table 2.1 illustrates some of the important function definitions.

Conclusion

This was an important section. Again, the details are less important than the concepts. Here the concept of passing a value from one function to another was introduced. You saw that this can be done using parameters that will represent a place holder for a value to be passed from one function to another. The second idea is using a special C command that will pass a value back to the calling function (`return()`).

In the next section, you will see a more useful application of passing values between functions. For now, test your understanding of this section by trying the following section review.

2.3 Section Review

1. State what is meant by the parameter of a function.
2. What is the difference between a formal parameter and an actual parameter? How must these be the same? How may these be different?
3. Explain the meaning of passing values between functions.
4. What is the difference between the coding of a function prototype and the head of a function declaration?
5. How is a value passed back to the calling function?

2.4 Using Functions, Part 2

Discussion

In the last section you were introduced to the power of passing values between functions. In this section you will see how to do more of this as well as some more useful items about functions. You will find that the backbone of the C language is its functions. You will spend a lot of time learning about functions. This section presents some more important information.

More Than One Argument

Program 2.6 illustrates the passing of **multiple arguments** between functions. What the program does is to use a function to calculate the capacitive reactance of a capacitor given the value of the capacitor and the frequency. The mathematical relationship is

$$X_c = 1/(2\pi f C)$$

Where

X_c = The reactance of the capacitor in ohms.
f = The frequency in hertz.
C = The value of the capacitor in farads.

Program 2.6

```
#include <stdio.h>

/* Function prototype. */

float capacitive_reactance(float capacitance, float frequency);
/* This is the function that computes the capacitive reactance
   for a given value of capacitor and frequency. */

main()
{
        float farads;           /* Value of the capacitor. */
        float hertz;            /* Value of the frequency. */
        float reactance;        /* Value of the reactance. */

        printf("Input value of the capacitor in farads => ");
        scanf("%f",&farads);
        printf("Input value of the frequency in hertz  => ");
        scanf("%f",&hertz);

        reactance = capacitive_reactance(farads, hertz);

        printf("The reactance of a %e farad capacitor\n",farads);
        printf("at a frequency of %e hertz is %e ohms\n",hertz, reactance);

        exit(0);
}
```

```
float capacitive_reactance(float capacitance, float frequency)
{
        float reactance;        /* The capacitive reactance. */
        float pi = 3.14159;

        reactance = 1/(2 * pi * frequency * capacitance);

        return(reactance);
}
```

Program Analysis

Program 2.6 uses two functions: `main()` and `capacitive_reactance()`. The feature of this program is that two values must be passed from `main()` to `capacitive_reactance()` (the value of the capacitor and the value of the frequency). This is accomplished by having two formal parameters for `capacitive_reactance()`:

```
float capacitive_reactance(float capacitance, float frequency);
```

As you can see from the prototype, these formal parameters are identified as `capacitance` and `frequency`. Both are declared as type `float`. The function itself will return a value, so it is also declared as a type `float`. Now look at the function definition:

```
float capacitive_reactance(float capacitance, float frequency)
{
        float reactance;   /* The capacitive reactance.  */
        float pi = 3.14159;

        reactance = 1/(2 * pi * frequency * capacitance);

        return(reactance);

}
```

Again, the function heading is exactly like its prototype—with the important exception that it does not end in a semicolon. The body of the function starts with the required { and then declares two identifiers. One, `reactance`, will represent the value of the computed capacitive reactance while the other is the value for pi. The calculation is then performed and the single value is returned back to the calling procedure.

Now, look at the function `main()`:

```
main()
{
        float farads;          /* Value of the capacitor. */
        float hertz;           /* Value of the frequency. */
        float reactance;       /* Value of the reactance. */

        printf("Input value of the capacitor in farads => ");
        scanf("%f",&farads);

        printf("Input value of the freqency in hertz => ");
        scanf("%f",&hertz);
```

```
        reactance = capacitive_reactance(farads, hertz);

        printf("The reactance of a %e farad capacitor\n",farads);
        printf("at a frequency of %e hertz is %e ohms\n",hertz, reactance);

        exit(0);
}
```

main() contains the type declarations of the three variables it will be using. Two, farads and hertz, will be entered by the program user through two separate scanf() functions. Each one is preceded by the necessary &. The third, reactance, will be used to display the resulting answer. After the user input is received, the defined function is called and the arguments are passed to it:

```
reactance = capacitive_reactance(farads, hertz);
```

This results in the variable identifier reactance receiving the result of the reactance calculation. This result is then echoed back to the monitor so the program user may observe the result.

Execution of Program 2.6 produces the following results (assume the program user has entered the value of a 1μf capacitor and a frequency of 5 kHz):

```
Input value of the capacitor in farads => 1e-6
Input value of the frequency in hertz => 5e3
The capacitive reactance => 3.183101e+001 ohms.
```

Calling More Than One Function

Recall that a C function may call more than one function. **Calling functions** is illustrated in Program 2.7. What the program does is to solve for both the capacitive reactance of a capacitor and the inductive reactance of an inductor where the inductive reactance is given by

$$X_L = 2\pi f L$$

Where

X_L = The inductive reactance in ohms.
f = The frequency in hertz.
L = The inductance in henrys.

Program 2.7

```
#include <stdio.h>

/* Function prototype. */

float capacitive_reactance(float capacitance, float frequency);
/* This function computes the capacitive reactance
   for a given value of capacitor and frequency. */

float inductive_reactance(float inductance, float frequency);
/* This function computes the inductive reactance
```

```
   for a given value of inductor and frequency. */

main()
{
        float farads;           /* Value of the capacitor. */
        float henrys;           /* Value of the inductor.  */
        float hertz;            /* Value of the frequency. */
        float reactance;        /* Value of the reactance. */

        printf("Input value of the capacitor in farads => ");
        scanf("%f",&farads);
        printf("Input value of the inductor in henrys => ");
        scanf("%f",&henrys);
        printf("Input value of the frequency in hertz  => ");
        scanf("%f",&hertz);

        reactance = capacitive_reactance(farads, hertz);

        printf("The reactance of a %e farad capacitor\n",farads);
        printf("at a frequency of %e hertz is %e ohms\n",hertz, reactance);

        reactance = inductive_reactance(henrys, hertz);

        printf("The reactance of a %e henry inductor\n",henrys);
        printf("at a frequency of %e hertz is %e ohms\n",hertz, reactance);

        exit(0);
}

float capacitive_reactance(float capacitance, float frequency)
{
        float reactance;        /* The capacitive reactance. */
        float pi = 3.14159;

        reactance = 1/(2 * pi * frequency * capacitance);

        return(reactance);
}

float inductive_reactance(float inductance, float frequency)
{
        float reactance;        /* The inductive reactance. */
        float pi = 3.14159;

        reactance = 2 * pi * frequency * inductance;

        return(reactance);
}
```

Program Analysis

The main difference in Program 2.7 is that there are now two functions other than `main()`. These are

```
float capacitive_reactance(float capacitance, float frequency);
```

and

```
float inductive_reactance(float inductance, float frequency);
```

Both of these functions have their separate definitions. As before, each function definition starts with the same heading as its prototype, with the required exception of the semicolon:

```
float capacitive_reactance(float capacitance, float frequency)
{
     float reactance;    /* The capacitive reactance.   */
     float pi = 3.14159;

     reactance = 1/(2 * pi * frequency * capacitance);

     return(reactance);
}

float inductive_reactance(float inductance, float frequency)
{
     float reactance;   /* The inductive reactance.   */
     float pi = 3.14159;

     reactance = 2 * pi * frequency * inductance;

     return(reactance);
}
```

It actually makes no difference in what order these functions appear in the program. Next, the calling to each of these functions is done by `main()`:

```
reactance = capacitive_reactance(farads, hertz);
```

and

```
reactance = inductive_reactance(henrys, hertz);
```

Notice that the same variable identifier `reactance` is used in both cases. This can be done because after each use its value is displayed:

```
reactance = capacitive_reactance(farads, hertz);

printf("The reactance of a %e farad capacitor\n"farads);
printf("at a frequency of %e hertz is %e ohms\n",hertz, reactance);

reactance = inductive_reactance(henrys, hertz);

printf("The reactance of a %e henry inductor\n",henrys);
printf("at a frequency of %e hertz is %e ohms\n",hertz, reactance);
```

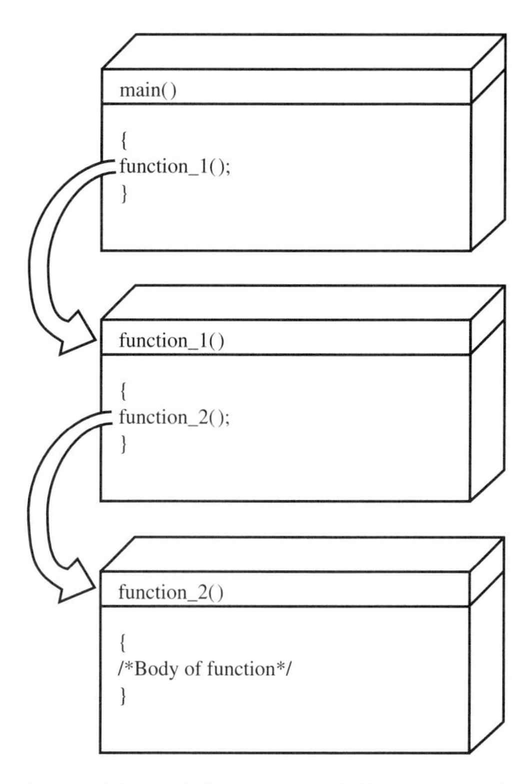

Figure 2.7 A Called Function Calling Another Function

Calling Functions from within a Function

A called function may also call another function. This concept is illustrated in Figure 2.7.

Program 2.8 illustrates the process. It solves for the impedance of a series LC circuit. The mathematical relationship is

$$Z = (X_L + X_C)$$

Where

Z = Circuit impedance in ohms.
X_c = Capacitive reactance in ohms.
X_L = Inductive reactance in ohms.

The impedance Z is based on the difference of the inductive and capacitive reactances. Both reactance values and the overall impedance can be calculated in separate functions. For now, the important point is to see how a called function in turn calls two other functions. Observe Program 2.8.

Program 2.8

```
#include <stdio.h>

/* Function prototype. */
```

```
float capacitive_reactance(float capacitance, float frequency);
/* This function computes the capacitive reactance
   for a given value of capacitor and frequency. */

float inductive_reactance(float inductance, float frequency);
/* This function computes the inductive reactance
   for a given value of inductor and frequency. */

float series_impedance(float capacitor, float inductor, float frequency);
/* This function computes the series impedance of an inductor and capacitor. */

main()
{
        float farads;            /* Value of the capacitor. */
        float henrys;            /* Value of the inductor.  */
        float hertz;             /* Value of the frequency. */
        float impedance;         /* Value of the impedance. */

        printf("Input value of the capacitor in farads => ");
        scanf("%f",&farads);
        printf("Input value of the inductor in henrys => ");
        scanf("%f",&henrys);
        printf("Input value of the frequency in hertz  => ");
        scanf("%f",&hertz);

        impedance = series_impedance(farads, henrys, hertz);

        printf("The impedance of the circuit is %e ohms\n",impedance);

        exit(0);
}

float capacitive_reactance(float capacitance, float frequency)
{
        float reactance;         /* The capacitive reactance. */
        float pi = 3.14159;

        reactance = 1/(2 * pi * frequency * capacitance);

        return(reactance);
}

float inductive_reactance(float inductance, float frequency)
{
        float reactance;         /* The inductive reactance. */
        float pi = 3.14159;

        reactance = 2 * pi * frequency * inductance;

        return(reactance);
}
```

```
float series_impedance(float capacitor, float inductor, float frequency)
{
        float cap_react;        /* Resulting capacitive reactance. */
        float ind_react;        /* Resulting inductive reactance.  */
        float impedance;        /* Resulting impedance.            */

        cap_react = capacitive_reactance(capacitor, frequency);
        ind_react = inductive_reactance(inductor, frequency);

        printf("The capacitive reactance => %e ohms\n", cap_react);
        printf("The inductive reactance  => %e ohms\n", ind_react);

        impedance = ind_react - cap_react;

        return(impedance);
}
```

Program Analysis

Program 2.8 illustrates that a called function may call other functions. Note that the order in which the functions are defined makes no difference in the program. The function called from `main()` is `series_impedance`. The definition for this function is located at the end of the program. This function then calls two other functions:

```
cap_react = capacitive_reactance(capacitor, frequency);
ind_react = inductive_reactance(inductor, frequency);
```

Note that each of these called functions returns its calculated value back to the calling function (`series_impedance`). These values are then used in the calculation of the impedance. This calculation requires taking the difference of the inductive and capacitive reactances. Note that the result will be negative when the capacitive reactance is larger than the inductive reactance.

The called function `series_impedance` then returns the value of the computation to its calling function `main()`. Here, the final value is displayed to the program user. Assuming that the program user entered a value of 1 μF for the capacitor, 1 mH for the value of the inductor, and a frequency of 5 kHz, then execution of the above program would yield:

```
Input value of the capacitor in farads => 1e-6
Input value of the inductor in henrys => 1e-3
Input value of the frequency in hertz => 5e3
The capacitive reactance => 3.183101e+001 ohms.
The inductive reactance => 3.141590e+001 ohms.
The impedance of the circuit is -4.151115e-001 ohms.
```

How Functions May Be Called

Figure 2.8 illustrates the different ways functions may call other functions. Notice that a function may also call itself. This is called **recursion**. The only restriction is that a function cannot be defined within the body of another function.

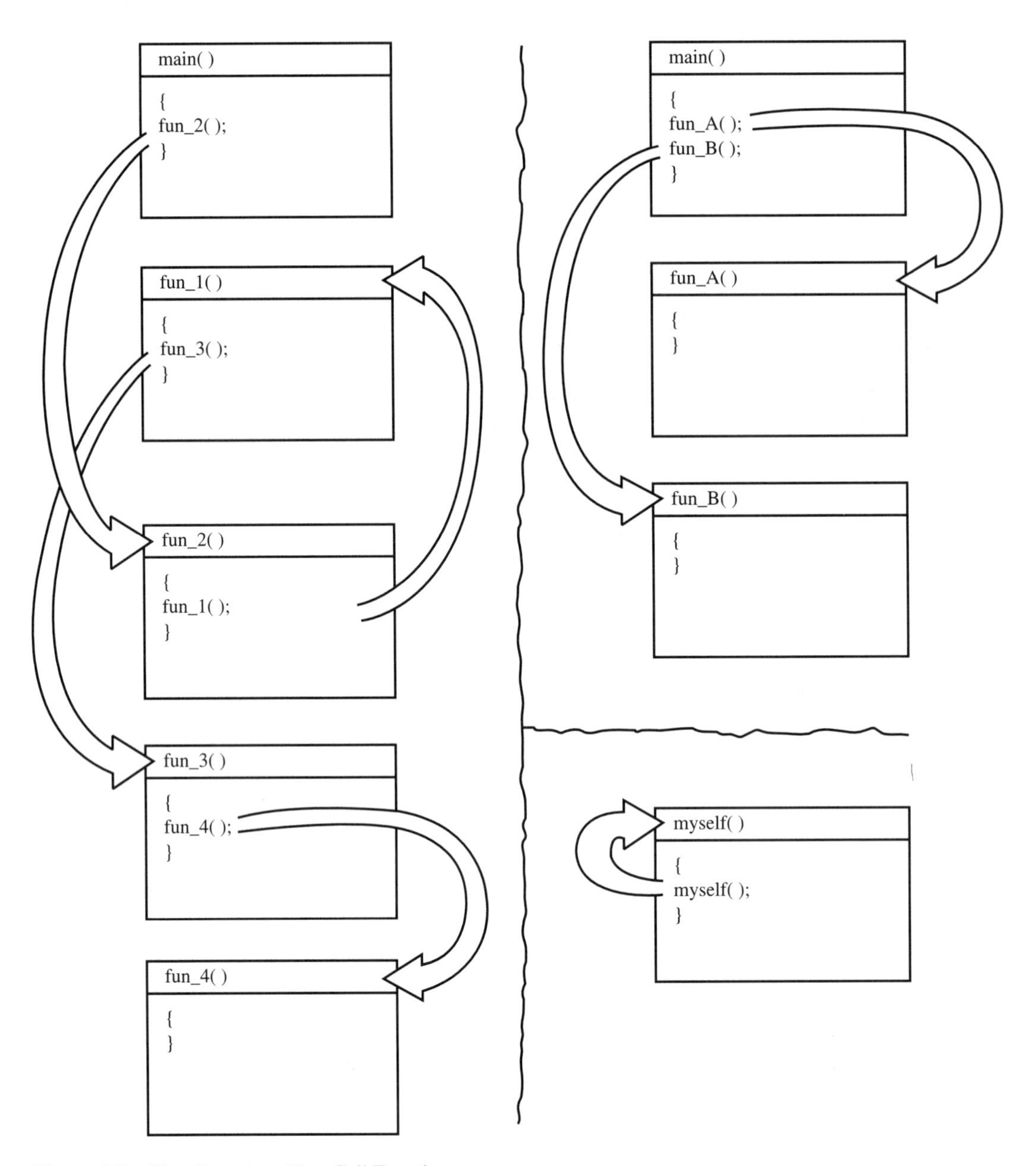

Figure 2.8 How Functions May Call Functions

Conclusion

In this section you learned a lot more about the structure of a C program. You saw that more than one value may be passed to a called function. You saw that a function may call more than one other function. The order of defining functions in the program makes no difference. You also saw that a called function may call another function and that a function may even call itself.

In the next section you will learn how to define items in your C program. As you will see, you may later store this information on your own disk for use by all of your programs. Check your understanding of this section by trying the following section review.

2.4 Section Review

1. May a function pass more than one value to a called function? Explain.
2. Is it possible for a function to call more than one function? Explain.
3. Does it make any difference in which order called functions are defined in the body of the C program?
4. Explain what is meant by a called function calling another function.
5. What is recursion?

2.5 Using #define

Discussion

In the last section you saw more of the power of functions. In this section you will see how you can define your own constants within your programs. Doing this makes your program more portable, easier to change, and easier to understand. Here you will see how to use this new and powerful feature of the C system.

Basic Idea

Look at Program 2.9. What it does is take the number 5 and square it. It does this by using the preprocessor command `#define` where the identifier `square` is defined with its parameter (`x`).

Program 2.9

```
#include <stdio.h>
#define square(x) x*x

main()
{
        float number;    /* Square of a number. */

        number = square(5);

        printf("The square of 5 is %f",number);
}
```

Program Analysis

Program 2.9 uses the preprocessor directive `#define`. This defines a preprocessor **macro**. A macro, in this sense, is simply a string of tokens that will be substituted for another string of tokens. As an example, in the preprocessor macro for Program 2.9

```
#define square(x) x*x
```

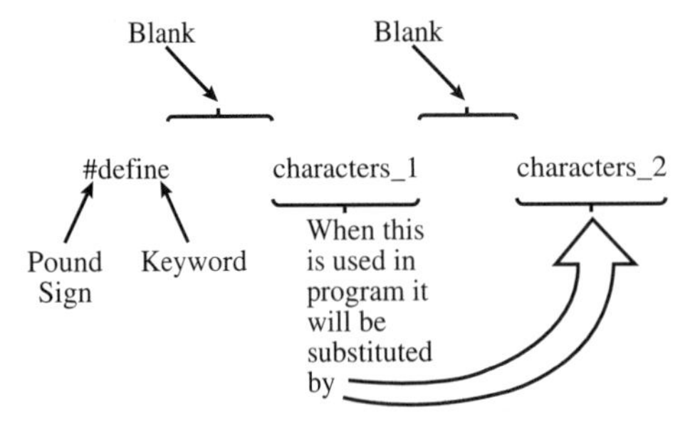

Figure 2.9 Construction of the `#define`

when the statement

```
square(x)
```

appears in the program, the compiler will actually substitute

```
x*x
```

Thus, in the above program, when the macro is used

```
number = square(5);
```

the processor actually substitutes

```
number = 5*5;
```

As you can see, the blanks on each side of the preprocessor directive serve to separate the tokens to be substituted. This is shown in Figure 2.9.

This preprocessor macro may now be used by any function anywhere in the program. Thus, to square any number (constant or variable) simply use

```
answer = square(number);
```

Defining Constants

Program 2.10 illustrates how to define a constant. In this program, the constant `PI` is defined.

This program computes the area of a circle. It uses the preprocessor directives of defining `PI` and `square(x)`. Thus, anywhere that `PI` appears in the program text, we can simply substitute the value 3.141592.

Program 2.10

```
#include <stdio.h>
#define square(x) x*x
#define PI 3.141592

main()
{
        float area;      /* Circle area.    */
```

```
        float radius;    /* Circle radius. */

        printf("Give me the radius => ");
        scanf("%f",&radius);

        area = PI * square(radius);

        printf("The area of a circle with a radius of %f\n",radius);
        printf("is %f square units.",area);
}
```

Program Analysis

Observe that Program 2.10 uses a preprocessor macro:

```
#define square(x) x*x
```

and defines a constant (in C, constants are written in uppercase):

```
#define PI 3.141592
```

Both statements start at the left side of the screen with the required pound sign (#) immediately followed by the reserved word (in lowercase letters) `define`. Then come the required space and the tokens you want defined, then another space and the tokens that are to be substituted. From now on, in any function you write in the program using these directives, the token `PI` will be substituted with `3.141592` and the token `square(x)` will be substituted with `x*x`.

Execution of Program 2.10 yields

```
Give me the radius => 3
The area of a circle with a radius of 3.000000
is 28.274334 square units.
```

Defining Operations

Mathematical operations themselves that use prior macros may be included in a new macro. This is illustrated in Program 2.11. Note that this time the formula for calculating the area of a circle is itself a preprocessor macro. Program 2.11 does exactly the same thing as Program 2.10; it just uses more macros.

Program 2.11

```
#include <stdio.h>
#define square(x) x*x
#define PI 3.141592
#define circle_area(r) PI*square(r)

main()
{
        float area;      /* Circle area.   */
        float radius;    /* Circle radius. */
```

```
        printf("Give me the radius => ");
        scanf("%f",&radius);

        area = circle_area(radius);

        printf("The area of a circle with a radius of %f\n",radius);
        printf("is %f square units.",area);
}
```

Again, execution of Program 2.11 will yield

```
Give me the radius => 3
The area of a circle with a radius of 3.000000
is 28.274334 square units.
```

Forms for #defines

It isn't necessary to use predefined macros for the circle area formula. Consider Program 2.12. Here there is simply one macro that defines `circle_area(r)`. The advantage to this is that there are fewer macro definitions. However, the disadvantage is that some flexibility is lost because the value `PI` is never given a macro definition. The purpose of this program is to show that such an arrangement can be done for the C precompiler.

Program 2.12

```
#include <stdio.h>
#define circle_area(r) 3.141592*r*r

main()
{
        float area;     /* Circle area.   */
        float radius;   /* Circle radius. */

        printf("Give me the radius => ");
        scanf("%f",&radius);

        area = circle_area(radius);

        printf("The area of a circle with a radius of %f\n",radius);
        printf("is %f square units.",area);
}
```

Revisiting an Old Program

Recall Program 2.6 in the previous section, which computed the reactance of a capacitor. That program can be rewritten so that the formula for capacitive reactance is given as a precompiler directive. Look at Program 2.13. It computes the value of the capacitive reactance. All that are needed are the value of the capacitor and the value of the frequency.

Program 2.13

```
#include <stdio.h>
#define PI 3.141592
#define capacitive_reactance(c,f) 1/(2*PI*f*c)

main()
{
        float capacitance;      /* Value of the capacitor in farads. */
        float frequency;        /* Value of the frequency in hertz.  */
        float cap_react;        /* Value of the capacitive reactance in ohms. */

        printf("Give me the capacitance => ");
        scanf("%f",&capacitance);
        printf("Give me the frequency => ");
        scanf("%f",&frequency);

        cap_react = capacitive_reactance(capacitance, frequency);

        printf("The reactance of a %e farad capacitor\n",capacitance);
        printf("with a frequency of %e hertz is %e ohms.",frequency, cap_react);
}
```

Program Analysis

The main feature of Program 2.13 is the use of a precompiler macro to define a complex formula:

```
#define capacitive_reactance(c,f) 1/(2*PI*f*c)
```

Note again that the tokens to be defined are one space from the `#define` directive. Note also the use of parameters `(c,f)`, which reserve two spaces for input by the arguments `c` and `f`. Following this is another space to indicate that what follows next is what is to be substituted: `1/(2*PI*f*c)`. A close look at this reveals that it is the formula for the capacitive reactance of a capacitor. This is exactly the same formula that was used in a separate function of a similar program presented in the previous section. The difference here is that now the same thing is accomplished by use of a preprocessor macro.

Assuming the program user entered a value of 1 μf for the capacitor and a frequency of 5 kHz, execution of Program 2.13 would yield

```
Give me the capacitance => 1e-6
Give me the frequency => 5e3
The reactance of a 1.000000e-006 farad capacitor
with a frequency of 5.000000e+003 hertz is 3.183101e+001 ohms.
```

Expanding the Concept

You can take the idea of defining complex formulas in precompiler macros a step further. Program 2.14 shows that macros with parameters may be used to define other macros with parameters. In this program, the series LC impedance formula is defined by using the predefined formulas for capacitive reactance (now shortened to `X_c`) and inductive reactance (`X_l`). There is one addition. Notice how the inductive reactance and capacitive reactance definitions are used to create the third definition `series impedance`.

Program 2.14

```
#include <stdio.h>
#include <math.h>
#define PI 3.141592
#define X_c(c,f) 1/(2*PI*f*c)
#define X_l(l,f) 2*PI*f*l
#define series_impedance(c,l,f) X_l(l,f) - X_c(c,f)

main()
{
        float capacitance;       /* Value of the capacitor in farads.    */
        float inductance;        /* Value of the inductor in henrys.     */
        float frequency;         /* Value of the frequency in hertz.     */
        float cap_react;         /* Value of the capacitive reactance in ohms. */
        float ind_react;         /* Value of the inductive reactance in ohms.  */
        float imped;             /* Value of the circuit impedance in ohms. */

        printf("Give me the capacitance => ");
        scanf("%f",&capacitance);
        printf("Give me the inductance => ");
        scanf("%f",&inductance);
        printf("Give me the frequency => ");
        scanf("%f",&frequency);

        imped = series_impedance(capacitance, inductance, frequency);

        printf("The impedance of a series LC circuit with a \n");
        printf("capacitance of %e farads, ", capacitance);
        printf("and an inductance of %e henrys\n", inductance);
        printf("is %e ohms.",imped);
}
```

Program Analysis

Program 2.14 does exactly the same thing as the one in the previous section that computed the value of series LC impedance. The difference with this one is that instead of using three different functions to define the formulas, this is done with the precompiler directives:

```
#define PI 3.141592
#define X_c(c,f) 1/(2*PI*f*c)
#define X_l(l,f) 2*PI*f*l
#define series_impedance(c,l,f) X_l(l,f) - X_c(c,f)
```

The order is important—`X_c` has to be defined before `series_impedance` because it is used there.

All that is now necessary is to use the defined statement

```
series_impedance(c,l,f)
```

any time you wish to compute the impedance of this type of circuit.

Assuming that the program user entered a value of 2 μf for the value of the capacitor and 1 mh for the inductor at a frequency of 5 kHz, this program will yield an output of

```
Give me the capacitance => 2e-6
Give me the inductance => 1e-3
Give me the frequency => 5e3
The impedance of a series LC circuit with a
capacitance of 2.000000e-006 farads
and an inductance of 1.000000e-003 henrys
is 1.550042e+001 ohms.
```

Including the #define

As you will see in the next section on header files, you may create a file of your own preprocessor macros and save it on your disk. You can then give the file a name (such as `mystuff.h`). This file could contain all of your necessary electronic formulas or other such formulas in your area of technical interest. Then, any C program you wish to write may use these formulas. All you need to do is to use the `#include` preprocessor directive.

Conclusion

This was an exciting section. Here you saw a whole new aspect of the C language—a valuable programming tool for creating large and complex technical programs. You will find that these preprocessor directives will become very familiar and useful. For now, check your understanding of this section by trying the following section review.

2.5 Section Review

1. What is a preprocessor directive?
2. Explain what is meant by the `#include` directive.
3. What is a macro?
4. How are constants usually defined in C?
5. May a parameter be used with the `#define`? Give an example.

2.6 Making Your Own Header Files

Discussion

In this section you will discover how to make your own **header files** (`.h`) in C. With these, you can store all of your own preprocessor macros. This means that you can have a library of information for your area of technology to use with any of your programs.

An Example

Look at Program 2.15.

Program 2.15

```
#include <stdio.h>
#define square(x) x*x
#define cube(x) x*x*x

main()
{
        float total;    /* Value of the square. */

        total = square(5);

        printf("The square of 5 is %f",total);
}
```

Program Analysis

Program 2.15 is similar to the one presented in the previous section. This program uses two `#define` statements:

```
#define square(x) x*x
#define cube(x) x*x*x
```

The first statement defines `square(x)` as `x*x` and the second defines `cube(x)` as `x*x*x`. The `main()` function uses the `square(x)` definition for demonstration purposes. Execution of the above program yields

```
The square of 5 is 25.000000
```

Saving Your #defines

Figure 2.10 illustrates the concept of taking the `#define` statements of Program 2.15 and saving them to a disk file.

As shown in Figure 2.10, you could now use the file by stating

```
#include "mystuff.h"
```

Note that the quotation marks are used instead of the <> symbols to enclose the name of the header file. This tells the C compiler to look for this header file on the active drive. Once you have done this, the previous program will appear as in Program 2.16.

Program 2.16

```
#include <stdio.h>
#include "mystuff.h"

main()
{
        float total;    /* Value of the square. */

        total = square(5);

        printf("The square of 5 is %f",total);
}
```

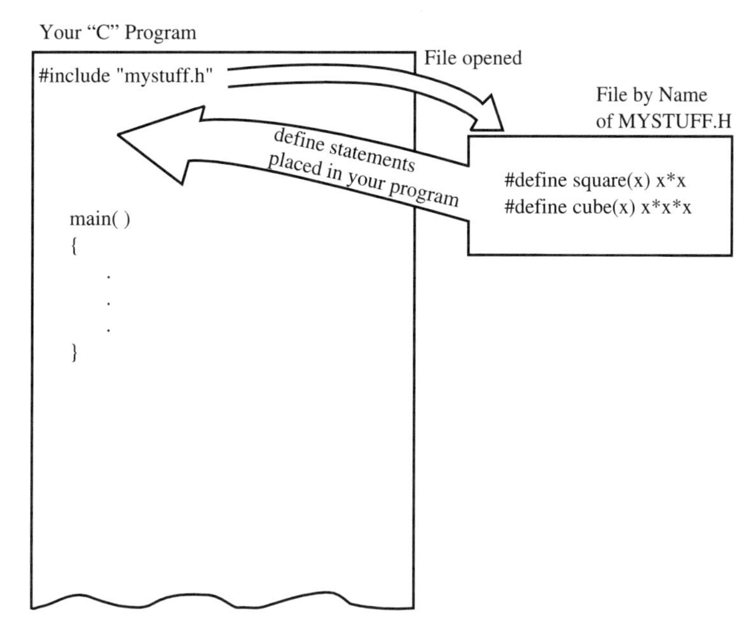

Figure 2.10 Concept of Saving the `#define` Statements

A Sample Program

Consider Program 2.17. It could be used by a technician at an integrated circuit manufacturing facility. Its `#define`'s refer to information about some of the elements used in the manufacturing process for integrated circuits.

Program 2.17

```
#include <stdio.h>
#define head    "Element       Symbol    Atomic Number     Atomic Weight\n"
#define Ge      "Germanium      Ge            32               72.60\n"
#define Si      "Silicon        Si            14               28.09\n"
#define Au      "Gold           Au            79              197.2\n"
#define Ag      "Silver         Ag            47              107.88\n"
#define As      "Arsenic        As            33               74.91\n"
#define Ge_Wt  72.60
#define Si_Wt  28.09
#define Au_Wt  197.2
#define Ag_Wt  107.88
#define As_Wt  74.91

main()
{
      float total;

      printf(head);
      printf(Ge);
```

```
    printf(Ag);
    total = Ge_Wt + Ag_Wt;
    puts("");
    printf("The combined atomic weights of germanium and silver is %f",total);
}
```

Program Analysis

Program 2.17 has `#include` statements that define strings:

```
#define head "Element      Symbol  Atomic Number  Atomic Weight\n"
#define Ge   "Germanium      Ge         32             72.60\n"
#define Si   "Silicon        Si         14             28.09\n"
#define Au   "Gold           Au         79            197.2\n"
#define Ag   "Silver         Ag         47            107.88\n"
#define As   "Arsenic        As         33             74.91\n"
```

As an example, the first `#define` defines the token `head` as the string (note the quotation marks):

```
"Element      Symbol  Atomic Number  Atomic Weight\n"
```

Note that the end of the string has the return to a new line command `\n`. What this means is that when a `printf()` uses the token `head`, the compiler will substitute this string instead. Thus

```
printf(head);
```

will produce

```
Element      Symbol  Atomic Number  Atomic Weight
```

upon program execution. The same is true of all the other tokens that have a string definition. For example,

```
printf(Ge);
```

produces

```
Germanium      Ge         32             72.60
```

upon program execution. Thus the source code of

```
printf(head);
printf(Ge);
```

produces

```
Element      Symbol  Atomic Number  Atomic Weight
Germanium      Ge         32             72.60
```

upon execution.

Now the technician may easily display information about specific atomic elements anywhere in the program simply by entering the symbol of the element in an output statement.

The remainder of the #define statements set numerical values for the tokens. These values represent the atomic weights of the corresponding element symbols.

```
#define Ge_Wt 72.60
#define Si_Wt 28.09
#define Au_Wt 197.2
#define Ag_Wt 107.88
#define As_Wt 74.91
```

Thus, the token Ge_Wt will have the numerical value 72.60 substituted. Assuming that total is a float type variable identifier, then

```
total = Ge_Wt + Ag_Wt;
```

will have the compiler actually substitute

```
total = 72.60 + 107.88;
```

and produce a value of 180.48 for the sum of the two atomic weights of these elements.

Execution of Program 2.17 produces

```
Element       Symbol  Atomic Number  Atomic Weight
Germanium       Ge          32            72.60
Silver          Ag          47           107.88

The combined atomic weights of germanium and silver is 180.480000
```

Making the Header File

To produce the header file that would contain all of the #define statements in Program 2.17, you should create a program similar to Program 2.17 and test all of the #define statements to ensure that they produce the expected result(s) in your program. The next step, after saving your program, is to eliminate all of the lines of your program except for the #define statements. Your program will now appear as shown in Program 2.18:

Program 2.18

```
#define head    "Element        Symbol    Atomic Number      Atomic Weight\n"
#define Ge      "Germanium        Ge           32                72.60\n"
#define Si      "Silicon          Si           14                28.09\n"
#define Au      "Gold             Au           79               197.2\n"
#define Ag      "Silver           Ag           47               107.88\n"
#define As      "Arsenic          As           33                74.91\n"
#define Ge_Wt  72.60
#define Si_Wt  28.09
#define Au_Wt  197.2
#define Ag_Wt  107.88
#define As_Wt  74.91
```

If your B: drive is the active drive, then save the #defines of Program 2.18 in a file called B:CHEM.H.

Using Your Header File

Program 2.19 illustrates the use of your new header file. Observe that it outputs exactly the same thing as did Program 2.17.

Program 2.19

```
#include <stdio.h>
#include "chem.h"

main()
{
        float total;

        printf(head);
        printf(Ge);
        printf(Ag);
        total = Ge_Wt + Ag_Wt;
        puts("");
        printf("The combined atomic weights of germanium and silver is %f",total);
}
```

The following reference will now invoke your header definitions:

```
#include "chem.h"
```

When C sees this (with the quotation marks instead of the <>), it will go to your disk, look for a file by that name, and then actually bring the material in that file into your program.

Token Pasting and String-izing Operators

The ANSI C standard includes the preprocessor operation that allows **token pasting**. The token pasting operator `##` allows you to join one token to another to create a third token. As an example, if you have used the following `#defines` in your program:

```
#define PI_1      3.14159
#define PI_2      6.28318
#define Value(x) PI_##x
```

then if Value(1) is declared, this will be replaced by `PI_1`, which has been defined as equal to 3.14159. In the same manner, Value(2) will be replaced by `PI_2`, which has been defined as equal to 6.28318.

The **string-izing** operator is the single # and allows you to make a string out of an operand with a # prefix. This is accomplished by placing the operand in quotes. As an example, if you have used the following `#define` at the beginning of your program:

```
#define present_value(x)  printf(#x " = %d",x)
```

then when `present_value(increment)` is declared, the preprocessor will generate the statement

```
printf("increment" " = %d",increment);
```

which reduces to

```
printf("increment = %d",increment)
```

Conclusion

As you can see, header files are a powerful extension of the C language system. As you develop more programs, your library of header files (developed by you for your particular area of technology) will become one of your most valuable programming tools. The other fact about header files is that they protect your source code. Another person may have a copy of your source code to look at, but without access to your header file code, they can do little to replicate your program.

Check your understanding of this section by trying the following section review.

2.6 Section Review

1. State the advantage of creating your own header files.
2. As presented in this section, what information may be contained in your header files?
3. What is the extension given to a C header file?
4. Explain how you may call your header file from a C program.

2.7 Troubleshooting Techniques

Discussion

In this section we will examine a very useful troubleshooting technique called **conditional compilation**. We will use a `#define` statement to control the display of *diagnostic* information. Since the example programs we have examined in this chapter performed calculations, a useful diagnostic would involve displaying the results of intermediate steps in a calculation. The `#define` directive controls the way another compiler directive called `#ifdef` does its job. `#ifdef` implements conditional compilation, where a select group of C statements are either compiled or not compiled (effectively ignored). Program 2.20 illustrates how conditional compilation is used.

Program 2.20

```
#include <stdio.h>

main()
{
        float R1,R2,numer,denom,Req;

        printf("Enter resistance #1 ===> ");
        scanf("%f",&R1);
        printf("Enter resistance #2 ===> ");
        scanf("%f",&R2);
        numer = R1 * R2;
        denom = R1 + R2;
        #ifdef TEST
                printf("numer = %f\n",numer);
```

```
        printf("denom = %f\n",denom);
    #endif
    Req = numer /  denom;
    printf("The equivalent of %f and %f in parallel is %f ohms.", R1, R2, Req);
```

The `#ifdef` statement

```
#ifdef TEST
```

checks to see if a variable called `TEST` is defined. If it is, the two `printf` statements

```
printf("numer = %f",numer);
printf("denom = %f",denom);
```

will be compiled.

If `TEST` is not defined, the `printf` statements will be ignored. Notice how `#endif` is used to indicate the end of a conditional compilation block.

The execution of Program 2.20 looks like this:

```
Enter resistance #1 ===> 10
Enter resistance #2 ===> 40
The equivalent of 10.000000 and 40.000000 in parallel is 8.000000 ohms.
```

It is obvious that the numerator and denominator values are not displayed. This is because there is no `#define` statement that defines the `TEST` variable.

Program 2.21 *does* define the `TEST` variable.

Program 2.21

```
#include <stdio.h>

#define TEST

main()
{
        float R1,R2,number,denom,Req;

        printf("Enter resistance #1 ===> ");
        scanf("%f",&R1);
        printf("Enter resistance #2 ===> ");
        scanf("%f",&R2);
        numer = R1 * R2;
        denom = R1 + R2;
        #ifdef TEST
                printf("numer = %f\n",numer);
                printf("denom = %f\n",denom);
        #endif
        Req = numer /  denom;
        printf("The equivalent of %f and %f in parallel is %f ohms.", R1, R2, Req);
```

When Program 2.21 executes, the results are:

```
Enter resistance #1 ===> 10
Enter resistance #2 ===> 40
numer = 400.000000
denom = 50.000000
The equivalent of 10.000000 and 40.000000 in parallel is 8.000000 ohms.
```

The intermediate calculations are now displayed. If the `#define TEST` statement is removed (or commented out), the results will go back to those of Program 2.20. This is the great advantage you get when using conditional compilation. Statements may be quickly placed in or out during a compilation, but always remain in the source file, to be reactivated at any time.

Conclusion

In this section you saw how conditional compilation can help troubleshoot a program under development. Check your understanding of this material with the following section review.

2.7 Section Review

1. Explain the operation of the `#ifdef` statement.
2. How is the `#define` statement used to control an `#ifdef`?
3. How is the end of a conditional block indicated?
4. What is accomplished by these two `#ifdef` statements?

```
#ifdef VARX
     #ifdef VARY
          printf("Both are defined.\n");
     #endif
#endif
```

2.8 Case Study: AC Series R-L Circuit

Discussion

This case study takes you step by step through the development of a C program that utilizes the new information presented in this chapter. Here you will see how decisions are made during the development of a very specialized program. As you increase your understanding of C and its versatility, you will come to see that there are many options available to you concerning how to create your source code. One of the powerful points of this language is this versatility of the source code.

Because of this, what is presented in this case study (as well as in the others) is an attempt to increase readability and understanding of what the program will do while still preserving the fundamental characteristics of the C language.

The Problem

Given the circuit shown in Figure 2.11, develop a C program that will compute the voltage drop across the resistor and the inductor.

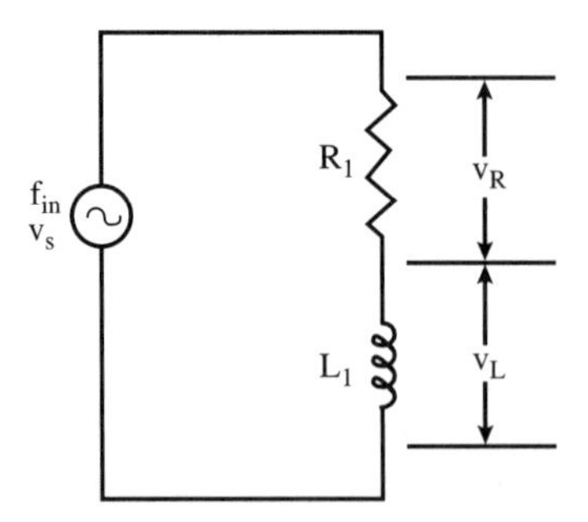

Figure 2.11 Circuit for Case Study

First Step—Stating the Problem

Recall from the case study in the last chapter that the first step in designing a computer program is to state the problem in writing. As a guide, the following areas are the minimum requirements:

- Purpose of the Program
- Required Input (source)
- Process on Input
- Required Output (destination)

The problem can be stated as follows:

- Purpose of the Program: Compute the voltage drop in volts across an inductor and a resistor in a series RL circuit.
- Required Input: Values of the following:
  - Inductor in henrys
  - Resistor in ohms
  - Source voltage in volts
  - Applied frequency in hertz

  (Source: keyboard)
- Process on Input: Compute the voltage across the inductor in volts. Compute the voltage across the resistor in volts.

$$X_L = 2\pi f L$$
$$Z = \sqrt{(R^2 + X_L{}^2)}$$
$$v_L = v_S(X_L/Z)$$
$$v_R = v_S(R/Z)$$

Where

X_L = Inductive reactance in ohms.
f = Applied frequency in hertz.
L = Value of inductor in henrys.
Z = Circuit impedance in ohms.
R = Value of resistor in ohms.
v_L = Voltage across inductor in volts.
v_R = Voltage across resistor in volts.
v_S = Voltage of the source in volts.

- Required Output: The value of the voltage across the resistor, v_R, and the value of the voltage across the inductor, v_L (displayed on monitor).

From this description, it should be clear to anyone what the program is required to do (and, just as important, what it will not do). The next step in the development of this program is to develop an algorithm. You can think of an algorithm as similar to a cake recipe. An algorithm is a step-by-step explanation of exactly what the program will do, written as concisely as possible (just like the cake recipe would be written).

Developing the Algorithm

If you have reduced the problem to writing, including the required four areas (program purpose, required input, process on input, and required output), then the algorithm may be developed straight from there. For most technology problems, the program will follow the order shown in the outline below:

The following is all commented (between `/*` and `*/`, except for the function prototypes) and placed in the programmer's block—sometimes referred to as the **prologue**.

I. Program Information
   A. Name of program
   B. Name of programmer
II. Program Explanation
   A. What the program will do
   B. What is required for input
   C. What process will be performed
   D. What the results will be
III. Describe All Functions
   A. Use function prototypes
   B. Explain purpose of each prototype
   C. Define all variables and constants

The following are the programming steps that are used by most technical programs:

IV. Explain Program to User
   A. Purpose of the program
   B. What the user is required to do
   C. What the program will do
   D. What the final results will be
V. Get Information from User
   A. Prompt user for input
   B. Acknowledge the input
VI. Perform the Process
   A. Do the process
VII. Display the Results
   A. Confirm what the user has entered
   B. Display all units
VIII. Ask for Program Repeat
   A. Ask if user wishes to repeat the program
   B. Give option to bypass instructions

The outline above will serve as the guide for this program.

First Coding

The first part of the program coding is to develop the programmer's block, compile it, save it to a disk file, and then make a printed copy. There is no output, but the purpose is to ensure that the documentation of the program has already started. This first step is shown in Program 2.22.

Program 2.22

```
#include <stdio.h>
/*****************************************************************/
/*                 RL Circuit Voltage Drops                      */
/*****************************************************************/
/*              Developed by: A. Good Programmer                 */
/*                                                               */
/*****************************************************************/
/*        This program will compute the voltage drop in          */
/*   volts across an inductor and a resistor in a series         */
/*   RL circuit.  The program user must enter the values of      */
/*   the inductor in henrys, resistor in ohms, applied           */
/*   frequency in hertz, and source voltage in volts.            */
/*---------------------------------------------------------------*/

void explain_program(void);
/*  This function explains the operation of the program to      */
/*  the program user.                                           */

void get_values(void);
/*  This function gets the values of the resistor, inductor,    */
/*  source voltage, and applied frequency, calculates and       */
/*  displays the results.                                       */

/*****************************************************************/

main()
{
}

void explain_program()
{
}

void get_values()
{
}
```

Program Analysis

Program 2.22 includes a program title and the developer's name. A brief description of the program follows. Note that all units of measurement are given in the program description. Also note that the programmer decided to use a single programming block that not only gets the values from the program user, but also does the calculations and displays the values. As the program development progresses, this may be too much coding in a single block, and the program may be more readable if this is divided into at least two blocks.

Entering the First Code

For this program, the next step is to do the coding for the user explanation block. After this is done, the program is compiled and executed. The block appears as shown below:

```
main()
{
        explain_program();     /* Explain program to user.  */
        exit(0);
 }
/*-----------------------------------------------------------*/
void explain_program()  /* Explain program to user.  */
{

        printf("\n\n This program will compute the voltage drop in\n");
        printf("volts across an inductor and a resistor in a series\n");
        printf("RL circuit. The program user must enter the values of\n");
        printf("the inductor in henrys, resistor in ohms, applied\n");
        printf("frequency in hertz, and source voltage in volts.\n\n");

}    /* End of Explain program to user.  */
```

Note that `main()` now contains a call to the function `explain_program`. This is included so that the output can be observed. It's important to see the use of the two `\n`'s at the beginning of the first `printf` function. This is good practice because it brings the first sentence down two spaces from the top of the monitor screen, thus making the text a little easier to read. Notice the ending comment `/* End of Explain program to user. */`. This may seem unnecessary for this small a function. However, it is a good practice to develop because it is a great aid in helping you find your way around larger programs.

Formula Decision

The program developer next has to decide if the formulas for the solution of this problem should be put into a separate function or into a header file. It was decided to create a header file for these formulas because they are formulas that will probably be used again in future programs.

The formulas were first developed using `#define` statements, as shown below:

```
#include <stdio.h>
#include <math.h>
#define PI              3.141592
#define square(x)       x*x
#define X_L(f,l)        2*PI*f*l
#define Z(f,l,r)        sqrt(square(X_L(f,l))+square(r))
#define V_L(f,l,r,v)    v*(X_L(f,l)/Z(f,l,r))
#define V_R(f,l,r,v)    v*(r/Z(f,l,r))
```

These are placed at the beginning of the program before the programmer's block. Observe that the `#include` files are used for input/output as well as math functions (such as `sqrt`). Essentially the programmer followed the formulas as they are stated in the written statement of the problem. Note that the `sqrt` function is defined in the `math.h` header file. See appendix B for a list of other functions defined in `math.h`.

Adding the Calculate and Display Block

Once the decision was made to use a header file for the formulas, it was then decided to use a separate program block to do the calculations and display the values.

The header files should not be made until the accuracy of the information to be included in them has been checked. For this program, this may be done with a desk-check of the program.

The combined results of all steps are shown in Program 2.23.

Program 2.23

```
#include <stdio.h>
#include <math.h>
#define PI 3.141592
#define square(x) x*x
#define X_L(f,l) 2*PI*f*l
#define Z(f,l,r) sqrt(square(X_L(f,l)+square(r)))
#define V_L(f,l,r,v) v*(X_L(f,l)/Z(f,l,r))
#define V_R(f,l,r,v) v*(r/Z(f,l,r))

/*************************************************************/
/*                RL Circuit Voltage Drops                  */
/*************************************************************/
/*            Developed by: A. Good Programmer              */
/*                                                          */
/*************************************************************/
/*        This program will compute the voltage drop in     */
/*   volts across an inductor and a resistor in a series    */
/*   RL circuit.  The program user must enter the values of */
/*   the inductor in henrys, resistor in ohms, applied      */
/*   frequency in hertz, and source voltage in volts.       */
/*************************************************************/
/*             Non-Standard Header Files Used:              */
/*----------------------------------------------------------*/
/*   serrl.h => Series RL, this file contains the formula   */
/*              for the inductor and resistor voltage drops */
/*              when the above values are given.            */
/*************************************************************/
/*                 Function prototypes                      */
/*----------------------------------------------------------*/

float calculate_and_display(float f, float l, float r, float v);
/*  f = Frequency in hertz.                                 */
```

The function `get_values` had a call to the `calculate_and_display` function. The actual parameters used by the function in getting information from the program user are

```
calculate_and_display(frequency,inductor,resistor,voltage_s);
```

The `calculate_and_display` function definition uses the same identical header as its prototype (with the exception that the semicolon is omitted). Notice how easy it is to code the calculations for these complex formulas:

```
float calculate_and_display(float f, float l, float r, float v)
{
        float inductor_v;     /* Value of voltage across inductor.  */
        float resistor_v;     /* Value of voltage across resistor.  */

        inductor_v = V_L(f,l,r,v);    /* Calculate inductor voltage.  */
        resistor_v = V_R(f,l,r,v);    /* Calculate resistor voltage.  */

        printf("\n\nFor a series LR circuit consisting of a\n");
        printf("%e henry inductor with a %e ohm resistor\n",l,r);
        printf("with an applied frequency of %e hertz\n",f);
        printf("and a source voltage of %e volts, the\n",v);
        printf("component voltage drops are: \n\n");
        printf("Inductor voltage => %e volts\n",inductor_v);
        printf("Resistor voltage => %e volts\n",resistor_v);

}    /* End of calculate_and_display.  */
```

The output of this program echoes the user input using E notation along with all units of measurement. The solutions are also displayed using E notation as well as the important units of measurement.

Final Program

Since the coding for giving the program user the option of repeating a program over again has not yet been presented, it was omitted from this one. The last step of the coding was to create a header file called `serrl.h` that contains the following information:

```
#define PI              3.141592
#define square(x)       x*x
#define X_L(f,l)        2*PI*f*l
#define Z(f,l,r)        sqrt(square(X_L(f,l))+square(r))
#define V_L(f,l,r,v)    v*(X_L(f,l)/Z(f,l,r))
#define V_R(f,l,r,v)    v*(r/Z(f,l,r))
```

Thus the top of the final program includes a listing of `#include` statements:

```
#include <stdio.h>
#include <math.h>
#include "serrl.h"
```

Conclusion

You have come a long way in your observation of the development of a C program. It's important that you apply the principles presented here to the programs that you develop. Careful planning and program documentation are important in serious programming. These steps are used by professional programmers who all know the old saying: "The sooner you start to enter program code, the longer it will take to debug the final program." There are usually good reasons for old sayings.

Check your understanding of this section by trying the following section review.

2.8 Section Review

1. State one of the main goals in the development of the case study for this section.
2. What should be included in the programmer's block?
3. Explain the reason for developing `#define` statements.
4. What is the last thing that is usually asked of the program user in a typical technology program?

Interactive Exercises

DIRECTIONS

These exercises require that you have access to a computer and software that supports C. They are provided here to give you valuable experience and most importantly immediate feedback on what the concepts and commands introduced in this chapter will do.

Exercises

1. Predict what Program 2.24 will do, then try it. Did you get a call to the next function? You should have.

Program 2.24

```
#include <stdio.h>

main()
{
        This_One();
}

void This_One()
{
        puts("Here I am.");
}
```

2. Predict what you think Program 2.25 will do, then try it. You should be able to call all of the functions.

Program 2.25

```
#include <stdio.h>

main()
```

```
{
        puts("This is main()");
        First_One();
}

void First_One()
{
        puts("This is First_One()");
        Second_One();
}

void Second_One()
{
        puts("This is Second_One()");
}
```

3. Program 2.26 is similar to Program 2.25 in that you will need to press Ctrl-Break to stop it. The difference here is that this function calls itself. This is an example of recursion.

Program 2.26

```
#include <stdio.h>

main()
{
        Recursive_One();
}

void Recursive_One()
{
        puts("This is an example of recursion.");
        Recursive_One();
}
```

4. C is such a versatile language that you can make it look like almost any other language you wish. Program 2.27 redefines C commands. These could be put in a header file called `newcode.h`. Try it to prove to yourself that it does work.

Program 2.27

```
#include <stdio.h>
#define START main(){
#define END }
#define WRITE puts

START
        WRITE("This is a C Program?");
END
```

Self-Test

DIRECTIONS

Use Program 2.23 developed in the case study section of this chapter to answer the following questions.

Questions

1. How many functions are defined in the program? Name them.
2. How many total functions are used in the program? Name those not named in question 1 above.
3. List the identifiers used for formal parameters in the program.
4. List the identifiers used for actual parameters in the program.
5. Which function passes variables to another function?
6. Explain how values are passed from one function to another using Program 2.23 as an example.
7. How many variable identifiers are used in the program? Name them.
8. What is the minimum change you would make in this program to display the value of the inductive reactance (`X_L`)?
9. Will the program accept user input in E notation? Explain.

End-of-Chapter Problems

General Concepts

Section 2.1

1. Explain what is meant by block structure.
2. Why is there no need for a `goto` statement in a C program?
3. Which type of block has the ability to go back and repeat a part of the program?
4. Which type of block has the ability to execute a different sequence of program code?

Section 2.2

5. In a C program, what gives the compiler specific information about the functions that will be defined in the program?
6. State the difference between the function prototype and the function header used to define the function.
7. How is a function called?
8. What C function causes normal program termination? What value is assigned to it to indicate normal termination?

Section 2.3

9. What type is assigned to a function when no value is to be returned by the function?
10. State what is used in a function to define the number and types of data that will be passed to the function when it is called.
11. What is used to return a value from the called function to the calling function?

Section 2.4

12. What is it called when a function calls itself?
13. May a function call more than one other function?
14. Does it make any difference in what order functions are called?
15. May a called function call another function?

Section 2.5

16. What is a preprocessor directive?
17. Name the preprocessor directive presented in this chapter.
18. Explain how constants are usually defined in C.
19. Give an example of how parameters may be used with the `#define`.

Section 2.6

20. Explain how you can make a file of your own `#define` statements for use in other programs.
21. How do you invoke your own header file?
22. What is the accepted extension given to C header files?
23. What is the conditional compiler directive used to begin a conditional block of cdde?
24. Why use conditional compilation?

Program Design

In developing the following C programs, use the method described in the case study section of this chapter. For each program you are assigned, document your design process and hand it in with your program. This process should include the design outline stating the purpose of the program, required input, process on the input, and required output as well as the program algorithm. Be sure to include all of the documentation in your final program. This should consist of, but not be limited to, the programmer's block, function prototypes, and a description of each function, as well as any formal parameters you may use.

25. Develop a C program that will compute the power delivered by a voltage source. The relationship is

$$P = I \times E$$

Where

P = The power delivered in watts.
I = The source current in amps.
E = The source voltage in volts.

Do not use any `#define` headers for this program. Use functions for each major part of the program. One function will explain the program to the user and the other will get the values and pass them on to the function that defines the power formula.

26. Modify the program developed in problem 25 so that the power formula is now included in a header file instead of being a separate function.
27. Create a C program that will compute the voltage drop across each resistor in a series circuit consisting of three resistors. The user input is to be the value of each resistor and the applied circuit voltage. The mathematical relationships are

$$R_T = R_1 + R_2 + R_3$$
$$I_T = V_S / R_T$$
$$V_N = I_T \times R_N$$

Where

R_T = Total circuit resistance in ohms.
R_1, R_2, R_3 = Value of each resistor in ohms.
I_T = Total circuit current in amps.
V_S = Applied circuit voltage.
V_N = Voltage drop across an individual resistor N in volts.
R_N = R_1, R_2, or R_3

Use only functions for this program; do not use any `#define` statements.

28. Redo the program created in problem 27 using `#define` statements. Also include the total circuit resistance, the total circuit current, and the voltage drop across each resistor as part of the output.
29. Expand the program in problem 27 to include the power dissipated by each resistor as well as the power dissipated by the voltage source. The relationship for resistor power dissipation is

$$P_{RN} = I_T^2 \times R_N$$

Where

P_{RN} = Power dissipated by individual resistor N in watts.
I_T = Total circuit current in amps.
R_N = Individual resistor value in ohms.

$$P_S = I_T \times V_S$$

Where

P_S = Power of the source.
V_S = Source of voltage.

30. Create a C program that will compute the volume of a magnetic ring or circular area. The volume is given by

$$V = 2\pi^2 R r^2$$

Where

V = The volume in square units.
R = Radius of the ring.
r = Radius of the cross-sectional area.

See Figure 2.12.

31. Develop a C program that will determine the number of bytes required to store an N by M array of elements. Compute for bit, byte, and double-byte element types.
32. Write a C program that will compute the volumes of three different rooms. Assume that each room is a perfect rectangle. User input is the height, width, and length of each room. The program is to return the volume of each separate room as well as the total volume of all three rooms.
33. Referring to a standard table of conversions, create a header file that will convert acres to square feet, and convert square feet to square meters as well as hectares. Use this file in a C program that demonstrates the correctness of the conversions.
34. Develop a C program that will compute the total bill of a hospital patient. The user inputs are
    1. Number of days in hospital
    2. Surgery cost

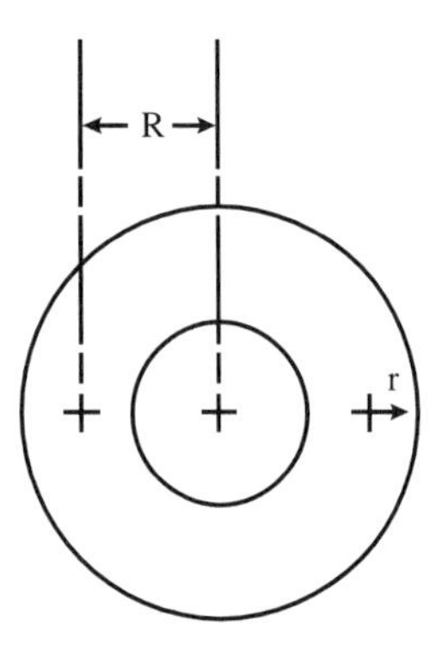

Figure 2.12 Magnetic Ring for Manufacturing Technology (used in Question 30)

3. Medication cost
4. Miscellaneous cost
5. Cost per day
6. Insurance deductible

The program must compute the following for insurance purposes:

1. Total cost
2. Total cost less insurance deductible
3. Total cost less cost of medication and deductible

35. Design a C program that will balance two different accounts consisting of five items each. The user inputs are

1. Number of widgets ordered
2. Number of widgets shipped
3. Shipping cost per widget
4. Cost of each widget
5. Handling charge per unit

The program output is to display

1. Number of widgets on back order
2. Total cost of widgets shipped
3. Total cost when all widgets are shipped

36. Show how this formula is implemented as a function:

$$X = \frac{B + \sqrt{(B^2 + 4AC)}}{2A}$$

3 Operations on Data and Decision Making

Objectives

This chapter gives you the opportunity to learn:

1. The meaning and use of relational operators.
2. The difference between an open branch and a closed branch.
3. Technical applications of open and closed branches.
4. Logical and bitwise operations used in C.
5. Applications of logical and relational operations to programming branches.
6. Methods of choosing one of several alternatives.
7. Methods of working with different data types and how to cast a type.
8 The use and methods of conditional compilation.
9. Special decision making used by C.
10. Program development from existing technical flow charts.

Key Terms

Relational Operators
TRUE
FALSE
Assignment
Open Branch
`if` Statement
Conditional Statement
Compound Statement
Closed Branch
`if...else` Statement
`if...else if...else`
Bit Manipulation
Boolean Operators
Bitwise Complement
Bitwise AND
Bitwise OR
Bitwise XOR
Bit Shifting
Logical AND
Logical OR

Logical NOT
Rank
Casting
lvalue
Program Stubs

Outline

Introduction

In this chapter you will learn how to develop C programs that make the computer look "smart." You will find out how to create programs that allow program decisions to be made. Doing this allows the program flow to branch in one direction or another depending on the input from a user or the results of a computation.

The chapter starts by explaining how quantities may be related. These relationships may then be used in the decision-making process of the program. You will then be introduced to the logic used by C and its symbolism. Here you will discover the simplicity and power contained in the TRUE and FALSE logic of a C program.

The chapter concludes with an interesting case study for troubleshooting a robot. When you finish with the material in this chapter, you will be well on your way to developing programs that solve and analyze technology problems.

3.1 Relational Operators

Discussion

This section introduces **relational operators**. As you will discover, such an operator shows the relationship between two quantities. You will use this information as a foundation for the material in the next section. There, you will learn how to create C programs that will exhibit decision-making capabilities.

Relational Operators

A relational operator is a symbol that indicates a relationship between two quantities. These quantities may be variables, constants, or functions. The important point about these relations is that they are either **TRUE** or **FALSE**—there is nothing in between!

The relational operators used in C are shown in Table 3.1.

Table 3.1 Relational Operators Used in C

Symbol	Meaning	TRUE Examples	FALSE Examples
>	Greater than.	5 > 3	3 > 5
		(3+8) > (5–2)	(12/6) > 18
>=	Greater than or equal to.	10 > = 10	3 > = 5
		(3*4) > = 8/2	8+5 > = 10*15
<	Less than.	3 < 5	5 < 3
		12/3 < 12*3	9–2 < 3+1
<=	Less than or equal to.	3 < = 15	15 < = 3
		18/6 < = 9/3	12+4 < = 12/4
==	Equal to.	5 == 5	10 == 5
		2+7 == 18/2	8–5 == 2+4
!=	Not equal to.	8 ! = 5	5 ! = 5
		8–5 ! = 2+4	24/6 ! = 12/3

For relational operators, the value returned for a TRUE condition is 1 whereas the value returned for a FALSE condition is a 0. This is illustrated by Program 3.1.

Program 3.1

```
#include <stdio.h>

main()
{
        float logic_value;      /* Numeric value of relational expression */

        printf("Logic values of the following relations:\n\n");

        logic_value = (3 > 5);
        printf("(3 > 5) is %f\n",logic_value);

        logic_value = (5 > 3);
        printf("(5 > 3) is %f\n",logic_value);

        logic_value = (3 >= 5);
        printf("(3 >= 5) is %f\n",logic_value);

        logic_value = (15 >= 3*5);
        printf("(15 >= 3*5) is %f\n",logic_value);

        logic_value = (8 < (10-2));
        printf("(8 < (10-2)) is %f\n",logic_value);

        logic_value = (2*3 < 24/3);
        printf("(2*3 < 24/3) is %f\n",logic_value);

        logic_value = (10 < 5);
```

```
        printf("(10 < 5) is %f\n",logic_value);

        logic_value = (24 <= 15);
        printf("(24 <= 15) is %f\n",logic_value);

        logic_value = (36/6 <= 2*3);
        printf("(36/6 <= 2*3) is %f\n",logic_value);

        logic_value = (8 == 8);
        printf("(8 == 8) is %f\n",logic_value);

        logic_value = (12+5 == 15);
        printf("(12+5 == 15) is %f\n",logic_value);

        logic_value = (8 != 5);
        printf("(8 != 5) is %f\n",logic_value);

        logic_value = (15 != 3*5);
        printf("(15 != 3*5) is %f\n",logic_value);
}
```

Program Analysis

Execution of Program 3.1 yields

```
Logic values of the following relations:
(3 > 5) is 0.000000
(5 > 3) is 1.000000
(3 >= 5) is 0.000000
(15 >= 3*5) is 1.000000
(8 < (10-2)) is 0.000000
(2*3 < 24/3) is 1.000000
(10 < 5) is 0.000000
(24 <= 15) is 0.000000
(36/6 <= 2*3) is 1.000000
(8 == 8) is 1.000000
(12+5 == 15) is 0.000000
(8 != 5) is 1.000000
(15 != 3*5) is 0.000000
```

As you can see from this output, a relational operation in C returns a value of either 1 or 0. If the operation is TRUE, a value of 1 is returned, and if the operation is FALSE, a value of 0 is returned.

To illustrate a portion of the program, consider the following program excerpt:

```
logic_value = (3 > 5);
printf("(3 > 5) is %f\n",logic_value);

logic_value = (5 > 3);
printf("(5 > 3) is %f\n",logic_value);
```

The variable `logic_value` has been declared a type `float`. You may wish to try this with other types. It is being set to the value of the relational operation (`3 > 5`). This statement is FALSE; therefore, the value returned will be a `0`. This value is displayed on the monitor, as a type `float (%f)`, as `0.000000`. In the following program line, the variable `logic_value` is being set to the value of the relational operation (`5 > 3`). This is a TRUE statement and hence the value returned will be `1`. This also is displayed as a type `float`, producing the display of `1.000000`.

The remainder of the program continues with the same type of process.

Equal To

It may at first seem strange that C uses the == (double equals or "equal-equal") to mean equal to. Understandably, many who are new to C may have thought that the use of the = (single equals) would mean equal to. The fact is that the = (single equals) does not mean the same in C as it does in ordinary math. What the = means in C is **assignment**. The concept of assignment is illustrated in Figure 3.1.

It's important to make the distinction between the assignment operator (=) and the equals operator (==) in C. The assignment operator takes the value on the right side of the

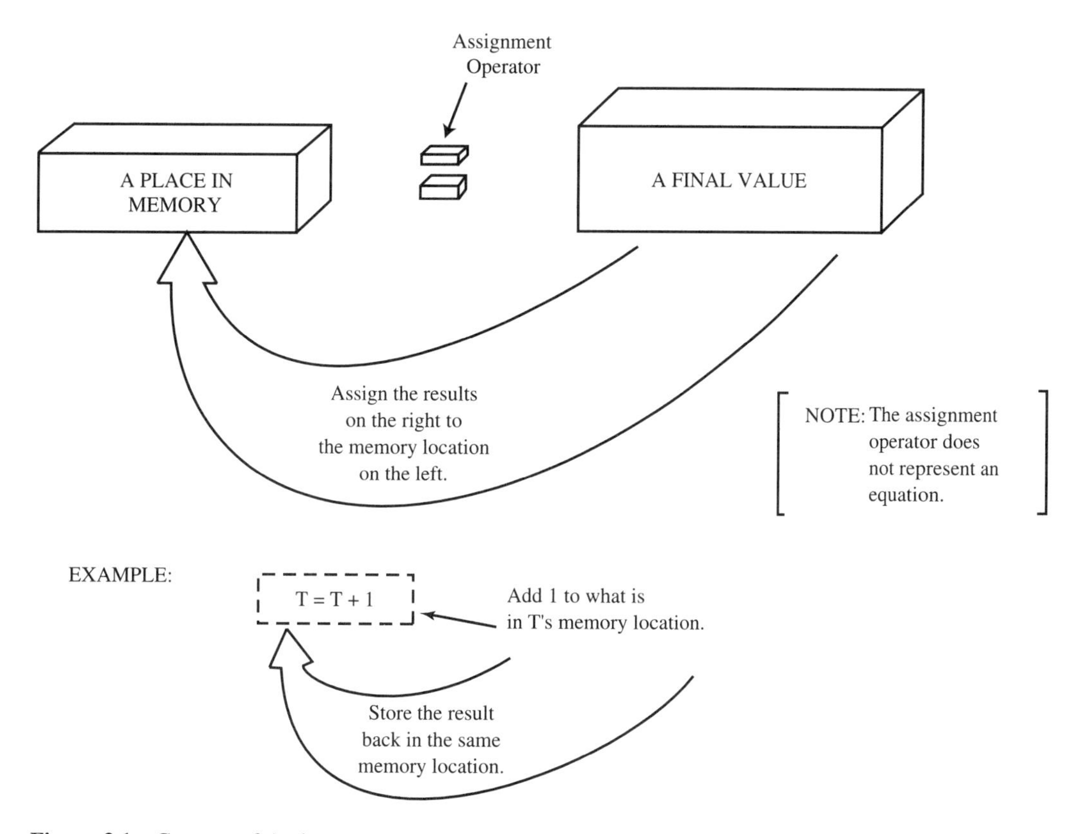

Figure 3.1 Concept of Assignment

assignment statement and puts it in the memory location of the variable on the left side of the assignment. The equals operator does something different. It simply compares the value of one memory location with the value of another memory location. No transfer of data from one memory location to the other takes place.

Check your understanding of relational operators by trying Example 3.1.

EXAMPLE 3.1

Assume that the following assignments have been made:

a = 5, b = 10, c = 5

State the final values of each of the following comparisons:

A. a == 5 C. a >= c
B. c == b D. b != a

Solution

A. a == 5 is TRUE, value is 1.
B. c == b is FALSE, value is 0.
C. a >= c is TRUE, value is 1.
D. b != a is TRUE, value is 1.

Conclusion

In this section you were introduced to relational operators used in C. You saw the symbols used for each of them and saw that there are only two possible conditions—TRUE or FALSE—and nothing in between. You also saw that a C relational operator actually returns a numerical value. The value returned for a TRUE operation is 1, and the value returned for a FALSE operation is 0. This chapter also made the important distinction between the assignment operator (=) and the equals operator (==).

In the next section you will discover how to use what you have learned here to develop programs that will exhibit decision-making capabilities. For now, test your understanding of this section by trying the following section review.

3.1 Section Review

1. What are relational operators?
2. State all of the relational operators used in C.
3. What are the two conditions that relational operators may have?
4. Explain what is meant by the statement that a relational operator in C returns a value.
5. State the difference between the C operation symbols = and ==.

3.2 The Open Branch

Discussion

This section will show you how to use the relational operators presented in the previous section. Here you will see how to code decision-making capabilities using C.

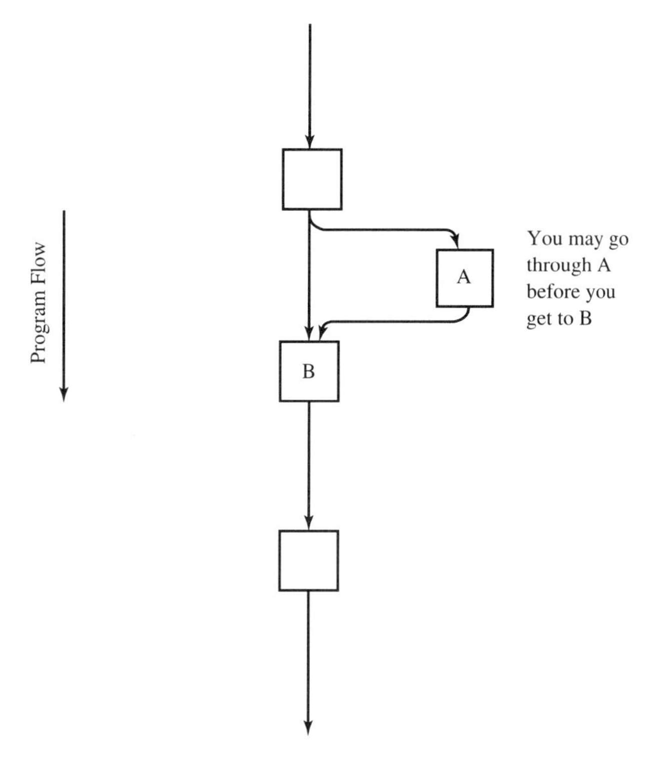

Figure 3.2 Concept of the Open Branch

The open branch is presented here. This is perhaps the simplest of the computer's decision-making capabilities. Even so, you will find this a powerful addition to your C programming skills.

Basic Idea

The basic idea of an **open branch** is illustrated in Figure 3.2.

There are two important points about the open branch. First, the flow of the program always goes forward to new information. Second, the option may or may not be used—but the remainder of the program is always executed. In the C programming language, the open branch is accomplished by the **`if`** statement.

The if Statement

The `if` statement in C is referred to as a **conditional statement** because its execution will depend on a specific condition. The form of the `if` statement is

```
if (expression) statement
```

What this means is that if `expression` is TRUE, then `statement` will be executed. If `expression` is FALSE, `statement` will not be executed. This is illustrated in Program 3.2.

Program 3.2

```
#include <stdio.h>
main()
{
        float number;   /* Value supplied by user. */

        printf("Give me a number from 1 to 10 => ");
        scanf("%f",&number);
        if (number > 5)
                printf("Your number is larger than 5.\n");

        printf("%f was the number you entered.",number);
}
```

Program Analysis

Program 3.2 asks the user to enter a number from 1 to 10. If the user enters a number from 1 to 5 (assume 3 was entered), the program will display

```
3 was the number you entered.
```

If the program user enters a number greater than 5 (say 7), then the program displays

```
Your number is larger than 5.
7 was the number you entered.
```

You can see the use of the `if` statement as

```
if (number > 5)
        printf("Your number is larger than 5.\n");
```

The expression is a relational one that compares the value of the user input to see if it is greater than 5. If it is, then the expression is TRUE and the statement

```
printf("Your number is larger than 5.\n");
```

will be executed.

If the relational operation inside the expression is FALSE, then the statement will not be executed.

Note how the `if` statement is indented from the rest of the program. Also note that there is no semicolon between the expression and the statement (no semicolon following `if (number > 5)`).

The important point to see here is that the program always goes forward and may or may not execute the conditional part of the program (depending on whether the value of the user input is greater than 5). However, regardless of the user input, the remainder of the program is always executed (the last `printf` function output is always displayed).

A Compound Statement

Program 3.2 illustrates the use of the `if` statement with only one statement following it. This is fine when all you need to output is a simple single statement. But suppose you need

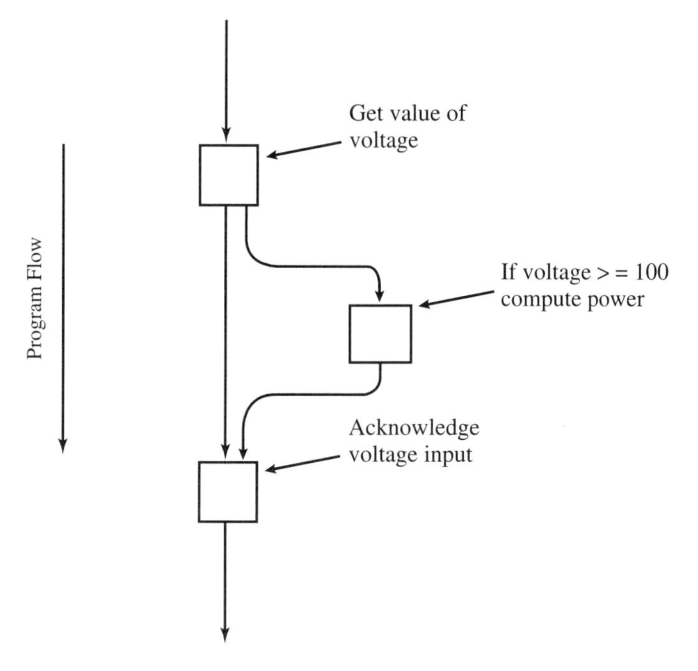

Figure 3.3 Optional Power Dissipation Computation

to develop a program that has more than one statement as a condition? For example, suppose your task is to develop a C program that takes the value of a voltage measurement across a resistor from the program user. The condition for the program is that if the voltage value is 100 volts or greater, then the program will compute the power dissipation in the resistor. If the voltage value is not equal to or greater than 100 volts, then the resistor power dissipation will not be computed. This task is illustrated in Figure 3.3.

When more than one statement must be included in the logical space reserved for a statement, then a **compound** statement may be used. This is nothing more than a series of C statements enclosed within braces (`{}`). Program 3.3 illustrates.

Program 3.3

```
#include <stdio.h>

main()
{
        float voltage;     /* Voltage measurement in volts. */
        float resistor;    /* Resistance value in ohms.     */
        float power;       /* Power calculation in watts.   */

        printf("Enter the voltage reading in volts => ");
        scanf("%f",&voltage);
        if (voltage >= 100.0)
        {
                printf("Voltage is equal to or greater than 100 V\n");
```

```
        printf("Please enter the resistor value => ");
        scanf("%f",&resistor);

        power = voltage*voltage/resistor;

        printf("The power dissipation is %f watts.\n",power);
    }
    printf("Input value of %f volts is acknowledged.",voltage);
    exit(0);
}
```

Figure 3.4 illustrates the logical construction of Program 3.3.

Program Analysis

Program 3.3 illustrates the use of a compound C statement:

```
{
        printf("Voltage is equal to or greater than 100 V\n");
        printf("Please enter the resistor value => ");
        scanf("%f",&resistor);

        power = voltage*voltage/resistor;

        printf("The power dissipation is %f watts.\n",power);
 }
```

Note that all of this part of the program is enclosed by braces (`{}`). What is between these braces is effectively another program block (just like the `{}` used in `main` or any other C function). This new program block is viewed by the `if` statement as another statement. This concept is illustrated in Figure 3.5.

So as far as the C program is concerned, the relational operation of

```
if (voltage >= 100.0)
```

will cause the execution of a single statement—which in this case is actually a compound statement.

Thus, in Program 3.3, if the program is executed and the user inputs a value of 25 volts across a 500 Ω resistor, the output will be

```
Enter the voltage reading in volts => 25
Input value of 25.000000 volts is acknowledged.
```

As you can see from this output, the `if` statement was not activated because the relation

```
voltage >= 100.0
```

is FALSE.

However, if the program user now enters a voltage reading of 125 volts across a 500 Ω resistor, the output will be

```
Enter the voltage reading in volts => 125
Voltage is equal to or greater than 100 V
Please enter the resistor value => 500
The power dissipation is 31.250000 watts
Input value of 125.000000 volts is acknowledged.
```

Observe from the preceding output that the last `printf` function is still executed.

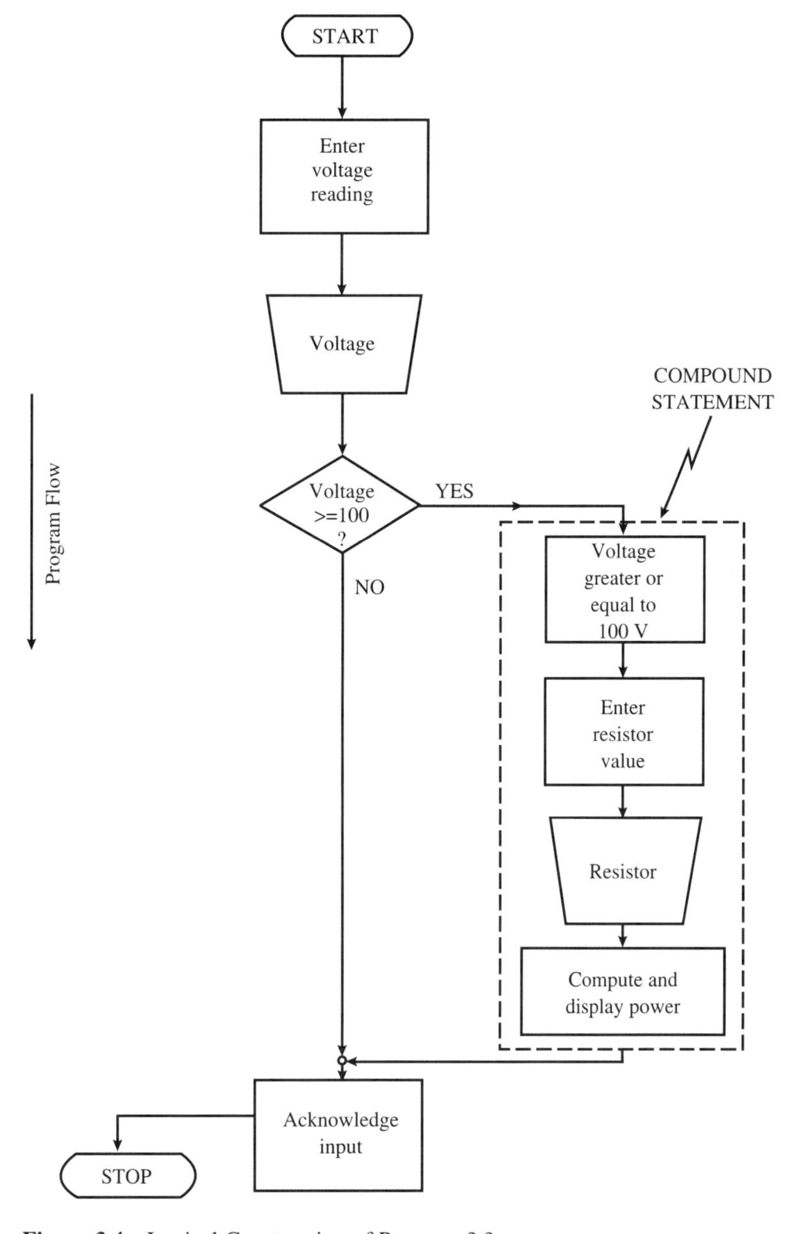

Figure 3.4 Logical Construction of Program 3.3

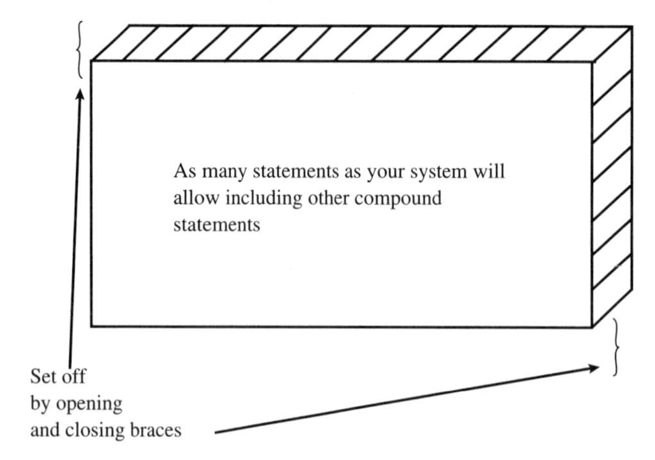

Figure 3.5 Concept of Viewing a Compound Statement

Calling a Function

Since the `if` statement can be used with a compound statement, it seems natural to ask if it can be used to call another C function (which could also contain some more `if` statements). The answer is yes it can. This is illustrated in Program 3.4.

Program 3.4

```
#include <stdio.h>

void power_calculation(float voltage);

main()
{
        float voltage;     /* Voltage measurement in volts. */

        printf("Enter the voltage reading in volts => ");
        scanf("%f",&voltage);

        if (voltage >= 100.0)
                power_calculation(voltage);

        printf("Input value of %f volts is acknowledged.",voltage);
        exit(0);
}

void power_calculation(float voltage)
{
        float resistor;    /* Resistance value in ohms.   */
        float power;       /* Power calculation in watts. */

        printf("Voltage is equal to or greater than 100 V\n");
        printf("Please enter the resistor value => ");
```

```
        scanf("%f",&resistor);

        power = voltage*voltage/resistor;

        printf("The power dissipation is %f watts.\n",power);
}
```

Program Analysis

Program 3.4 illustrates good programming practice in that it is divided into two separate parts. This doesn't mean that the more parts a program has the better it is. It means that each distinct task should be distinguished within the program. This program is divided by putting the optional power dissipation calculation in a separate function of its own. Now the `if` statement in the body of `main()` simply becomes

```
if (voltage >= 100.0)
   power_calculation(voltage);
```

This is much easier to read. It means that if the voltage value entered by the program user is greater than or equal to 100 volts, then do a power calculation using the voltage value as a part of that calculation. Here, you can see that the value of the voltage is passed to the function `power_calculation`. The function prototype at the beginning of the program

```
void power_calculation(float voltage);
```

tells the compiler to expect a `void` function (it will not return any value) that has one argument of type `float`. The function definition header

```
void power_calculation(float voltage)
```

is identical to the prototype (except, of course, that there is no ending semicolon). The declarations of the resistor value and power value are now in the function that uses them (`main()` is no longer cluttered with them):

```
float resistor;    /* Resistance value in ohms.    */
float power;       /* Power calculation in watts. */
```

The output of this program is exactly the same as the output of Program 3.3. The difference is that this program is easier to read and debug.

Conclusion

This section demonstrated the use of the `if` statement as a conditional C statement in an open branch. Here you saw that the `if` statement consists of an expression and a statement. If the expression is TRUE, the statement will be executed. If the expression is FALSE, the statement will not be executed. In either case, the remainder of the program is always executed.

You were also introduced to the compound statement. In C, this consists of using braces `{}` to set off more than one statement. Doing this allows many different operations to be parts of a conditional statement. The practice of using the `if` statement to call another C function was also presented here.

In the next section, you will be introduced to the closed branch. For now, test your understanding of this section by trying the following section review.

3.2 Section Review

1. Describe an open branch.
2. Explain the operation of the `if` statement.
3. What is a compound statement?
4. May an `if` statement call a function? Explain.

3.3 The Closed Branch

Discussion

This section demonstrates the action of the closed branch in C. The **closed branch** extends the decision-making capabilities of the language. As you will see, the closed branch statement in C is very similar to the open branch statement.

Basic Idea

The basic idea of a *closed branch* is illustrated in Figure 3.6.

There are two important points about the closed branch. First, the flow of the program always goes forward to new information. Second, the program will do one of two options (not both) and then proceed with the rest of the program. In C, the closed branch is accomplished by the **`if...else`** statement.

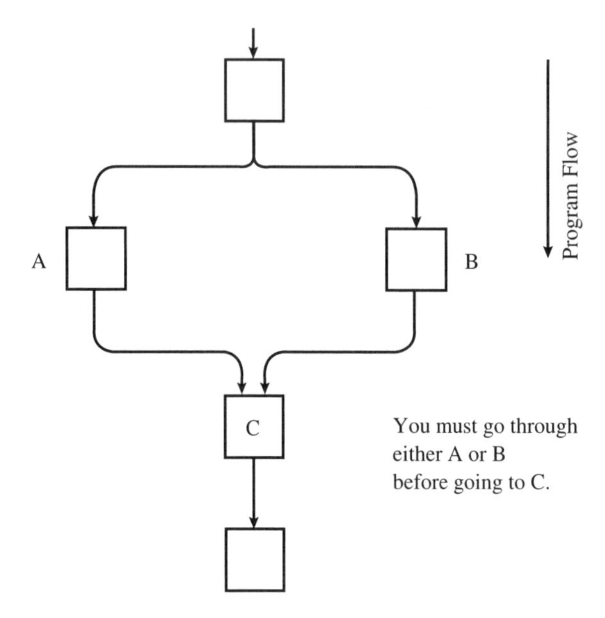

Figure 3.6 Concept of a Closed Branch

The if...else Statement

The `if...else` statement in C is another conditional statement. It differs from the `if` statement in that the `if` statement represents an open branch whereas the `if...else` statement represents a closed branch. The form of this statement is

if (expression) $statement_1$ else $statement_2$

What this means is that if `expression` is TRUE, then $statement_1$ will be executed and $statement_2$ will not be executed. If, on the other hand, `expression` is FALSE, then $statement_1$ will not be executed and $statement_2$ will be executed. This is illustrated in Program 3.5.

Program 3.5

```
#include <stdio.h>

main()
{
        float number;     /* User number value. */

        printf("\n\nGive me a number from 1 to 10 => ");
        scanf("%f",&number);

        if (number > 5.0)
                printf("Your number is greater than 5.\n");
        else
                printf("Your number is less than or equal to 5.\n");

        printf("The value of your number was %f",number);
}
```

Figure 3.7 illustrates the construction of Program 3.5.

Program Analysis

Note the construction of the `if...else` in Program 3.5.

```
if (number > 5.0)
        printf("Your number is greater than 5.\n");
else
        printf("Your number is less than or equal to 5.\n");
```

The general form of the `if...else` statement presents itself with the expressions

```
if (number > 5.0)
```

and

```
printf("Your number is greater than 5.\n");
```

as $statement_1$, and

```
printf("Your number is less than or equal to 5.\n");
```

as $\texttt{statement}_2$.

This `if...else` combination forces a selection between one of two `printf` functions. If the number entered by the program user is greater than 5, then the expression (`number > 5`) will be TRUE and $\texttt{statement}_1$ will be executed, yielding

```
Your number is greater than 5.
```

and the program will skip the `printf` function following the `else`.

If the program user inputs a 4, then the expression (`number > 5.0`) will be FALSE and $\texttt{statement}_1$ will not be executed. However, now $\texttt{statement}_2$ following the `else` will be executed.

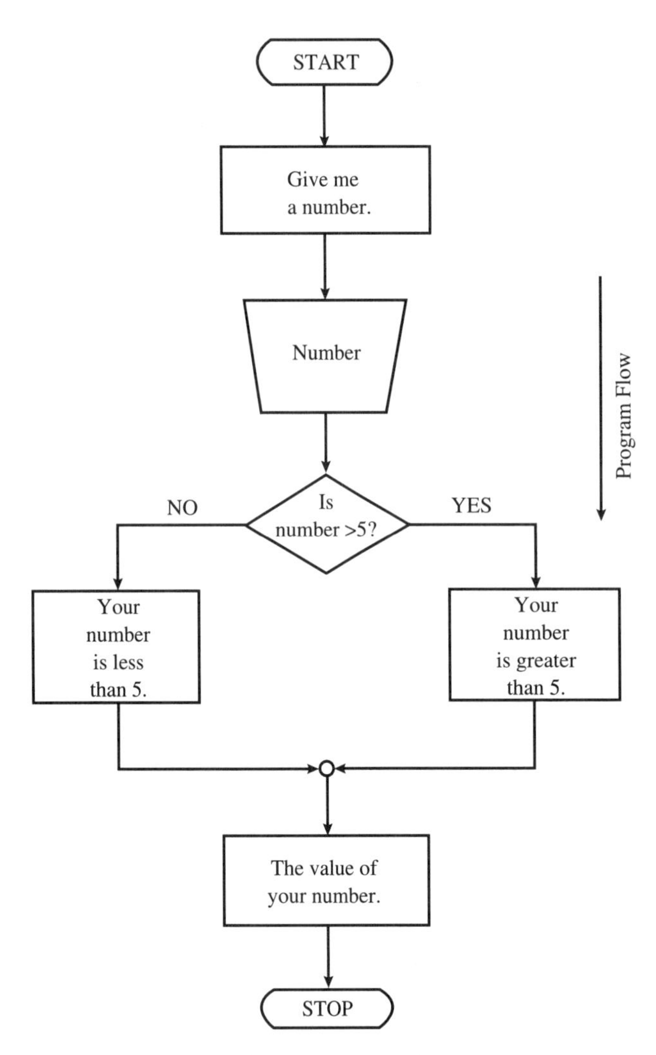

Figure 3.7 Construction of Program 3.5

In either case, program flow always goes forward and the final `printf` function is always executed:

```
printf("The value of your number was %f",number);
```

Compound if...else

You can add even more power to your program's decision-making capabilities through the use of compound `if...else` statements. This is illustrated in Program 3.6. The program will compute the area of a square or the area of a circle. The program user selects which computation is to be performed.

Program 3.6

```
#include <stdio.h>
#define PI 3.141592

main()
{
        float selection;    /* User selection.           */
        float length;       /* Length of side or radius. */
        float area;         /* Area in square units.     */

        printf("\n\nThis program will compute the area of\n");
        printf("a square or the area of a circle.\n");

        printf("\nSelect by number:\n");
        printf("1] Area of circle. 2] Area of square.\n");
        printf("Your selection (1 or 2) => ");
        scanf("%f",&selection);

        if (selection == 1)
        {
               printf("Give me the length of the circle radius => ");
               scanf("%f",&length);

               area = PI*length*length;
               printf("A circle of radius %f has an area of ",length);
               printf("%f square units.",area);
        }
        else
        if (selection == 2)
        {
               printf("Give me the length of one side of the square => ");
               scanf("%f",&length);

               area = length*length;
               printf("A square of length %f has an area of ",length);
               printf("%f square units.",area);
        }
```

```
        else
        {
                printf("That was not one of the selections.\n");
                printf("You must run the program again and\n");
                printf("select either a 1 or a 2.\n");
        }

        printf("\n\nThis concludes the program to calculate\n");
        printf("the area of a circle or a square.");

        exit(0);
}
```

Figure 3.8 illustrates the construction of Program 3.6.

Program Analysis

As you can see from Figure 3.8, this program has three options, as indicated by the **`if...else if...else`** construction. This can be viewed as

`if` ($\texttt{expression}_1$) $\texttt{statement}_1$ `else if` ($\texttt{expression}_2$) $\texttt{statement}_2$ `else` $\texttt{statement}_3$.

What this means is if $\texttt{expression}_1$ is TRUE, then $\texttt{statement}_1$ will be executed and none of the other statements. If $\texttt{expression}_2$ is TRUE, then $\texttt{statement}_2$ will be executed and none of the others. If neither $\texttt{statement}_1$ nor $\texttt{statement}_2$ is TRUE, then $\texttt{statement}_3$ will be executed. This is coded in Program 3.6 as

```
if (selection == 1)
 {

    [Body of compound statement to compute area of circle.]

 }
else
if (selection == 2)
 {

    [Body of compound statement to compute area of square.]
 }
else
 {

    [Body of compound statement to give message to user.]
 }
```

Notice that there are no semicolons between the `else if` or after any of the `if` or `else` keywords.

For this program, the program user enters a 1 or a 2 to select which computation will be performed. If a 1 is entered, the compound statements for the computation of the area of a circle are executed. If a 2 is entered, the compound statements for the computation of

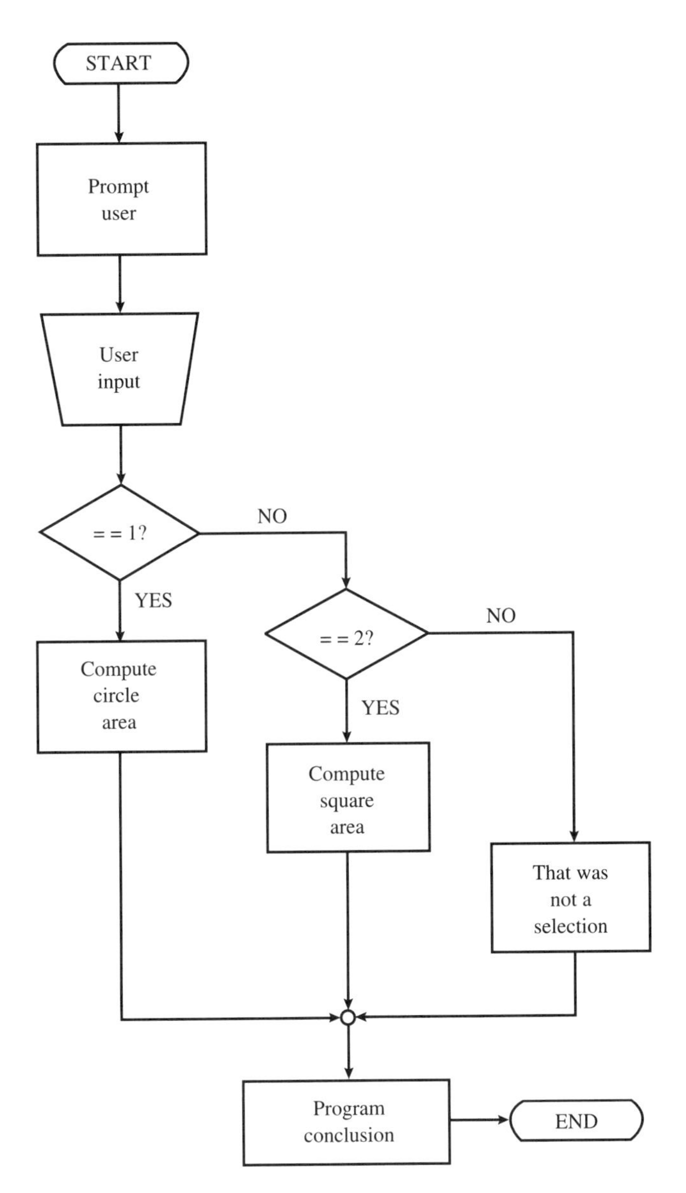

Figure 3.8 Construction of Program 3.6

the area of a square are executed. If a value other than 1 or 2 is entered, the statement following the last `else` is executed.

Another Way

A more structured way of presenting the same program is shown in Program 3.7. Note that now all of the compound statements are placed in their own functions. Also notice that both formulas have now been defined using the `#define` directive at the head of the program.

Program 3.7

```
#include <stdio.h>
#define PI 3.141592           /* The constant pi.  */
#define square(x) x*x         /* Area of a square. */
#define circle(r) PI*r*r      /* Area of a circle. */

void user_selection(void);    /* Get selection from user.          */
void circle_data(void);       /* Get circle radius and compute.    */
void square_data(void);       /* Get square side and compute.      */
void wrong_selection(void);   /* Notify user of wrong selection.   */

main()
{

        printf("\n\nThis program will compute the area of\n");
        printf("a square or the area of a circle.\n");

        user_selection();    /* Get selection from user. */

        printf("\n\nThis concludes the program to calculate\n");
        printf("the area of a circle or a square.");

        exit(0);
}

/* ------------------------------------------------------------------ */

void user_selection(void)   /* Get selection from user. */
{
        float selection;    /* User selection.          */

        printf("\nSelect by number:\n");
        printf("1] Area of circle. 2] Area of square.\n");
        printf("Your selection (1 or 2) => ");
        scanf("%f",&selection);

        if (selection == 1)
               circle_data();
        else
        if (selection == 2)
               square_data();
        else
               wrong_selection();
}

/* ------------------------------------------------------------------ */

void circle_data(void)      /* Get circle radius and compute. */
{
        float radius;       /* Radius of the circle.          */
```

```
        float area;          /* Circle area in square units.   */

        printf("Give me the length of the circle radius => ");
        scanf("%f",&radius);

        area = circle(radius);

        printf("A circle of radius %f has an area of ",radius);
        printf("%f square units.",area);
}

/* ------------------------------------------------------------ */

void square_data(void)     /* Get square side and compute.       */
{
        float side;          /* Side of the square.                */
        float area;          /* Area of the square in square units. */

        printf("Give me the length of one side of the square => ");
        scanf("%f",&side);

        area = square(side);

        printf("A square of length %f has an area of ",side);
        printf("%f square units.",area);
}

/* ------------------------------------------------------------ */

void wrong_selection(void)   /* Notify user of wrong selection. */
{
        printf("That was not one of the selections.\n");
        printf("You must run the program again and\n");
        printf("select either a 1 or a 2.\n");
}

/* ------------------------------------------------------------ */
```

Figure 3.9 illustrates the structure of Program 3.7.

As shown in the figure, the structure of the program has now been divided into function blocks.

Program Analysis

The program starts with the `#define` directives:

```
#define PI 3.141592          /* The constant pi.   */
#define square(x) x*x        /* Area of a square.  */
#define circle(r) PI*r*r     /* Area of a circle.  */
```

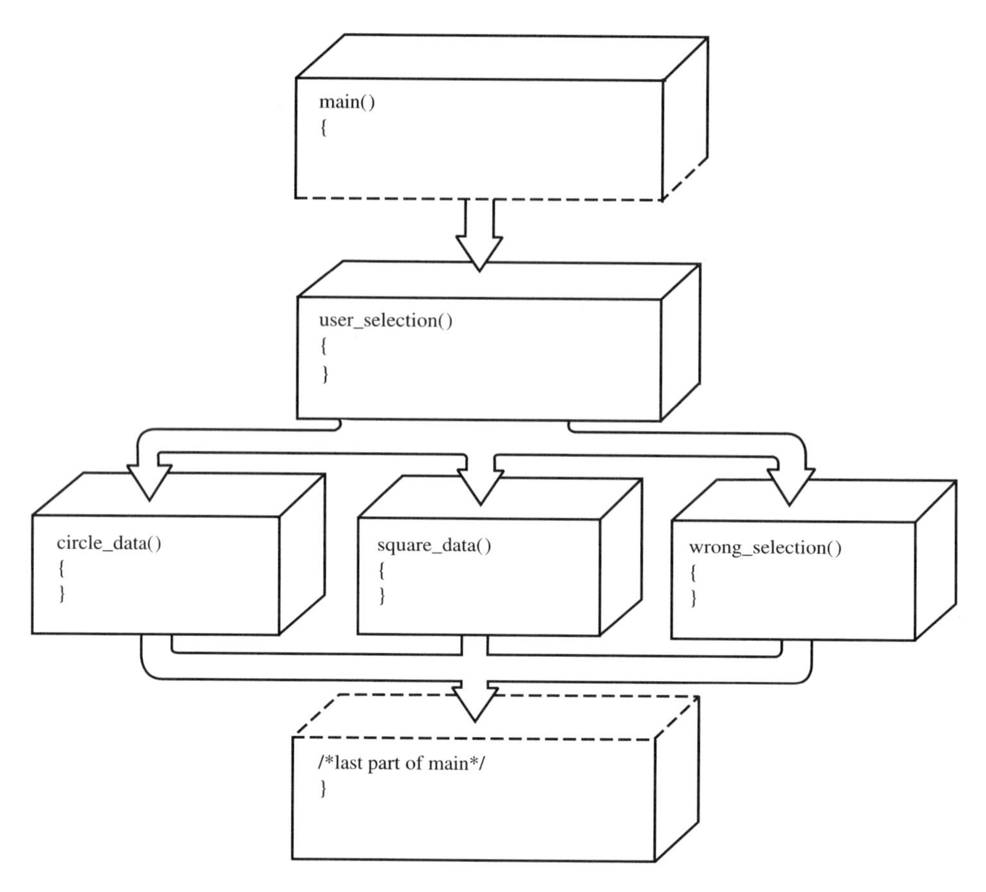

Figure 3.9 Structure of Program 3.7

Note that each `#define` is commented to exactly describe the purpose of the definition. Next, the function prototypes are presented:

```
void user_selection(void);    /* Get selection from user.          */
void circle_data(void);       /* Get circle radius and compute.    */
void square_data(void);       /* Get square side and compute.      */
void wrong_selection(void);   /* Notify user of wrong selection. */
```

Each prototype states its type and the type of any arguments. Each one is commented—and this same comment will be used on the header of the function definition in the body of the program. Next follows function `main()` with its simple coding:

```
main()
{

        printf("\n\nThis program will compute the area of\n");
        printf("a square or the area of a circle.\n");

        user_selection();   /* Get selection from user.   */
```

```
        printf("\n\nThis concludes the program to calculate \n");
        printf("the area of a circle or a square.");

       exit(0);

}
```

It is now very easy to read through `main()` and get the idea of what the program is to do and how it is going to do it. Note that only one function is called `user_selection()`. This function is now also easy to follow. It contains the necessary `if...else if...else` statements:

```
void user_selection(void) /* Get selection from user. */
{
        float selection;    /*  User selection.   */

        printf("\nSelect by number:\n");
        printf("1] Area of circle. 2] Area of square.\n");
        printf("Your selection (1 or 2) => ");
        scanf("%f",&selection);

        if (selection == 1)
             circle_data();
        else
        if (selection == 2)
             square_data();
        else
             wrong_selection();

}
```

In this function, it is clear that the user will be asked to choose between one of two calculations. The `if...else if...else` statements that follow simply call functions. Thus, it can be seen that `circle_data` or `square_data` or `wrong_selection` will be called by this function. Also note that the names of these functions are descriptive of what they do.

The other functions simply get the information, do the appropriate calculation, and display a corresponding message on the monitor. The action of this program, from a user standpoint, will be no different from the action of Program 3.6.

Conclusion

This section presented the powerful decision-making C statement `if...else if...else`. This feature is used very often in technical programming. You will see many examples of it in this and future chapters. In the next section you will be introduced to another powerful C decision-making structure. For now, test your understanding of this section by trying the following section review.

3.3 Section Review

1. Explain the meaning of a closed branch.
2. What is the difference between the `if` and the `if...else` statements?

3. May compound statements be used with the `if...else`? Explain.
4. May function calls be used with the `if...else`? Explain.
5. Describe the action of the `if...else if...else` statement in C.

3.4 Bitwise Boolean Operations

Discussion

This section presents program excerpts demonstrating the effect of bit manipulation in C. As you will see, **bit manipulation** is a method of working at the bit level with your machine. There are many applications for this low-level work such as in the area of hardware interfacing and using the computer as a controlling device for other systems.

Bit Manipulation

When doing bit manipulation, you are working with the individual bits of data within the computer. With bit manipulation you should think of all stored data in its binary form of 1s and 0s. Doing this is a great aid to understanding what is to follow. Think of the binary 1 as a Boolean TRUE and the binary 0 as a Boolean FALSE. Table 3.2 illustrates the meaning of various **Boolean operators**. Some of these will already be familiar to you.

The following examples give you some practice with each of the C bitwise operators presented in Table 3.2.

Bitwise Complementing

To get the **bitwise complement** of a number, convert the number to its binary equivalent, then change each 1 to a 0 and each 0 to a 1 (the same as taking the ones complement). Convert the resulting binary number back to the base of the original number. The resulting value is the bitwise complement of the number.

Some examples of bitwise complementation are shown in Example 3.2.

EXAMPLE 3.2

Determine the bitwise complements of the following hex numbers:

A. 00_{16} B. FF_{16} C. $A3_{16}$

Solution

A. Convert to binary: 00_{16} = $0000\ 0000_2$
Take ones complement: $1111\ 1111_2$
Convert back: F F = FF_{16}
Thus, the bitwise complement of 00_{16} is FF_{16}

B. Convert to binary: FF_{16} = $1111\ 1111_2$
Take ones complement: $0000\ 0000_2$
Convert back: 0 0 = 00_{16}
Thus, the bitwise complement of $FF_{16} = 00_{16}$

C. Convert to binary: $A3_{16}$ = $1010\ 0011_2$
Take ones complement: $0101\ 1100_2$
Convert back: 5 C = $5C_{16}$
Thus, the bitwise complement of $A3_{16}$ = $5C_{16}$

Table 3.2 Boolean Operations

Operation	Bitwise Operator	Meaning	With Bits
Bitwise COMPLEMENT	~	Change bit to its opposite	~1 = 0 ~0 = 1
Bitwise AND	&	The result is 1 if both bits are 1	0 & 0 = 0 0 & 1 = 0 1 & 0 = 0 1 & 1 = 1
Bitwise OR	|	The result is 0 if both bits are 0	0 | 0 = 0 0 | 1 = 1 1 | 0 = 1 1 | 1 = 1
Bitwise exclusive OR	^	The result is 1 if both bits are different	0 ^ 0 = 0 0 ^ 1 = 1 1 ^ 0 = 1 1 ^ 1 = 0

Program 3.8 is a program excerpt that illustrates the action of the bitwise complement.

Program 3.8

```
#include <stdio.h>

main()
{
        int value;

        printf("Input a hex number from 00 to FF =>");
        scanf("%X",&value);
        printf("The bitwise complement is => %X",~value);
}
```

Assuming the program user enters the value A3, execution of Program 3.8 yields

```
Input a hex number from 00 to FF => A3
The bitwise complement is => 5C
```

Note that the bitwise complement operator ~ is used on the variable `value`. This is done in the second `printf` function:

```
printf("The bitwise complement is => %X",~value);
```

Bitwise ANDing

To get the **bitwise AND** of a number, convert the number to its binary equivalent, then AND each corresponding bit of the two resulting binary numbers. Convert the resulting binary number back to the base of the original number. The resulting value is the bitwise ANDing of the two numbers.

Example 3.3 demonstrates bitwise ANDing.

EXAMPLE 3.3

Determine the hex results of bitwise ANDing the following hex numbers:

A. $00_{16}\&FF_{16}$ B. $FF_{16}\&A5_{16}$ C. $D3_{16}\&8E_{16}$

Solution

A. Convert to binary: $00_{16} = 0000\ 0000_2$
$FF_{16} = 1111\ 1111_2$
AND each bit pair: $0000\ 0000_2$
Convert back: 0 0 $= 00_{16}$
Thus, the bitwise ANDing of 00_{16} and FF_{16} is 00_{16}

B. Convert to binary: $FF_{16} = 1111\ 1111_2$
$A5_{16} = 1010\ 0101_2$
AND each bit pair: $1010\ 0101_2$
Convert back: A 5 $= A5_{16}$
Thus, the bitwise ANDing of FF_{16} and $A5_{16} = A5_{16}$

C. Convert to binary: $D3_{16} = 1101\ 0011_2$
$8E_{16} = 1000\ 1110_2$
AND each bit pair: $1000\ 0010_2$
Convert back: 8 2 $= 82_{16}$
Thus, the bitwise ANDing of $D3_{16}$ and $8E_{16} = 82_{16}$

Program 3.9 illustrates the action of bitwise ANDing.

Program 3.9

```
#include <stdio.h>

main()
{
        int value1;

        int value2;

        printf("Input a hex number from 00 to FF => ");
        scanf("%X",&value1);
        printf("Input a hex number to be bitwise ANDed => ");
        scanf("%X",&value2);

        printf("Bitwise ANDing of %X and %X produces => ",value1,value2);
        printf("%X",value1 & value2);
}
```

Assuming that the program user enters the values of D3 and 8E, execution of the above program produces

```
Input a hex number from 00 to FF => D3
Input a hex number to be bitwise ANDed => 8E
Bitwise ANDing of D3 and 8E produces => 82
```

Bitwise ORing

To get the **bitwise OR** of a number, convert the number to its binary equivalent, then OR each corresponding bit of the two resulting binary numbers. Convert the resulting binary number back to the base of the original number. The resulting value is the bitwise ORing of the two numbers.

Example 3.4 illustrates the bitwise OR operation.

EXAMPLE 3.4

Determine the hex results of bitwise ORing the following hex numbers:

A. $00_{16} \mid FF_{16}$ B. $FF_{16} \mid A5_{16}$ C. $D3_{16} \mid 8E_{16}$

Solution

A. Convert to binary: $00_{16} = 0000\ 0000_2$
$FF_{16} = 1111\ 1111_2$
OR each bit pair: $1111\ 1111_2$
Convert back: F F = FF_{16}
Thus, the bitwise ORing of 00_{16} and FF_{16} is FF_{16}

B. Convert to binary: $FF_{16} = 1111\ 1111_2$
$A5_{16} = 1010\ 0101_2$
OR each bit pair: $1111\ 1111_2$
Convert back: F F = FF_{16}
Thus, the bitwise ORing of FF_{16} and $A5_{16}$ is FF_{16}

C. Convert to binary: $D3_{16} = 1101\ 0011_2$
$8E_{16} = 1000\ 1110_2$
OR each bit pair: $1101\ 1111_2$
Convert back: D F = DF_{16}
Thus, the bitwise ORing of $D3_{16}$ and $8E_{16}$ is DF_{16}

Program 3.10 illustrates the action of bitwise ORing.

Program 3.10

```
#include <stdio.h>

main()
{
        int value1;
        int value2;

        printf("Input a hex number from 00 to FF => ");
        scanf("%X",&value1);
        printf("Input a hex number to be bitwise ORed => ");
        scanf("%X",&value2);
        printf("Bitwise ORing of %X and %X produces => ",value1,value2);
        printf("%X",value1 | value2);
}
```

Assuming that the program user enters the values of D3 and 8E, execution of Program 3.10 produces

```
Input a hex number from 00 to FF => D3
Input a hex number to be bitwise ORed => 8E
Bitwise ORing of D3 and 8E produces => DF
```

Bitwise XORing

To get the **bitwise XOR** (exclusive OR) of a number, convert the number to its binary equivalent, then XOR each corresponding bit of the two resulting binary numbers. Convert the resulting binary number back to the base of the original number. The resulting value is the bitwise XORing of the two numbers. This is shown in Example 3.5.

EXAMPLE 3.5

Determine the hex results of bitwise XORing the following hex numbers:

A. $00_{16} \wedge FF_{16}$ B. $FF_{16} \wedge A5_{16}$ C. $D3_{16} \wedge 8E_{16}$

Solution

A. Convert to binary: $00_{16} = 0000\ 0000_2$
$FF_{16} = 1111\ 1111_2$
XOR each bit pair: $1111\ 1111$
Convert back: F F = FF_{16}
Thus, the bitwise XORing of 00_{16} and FF_{16} is FF_{16}

B. Convert to binary: $FF_{16} = 1111\ 1111_2$
$A5_{16} = 1010\ 0101_2$
XOR each bit pair: $0101\ 1010_2$
Convert back: 5 A = $5A_{16}$
Thus, the bitwise XORing of FF_{16} and $A5_{16} = 5A_{16}$

C. Convert to binary: $D3_{16} = 1101\ 0011_2$
$8E_{16} = 1000\ 1110_2$
XOR each bit pair: $0101\ 1101_2$
Convert back: 5 D = $5D_{16}$
Thus, the bitwise XORing of $D3_{16}$ and $8E_{16} = 5D_{16}$

Program 3.11 illustrates the action of bitwise XORing.

Program 3.11

```
#include <stdio.h>

main()
{
        int value1;
        int value2;

        printf("Input a hex number from 00 to FF => ");
        scanf("%X",&value1);
        printf("Input a hex number to be bitwise XORed => ");
        scanf("%X",&value2);
```

```
        printf("Bitwise XORing of %X and %x produces => ",value1,value2);
        printf("%X",value1 ^ value2);
}
```

Assuming that the program user enters the values of D3 and 8E, execution of Program 3.11 produces

```
Input a hex number from 0 to FF => D3
Input a hex number to be bitwise XORed => 8E
Bitwise XORing of D3 and 8E produces => 5D
```

Shifting Bits

C also allows for shifting bits left or right. This is accomplished with the shift left operator (<<) or the shift right operator (>>). To determine the result of a bitwise shift, convert the value to binary, then shift the binary bits the required number of bits in the indicated direction. Convert the resulting binary number back to the base of the original value. Example 3.6 illustrates the result of **bit shifting**.

EXAMPLE 3.6

Determine the hex value of the following hex numbers when shifted to the right or left the number of bits indicated:

A. FF_{16}<<2   B. $5C_{16}$>> 3   C. 05_{16}<<4

Solution

A. Convert to binary => FF_{16} = $1111\ 1111_2$
   Shift left 2 bits => $11\ 1111\ 1100_2$
   Convert back to hex => 3 F C = $3FC_{16}$
   Thus FF_{16}<<2 is FC_{16}

B. Convert to binary => $5C_{16}$ = $0101\ 1100_2$
   Shift right 3 bits => $0000\ 1011_2$
   Convert back to hex => 0 B = $0B_{16}$
   Thus $5C_{16}$>>3 is $0B_{16}$

C. Convert to binary => 05_{16} = $0000\ 0101_2$
   Shift left 4 bits => $0101\ 0000_2$
   Convert back to hex => 5 0 = 50_{16}
   Thus 05_{16}<<4 is 50_{16}

Program 3.12 illustrates a shift left operation.

Program 3.12

```
#include <stdio.h>

main()
{
        int value1;
        int shift_left;
```

```
        printf("Input a hex number from 00 to FF => ");
        scanf("%X",&value1);
        printf("Input a number of bits to be left shifted => ");
        scanf("%X",&shift_left);
        printf("Shifting %X %d places to the left produces => "
                ,value1,shift_left);
        printf("%X",value1 << shift_left);
}
```

Assuming that the program user enters the value 5C to be shifted to the left 3 bits, execution of Program 3.12 yields

```
Input a hex number from 00 to FF => 5C
Input a number of bits to be left shifted => 3
Shifting 5C 3 places to the left produces => 2E0
```

Conclusion

This section presented the concepts of bitwise operations available with C. In this section you saw how to perform bitwise Boolean operations as well as bitwise shifting. The information presented here was in the form of several small programs which were developed to demonstrate these concepts. Test your understanding of this section by trying the following section review.

3.4 Section Review

1. Define what is meant by a bitwise complement in C.
2. What is the meaning of the bitwise AND operation?
3. Explain what is meant by the bitwise OR operation.
4. Define what is meant by a bitwise XOR operation.
5. What is meant by a bitwise shift in C?

3.5 Logical Operation

Discussion

In this section you will learn about the logical operators used in C. Here the information presented in the first part of this chapter will be applied. By using C's logical operators, your programs will have increased decision-making capabilities.

Logical AND

The **logical AND** operation in C is expressed as

$(\texttt{expression}_1)\&\&(\texttt{expression}_2)$

The above operation will be evaluated TRUE only if $\texttt{expression}_1$ is TRUE *and* $\texttt{expression}_2$ is TRUE; otherwise the operation will be evaluated as FALSE. Keep in mind that in C a FALSE evaluation is actually a 0, whereas a TRUE evaluation is actually a non-zero value. Table 3.3 summarizes the AND operation.

Table 3.3 The AND Operation

$expression_1$	$expression_2$	**Result**
FALSE	FALSE	FALSE
FALSE	TRUE	FALSE
TRUE	FALSE	FALSE
TRUE	TRUE	TRUE

Observe that the double `&&` is used to represent this operation. No spaces are allowed between these symbols, although spaces are allowed to the left and right of this double symbol.

Program 3.13 illustrates the use of the logical AND operation in C.

Program 3.13

```
#include <stdio.h>

main()
{
        float result;    /* Result of logical expression. */

        result = 0 && 0;
        printf("0 && 0 = %f\n",result);

        result = 0 && 1;
        printf("0 && 1 = %f\n",result);

        result = 1 && 0;
        printf("1 && 0 = %f\n",result);

        result = 1 && 1;
        printf("1 && 1 = %f\n",result);
}
```

Figure 3.10 illustrates the operation of Program 3.13.

Program Analysis

Execution of Program 3.13 yields

```
0 && 0 = 0.000000
0 && 1 = 0.000000
1 && 0 = 0.000000
1 && 1 = 1.000000
```

Notice the variable `result` is of type `float`. You should try this with other data types as well. It is then used to store the result of each of the AND operations.

The first AND operation

```
result = 0 && 0;
```

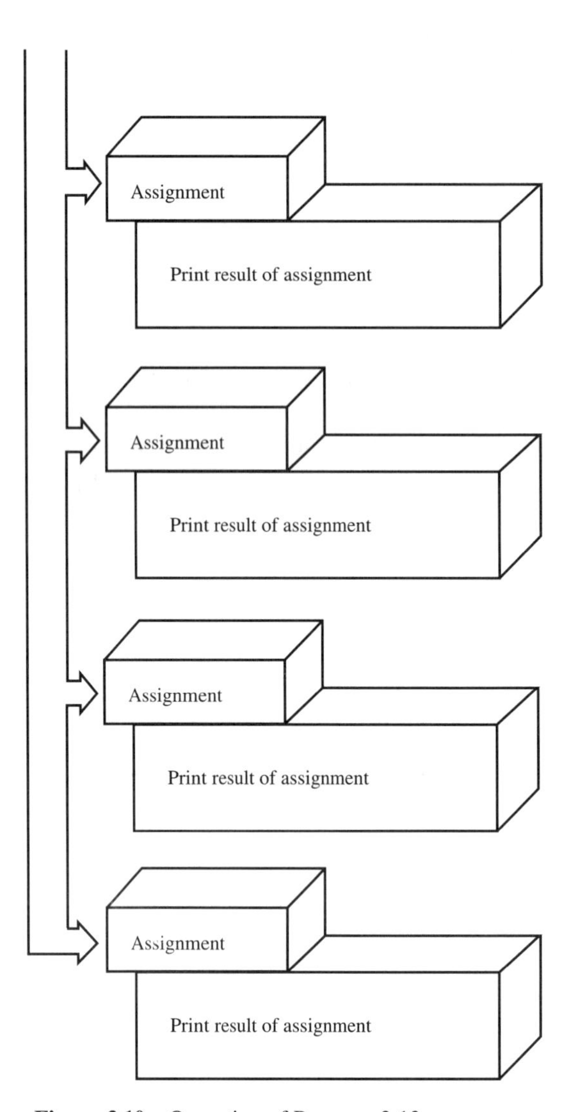

Figure 3.10 Operation of Program 3.13

causes two FALSE values (represented in C as `0`) to be ANDed. Since the result of this is FALSE, the value of the variable `result` will be `0`.

The next AND operation

```
result = 0 && 1;
```

causes a FALSE and a TRUE (represented in C as a `0` and a `1`) to be evaluated. Since the result of this operation is FALSE, the value of the variable `result` will again be `0`.

The third AND operation

```
result = 1 && 0;
```

causes a TRUE and a FALSE (represented in C as a `1` and a `0`) to be evaluated. This is another operation that is FALSE, and the value of `result` is once again `0`.

The last operation is the only one that evaluates to a TRUE:

```
result = 1 && 1;
```

Here, a TRUE is being ANDed with a TRUE (represented as a `1` and a `1`). The result of this operation is TRUE, and the value of `result` will be evaluated to a `1`.

The OR Operation

The **logical OR** operation in C is expressed as

$(\texttt{expression}_1)$ `||` $(\texttt{expression}_2)$

This operation will be evaluated FALSE only if $\texttt{expression}_1$ is FALSE *and* $\texttt{expression}_2$ is FALSE; otherwise the operation will be evaluated as TRUE. As before, in C a FALSE evaluation is actually a `0` whereas a TRUE evaluation is actually a non-zero value. Table 3.4 summarizes the OR operation.

Observe that the double || is used to represent this operation. No spaces are allowed between these symbols, although spaces are allowed to the left and right of this double symbol.

Program 3.14 illustrates the use of the OR operation in C.

Program 3.14

```
#include <stdio.h>

main()
{
        float result;    /* Result of logical expression. */

        result = 0 || 0;
        printf("0 || 0 = %f\n",result);

        result = 0 || 1;
        printf("0 || 1 = %f\n",result);

        result = 1 || 0;
        printf("1 || 0 = %f\n",result);

        result = 1 || 1;
        printf("1 || 1 = %f\n",result);
}
```

Table 3.4 The OR Operation

$\textbf{expression}_1$	$\textbf{expression}_2$	**Result**
FALSE	FALSE	FALSE
FALSE	TRUE	TRUE
TRUE	FALSE	TRUE
TRUE	TRUE	TRUE

Program Analysis

Execution of Program 3.14 yields

```
0 || 0 = 0.000000
0 || 1 = 1.000000
1 || 0 = 1.000000
1 || 1 = 1.000000
```

Notice that the variable `result` in this case is of type `float`. Again, other types could have been used. It is then used to store the result of each of the OR operations.

The first OR operation

```
result = 0 || 0;
```

causes two FALSE values (represented in C as `0`) to be ORed. Since the result of this is FALSE, the value of the variable `result` will be `0`.

The next OR operation

```
result = 0 || 1;
```

causes a FALSE or a TRUE (represented in C as a `0` or a `1`) to be evaluated. Since the result of this operation is TRUE, the value of the variable `result` will now be `1`.

The third OR operation

```
result = 1 || 0;
```

causes a TRUE and a FALSE (represented in C as a `1` or a `0`) to be evaluated. This is another operation that is TRUE, and the value of `result` is once again `1`.

The last operation also evaluates to a TRUE.

```
result = 1 || 1;
```

Here, a TRUE is being ORed with a TRUE (represented as a `1` and a `1`). The result of this operation is TRUE, and the value of `result` will be evaluated to a `1`.

Relational and Logical Operations

Relational operations may be used with logical operations. This can be done because a relational operation returns a TRUE or FALSE condition. These are the same conditions used by logical operators. Table 3.5 shows the order of *precedence* for the C operators

Table 3.5 C Precedence

Operators	Name
!	Logical Not
* /	Multiplication and Division
+ -	Addition and Subtraction
< <= >= >	Less, Less or Equal, Greater or Equal, Greater
== !=	Equal, Not Equal
&&	Logical AND
||	Logical OR

presented up to this point. This means that the ! is evaluated before the *, which is evaluated before the <, and so on.

There is a new logical operator shown in Table 3.5, called the **logical NOT** and represented by the !. You will see an example of this shortly. For now, realize that what Table 3.5 shows is which of these operations will be done before others in a program line that contains more than one of them. The use of this table and of the logical NOT is illustrated in the following example.

EXAMPLE 3.7

Determine the results of each of the following expressions (will they be TRUE or FALSE?):

A. (5 == 5) || (6 == 7)
B. (5 == 8)&&(6 != 7)
C. (8 >= 5)&&(!(5 <= 2))

Solution

A. Analyze the expression by first determining the logical value of the operation in the innermost parentheses. For this one, (5==5) is TRUE. It now makes no difference what the condition of the following operation is because of the OR. The final result of an OR will always be TRUE as long as any one member of the OR expression is TRUE. This analysis method is the same used by the C compiler. If it determines that the first part of a || is TRUE, then it will not bother to check the remainder of the expression. If it finds that the first part is FALSE, then the second part will be evaluated to test its final condition.
B. Using the same analysis method, you see that the expression (5 == 8) is FALSE. Since this is an AND expression, you do not need to evaluate the (6 != 7) (which is TRUE). The reason for this is that any AND that has one member FALSE will always be FALSE. Again, this is how C evaluates a && expression. If the first part is FALSE, it does not go on to evaluate the second part. The only time the second part will be evaluated is when the first part is TRUE. Thus, the result of this expression is FALSE.
C. This expression uses the NOT (!) logical operator. Therefore !TRUE means FALSE (NOT TRUE is FALSE), and !FALSE means TRUE (NOT FALSE is TRUE). Evaluation of the first part: (8>=5) is TRUE. Since this is a && operation, it's now necessary to evaluate the next part: !(5 <= 2). Because (5 <= 2) is FALSE, NOT(5 <= 2) is TRUE. Because both sides of the AND are TRUE, the final result is also TRUE.

Program 3.15 shows an example of using relational and logical operations together. The program gets a number from the program user and then checks to see if its value is between 1 and 10. If it is, a message is printed to the screen.

Program 3.15

```
#include <stdio.h>

main()
{
        float number;      /* User input number. */

        printf("\n\nGive me a number from 1 to 100 => ");
        scanf("%f",&number);

        if ((number >= 1.0)&&(number <= 10.0))
               printf("You gave me a number between 1 and 10.");
}
```

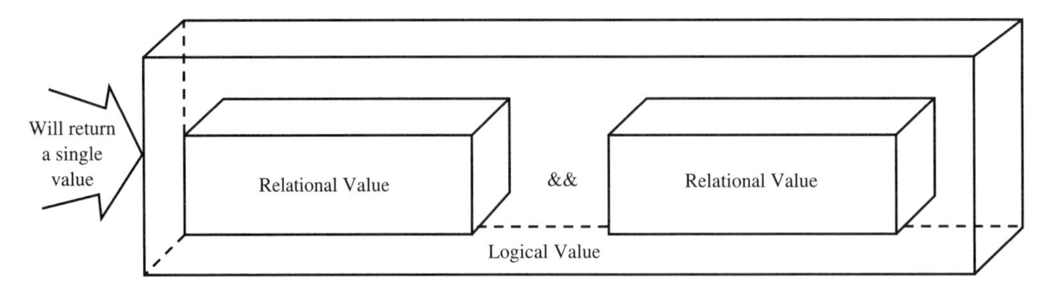

Figure 3.11 Relational and Logical Operation

Program Analysis

Note the program line that combines a relational and a logical expression:

```
if ((number >= 1.0)&&(number <= 10.0))
```

This statement can be thought of as shown in Figure 3.11.

The parentheses around the relational operations may be omitted because the >= and the <= have higher precedence than the `&&`. However, there is no harm in putting them there, and they should be used if they improve clarity.

Compounding the Logic

Any combination of relational and logical operations may be used. For example, Program 3.16 uses a complex logical operation consisting of OR as well as AND with relational operations. The program checks the user input of a number between 1 and 100 to see if the number entered is in the top or bottom 10%.

Program 3.16

```
#include <stdio.h>

main()
{
        float number;     /* User input number. */

        printf("\n\nGive me a number from 1 to 100 => ");
        scanf("%f",&number);

        if (((number >= 1.0)&&(number <= 10.0))||((number >= 90.0)&&(number <= 100.0)))
                printf("You gave me a number in the top or bottom 10%%.");
}
```

Program Analysis

The key to Program 3.16 is the statement

```
if (((number>=1.0)&&(number<=10.0))||((number>=90.0)&&(number<=100.0)))
```

An analysis of this statement is shown in Figure 3.12.

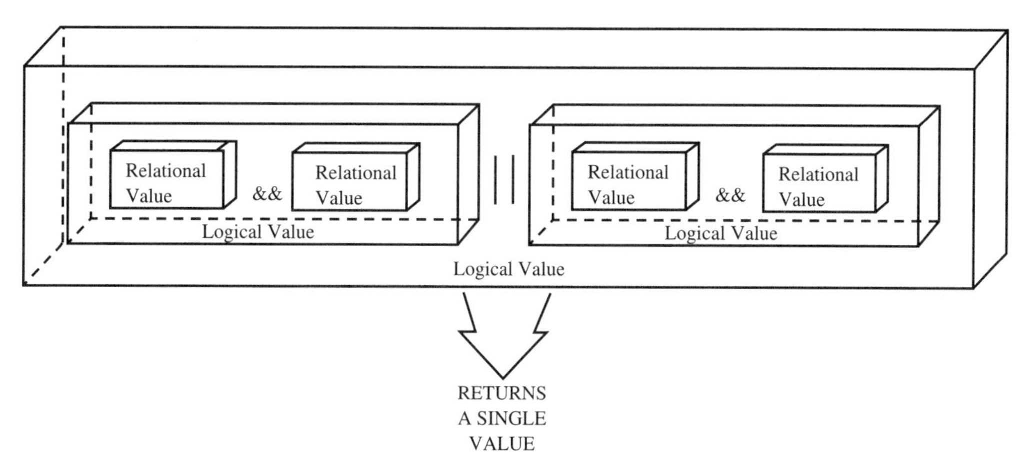

Figure 3.12 Analysis of Another Compound Operation

Again, parentheses are used around the relational operations to help read the logic of what the program is to do.

Application Program

Program 3.17 illustrates a technical application using relational and logical operations. The program reads the temperature of a process. In this program, this is entered by the program user. However, it could also be done directly from a temperature probe interfaced with the computer. The result of the program would then depend on the temperature reading of the probe. The logic of this program is illustrated in Figure 3.13.

Program 3.17

```
#include <stdio.h>

main()
{
        float temp;     /* Temperature reading. */

        printf("\n\nGive me the temperature reading => ");
        scanf("%f",&temp);

        if ((temp >= 100.0)&&(temp <= 120.0))
               printf("Temperature OK, continue process.");

        if (temp < 100.0)
               printf("Temperature too low, increase energy.");

        if ((temp > 120.0)&&(temp <= 150.0))
               printf("Temperature too high, decrease energy.");

        if (temp > 150.0)
               printf("Danger! Shut down systems.");
}
```

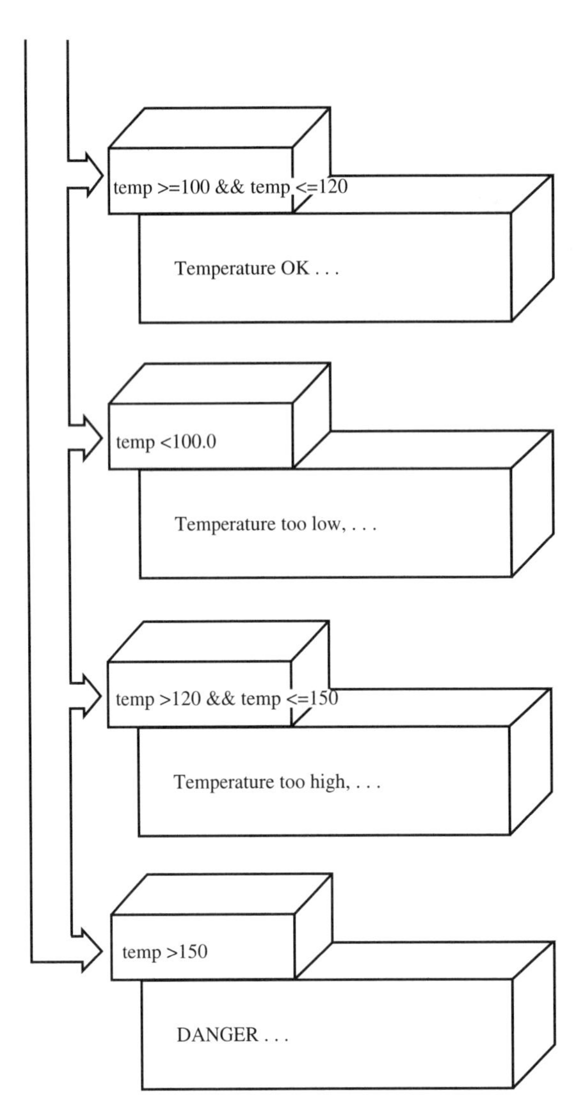

Figure 3.13 Temperature Probe Program

Conclusion

This section demonstrated the power of combining relational and logical operations. You saw some very simple examples as well as a more complex technical application. In the next section, you will learn about another powerful decision-making tool available in the C language. First check your understanding of this section by trying the following section review.

3.5 Section Review

1. Describe the operation of the logical AND. What symbol is used to represent this in C?
2. Describe the operation of the logical OR. What symbol is used to represent this in C?

3. Describe the operation of the logical NOT.
4. Give an example of combining a relational and a logical operation.
5. State how C evaluates the AND operation and the OR operation.

3.6 Conversion and Type Casting

Discussion

This section presents information regarding what may seem to be a small detail. You may not have had to pay attention to it up to this point, but this "detail" could become very important in more complex C programs.

Data Types

You will recall that every piece of data in C has a type. It may be an `int`, `char`, or other type. The C language allows you to add a type `int` to a type `float`. Doing this is called *mixing types*.

When data types are mixed in an expression, the compiler will convert all of the variables to a single compatible type. Then the operation will be carried out. This is done according to the **rank** of the data type. Variables of a lower rank are converted to the type of variables with a higher rank. The ranking of variables is

Low Rank <= char, int, long, float, double => High Rank

This means if an operation of a `char` and `int` takes place, the resulting value will be of type `int`. If an operation with a `float` and an `int` takes place, the resulting output will be of type `float`. A good rule to follow is not to mix data types. If it is necessary to do so, use the method of **type casting**.

Type Casting

Type casting provides a method for converting a variable to a particular type. To do this, simply precede the variable by the desired type. As an example, if `value` were originally declared as a type `int`, it is converted to a type `float` as follows:

```
result = (float)value;
```

You should be cautious when demoting a variable from a high rank to a lower rank. In doing this, you are asking a larger value to fit into a smaller value, and this usually results in corrupted data.

You cannot use the cast on a type `void`. You can cast any type to `void`, but you cannot cast a type `void` to any other type.

lvalue

There are times in programming when you will see the term **lvalue**. It literally means left-handed value because an assignment operation assigns the value of the right-hand operand to the memory location indicated by the left-hand operand. As an example: `value = 3 + 5;` is a legal statement in C; however, `3 + 5 = value;` is not a legal statement in C because the left-hand operand is not an lvalue (does not represent a unique memory location). What this means is that the left-hand operand of an assignment operation must be an expression that refers to a modifiable memory location, and expressions that refer to memory locations in this manner are called lvalue expressions.

Conclusion

This section presented some important details about C. Here you learned about mixed data types and how they are converted in C. You also saw how to cast to a different data type, and you discovered the meaning of lvalue. Check your understanding of this section by trying the following section review.

3.6 Section Review

1. Explain what is meant by mixing data types.
2. What happens if a type `int` is added to a type `float`?
3. State the rule for mixed data types in C.
4. What is meant by a cast in C?
5. What is an lvalue expression?

3.7 The C switch

Discussion

This section presents a method used by C to make one selection when there are several choices to be made. Here you will see one of the most powerful decision-making commands in the C language. First, the need for this command will be demonstrated. Then the command will be defined and applied to a technology problem. You will also see the need for using `char` or `int` types as presented in the previous section.

Decision Revisited

Program 3.18 is an example of having one selection from several different alternatives. What the program does is to display one of three forms of Ohm's law depending on the choice made by the program user. In the program, the program user selects one of three letters, and the corresponding formula is displayed on the monitor.

Program 3.18

```
#include <stdio.h>

main()
{
	char selection;    /* Item to be selected by program user. */

	printf("\n\nSelect the form of Ohm's law needed by letter:\n");
	printf("A] Voltage B] Current C] Resistance\n");
	printf("Your selection (A, B, or C) => ");
	scanf("%c",&selection);
```

```
        if (selection == 'A')
                printf("V = I*R");

        else
        if (selection == 'B')
                printf("I = V/R");

        else
        if (selection == 'C')
                printf("R = V/I");

        else
        printf("That was not one of the proper selections.");
}
```

Program Analysis

Figure 3.14 illustrates the logic of Program 3.18.

As shown in the figure, the program allows the program user three choices. This is achieved in the program by the use of the `if...else if...else` in C. Note the use of the `char` variable as the selection variable. The program user must enter one of the three uppercase letters A, B, or C.

The selection is made by the relational statements. As an example,

```
if (selection == 'A')
        printf("V = I*R");
```

If `selection` equals the character A, the `printf()` function will be executed. If none of the correct selections is made, the program will default to the last `else`:

```
else
        printf("That was not one of the proper selections.");
```

This type of program, where there is a selection from a list of different possibilities, is so common that a special C statement is used for it.

How to Switch

In C the `switch` statement is an easier way to code multiple `if...else if...else` statements. The `switch` statement has the form shown below:

```
switch (expression1)
{
        case constant-expression : (expression1)
        default : (expression2)
}
```

Where

`switch` = Reserved word indicating that a switch statement is about to take place.

$expression_1$ = Any legal C expression.

{ Defines the beginning of the `switch` body.

`case` = Reserved word indicating that what follows is the constant-expression required for a match.

`constant-expression` Identifies what is required for a match. This must be of the same data type as $\texttt{expression}_1$ (sometimes called the case label).

`default` = C reserved word indicating the option to be exercised if no match is made.

`}` = Defines the ending of the `switch` body.

The construction of the `switch` statement requires the introduction of two other C keywords: `case` and `break`. Essentially the `switch` statement identifies a variable whose value will determine which `case` will be activated. The `break` lets C know when the selected program code is to end. Program 3.19 does exactly the same thing as Program 3.18 did; the program user gets to select one of three forms of Ohm's law. The difference is that now the `switch` statement is used in place of the `if...else if...else` statements.

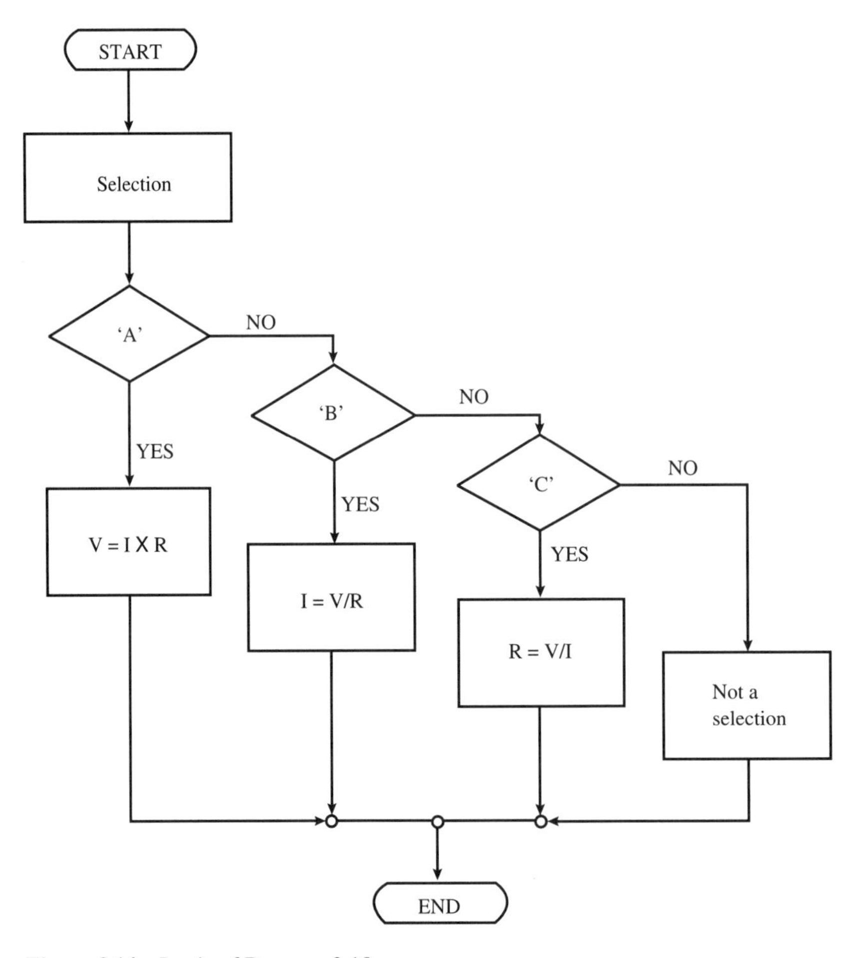

Figure 3.14 Logic of Program 3.18

Program 3.19

```
#include <stdio.h>

main()
{
        char selection;    /* Item to be selected by program user. */

        printf("\n\nSelect the form of Ohm's law needed by letter:\n");
        printf("A] Voltage B] Current C] Resistance\n");
        printf("Your selection (A, B, or C) => ");
        scanf("%c",&selection);

        switch(selection)
        {
                case 'A' : printf("V = I*R");
                           break;
                case 'B' : printf("I = V/R");
                           break;
                case 'C' : printf("R = V/I");
                           break;
                default  : printf("That was not one of the proper selections.");
        } /* End of switch. */
}
```

Program Analysis

It should be clear from Program 3.19 that it is easier to read program code that uses the `switch` statement instead of multiple `if...else if...else` statements. Observe the following extract from Program 3.19:

```
switch(selection)
{
  case 'A' : printf("V = I*R");
             break;
  case 'B' : printf("I = V/R");
             break;
  case 'C' : printf("R = V/I");
             break;
  default  : printf("That was not one of the proper selections.");
} /* End of switch. */
```

Note that the `switch` statement is followed by the `char` variable selection. This means that each of the `case` labels must be of type `char` (and they are: `'A'`, `'B'`, and `'C'`). Note also the use of the opening `{` and closing `}`. These are both brought out to the left so that it is clear exactly where the body of the `switch` statement begins and ends. One other important item is the comment following the closing `}`. It is there to remind you that this is the end of the body of a `switch` statement. This is a good habit to get into—especially when you have programs with long selection lists. Note that the `break` statement is aligned so that it clearly identifies where the execution of the selection option will terminate.

Again, if none of the correct inputs is selected, the program will `default` to

```
default   : printf("That was not one of the proper selections.");
```

Compound statements may also be used with the `switch`. This is presented below.

Compounding the switch

You may use compound statements with the C `switch`. This is illustrated in Program 3.20. This program is an expansion of the previous ones in this section. Now the program user may actually perform a calculation with the selected form of Ohm's law.

Program 3.20

```
#include <stdio.h>

main()
{
        char selection;     /* Item to be selected by program user. */
        float voltage;      /* Circuit voltage in volts.            */
        float current;      /* Circuit current in amps.             */
        float resistance;   /* Circuit resistance in ohms.          */

        printf("\n\nSelect the form of Ohm's law needed by letter:\n");
        printf("A] Voltage B] Current C] Resistance\n");
        printf("Your selection (A, B, or C) => ");
        scanf("%c",&selection);

        switch(selection)
        {
               case 'A' : {  /* Solve for voltage. */
                             printf("Input the current in amperes => ");
                             scanf("%f",&current);
                             printf("Value of the resistance in ohms => ");
                             scanf("%f",&resistance);
                             voltage = current*resistance;
                             printf("The voltage is %f volts.",voltage);
                          }
                          break;
               case 'B' : {  /* Solve for current. */
                             printf("Input the voltage in volts => ");
                             scanf("%f",&voltage);
                             printf("Value of the resistance in ohms => ");
                             scanf("%f",&resistance);
                             current = voltage/resistance;
                             printf("The current is %f amperes.",current);
                          }
                          break;
               case 'C' : {  /* Solve for the resistance. */
                             printf("Input the voltage in volts => ");
                             scanf("%f",&voltage);
```

```
                        printf("Value of the current in amperes => ");
                        scanf("%f",&current);
                        resistance = voltage/current;
                        printf("The resistance is %f ohms.",resistance);
                      }
                      break;
        default     : printf("That was not a correct selection.\n");
                      printf("Please go back and select A, B, or C");
    } /* End of switch. */
}
```

The general structure of Program 3.20 is shown in Figure 3.15.

Observe the block structure used in the C `switch` statement. This structure makes it easier to read and understand the program code. Notice that the compound statement is indented and how its beginning and closing are clearly defined with the `{` and `}`. Also note the location of the `break`. It lets you clearly see where the body of the option ends. Again, structure makes very little difference to the program user. Program structure is for you and others who will be modifying or trying to understand your program.

Program Analysis

The important characteristic of Program 3.20 is the use of a compound C statement with the `switch` statement.

```
switch(selection)
 {
  case 'A' : {    /*  Solve for voltage.  */
                  printf("Input the current in amperes => ");
                  scanf("%f",&current);
                  printf("Value of the resistance in ohms => ");
                  scanf("%f",&resistance);
                  voltage = current*resistance;
                  printf("The voltage is %f volts.",voltage);
             }
             break;
```

If the selection above is made (`selection == 'A'`), then the whole body of statements defined by the compound statement between the opening `{` and the closing `}` will be executed. The concept of the `switch` statement here is the same as in the previous program. The main difference is the use of a compound statement.

As you might suspect, the `switch` statement also allows you to call other functions.

Functional Switching

Program 3.21 illustrates another program structure using the `switch` statement. This program does exactly the same thing as the previous one. It allows the program user to select any one form of Ohm's law and then permits a calculation. The difference in this program structure is that now each of the compound statements is replaced by a function call. In turn, each of these function definitions contains the statements needed to implement the user selection.

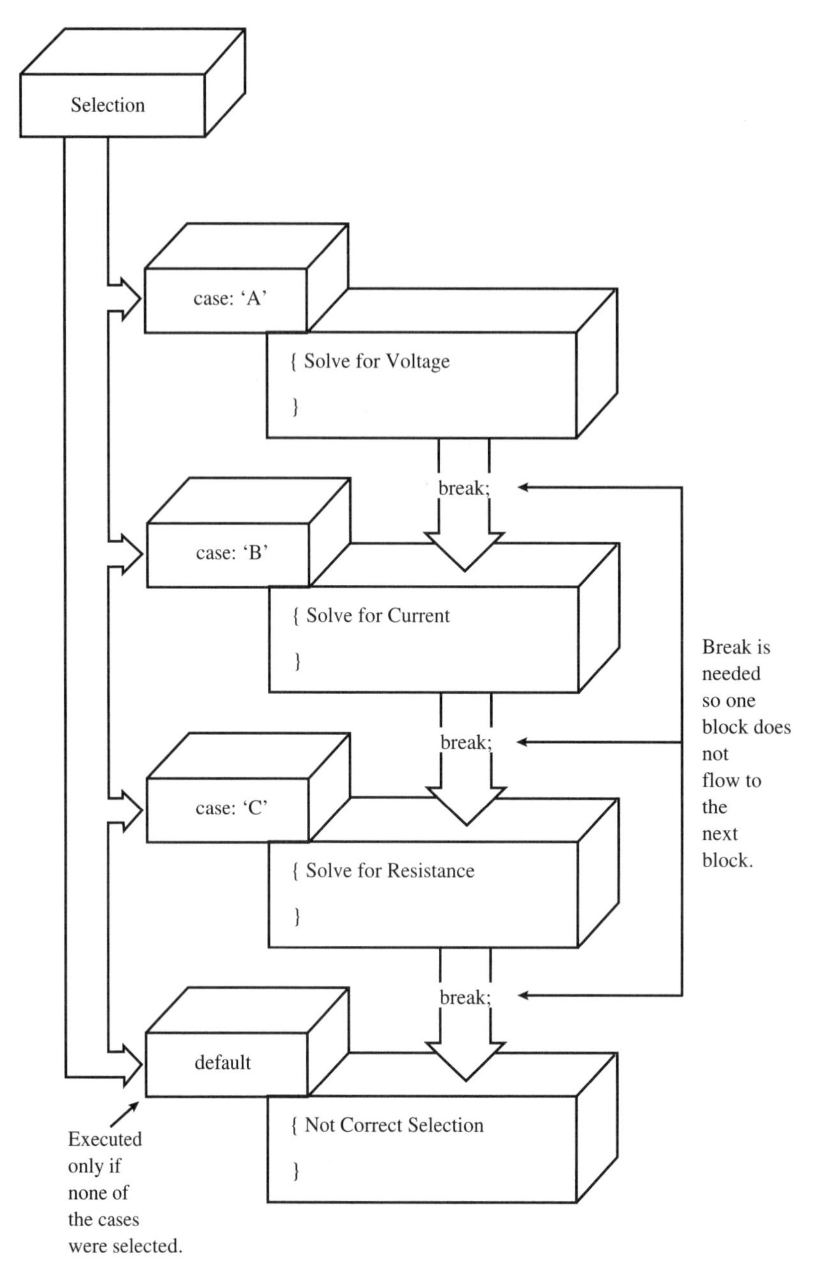

Figure 3.15 General Structure of Program 3.20

Program 3.21

```
#include <stdio.h>

void voltage_solution(void);      /* Solve for voltage in volts.   */
void current_solution(void);      /* Solve for current in amperes. */
```

```
void resistance_solution(void);   /* Solve for resistance in ohms. */

main()
{
        char selection;    /* Item to be selected by program user. */
        float voltage;     /* Circuit voltage in volts.            */
        float current;     /* Circuit current in amperes.          */
        float resistance;  /* Circuit resistance in ohms.          */

        printf("\n\nSelect the form of Ohm's law needed by letter:\n");
        printf("A] Voltage B] Current C] Resistance\n");
        printf("Your selection (A, B, or C) => ");
        scanf("%c",&selection);

        switch(selection)
        {
               case 'A' : voltage_solution();
                          break;
               case 'B' : current_solution();
                          break;
               case 'C' : resistance_solution();
                          break;
        }   /* End of switch. */
}

/* ------------------------------------------------------------ */

void voltage_solution(void)   /* Solve for voltage in volts. */
{
        float resistance;   /* Value of circuit resistance in ohms.   */
        float voltage;      /* Value of circuit voltage in volts.     */
        float current;      /* Value of circuit current in amperes.   */

        printf("Input the current in amperes => ");
        scanf("%f",&current);
        printf("Value of the resistance in ohms => ");
        scanf("%f",&resistance);
        voltage = current*resistance;
        printf("The voltage is %f volts.",voltage);
}

/* ------------------------------------------------------------ */

void current_solution(void)   /* Solve for current in amperes. */
{
        float resistance;   /* Value of circuit resistance in ohms.   */
        float voltage;      /* Value of circuit voltage in volts.     */
        float current;      /* Value of circuit current in amperes.   */

        printf("Input the voltage in volts => ");
        scanf("%f",&voltage);
```

```
        printf("Value of the resistance in ohms => ");
        scanf("%f",&resistance);
        current = voltage/resistance;
        printf("The current is %f amperes.",current);
}

/* ------------------------------------------------------------------ */

void resistance_solution(void)   /* Solve for resistance in ohms. */
{
        float resistance;    /* Value of circuit resistance in ohms.    */
        float voltage;       /* Value of circuit voltage in volts.      */
        float current;       /* Value of circuit current in amperes.    */

        printf("Input the voltage in volts => ");
        scanf("%f",&voltage);
        printf("Value of the current in amperes => ");
        scanf("%f",&current);
        resistance = voltage/current;
        printf("The resistance is %f ohms.",resistance);
}
```

Program Analysis

Note how easy it now is to read the `switch` statement:

```
switch(selection)
{
        case 'A' : voltage_solution();
                   break;
        case 'B' : current_solution();
                   break;
        case 'C' : resistance_solution();
                   break;
}  /*  End of switch.  */
```

This is much cleaner than the compound statements used in Program 3.20; as such, this is the preferred method. If the actual function definition is needed for any one of the called functions, it is only necessary to scan down the program and find the proper one.

What May Be Switched

In the C `switch`, $\texttt{expression}_1$ may be of type `int` as well as `char`. A type `float` is not permitted as an $\texttt{expression}_1$ type in a `switch` statement.

Switching within Switches

You may have multiple `switch` statements within other `switch` statements. This concept is illustrated in Figure 3.16.

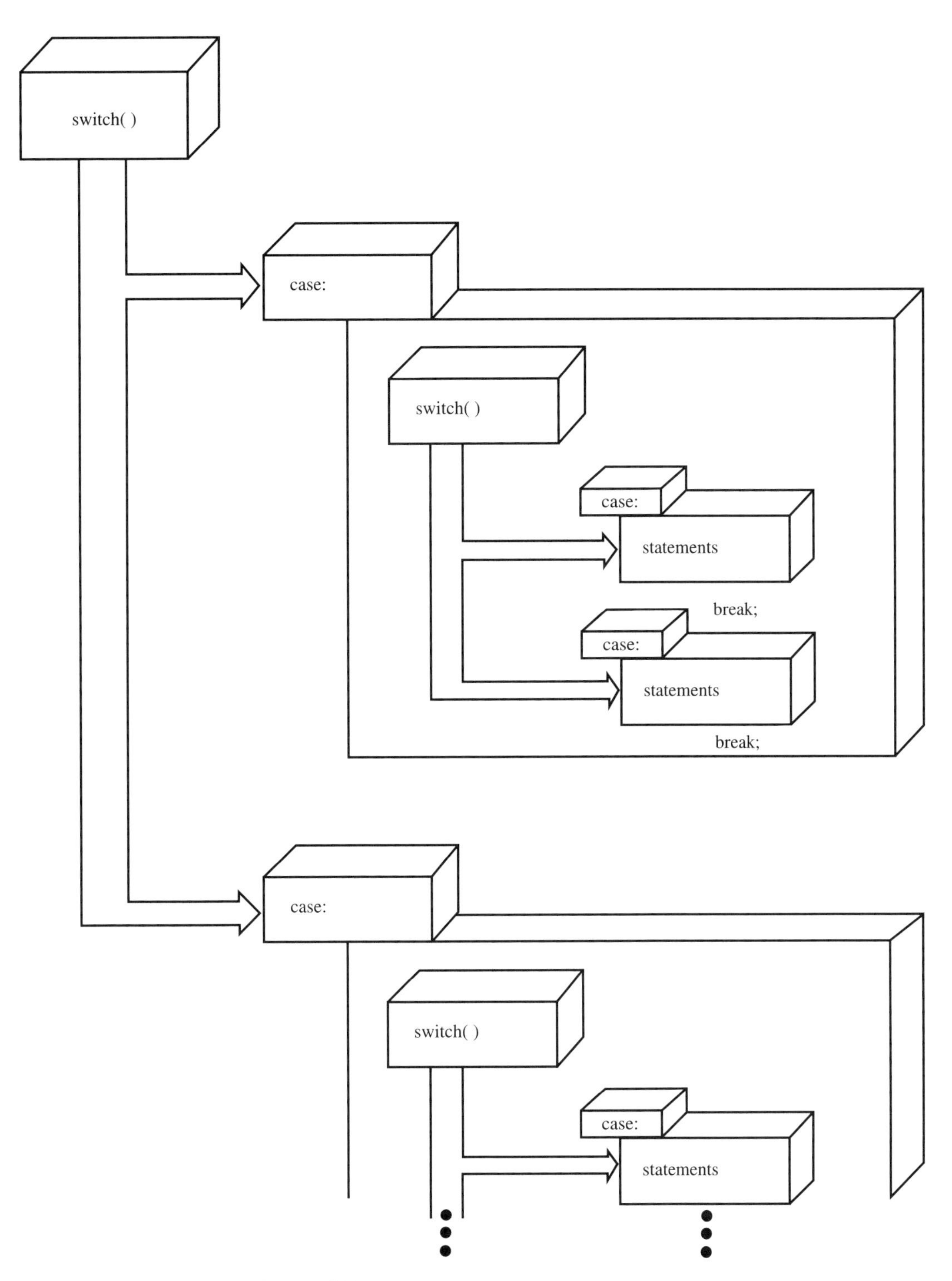

Figure 3.16 Concept of Switches within a `switch`

Conclusion

This section presented the powerful C `switch` statement. You will find that the correct writing of technical programs will require the use of the `switch` statement. Here you were introduced to the `switch` structure and several different ways of applying this general structure. You were also shown a suggested program structure both for single and compound statements. The last program in this section presented the concept of function calls from the `switch` statement.

There is one more C decision-making statement. This is presented in the next section. For now, test your understanding of this section by trying the following section review.

3.7 Section Review

1. State the purpose of the `switch` statement in C.
2. What other keywords must be used with the `switch` statement?
3. State the purpose of the keyword `default` in the `switch` statement.
4. May function calls be made from a `switch` statement? Explain.

3.8 One More switch and the Conditional Operator

Discussion

This section will present one more application of the C `switch`. Here you will also be introduced to the final decision-making process used in C. This section will conclude the discussion of the important decision-making capabilities of the C language.

One More switch

Recall from the last section that the C `switch` has the form

```
switch(expression)
{
    case label : statement;
                 break;
}
```

As you will see, the `break` is important in determining the logic of the `switch`. This is illustrated by Program 3.22. This program displays formulas required to compute the power delivered by the power source in a series circuit consisting of three resistors. The program user only needs to enter what conditions are known about the circuit. Note where the `break` has been omitted from the C `switch`.

Program 3.22

```
#include <stdio.h>

main()
{
        char selection;     /* User input selection. */
```

```
    printf("This program will show the formulas necessary\n");
    printf("to compute the power delivered by a voltage source\n");
    printf("to a series circuit consisting of three resistors\n");
    printf("when the value of the source voltage is known.\n");
    printf("\n\nSelect by letter:\n");
    printf("A] Resistor values known.   B] Total resistance known.\n");
    printf("C] Total current known.\n");
    printf("Your selection => ");
    scanf("%c",&selection);

    switch(selection)
    {
           case 'A' : printf("Rt = R1 + R2 + R3\n");
           case 'B' : printf("It = Vt/Rt\n");
           case 'C' : printf("Pt = Vt*It\n");
                      break;
           default  : printf("That was not a correct selection!");
    }  /* End of switch. */
}
```

Program Analysis

What follows is the monitor display for four different responses to Program 3.22 by the program user:

1. Program user enters the letter A:

```
Your selection => A
Rt = R1 + R2 + R3
It = Vt/Rt
Pt = Vt*It
```

2. Program user enters the letter B:

```
Your selection => B
It = Vt/Rt
Pt = Vt*It
```

3. Program user enters the letter C:

```
Your selection => C
Pt = Vt*It
```

4. Program user enters the letter T:

```
That was not a correct selection!
```

What has happened here is that the `break` between the labels in the C `switch` has been omitted:

```
switch(selection)
{
  case 'A' : printf("Rt = R1 + R2 + R3\n");
  case 'B' : printf("It = Vt/Rt\n");
```

```
  case 'C' : printf("Pt = Vt*It\n");
             break;
  default  : printf("That was not a correct selection!");
}  /* End of switch. */
```

Observe from the indicated output that if there is a match with `case 'A'`, then all remaining statements are executed up to the first `break`. If the `break` were not present, then the `default` statement would be executed as well. As you can see from the above discussion, wherever a match is made, every statement from that point on is executed up to the first `break` or the end of the `switch` statement. This concept is illustrated in Figure 3.17.

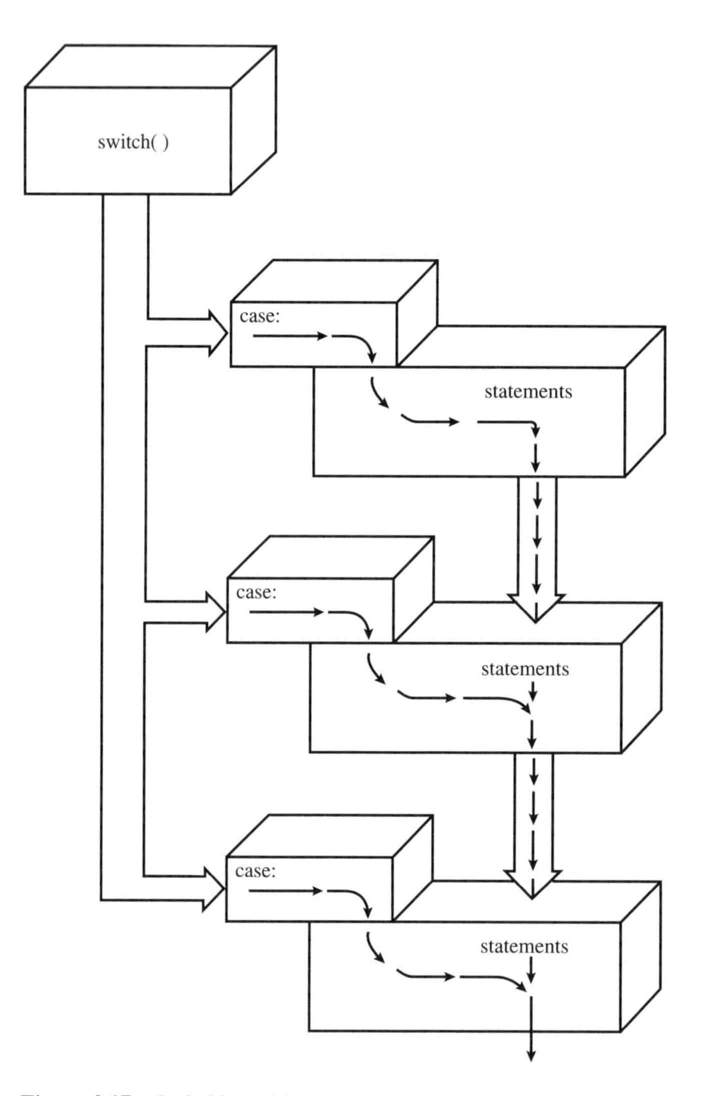

Figure 3.17 Switching without a `break`

The Conditional Operator

The last decision maker in C is the conditional operator. This has the form of an `if...else` statement. It is

$expression_1$? $expression_2$: $expression_3$

What happens is that when $expression_1$ is TRUE (any value other than zero), the whole operation becomes the value of $expression_2$. If, on the other hand, $expression_1$ is FALSE (equal to zero), the whole operation becomes the value of $expression_3$. A simple illustration is presented in Program 3.23.

Program 3.23

```
#include <stdio.h>

main()
{
        int selection;     /* User input selection */

        printf("Enter a 1 or a 0 => ");
        scanf("%d",&selection);

        selection ? printf("A one.") : printf("A zero.");
}
```

Program Analysis

In Program 3.23, if the program user inputs a `0`, the second `printf` will be evaluated, and the monitor will display

```
A zero.
```

If the program user inputs a `1` (or any non-zero value), the first `printf` will be evaluated, and the monitor will display

```
A one.
```

What has happened here is that the condition of `selection` determines which of the following two expressions will be evaluated. Actually, any value entered by the user that is not equal to zero will cause evaluation of the first expression. This means that if the program user entered the value of `12`, the first expression would be evaluated and the monitor would display

```
A one.
```

Conditional Operator Application

For an application of the conditional operator, consider the electrical circuit shown in Figure 3.18.

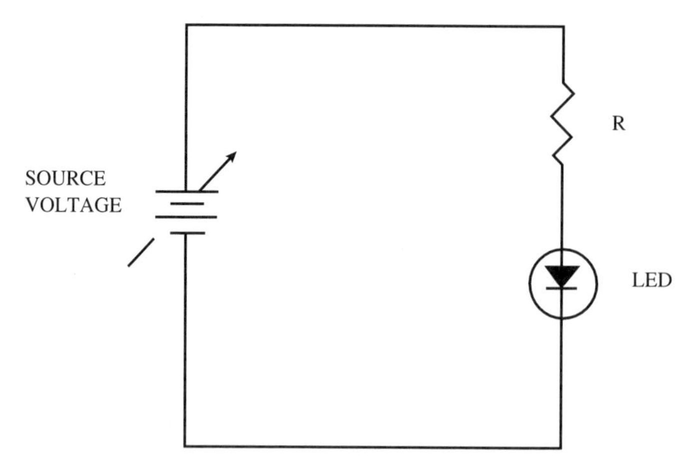

Figure 3.18 Application Circuit for Conditional Operator

In this particular circuit, the device called the LED (light emitting diode) will not conduct current until the voltage applied across it is greater than 2.3 volts. Once this happens, further increasing the source voltage will not significantly change the voltage across the LED, and the remaining voltage will be dropped across the resistor. To calculate the circuit current, the voltage across the resistor is divided by the value of the resistor. This type of problem is easily solved by the C conditional operator, as illustrated in Program 3.24.

Program 3.24

```
#include <stdio.h>

main()
{
        float led_voltage;          /* Voltage across LED in volts.      */
        float resistor_voltage;     /* Voltage across resistor in volts. */
        float source_voltage;       /* Voltage of the source in volts.   */
        float circuit_current;      /* Current in the LED in amperes.    */
        float resistor_value;       /* Value of resistor in ohms.        */

        printf("\n\nEnter the source voltage in volts => ");
        scanf("%f",&source_voltage);
        printf("Enter value of resistor in ohms => ");
        scanf("%f",&resistor_value);

        led_voltage = (source_voltage < 2.3) ? source_voltage : 2.3;
        resistor_voltage = source_voltage - led_voltage;
        circuit_current = resistor_voltage/resistor_value;

        printf("Total circuit current is %f amperes.",circuit_current);
}
```

Program Analysis

Assume that the user of Program 3.24 enters a value of 2 volts. This means that the expression

```
(source_voltage < 2.3)
```

will be TRUE. Thus `led_voltage` will be evaluated to `source_voltage` (which, as entered by the program user, is 2 volts).

Next, the evaluation

```
resistor_voltage = source_voltage - led_voltage;
```

becomes

```
resistor_voltage = 2 - 2
```

and the resistor voltage is zero (as it should be under these conditions). Now, when the next expression

```
circuit_current = resistor_voltage/resistor_value
```

is evaluated, the evaluation becomes

```
circuit_current = 0/resistor_value
```

which evaluates to 0. Again this is as it should be, because if the voltage across the LED is not more than 2.3 volts, there will not be any current flow in the circuit.

Next, assume the program user enters a value greater than 2.3 volts—say, 5 volts—for the source voltage. Note that the calculations will show a circuit current (as they should). The expression

```
(source_voltage < 2.3)
```

will be FALSE. Thus `led_voltage` will now be evaluated to 2.3 volts (the second expression on the right of the : in the ? operator):

```
(source_voltage < 2.3) ? source_voltage : 2.3
```

Now the following calculation becomes

```
resistor_voltage = 5 - 2.3 = 2.7
```

and the circuit current will now have a value different from zero. The exact value will be determined by the value of the resistor entered by the program user.

Conclusion

This section demonstrated how to use the C `switch` with variations of the `break`. You also saw an application of this concept. The conditional operator was also presented here along with an important application. Check your understanding of this section by trying the following section review.

3.8 Section Review

1. State the effect of omitting the `break` in a C `switch`.
2. What do you need to do in order to make sure that `default` is always executed in a C `switch`?
3. Explain the operation of the conditional operator.
4. What values make the conditional operator TRUE? What values make it FALSE?

3.9 Example Programs

This section contains six additional programs meant to highlight branch and logical operations. Although they are simple in nature, each performs a useful task.

Program 3.25 illustrates how a two-input binary *truth table* can be generated.

Program 3.25

```
#include <stdio.h>

void eval(int a,int b);

main()
{
        printf("Truth Table\n\n");
        printf("  A  B  F\n\n");
        eval(0,0);
        eval(0,1);
        eval(1,0);
        eval(1,1);
}

void eval(int a,int b)
{
        int f;     /* Used in Boolean computation. */
        f = (a && b)||(a && !b);
        printf("  %i  %i  %i\n",a,b,f);
}
```

The logical expression for the output of the truth table is computed in the `eval` function. The Boolean expression being evaluated is:

$$F = AB + A\overline{B}$$

which is read as "F equals (A and B) or (A and not B)." Examine how this expression is coded in C.

The output of the program is as follows:

```
Truth Table

  A  B  F

  0  0  0
```

```
0 1 0
1 0 1
1 1 1
```

Try modifying the program to produce other two-input truth tables, or even three-input truth tables.

Program 3.26 shows how a decimal integer supplied by the user can be converted into binary. Note the use of recursion in the `tobin` function.

Program 3.26

```
#include <stdio.h>

void tobin(int n);

main()
{
        int n;    /* User input number. */

        printf("Enter an integer from 1 to 255 => ");
        scanf("%i",&n);
        tobin(n);
}

void tobin(int n)
{
        int m;     /* Used in remainder computation. */

        if (n != 0)
        {
                m = n / 2;
                tobin(m);
                printf("%i  ",n - 2*m);
        }
}
```

Note the output of the following sample execution:

```
Enter an integer from 1 to 255 => 100
1 1 0 0 1 0 0
```

It is easy to verify that the binary output `1 1 0 0 1 0 0` does equal 100 decimal. The three 1s in the output have weights of 4, 32, and 64, which do add up to 100.

The program uses a divide-by-2 conversion technique that always halts, thus making the use of recursion possible. Can you determine the way the recursion is stopped?

Conversion from decimal integer to hexadecimal is the subject of Program 3.27. The number entered by the user is divided into two halves, with each 4-bit half containing a value from 0 to 15. The values 10 through 15 are used in a `switch` statement to display the appropriate A to F hexadecimal character.

Program 3.27

```
#include <stdio.h>

void tohex(int n);

main()
{
        int n;          /* User input number. */
        int a,b;        /* Temporary storage. */

        printf("Enter an integer from 0 to 255 => ");
        scanf("%i",&n);
        a = n / 16;         /* Get upper 4-bit value. */
        b = n % 16;         /* Get lower 4-bit value. */
        tohex(a);
        tohex(b);
}

void tohex(int n)
{
        if ((n >= 0) && (n <= 9))
                printf("%i",n);
        else
        {
                switch(n)
                {
                        case 10 : printf("A"); break;
                        case 11 : printf("B"); break;
                        case 12 : printf("C"); break;
                        case 13 : printf("D"); break;
                        case 14 : printf("E"); break;
                        case 15 : printf("F"); break;
                        default : printf("Error!\n");
                }
        }
}
```

Two executions are shown to illustrate the hexadecimal conversion performed with the aid of the `tohex()` function.

```
Enter an integer from 0 to 255 => 100
64
```

In the first execution, 100 decimal is converted into 64 hexadecimal. This can be checked as follows: 6*16 + 4*1 = 100. This conversion, however, does not utilize the `switch` statement because neither 4-bit half of the number is greater than 9. In the second execution the `switch` statement is needed to convert the upper 4-bit half of the input number 200.

```
Enter an integer from 0 to 255 => 200
C8
```

Note that C8 hexadecimal equals C*16 + 8*1, which is also 12*16 + 8*1, giving 200.

Note that Program 3.27 deals only with 8-bit numbers. Is it possible to convert 16-bit numbers?

Recursion is used again in Program 3.28 to show how **Euclid's** algorithm (circa 300 B.C.) is implemented.

Program 3.28

```
#include <stdio.h>

int euclid(int a, int b);

main()
{
        int m;    /* First user input number.  */
        int n;    /* Second user input number. */

        printf("Enter the first number => ");
        scanf("%i",&m);
        printf("Enter the second number => ");
        scanf("%i",&n);
        printf("\nThe GCD of %i and %i is %i",m,n,euclid(m,n));
}

int euclid(int a, int b)
{
        if (b == 0)
               return a;
        else
               return euclid(b,a % b);
}
```

Euclid's algorithm is used to find the *greatest common divisor (GCD)* of two integers. For example, what is the GCD of 40 and 100? The factors of both numbers are:

```
 40:  1  2  4  5  8   10  20  40
100:  1  2  4  5  10  20  25  50  100
```

The highest number both lists have in common is 20. The `euclid` function finds this number recursively using remainder arithmetic (e.g., the `%` operator). Let us verify the GCD for 40 and 100 with a sample execution.

```
Enter the first number => 40
Enter the second number => 100

The GCD of 40 and 100 is 20
```

Try these pairs of numbers as well: 9 and 27, 9 and 28, 36 and 544.

The next program in this section deals with an important topic to all computer users: *memory speed.* Program 3.29 performs calculations for two common types of memory systems, cache memory and interleaved memory.

Program 3.29

```
#include <stdio.h>

float Cacher(void);
float RAMer(void);

main()
{
        char choice;    /* User selection. */
        float Tacc;     /* Access time in nanoseconds. */

        printf("Which would you like to compute?\n");
        printf("1] Average Cache Access Time\n");
        printf("2] Interleaved RAM Access Time\n");
        printf("Choice ? ");
        scanf("%c",&choice);
        switch(choice)
        {
                case '1' : Tacc = Cacher(); break;
                case '2' : Tacc = RAMer(); break;
                default  : printf("Sorry. That choice is invalid.");
        }
        if ((choice == '1')||(choice == '2'))
                printf("\nThe access time is %f nanoseconds.",Tacc);
}

float Cacher()
{
        float cat;   /* Cache access time. */
        float ram;   /* RAM access time. */
        float hit;   /* hit ratio. */

        printf("Enter the Cache access time (in nanoseconds) => ");
        scanf("%f",&cat);
        printf("Enter the RAM access time (in nanoseconds) => ");
        scanf("%f",&ram);
        printf("Enter the hit ratio => ");
        scanf("%f",&hit);
        return ((hit*cat) + (1.0 - hit)*(cat + ram));
}

float RAMer()
{
        float n;     /* Number of RAM modules. */
        float ram;   /* RAM access time. */

        printf("Enter the number of RAM modules => ");
        scanf("%f",&n);
        printf("Enter the RAM access time (in nanoseconds) => ");
        scanf("%f",&ram);
        return (ram / n);
}
```

In a *cache* memory system, a special high-speed memory called a cache is used. Cache memories are typically ten times faster than standard Random Access Memory (RAM). Using cache memory together with RAM leads to an average memory access time that lies somewhere between the fast cache and the slower RAM. The average access time is determined by a third parameter, the *hit ratio*. A hit ratio of 0.85 means that data is found in the cache 85% of the time and in RAM the other 15%. Thus, the average access time will be a blend of the cache and RAM access times. Suppose that the cache access time is 10 nanoseconds (nS) and the RAM access time is 70 nS. Executed with these values, Program 3.29 shows the following:

```
Which would you like to compute?
1] Average Cache Access Time
2] Interleaved RAM Access Time
Choice ? 1
Enter the Cache access time (in nanoseconds) => 10
Enter the RAM access time (in nanoseconds) => 70
Enter the hit ratio => .85

The access time is 20.499998 nanoseconds.
```

Notice that 20 nS lies within the range of 10 to 70 nS. What does a hit ratio of 0.92 give?

The other common type of memory system utilizes *interleaved* memory. In this memory system, a number of identical memory modules are connected in such a way that they are accessed simultaneously, with the output of each module associated with a series of consecutive memory locations. For example, an interleaved memory composed of eight modules would access locations X through $X + 7$ for each address X supplied to them. Since the modules are accessed simultaneously, the average access time becomes the access time of a single module divided by the number of modules. Program 3.29 illustrates this principle.

```
Which would you like to compute?
1] Average Cache Access Time
2] Interleaved RAM Access Time
Choice ? 2
Enter the number of RAM modules => 8
Enter the RAM access time (in nanoseconds) => 70

The access time is 8.750000 nanoseconds.
```

Use the program to determine how many 80 nS memory modules are required to get an average access time of 16 nS.

Our last example program revisits the bitwise operation AND. In the ASCII representation of ‘a’ we use the value 61 hexadecimal (0x61). An uppercase ‘A’ has the value 41 hexadecimal. If you examine the binary patterns for each letter, what do you see?

```
7 -- bit --   0

0 1 1 0 0 0 0 1  -  0x61  -  'a'
0 1 0 0 0 0 0 1  -  0x41  -  'A'
```

Notice that bit 5 is the only bit that is different in the uppercase and lowercase ASCII representations for these two letters. In general, the code for *any* lowercase letter can be converted into uppercase by simply clearing bit 5. This is easily done by bitwise ANDing the ASCII code with the pattern 0xdf, which contains a 0 in the bit 5 position.

Program 3.30 uses this technique to perform lowercase to uppercase conversion.

Program 3.30

```
#include <stdio.h>

main()
{
        int n1 = 0x68, n2 = 0x69;
        printf("%c%c\n",n1,n2);
        n1 &= 0xdf;
        printf("%c%c\n",n1,n2);
        n2 &= 0xdf;
        printf("%c%c\n",n1,n2);
}
```

Program 3.30 gives the following results during execution:

```
hi
Hi
HI
```

What would happen if some other ASCII code (such as '0' or '=') was ANDed this way?

Conclusion

In this section we examined many different applications requiring the use of mathematical, bitwise, and Boolean operations. Check your understanding of this material with the following section review.

3.9 Section Review

1. What happens if one of the numbers passed to `eval()` in Program 3.25 is negative, as in `eval(0,-1)`?
2. What changes must be made to the `tobin()` function of Program 3.26 to allow numbers as large as 65535 to be converted?
3. What makes a function recursive?
4. Show the C statement or statements needed to determine if the integer variable `xyz` contains an even number. If `xyz` contains an even number, display "even" on the screen, otherwise display "odd."

3.10 Troubleshooting Techniques

Discussion

In this section we will examine many of the common errors made by new C programmers when writing statements that contain the new operators covered in this chapter. In general, these errors are the result of improper use of the relational, bitwise, and Boolean operators.

Using = Instead of ==

It is very easy to make the following mistake when writing a conditional statement:

```
if (var = 5)
    printf("Ok\n");
```

The problem is the use of the = operator instead of the == operator to compare `var` with 5. In fact, the value of `var` is changed to 5 during execution of the statement. The `if` statement will interpret this non-zero conditional value as true, and print `Ok`.

The correct `if` statement is as follows:

```
if (var == 5)
    printf("Ok\n");
```

Notice that the == operator is being used to compare `var` with 5. This `if` statement will only print `Ok` when `var` equals 5.

It is very easy to skip over this error when troubleshooting a program because it looks correct at first glance.

Using && Instead of &

Another hard-to-catch mistake involves the use of bitwise and Boolean operators at the wrong time. For example, suppose that we are trying to determine when the most significant bit of an 8-bit value goes high. The MSB has a decimal value of 128 (10000000 binary), so ANDing the 8-bit value with 128 results in either 128 or zero. Getting 128 means the MSB was high. Getting zero means it was low. Examine the `if` statement used to perform this check:

```
if (abc && 128)
    printf("MSB is high\n");
```

Unfortunately, this `if` statement will print `MSB is high` for *every* value of `abc` from 1 to 255. This is because the Boolean AND operator `&&` interprets any non-zero value as a true result. So, values of `abc` from 1 to 255 will all produce true results to the `&&` condition.

The correct way to check the MSB is like this:

```
if (abc & 128)
    printf("MSB is high");
```

The bitwise AND operator `&` will only produce a true result if the MSB of `abc` is high.

Missing Breaks in a switch

It is also common to leave `break` statements out of a `switch`, which leads to unpredictable results. Consider this `switch` statement:

```
switch(abc)
{
    case 0: printf("abc equals zero\n");
        abc++;
    case 1: abc++;
    default: abc = 0;
}
```

It is the intention of the programmer to execute the `printf` statement when `abc` equals 0 and then increment `abc` to 1. If `abc` equals 1 it should be incremented to 2. Otherwise, `abc` should be loaded with zero. Now, what happens if `abc` does equal zero when the `switch` is executed? The first `case` statement is matched and the `printf` executes and `abc` is incremented to 1. Then, since there is no `break` statement, the next `case` statement is also executed (making `abc` equal 2), as is the `default` statement, which sets `abc` back to zero.

When `break` statements are used, only a single `case` statement (or the `default`) will execute:

```
switch(abc)
{
     case 0: printf("abc equals zero\n");
             abc++;
             break;
     case 1: abc++;
             break;
     default: abc = 0;
}
```

Now, `abc` will be updated as the programmer intended.

Missing Braces

A program that looks like it is structured may not always execute as expected. For instance, the programmer who wrote the following statements used tab columns to add structure:

```
if (abc == 5)
      printf("abc equals 5\n");
      abc = 0;
abc++;
printf("Now abc equals %d\n",abc);
```

It *looks* like the programmer wants the variable `abc` to be set to zero after the `printf` statement executes, when `abc` equals 5. In actual use, `abc` will *always* be set to zero because it is not part of the `if` statement. The programmer forgot to use braces to surround the body of the statements used in the `if` statement.

When the braces are added, `abc` is only set to zero when it initially equals 5, instead of every time:

```
if (abc == 5)
{
      printf("abc equals 5\n");
      abc = 0;
}
abc++;
printf("Now abc equals %d\n",abc);
```

Recall that braces are optional when the body only contains one statement. It is good practice to always used them, however, if only to save yourself the trouble of remembering to add them if you add more statements to the body of the `if` statement.

tion, program explanation, and function descriptions. This information is contained in the first pass of the program.

First Development Stage

Program 3.31 shows the first step in the development of the hypothetical robot program.

Program 3.31

```
/*****************************************************************/
/*                    Robot Troubleshooter                       */
/*****************************************************************/
/*                Developed by Robert Shooter                    */
/*                                                               */
/*****************************************************************/
/*    This program will take the user step by step through the   */
/*    analysis and troubleshooting of a hypothetical robot. The  */
/*    program also demonstrates the various kinds of decision    */
/*    making available with the C language.                      */
/*****************************************************************/
/*               Non-Standard Header Files Used:                 */
/*---------------------------------------------------------------*/
/*                           None                                */
/*****************************************************************/
/*                    Function Prototypes                        */
/*---------------------------------------------------------------*/

void explain_program(void);

/* This function explains the operation of the program to the user. */
/*****************************************************************/

main()
{
        explain_program();    /* Explain program to user.   */
        exit(0);
}

/*---------------------------------------------------------------*/

void explain_program(void)    /* Explain program to user.   */
{
        printf("\n\nThis program represents the troubleshooting\n");
        printf("of a hypothetical robot system.\n");
        printf("\nThe information you enter will simulate actual\n");
        printf("measurements you would make on this robot.\n");
        printf("\nThe program will instruct you as to what measurement\n");
        printf("you are to enter next.\n");
}
```

Program Analysis

The programmer's block clearly gives the name of the program, date (if required), name of developer, and purpose of the program. The only function prototype is the one for the function, which will be used to explain the purpose of the program to the program user. This is enough to compile and execute. You will see nothing more than the instructions explaining what the program will do. However, it is now documented as a part of the program exactly what it is to do. There is no longer any reason to refer back to written documentation for this information. As the program is developed further, all the programmer now needs to do is look at the programmer's block to be reminded of what is needed.

Adding Some Structure

Program 3.32 illustrates the next step in the development of this program.

Program 3.32

```
#include <stdio.h>

/********************************************************************/
/*                        Robot Troubleshooter                      */
/********************************************************************/
/*                     Developed by Robert Shooter                  */
/*                                                                  */
/********************************************************************/
/*      This program will take the user step by step through the    */
/*      analysis and troubleshooting of a hypothetical robot. The   */
/*      program also demonstrates the various kinds of decision     */
/*      making available with the C language.                       */
/********************************************************************/
/*                   Non-Standard Header Files Used:                */
/*------------------------------------------------------------------*/
/*                              None                                */
/********************************************************************/
/*                        Function Prototypes                       */
/*------------------------------------------------------------------*/

void explain_program(void);

/* This function explains the operation of the program to the user. */
/*------------------------------------------------------------------*/

float arm(void);

/* This function contains the arm service routine for the robot.    */
/*------------------------------------------------------------------*/

void power_unit(void);

/* This routine is the power unit service routine. It is a final    */
/* testing function.                                                */
```

```
/*------------------------------------------------------------------*/

void light_check(void);

/* This function presents instructions on how to check the status   */
/* light indicator.                                                 */
/*------------------------------------------------------------------*/

void arm_drive_disconnect(void);

/* This function presents instructions on the testing of the arm    */
/* drive disconnect unit.                                           */
/*------------------------------------------------------------------*/

main()
{
        float measurement;     /* Measurement conducted by program user.  *//

        explain_program();     /* Explain program to user.    */
        measurement = arm();   /* Arm measurement value.      */

        if (measurement > 35) power_unit();
        else
        if ((measurement <= 35)&&(measurement >= 30)) light_check();
        else
        if (measurement < 30) arm_drive_disconnect();

        exit(0);
}

/*------------------------------------------------------------------*/

void explain_program()   /* Explain program to user.   */
{
        printf("\n\nThis program represents the troubleshooting\n");
        printf("of a hypothetical robot system.\n");
        printf("\nThe information you enter will simulate actual\n");
        printf("measurements you would make on this robot.\n");
        printf("\nThe program will instruct you as to what measurement\n");
        printf("you are to enter next.\n");
}

/*------------------------------------------------------------------*/

float arm()   /* Do arm service routine.  */
{
        float measurement;  /* User measurement in volts.  */

        printf("\nMeasure the voltage at test point #1\n");
        printf("Your measurement in volts => ");
        scanf("%f",&measurement);
```

```
        return(measurement);
}

/*-----------------------------------------------------------------*/

void power_unit()  /* Power unit service instructions. */
{
        printf("\nRefer to power unit test in service manual\n");
        printf("and replace arm power unit.");
}

/*-----------------------------------------------------------------*/

void light_check()   /* Checking the status light indicator. */
{
        printf("\nThis is the status light indicator check.");
}

/*-----------------------------------------------------------------*/

void arm_drive_disconnect()  /* Arm drive instructions. */
{
        printf("\nThese are the arm drive disconnect instructions.");
}

/*-----------------------------------------------------------------*/
```

Program Analysis

Observe the new structure of `main()`. Essentially its logic reads the same as the first part of the troubleshooting flow chart. In the flow chart, the first measurement to be made is the voltage at TP#1. From this measurement, the program can branch in one of three directions depending on the value of the voltage. After calling `explain_program`, `main()` sets the variable `measurement` equal to the value of the function `arm()`. This function has already been identified as a type `float` prototype. This means it will `return()` a value. This value will be passed on to `measurement`, and the rest of the program will be determined by the value of `measurement`. This will be done by the series of `if...else if...else` statements that follows.

```
main()
{
      float measurement;   /* Measurement conducted by program user. */

      explain_program();      /* Explain program to user.  */
      measurement = arm();    /* Arm measurement value.    */

      if (measurement > 35) power_unit();
      else
      if ((measurement <= 35)&&(measurement >= 30)) light_check();
      else
```

```
    if (measurement < 30) arm_drive_disconnect();
    exit(0);

}
```

Thus, reading the code of the function above gives you the overall structure of this program. The point is, you don't have to read through a lot of program code in order to get the idea of what the structure of the program is or how the program does what it is supposed to do. The details of the code are there in the function definitions if and when you need them.

All of the new functions have their function prototypes in the programmer's block along with the explanations of what they are to do.

```
/*                       Function Prototypes                          */
/*--------------------------------------------------------------------*/
void explain_program(void);
/* This function explains the operation of the program to the user.*/
/*--------------------------------------------------------------------*/
float arm(void);
/* This function contains the arm service routine for the robot.   */
/*--------------------------------------------------------------------*/
void power_unit(void);
/* This function is the power unit service routine. It is a final  */
/* testing function.                                                */
/*--------------------------------------------------------------------*/
void light_check (void);
/* This function presents instructions on how to check the status  */
/* light indicator.                                                 */
/*--------------------------------------------------------------------*/
void arm_drive_disconnect(void);
/* This function presents instructions on the testing of the arm   */
/* drive disconnect unit.                                           */
/*--------------------------------------------------------------------*/
```

This type of documentation makes it easy to come back to the program at any time and see exactly how many functions are used in the program, what type of function each is, and exactly what each is supposed to do. This makes program modification, debugging, and maintenance much easier and less frustrating tasks.

Note that not all of the functions are completed. There is just enough information there to let you know that the program is working up to this point.

```
/*--------------------------------------------------------------------*/
float arm(void)     /*  Do arm service routine.  */
{
     float measurement;  /* User measurement in volts. */

     printf("\nMeasure the voltage at test point #1\n");
     printf("Your measurement in volts => ");
     scanf("%f",&measurement);

     return(measurement);
```

```
}

/*-------------------------------------------------------------*/

void power_unit()     /*  Power unit service instructions.  */
{

     printf("\nRefer to power unit test in service manual\n");
     printf("and replace arm power unit.");

}

/*-------------------------------------------------------------*/

void light_check()    /*  Checking the status light indicator.  */
{

      printf("\nThis is the status light indicator check.");

}

/*-------------------------------------------------------------*/

void arm_drive_disconnect()    /* Arm drive instructions.  */
{

      printf("\nThese are the arm drive disconnect instructions.");

}

/*-------------------------------------------------------------*/
```

Programmers frequently put only part of the information in a program—just enough to test the overall flow of the program—to catch bugs during the development of the program. Such undeveloped functions are called **program stubs.**

Final Program

Program 3.33 represents the final development of the program for the hypothetical robot troubleshooter. Note that the decision-making statements used include most of the techniques presented in this chapter.

Program 3.33

```
#include <stdio.h>

/****************************************************************/
/*                    Robot Troubleshooter                      */
/****************************************************************/
/*                 Developed by Robert Shooter                  */
/*                                                              */
```

```
/*****************************************************************/
/*     This program will take the user step by step through the     */
/*     analysis and troubleshooting of a hypothetical robot. The    */
/*     program also demonstrates the various kinds of decision      */
/*     making available with the C language.                        */
/*****************************************************************/
/*                  Non-Standard Header Files Used:                 */
/*-----------------------------------------------------------------*/
/*                              None                                */
/*****************************************************************/
/*                        Function Prototypes                       */
/*-----------------------------------------------------------------*/

void explain_program(void);

/* This function explains the operation of the program to the user. */
/*-----------------------------------------------------------------*/

float arm(void);

/* This function contains the arm service routine for the robot.    */
/*-----------------------------------------------------------------*/

void power_unit(void);

/* This routine is the power unit service routine. It is a final    */
/* testing function.                                                */
/*-----------------------------------------------------------------*/

void light_check(void);

/* This function presents instructions on how to check the status   */
/* light indicator.                                                 */
/*-----------------------------------------------------------------*/

void arm_drive_disconnect(void);

/* This function presents instructions on the testing of the arm    */
/* drive disconnect unit.                                           */
/*-----------------------------------------------------------------*/

main()
{
        float measurement;      /* Measurement conducted by program user.  */

        explain_program();      /* Explain program to user.   */
        measurement = arm();    /* Arm measurement value.     */

        if (measurement > 35) power_unit();
        else
        if ((measurement <= 35)&&(measurement >= 30)) light_check();
```

```
        else
        if (measurement < 30) arm_drive_disconnect();

        exit(0);
} /* End of main. */

/*-----------------------------------------------------------------*/

void explain_program(void)   /* Explain program to user.   */
{
        printf("\n\nThis program represents the troubleshooting\n");
        printf("of a hypothetical robot system.\n");
        printf("\nThe information you enter will simulate actual\n");
        printf("measurements you would make on this robot.\n");
        printf("\nThe program will instruct you as to what measurement\n");
        printf("you are to enter next.\n");
} /* End of explain_program. */

/*-----------------------------------------------------------------*/

float arm(void)   /* Do arm service routine.  */
{
        float measurement;  /* User measurement in volts.  */

        printf("\nMeasure the voltage at test point #1\n");
        printf("Your measurement in volts => ");
        scanf("%f",&measurement);

        return(measurement);
} /* End of arm. */

/*-----------------------------------------------------------------*/

void power_unit()  /* Power unit service instructions. */
{
        printf("\nRefer to power unit test in service manual\n");
        printf("and replace arm power unit.");
} /* End of power_unit. */

/*-----------------------------------------------------------------*/

void light_check()   /* Checking the status light indicator. */
{
        int light_status;  /* User input of light status. */

        printf("\nInput the condition of the status light:\n");
        printf("1] Red  2] Green  3] Off \n");
        printf("Enter number => ");
        scanf("%d",&light_status);
```

```
        switch(light_status)
        {
                case 1 :  printf("Disconnect power and replace fuse F1.\n");
                          break;
                case 2 :  printf("Disconnect power and replace arm drive board.\n");
                          break;
                case 3 :  printf("Replace bulb.\n");
                          break;
                default : printf("That was not a correct selection!\n");
        } /* End of switch. */
        printf("\nRepeat service routine.\n");
} /* End of light_check. */

/*-----------------------------------------------------------------*/

void arm_drive_disconnect()   /* Arm drive instructions. */
{
        float measurement;    /* Measurement made by user. */

        printf("Disconnect arm drive and remeasure voltage at TP#1.\n");
        printf("Your measurement in volts => ");
        scanf("%f",&measurement);

        measurement = (measurement > 30)? 30 : measurement;

        if (measurement < 30)
        {
                printf("Replace arm power unit. Refer to\n");
                printf("service manual page 183, then repeat\n");
                printf("SVS testing.");
        }
        else
        if (measurement == 30)
        {
                printf("Replace arm drive circuit board. Refer to\n");
                printf("service manual page 235, then repeat service\n");
                printf("testing.");
        }
} /* End of arm_drive_disconnect. */

/*-----------------------------------------------------------------*/
```

Program Analysis

A complete analysis of Program 3.33 is left for the Self-Test of this chapter. This will help you test your understanding of the material presented in this chapter as well as your ability to apply what you have learned.

Conclusion

This section presented a case study that used most all of the decision-making capabilities of the C language. Here you were again taken step by step through the development of a major program for an area of technology. Check your understanding of this section by trying the following section review.

3.11 Section Review

1. State the first step in the development of any program.
2. Explain the purpose of a troubleshooting flow chart.
3. Define a program stub.
4. For the program developed in this case study, what does the user input represent?

Interactive Exercises

DIRECTIONS

These exercises require that you have access to a computer and software that supports C. They are provided here to give you valuable experience and most importantly immediate feedback on what the concepts and commands introduced in this chapter will do.

Exercises

1. Keep in mind that relational statements in C all return a value. Look at Program 3.34. It works. Try it and see what is displayed.

Program 3.34

```
#include <stdio.h>

main()
{
        printf("What is this => %d",(5 > 3));
}
```

2. Program 3.35 uses bitwise operators. Note that the input is to the base 10 whereas the output is in hex. As before, predict what the output will be, then test it.

Program 3.35

```
#include <stdio.h>

main()
{
        char bit1;
        char bit2;

        bit1 = 5;
        bit2 = 10;

        printf("%X\n",bit1&bit2);
```

```
        printf("%X\n",bit1|bit2);
        printf("%X\n",bit1^bit2);
        printf("%X\n",~bit2);

}
```

3. Program 3.36 uses a logical statement. Remember, these too return a value. Predict what the following output will be, and then try it.

Program 3.36

```
#include <stdio.h>

main()
{
        printf("What is this => %d",(1&&1));
}
```

4. In C, any value other than `0` is considered to be TRUE. To demonstrate this, try Program 3.37.

Program 3.37

```
#include <stdio.h>

main()
{
        printf("What is this => %d",(1&&23));
}
```

5. Program 3.38 presents a demonstration with the logical OR. Predict what the program will do and then try it.

Program 3.38

```
#include <stdio.h>

main()
{
        printf("What is this => %d",(1||1));
}
```

6. Program 3.39 again emphasizes that any value other than `0` is considered TRUE in C. The program illustrates this fact with a logical OR statement.

Program 3.39

```
#include <stdio.h>

main()
{
        printf("What is this => %d",(0||38));
}
```

7. Program 3.40 gives you an opportunity to test your ability to predict the outcome of a complex logical statement. Try to predict the outcome, and then try the program.

Program 3.40

```
#include <stdio.h>

main()
{
        int logic_variable;

        logic_variable = (5 < 7)&&((8 > 5)||(7 > 3));
        printf("What is this => %d",logic_variable);
}
```

8. What do you think of the way Program 3.41 is written? Do you find that it is easier to read and that it makes it easier to predict the outcome?

Program 3.41

```
#include <stdio.h>
#define TRUE 1
#define FALSE 0
#define AND &&
#define OR ||
#define NOT !

main()
{
        int logic_value;

        logic_value = TRUE AND TRUE;

        if (logic_value) printf("TRUE");
        else printf("FALSE");
}
```

9. Read through Program 3.42 and see if you find it easier to read the C `switch` statements. Predict the outcome, and then try it.

Program 3.42

```
#include <stdio.h>
#define TRUE 1
#define FALSE 0
#define AND &&
#define OR ||
#define NOT !

main()
{
        int logic_value;
```

```
        logic_value = TRUE OR NOT TRUE;

        switch(logic_value)
        {
                case FALSE :  printf("FALSE");
                              break;
                case TRUE  :  printf("TRUE");
        }
}
```

10. Read through Program 3.43. Then try it.

Program 3.43

```
#include <stdio.h>
#define TRUE 1
#define FALSE 0
#define AND &&
#define OR ||
#define NOT !
#define Check_This_Out switch
#define Stop break;
#define Is_It case

main()
{
        int logic_value;

        logic_value = NOT(TRUE OR NOT TRUE) AND TRUE;

        Check_This_Out(logic_value)
        {
                Is_It FALSE :  printf("FALSE");
                               break;
                Is_It TRUE  :  printf("TRUE");
        }
}
```

Self-Test

DIRECTIONS

Use Program 3.33 developed in the case study section of this chapter to answer the following questions.

Questions

1. How many functions are defined in the program? Name them.
2. Are there any functions that contain open branches? If so, which one(s)?
3. Are there any functions that contain closed branches? If so, which one(s)?

4. Which function(s) use(s) the C `switch`?
5. State the meaning of

```
measurement = (measurement > 30) ? 30 : measurement;
```

6. How many variable identifiers are used in the program?
7. Which functions return values?
8. Of the functions that return values, what value(s) is (are) returned?
9. State why the variable `light_status` is not of type `float`.

End-of-Chapter Problems

General Concepts

Section 3.1

1. Write the six relational operators presented in this chapter. State what each means.
2. What is the numerical value used by C to represent a FALSE? To represent a TRUE?
3. If the == symbol in C means equal to, what does the = sign mean in C?
4. Indicate the numerical values of the following relational operations, given that $A = 0$, $B = 3$, and $C = 8$:
   A. !A
   B. B == B
   C. C > A

Section 3.2

5. Explain the concept of an open branch.
6. State the C statement that is used for the open branch.
7. What is meant by a compound statement in C?
8. Can a function be called as a branch option? Explain.

Section 3.3

9. Explain the concept of a closed branch.
10. State the C statement that is used for the closed branch.
11. What is a compound `if...else` in C?
12. May compound statements and function calls be made from the `if...else if...else` statement?

Section 3.4

13. Determine the bitwise complements of the following hexadecimal values:
   A. $\mathtt{C}_{16}$ B. $\mathtt{FE}_{16}$ C. $\mathtt{50}_{16}$
14. Find the results of bitwise ANDing the following hexadecimal values:
   A. 3_{16} & 5_{16} B. E_{16} & E_{16} C. F_{16} & 7_{16}
15. Solve for the bitwise ORing of the following hexadecimal values:
   A. 3_{16} | 5_{16} B. A_{16} | 5_{16} C 0_{16} | F_{16}
16. Determine the bitwise XORing of the following hexadecimal values:
   A. 3_{16} ^ 5_{16} B. A_{16} ^ 5_{16} C. F_{16} ^ F_{16}

Section 3.5

17. What is meant by a logical operation?
18. State what is meant by the logical AND operation. What symbol is used to represent this in C?

19. State what is meant by the logical OR operation. What symbol is used to represent this in C?
20. State what is meant by the logical NOT operation. What symbol is used to represent this in C?

Section 3.6

21. Is it legal in C to add a type `int` to a type `char`? What is this called?
22. What resulting type would the addition given in problem 21 produce?
23. State the purpose of a cast in C.
24. What is an lvalue expression?

Section 3.7

25. State when the C `switch` is used.
26. Identify the purpose of the keyword `case` in a C `switch`.
27. Identify the purpose of the keyword `break` in a C `switch`.
28. Can compound statements and function calls be used as part of a C switch?

Section 3.8

29. State the purpose of the keyword `default` in a C `switch`.
30. Explain the operation of the conditional operator in C.

Section 3.9

31. Is it possible to generate a truth table using recursion?
32. How should the conversion techniques presented be changed to allow conversion to any base?
33. Is it possible to use recursion to display all factors of a number?

Section 3.10

34. What happens when `=` is used in place of `==`?
35. What is logically different about `abc & def` versus `abc && def`?

Section 3.11

36. State what a troubleshooting flow chart does for a technician.
37. What property of the C language allows the information in a troubleshooting flow chart to be developed into an interactive computer program?

Program Design

In developing the following C programs, use the method described in the case study section of Chapter 2 and used again in the case study section of this chapter. For each program you are assigned, document your design process and hand it in with your program. This documentation should include the design outline that states the purpose of the program, required input, process on the input, and required output as well as the program algorithm. Be sure to include all of the documentation in your final program. This should consist of, but not be limited to, the programmer's block, function prototypes, and a description of each function as well as any formal arguments you may use.

38. Develop a C program that will compute the power dissipation of a resistor where the user input is the value of the resistor and the current in the resistor. The program is to warn the user when the power dissipation is above 1 watt. The power dissipation of a resistor is:

$$P = I^2 R$$

Where

P = Power dissipation in watts.
I = Current in amps.
R = Resistance in ohms.

39. Create a C program that will convert an input number into the resistor color code. The relationship is as follows:

0 = Black	3 = Orange	6 = Blue	9 = White
1 = Brown	4 = Yellow	7 = Violet	
2 = Red	5 = Green	8 = Gray	

The program user inputs the number, and the corresponding color is displayed.

40. Write a C program that determines standard capacitor values between which the value of a given capacitor lies. The program user inputs a capacitor value and the program will indicate between which two standard values it belongs. For the sake of simplicity, assume that the standard capacitor values to be used by the program are as follows:

0.001μF	0.0015μF	0.0022μF	0.0033μF
0.1μF	0.15μF	0.22μF	0.33μF
10μF	15μF	22μF	33μf
1000μF	1500μF	2200μF	3300μF

41. Create a C program in which the prices of the following items have already been entered:

Soap = $12.50 Asphalt = $27.59 Glue = $2.33 Gum = $0.57

The program user only needs to select the item by number and indicate the quantity. The program computes the total cost.

42. Develop a C program that demonstrates to the program user all forms of branching used by C.
43. Write a C program that will generate a three-input truth table for the following Boolean expression:

$$F = \overline{A}B + AC + A\overline{B}C$$

44. Write a C program that converts a user input number into octal.
45. Write a C program that determines whether an integer is even or odd.
46. Write a C program that determines whether an integer, when converted into binary, has odd or even parity.
47. Write a C program that will compute the average instruction execution time for a microprocessor. The user inputs the average number of clock cycles per instruction, and the clock frequency (in hertz). For example, five clock cycles per instruction at a clock speed of 2 megahertz gives 2.5 microseconds per instruction. The program must convert the result into milliseconds, microseconds, or nanoseconds, whichever is closest.
48. Create a C program where the program user selects the name of a figure and the program presents the formula that is used for calculating its area. Use a rectangle, triangle, circle, and parallelogram as the figure choices.
49. Develop a C program that will simulate computer aided troubleshooting of a hypothetical diesel engine. The troubleshooting chart is shown in Figure 3.20.
50. Create a C program that will allow the program user to convert from or to metric. The user can then choose length, volume, or weight for the conversion.
51. Develop a C program that will compute the weight of a machined part. The program user can select from three forms—cylinder, rectangle, or cone—and can select from three materials—copper, aluminum, or steel. The program must automatically compute the volume and use the density of the selected material to compute the total weight.

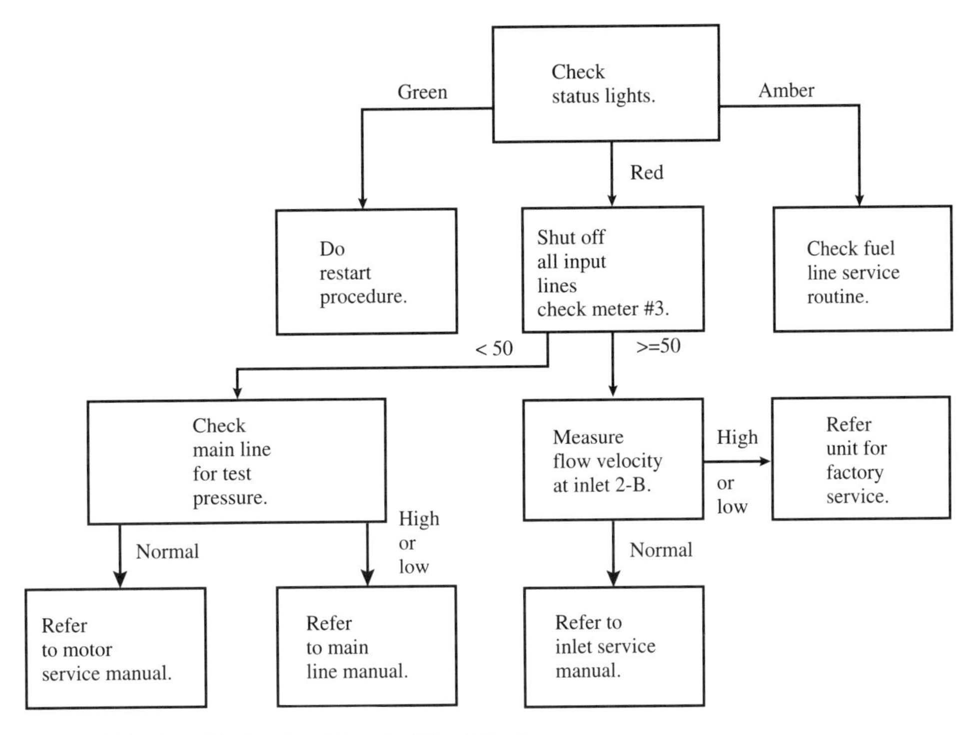

Figure 3.20 Troubleshooting Chart for Diesel Engine

4 Looping and Recursion

Objectives

This chapter gives you the opportunity to learn:

1. The purpose of program loops.
2. Different types of loops.
3. Loops used by C.
4. The structure and coding of C loops.
5. Practical applications of loops.
6. The nesting of loops.
7. Debugging techniques used in C.
8. The development of a complex technical program that requires the use of loops.
9. The details of recursion.

Key Terms

`for` Loop
Compound Statement
Post-Incrementing
Pre-Incrementing
Post-Decrementing
Pre-Decrementing
Comma Operator
Sequential-Evaluation Operator
`while` Loop
Conditional Loop
Sentinel Loop
`do while` Loop
Nested Loops
Recursion
Run Time Stack
Run Time Error
Tracing
Auto Debug Function

Outline

Introduction

Recall that three different types of program blocks are all you need to solve for any programming logic, no matter how complex. These are the sequential (or action), branch, and loop blocks. You have already studied branch blocks in the previous chapter and have worked with sequential blocks throughout the previous chapters. Once you have completed this chapter you will have used all three kinds of blocks.

You will discover that one of the powers of using the computer for the analysis and solution of technical problems comes from the use of the loop block. Using the loop block allows you to quickly test conditions for a range of values. By doing this, you can easily spot minimum or maximum values. You can also see the direction of change for many different conditions. An introduction to recursion is presented to complete the concept of looping.

4.1 The for Loop

Discussion

This section shows you how to develop a C program that will allow you to do something an exact number of times. This has many useful applications in the world of technology. Here you are no longer limited to solving a problem only once for a single answer; you will now be able to solve a problem any number of times with different values each time.

What the for Loop Looks Like

The C **`for` loop** contains four major parts:

1. The value at which the loop starts.
2. The condition under which the loop is to continue.
3. The changes that are to take place for each loop.
4. What is done during each loop.

These parts are put together in the C `for` loop as follows:

```
for(initial-expression; conditional-expression; loop-expression)
      {loop instructions};
```

The use of the C `for` loop is shown in Program 4.1. The program increments the value of a variable called `time` by one each time through the loop. The variable starts with a value of 1, and the loop continues to repeat as long as the variable `time` is less than or equal to 5. As soon as `time` is larger than 5, the program breaks out of the loop.

Program 4.1

```
#include <stdio.h>

main()
{
        int time;        /* Counter variable. */

        /* The loop starts here. */

        for(time = 1; time <= 5; time = time + 1)
                printf("Value of time = %d\n",time);

        /* The loop stops here. */

        printf("This is the end of the loop.");
}
```

Execution of Program 4.1 produces

```
Value of time = 1
Value of time = 2
Value of time = 3
Value of time = 4
Value of time = 5
This is the end of the loop.
```

Program Analysis

The C `for` loop starts with the keyword `for` followed by the loop expression. The loop expression is divided into four parts:

- The value at which the loop starts (the initial expression)
- The condition under which the loop is to continue (the conditional expression)
- What changes are to take place for each loop (the loop expression)
- What is done during each loop (the loop instructions)

In the case of this program, the C `for` loop is

```
for(time = 1; time <= 5; time = time + 1)
        printf("Value of time = %d\n",time);
```

This means the value at which the loop starts is `time = 1`, the conditions under which the loop is to continue are `time <= 5`, and the the changes that are to take place are `time = time + 1`. The structure of the C `for` loop is shown in Figure 4.1.

for Loop Facts

Note that the `for` loop expression is not terminated with a semicolon. This is because the whole combination of the keyword `for`, the loop expression, and the statement making up the body of the loop are one single C statement:

```
for(time = 1; time <= 5; time = time + 1)
        printf("Value of time = %d\n",time);
```

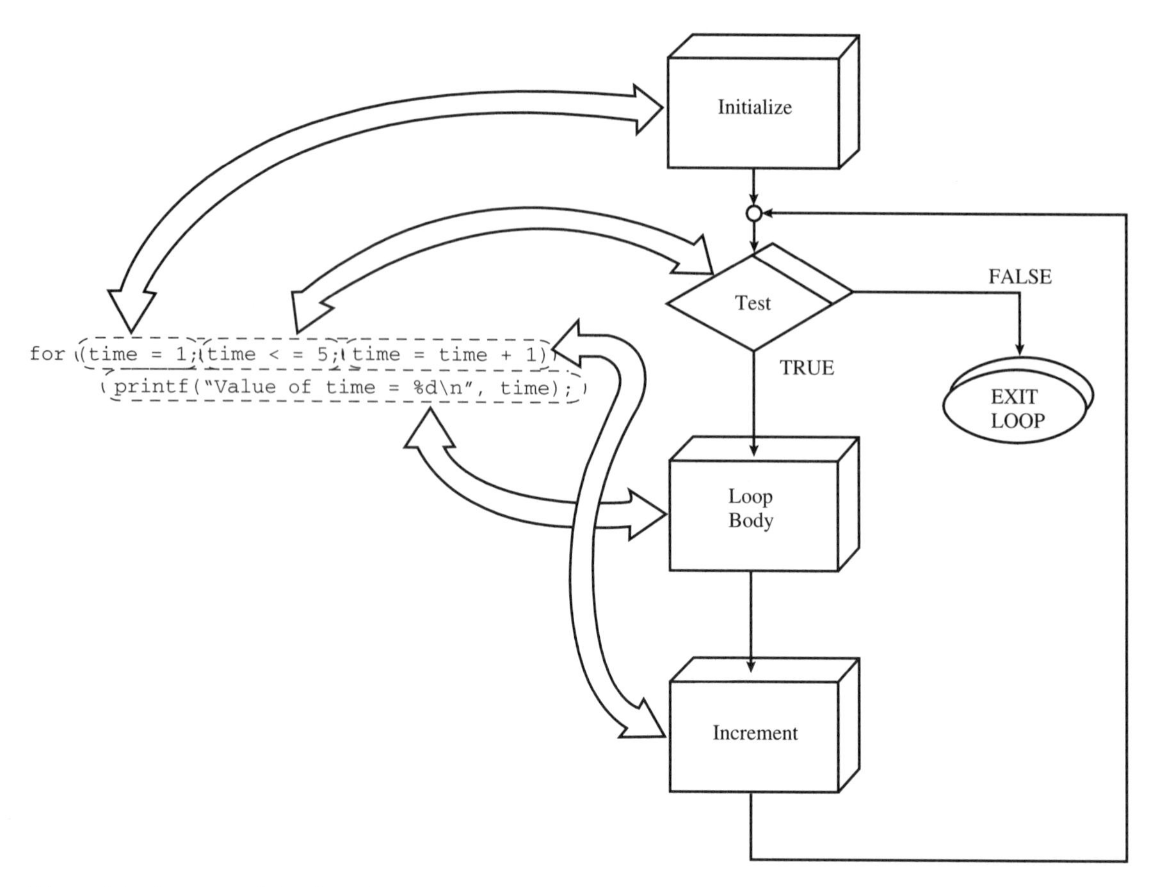

Figure 4.1 Structure of the C `for` Loop

More Than One Statement

You can have more than one statement in a C `for` loop. To do this, you need to indicate the beginning of the loop statements with an opening `{` and the end of these statements with a closing `}`. This is often called a **compound statement** and is illustrated in Program 4.2. The program computes the distance a body falls in feet per second, for the first 5 seconds of free fall as given by the equation

$$S = 1/2\ at^2$$

Where

S = The distance in feet.
a = Acceleration due to gravity (32 ft/sec^2).
t = Time in seconds.

Program 4.2

```
#include <stdio.h>
#define a 32.0

main()
```

```
{
        int time;        /* Counter variable. */
        int distance;    /* Distance covered by the falling body. */

        /* The loop starts here. */

        for(time = 1; time <= 5; time = time + 1)
        {
                distance = 0.5*a*time*time;
                printf("Distance at the end of %d seconds is %d feet.\n"
                        ,time,distance);
        }
        /* The loop stops here. */

        printf("This is the end of the loop.");
}
```

Execution of Program 4.2 produces

```
Distance at the end of 1 seconds is 16 feet.
Distance at the end of 2 seconds is 64 feet.
Distance at the end of 3 seconds is 144 feet.
Distance at the end of 4 seconds is 256 feet.
Distance at the end of 5 seconds is 400 feet.
This is the end of the loop.
```

Note that each of the statements within the `{ }` ends with a semicolon. This is because each is a complete statement. Also note that the ending brace of the compound statement does not require a semicolon. This is no different from other compound statements.

The Increment and Decrement Operators

C offers a shorthand notation for a common programming operation. This is the ability to increment or to decrement a value. The C `for` loop of the previous program uses `time = time + 1` to increment the variable `time`. In C, this could have been condensed to `time++`. The meanings of the increment and decrement operators are given in Table 4.1.

Table 4.1 Increment and Decrement Operators

Operator	**Meaning**
X++	Increment *X* after any operation with it (called **post-incrementing**).
++*X*	Increment *X* before any operation with it (called **pre-incrementing**).
X– –	Decrement *X* after any operation with it (called **post-decrementing**).
– –*X*	Decrement *X* before any operation with it (called **pre-decrementing**).

Program 4.3 illustrates the post-increment and pre-increment operators in action.

Program 4.3

```
#include <stdio.h>

main()
{
        int count;              /* Program counter.           */
        int number1 = 0;        /* Number to post-increment. */
        int number2 = 0;        /* Number to pre-increment.   */

        for(count = 1; count <= 5; count++)
        {
                printf("Post-incrementing number = %d",number1++);
                printf("Pre-incrementing number = %d\n",++number2);
        }
}
```

Execution of Program 4.3 produces

```
Post-increment number = 0 Pre-increment number = 1
Post-increment number = 1 Pre-increment number = 2
Post-increment number = 2 Pre-increment number = 3
Post-increment number = 3 Pre-increment number = 4
Post-increment number = 4 Pre-increment number = 5
```

Note that the `number++` does not get incremented until after it is used. However, the `++number` is incremented before it is used. Observe that both numbers are initially set to 0 in the declaration part of the program.

EXAMPLE 4.1

For the given values, determine the results of the following operations:

`i = 3 c = 10`

A. `x = i++` C. `x = c++` E. `x = i--+++c`
B. `x = ++i` D. `x = --c + 2`

Solution

A. `x = i++ = 3++=>` increment after process thus `x = 3`
B. `x = ++i = ++3 =>` increment before process thus `x = 4`
C. `x = c++ = 10++ =>` increment after process thus `x = 10`
D. `x = --c +2 = --10 + 2 = 9 + 2 = 11`
E. `x = i--+++c = 3--+++10 = 3 + 11 = 14`

The results of the operations in Example 4.1 are cumulative, as shown in Program 4.4.

Program 4.4

```
#include <stdio.h>

main()
{
```

```
        int i = 3;
        int c = 10;

        printf("i = %d\n",i);
        printf("c = %d\n",c);
        printf("x = i++ => %d\n",i++);
        printf("x = ++i => %d\n",++i);
        printf("x = c++ => %d\n",c++);
        printf("x = --c + 2 => %d\n",(--c + 2));
        printf("x = i-- + ++c => %d\n",(i-- + ++c));
}
```

Execution of Program 4.4 yields

```
i = 3
c = 10
x = i++ => 3
x = ++i => 5
x = c++ => 10
x = --c + 2 => 12
x = i-- + ++c => 16
```

Program Analysis

For Program 4.4, the result of each action is cumulative. This means that when `x = i++ => 3` is completed, the variable `i` now equals 4 (it gets incremented after the operation). Thus, when the next operation happens `(x = ++i => 5)`, the incrementing takes place before the operation, and since `i` now has a value of 4, the increment before the operation makes it a 5.

In the `--c + 2` operation, the variable `c` starts with the value of 11 (it was incremented after the previous operation, `x = c++ => 10`). Since this is a decrement before the operation, `--c` causes the value of `c` to become 10; and thus 10 + 2 = 12.

For the last operation, `x = i-- + ++c => 16`, `i` starts with its previous value from its last operation (5), and `c` has a value of 10 from its previous operation. Since `++c` causes an increment before operation, this makes it an 11, and 5 + 11 = 16.

The Comma Operator

As a general rule, two C expressions may be separated by a **comma operator**:

$\texttt{expression}_1\texttt{, expression}_2$

This means that in a C `for` loop, you may have more than one expression within the `for` loop expression. This means that you could initialize two or more variables at the same time:

```
for(count = 0, value = 2; count <= 3, count++)
```

Note that two variables, separated by a comma, are being initialized: `count` and `value`. You could have also used multiple statements in the increment part of the `for` statement. However, only one expression is allowed in the test part of the expression. Items separated by the comma operator are evaluated from left to right. The comma operator is sometimes referred to as the **sequential-evaluation operator**.

Conclusion

This section presented the operation of the C `for` loop. Here you saw the main elements of the C `for` along with some examples of how it may be used. You were also introduced to the C increment and decrement operators as well as the comma operator. Check your understanding of this section by trying the following section review.

4.1 Section Review

1. Name the four major parts of the C `for` loop.
2. Can you have more than one statement in a C `for` loop? Explain.
3. Explain the meaning of `++Y`.
4. What is the comma operator?

4.2 The while Loop

Discussion

This section will introduce you to another kind of loop structure offered by C. It is called the **`while` loop**. As you will see, this loop has the same elements as the C `for` loop. The difference is that its elements are distributed throughout the program. You will find that it's best to use the `while` loop in situations where you don't know ahead of time how many times the loop will be repeated (such as creating a loop in your program that lets the program user automatically repeat the program).

Structure of the while Loop

The structure of the C `while` loop is

```
while(expression)
        statement;
```

The statement may be a single statement or a compound statement (enclosed in braces `{}`). The statement is executed zero or more times until the expression becomes FALSE.

In the operation of the C `while` loop, `expression` is first evaluated. If this evaluation is FALSE (`0`), then `statement` is never executed, and control passes from the `while` statement to the rest of the program. If the evaluation is TRUE (not zero), then `statement` is executed and the process is repeated again.

Program 4.5 illustrates the C `while` loop. The program does the same thing that Program 4.1 did. It evaluates a variable called `time` for five different values.

Program 4.5

```
#include <stdio.h>

main()
{
```

```
        int time = 1;         /* Counter variable. */

        while(time <= 5)
        {
                printf("Value of time = %d\n",time);
                time++;
        } /* End of while. */

        printf("End of the loop.");
}
```

Execution of Program 4.5 produces

```
Value of time = 1
Value of time = 2
Value of time = 3
Value of time = 4
Value of time = 5
End of the loop.
```

Program Analysis

The operation of the `while` loop does the test before execution. The logic diagram for Program 4.5 is illustrated in Figure 4.2.

As long as the `while` condition is TRUE (meaning not zero), the statements following it will be executed. In Program 4.5, it was first necessary to initialize the counting variable `time` to a value of 1:

```
int time = 1;    /* Counter variable. */

while(time <= 5)
```

This was done to ensure its starting value. The `while` will be TRUE as long as the counter `time` is less than or equal to 5.

It is also necessary to change the value of the counter while in the loop. If this is not done, the loop will repeat itself forever. The counter `time` is incremented within the body of the loop:

```
{
        printf("Value of time = %d\n", time);
        time++;
} /* End of while. */
```

Observe that this is a compound statement containing two statements enclosed by the opening `{` and closing `}`.

while Application

An application of the C `while` is illustrated in Program 4.6. C `while` loops are more appropriate than C `for` loops when you do not know when the condition that terminates the loop will occur.

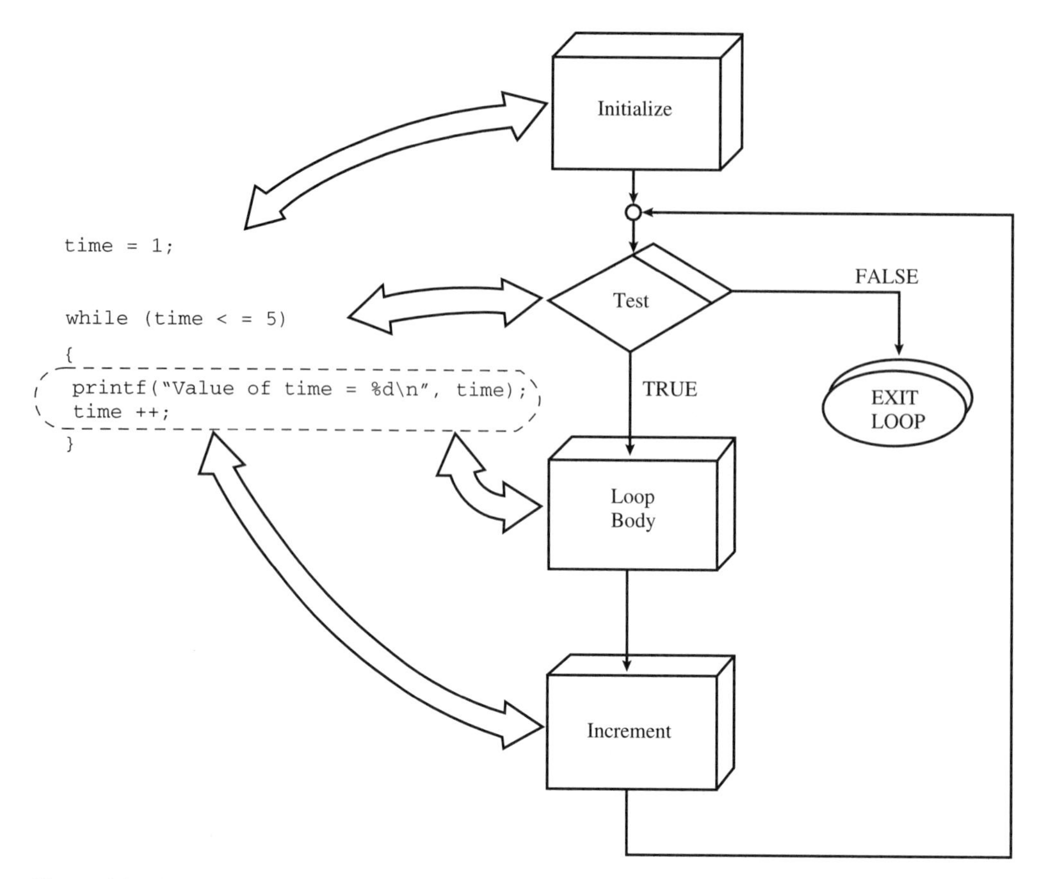

Figure 4.2 Structure of `while` Loop

A new C function is used here called `getche()`. This function gets a single character from the keyboard and echoes it to the screen. It is a useful function whenever a single character is needed for the program.

Program 4.6 simply checks the user input to see if the body of the program is to be repeated. Note that to terminate the loop, the program user must input a capital N.

Program 4.6

```
#include <stdio.h>

main()
{
        char answer = 'Y';      /* Response of program user. */

        while(answer != 'N')
        {
                printf("This is the body of the program.\n");
                printf("Do you want to repeat this program (Y/N) => ");
                answer = getche();
```

```
        printf("\n");
    } /* End of while. */

    printf("Thanks for using the program.");
}
```

A sample output of the program is

```
This is the body of the program.
Do you want to repeat this program (Y/N) => Y
This is the body of the program.
Do you want to repeat this program (Y/N) => n
This is the body of the program.
Do you want to repeat this program (Y/N) => N
Thanks for using the program.
```

Program Analysis

First the character variable is declared and initialized to Y. This is done to ensure that it does not come up with the value of N all by itself (a remote chance, but possible).

```
char answer = 'Y';    /* Response of program user. */
```

Next the condition for continuing the loop is given:

```
while(answer != 'N')
```

This means that as long as the character variable `answer` does not equal N, the body of the loop will be activated:

```
{
        printf("This is the body of the program.\n");
        printf("Do you want to repeat this program (Y/N) => ");
        answer = getche();
        printf("\n");
}  /* End of while.  */
```

Once the condition for the C `while` is FALSE (meaning that `answer != 'N'` is FALSE), the program continues on to the next statement:

```
printf("Thanks for using the program.");
```

Functions Inside the while

You can place a C function inside a C `while` loop expression. This practice is often done to reduce the amount of source code. As an example,

```
while(getche() != '\r')
```

will allow the program user to input characters and have them echoed to the screen until the carriage return is pressed. What happens here is that the `getche()` function is activated first, and then the evaluation is made. You should be aware of this practice.

Conclusion

The unique characteristic of the C `while` loop is that the loop condition is tested *before* the loop is executed. This type of loop should be used in situations when you do not know ahead of time how many times the loop is to be repeated (sometimes called a **conditional loop** or a **sentinel loop**).

Check your understanding of this section by trying the following section review.

4.2 Section Review

1. State the construction of the C `while` loop.
2. Under what conditions will a `while` loop be repeated?
3. When is the loop condition tested in a C `while` loop?
4. What is a good use of a `while` loop?

4.3 The do while Loop

Discussion

The last of the three C loop types is the C **`do while` loop**. As you will see, this loop structure is similar to the C `while` loop, the difference being that the test condition is evaluated *after* the loop is executed. Recall that the C `while` loop tests the condition *before* the loop is executed.

What the do while Loop Looks Like

The C `do while` statement has the form

```
do
  statement
while(expression);
```

where `statement` may be a single or a compound statement. `statement` is executed one or more times until `expression` becomes FALSE (a value of `0`). Execution is done by first executing `statement`, then testing `expression`. If `expression` is FALSE, the `do` statement terminates, and control passes to the next statement in the program. Otherwise, if `expression` is TRUE, `statement` is repeated, and the process starts over again.

Using the C do while

Program 4.7 illustrates the action of the C `do` loop. Note that it is a counting loop. The counting variable is `time`. The program simply increments the counter from 1 to 5 in steps of 1.

Program 4.7

```
#include <stdio.h>

main()
{
        int time;        /* Counter variable. */
```

```
        time = 1;

        do
        {
                printf("Value of time = %d\n",time);
                time++;
        }while(time <= 5);

        printf("End of the loop. ");
}
```

Program output is

```
Value of time = 1
Value of time = 2
Value of time = 3
Value of time = 4
Value of time = 5
End of the loop.
```

Note that the output is no different from the similar program using the C `while` loop (Program 4.5). The program logic is different in that the action part of the loop is done before the test.

Program Analysis

The counter variable is first initialized to a 1 as before:

```
time = 1;
```

Now the program gets into the C `do`:

```
do
{
        printf("Value of time = %d\n",time);
        time++;
}
```

Observe that a compound statement follows the keyword `do`. This is the action part of the loop. What must immediately follow it is the condition for doing the action part again. This is achieved by the `while` statement:

```
while(time <= 5);
```

This says that while the counter `time` is less than or equal to 5, the loop will continue. The construction of the C `do while` loop is illustrated in Figure 4.3.

Loop Details

To emphasize that a C `do` loop is always executed at least once, look at Program 4.8. Here the `while` is made FALSE (by setting its argument to zero).

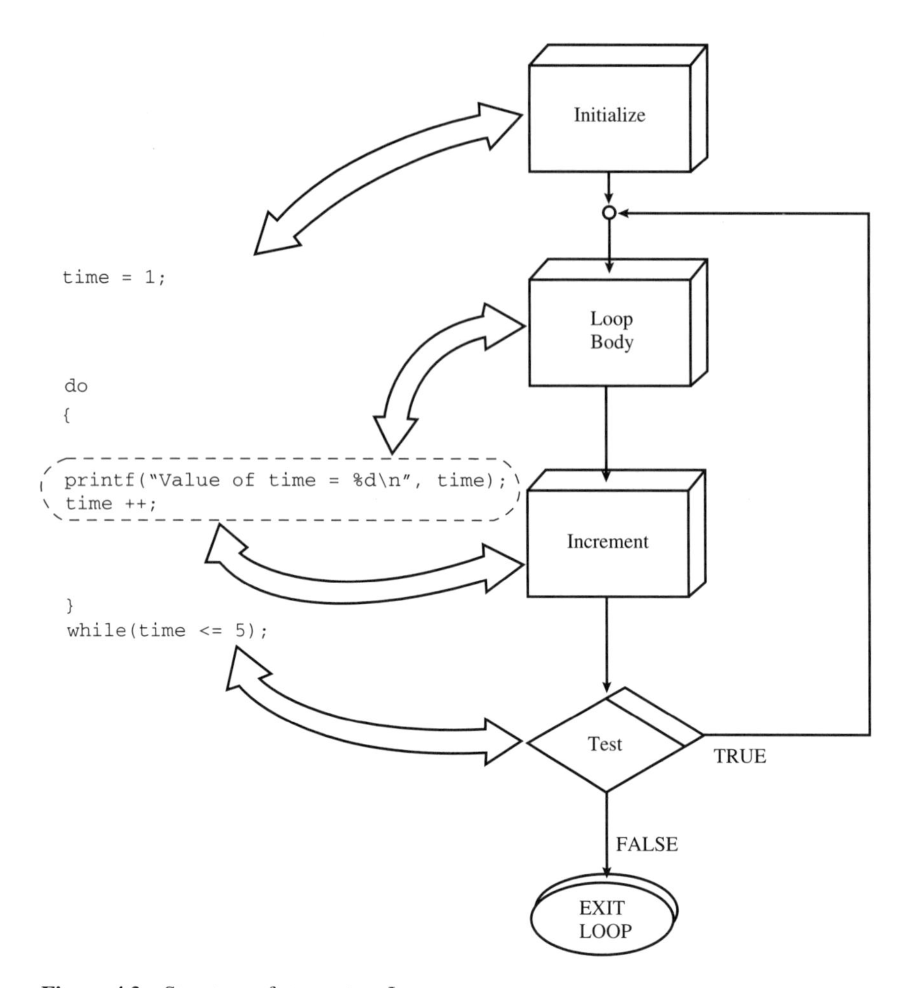

Figure 4.3 Structure of `do while` Loop

Program 4.8

```
#include <stdio.h>

main()
{
        do
                printf("This always happens at least once...");

        while(0);

        printf("End of the loop.");
}
```

The output of Program 4.8 is:

```
This always happens at least once...
End of the loop.
```

Now look at Program 4.9. This is a C `while` loop. Again, however, the `while` argument is set to FALSE. Since for a C `while` the test is always made first, the action part of the loop is never executed!

Program 4.9

```
#include <stdio.h>

main()
{
        while(0)
                printf("This will not happen if test is FALSE...");

        printf("End of the loop.");
}
```

Sentinel Loops

Both the C `do` and the C `while` loops may be used as sentinel loops. Program 4.10 illustrates the C `do` used as a sentinel loop. Here the program user is asked to enter a value that is less than 5. The program tests to see if this happened. If not, it warns the program user and repeats the loop so that a value within the requested range may be entered.

Program 4.10

```
#include <stdio.h>

main()
{
        float input;     /* User input number. */

        do
        {
                printf("Input a number less than 5 => ");
                scanf("%f",&input);
                if (input >= 5)
                        printf("That value is too large, try again.\n");
        }
        while(input >= 5);

        printf("Thank you...");
}
```

A sample execution of the above program could produce

```
Input a number less than 5 => 7
That value is too large, try again.
Input a number less than 5 => 25
```

```
That value is too large, try again.
Input a number less than 5 => 3
Thank you...
```

Program Analysis

There is no counter in Program 4.10 because it is a sentinel loop. The program executes the C `do` first:

```
do
{
        printf("Input a number less than 5 => ");
        scanf("%f",&input);
```

This part of the program waits for a user input. The next part of the program starts with a C `if`:

```
        if(input >= 5)
                printf("That value is too large, try again.\n");
}
```

This part of the program is necessary in order to prompt the program user to the fact that the input is not acceptable (if it is 5 or larger) and that another attempt to input a value within the correct range is needed. If a value within the correct range is entered, then the `printf` function within the loop is never evaluated.

The C `do while` loop ends with the required `while` test:

```
while(input >= 5)
```

Loop Comparisons

Both the C `while` and the C `do while` loops can be used when you do not know ahead of time how many times a loop is to be repeated. When choosing between the two, it is generally considered good programming practice to use the C `while` over the C `do while`. The reason for this is that the C `while` will first test the condition before executing the loop. This is analogous to testing the temperature of the water before you dive in.

Conclusion

Here you were introduced to the last of the three different loop structures available in C. Here you saw the C `do while` loop and how it compares to the C `while` loop. You saw that with the C `do while`, the loop is tested after the execution of the loop statement. Check your understanding of this section by trying the following section review.

4.3 Section Review

1. State the construction of the C `do while` loop.
2. Under what conditions will the `do while` loop be repeated?
3. When is the loop condition tested in a C `do while` loop?
4. Which is preferred, the `do while` or the `while`? Explain.

4.4 Nested Loops

Discussion

In this section you will be introduced to the concept of having one loop inside another. Being able to do this gives you great flexibility when developing C programs to solve technical problems. Here you will see methods to help you structure nested loops to make them easier to read and less prone to error.

Basic Idea

Program 4.11 contains a loop within a loop. This is known as a **nested** loop. The program has two counters and displays the value of each of these counters. Both loops are C `for` loops, and each uses a different counter.

Program 4.11

```
#include <stdio.h>

main()
{
        int outside_counter;     /* Counter for outside loop. */
        int inside_counter;      /* Counter for inside loop.  */

        for(outside_counter = 0; outside_counter <= 3; outside_counter++)
        {
                printf("Start of outside loop.\n");
                printf("Outside loop counter => %d\n\n",outside_counter);

                for(inside_counter = 0; inside_counter <= 3; inside_counter++)
                {
                        printf("Inside loop counter => %d\n",inside_counter);
                } /* End of inside loop. */

                printf("End of outside loop.\n\n");
        } /* End of outside loop. */
}
```

Execution of Program 4.11 produces

```
Start of outside loop.
Outside loop counter => 0

Inside loop counter => 0
Inside loop counter => 1
Inside loop counter => 2
Inside loop counter => 3

End of outside loop.

Start of outside loop.
```

```
Outside loop counter => 1

Inside loop counter => 0
Inside loop counter => 1
Inside loop counter => 2
Inside loop counter => 3

End of outside loop.

Start of outside loop.
Outside loop counter => 2

Inside loop counter => 0
Inside loop counter => 1
Inside loop counter => 2
Inside loop counter => 3

End of outside loop.

Start of outside loop.
Outside loop counter => 3

Inside loop counter => 0
Inside loop counter => 1
Inside loop counter => 2
Inside loop counter => 3

End of outside loop.
```

Program Analysis

The outside loop is the first C `for` in the program:

```
for(outside_counter = 0; outside_counter <= 3; outside_counter++)
{
        printf("Start of outside loop.\n");
        printf("Outside loop counter => %d\n\n",outside_counter);
```

The loop counter starts at 0 and on its first pass activates the nested loop:

```
for(inside_counter = 0; inside_counter <= 3; inside_counter++)
{
      printf("Inside loop counter => %d\n",inside_counter);
} /* End of inside loop. */
```

The nested loop is also a C `for` loop, and its counter starts at 0. Once this loop is initialized, it will repeat itself until the condition of its loop is met. This will happen when `inside_counter` is no longer less than or equal to 3. When this happens, the loop quits and program flow goes back to the remainder of the outer loop:

```
        printf("End of outside loop.\n\n");
} /* End of outside loop.  */
```

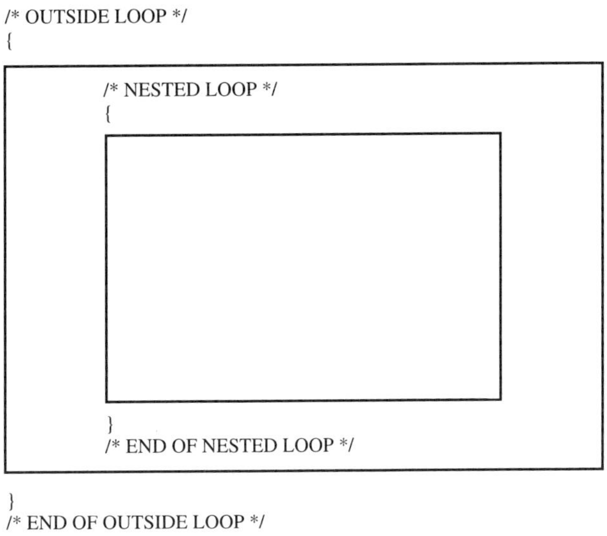

```
/* OUTSIDE LOOP */
{
        /* NESTED LOOP */
        {

        }
        /* END OF NESTED LOOP */
}
/* END OF OUTSIDE LOOP */
```

Figure 4.4 Recommended Structure of Nested Loops

Here the outer loop terminates and goes back to its beginning, incrementing its own counter by one. Thus, the inner loop increments itself from 0 to 3 for each single increment of the outer loop.

Nested Loop Structure

Figure 4.4 shows the recommended structure of nested loops.

As you can see from the figure, loops should be nested in such a manner as to encourage understanding of where the loops begin and end. This structure should include comments to make it clear to which loop the closing `}` of each loop belongs. Doing this will help minimize programming errors.

Nesting with Different Loops

You can nest as many levels of loops as your hardware will allow. However, when you nest several levels and different types of loops, you should be careful because the results may not be exactly what you expect. Consider, for example, the nested loops in Program 4.12. This program has three nested loops. The outer loop is a C `for`, the next inner loop is a C `while`, and the innermost loop is a C `do`. Look through the program and see if you can predict what the results will be. Then look at the actual output presented in the text for what may be some surprising results.

Program 4.12

```
#include <stdio.h>

main()
{
```

```
    int counter_1 = 0;  /* Counter for the first loop.  */
    int counter_2 = 0;  /* Counter for the second loop. */
    int counter_3 = 0;  /* Counter for the third loop.  */

    for(counter_1 = 0; counter_1 <= 2; counter_1++)
    {
        while(counter_2++ <= 3)
        {
            do
            {
                printf("counter_1 = %d\n",counter_1);
                printf("counter_2 = %d\n",counter_2);
                printf("counter_3 = %d\n",counter_3);
            }
            while(counter_3++ <= 3); /* End of third loop. */
        } /* End of second loop. */
    } /* End of first loop. */
}
```

When executed, Program 4.12 produces

```
Counter_1 = 0
Counter_2 = 1
Counter_3 = 0
Counter_1 = 0
Counter_2 = 1
Counter_3 = 1
Counter_1 = 0
Counter_2 = 1
Counter_3 = 2
Counter_1 = 0
Counter_2 = 1
Counter_3 = 3
Counter_1 = 0
Counter_2 = 1
Counter_3 = 4
Counter_1 = 0
Counter_2 = 2
Counter_3 = 5
Counter_1 = 0
Counter_2 = 3
Counter_3 = 6
Counter_1 = 0
Counter_2 = 4
Counter_3 = 7
```

Program Analysis

The results of Program 4.12 may not be what you expected. A close analysis of the program follows.

First, all three counter variables are initialized to 0:

```
int counter_1 = 0;      /* Counter for first loop.  */
int counter_2 = 0;      /* Counter for second loop. */
int counter_3 = 0;      /* Counter for third loop.  */
```

Next, the outer C `for` loop is activated:

```
for(counter_1 = 0; counter_1 <= 2; counter_1++)
{
```

The condition for this loop is that it will continue as long as its counter is equal to or less than 2. For now, its counter has been set to 0 for the first pass.

Next, the inner loop is activated (it is the first instruction of the active outer loop):

```
while(counter_2++ <= 3)
{
```

This is a C `while` loop. Its counter is being tested and then incremented to a 1. Since its test condition is TRUE, it will be active and activate its first instruction, which is the beginning of a C `do`:

```
do
{
        printf("counter_1 = %d\n",counter_1);
        printf("counter_2 = %d\n",counter_2);
        printf("counter_3 = %d\n",counter_3);
}
while(counter_3++ <= 3); /* End of third loop.  */
```

Recall that a C `do` loop does its loop before any test for a condition. Thus, no matter what the condition of the counter for the C `do`, all of the `printf` functions will be activated, and the program will display the values for each counter, which are

```
counter_1 = 0
counter_2 = 1
counter_3 = 0
```

The reason why the counter for the third loop is still a 0 is because it hasn't yet been incremented. It won't be until the ending `while`, where the counter is incremented and tested.

```
while(counter_3++ <= 3); /* End of third loop. */
```

Since the test is still TRUE, this inner C `do` loop will repeat itself until its counter has reached a count of 4. Then it will fall out of the loop and go to the next level, which is the C `while`.

Here, the second counter gets incremented. (Note that the first counter is not incremented because the program has only fallen out to the next highest level of loops—which is from the innermost to the next level. This next level has not yet completed its counting condition, so it is still active.)

Since this C `while` loop is still active, it activates its first instruction, which is the C `do`. Even though the C `do` counter has met its count, it is still activated because a C `do` always gets done every time it is active (no matter what its condition is). Thus, its counter is now forced to be incremented again! This is what eventually brings `counter_3` up to the surprising value of 7! This is another reason why the C `while` loop is preferred over the C `do`.

Once the C `while` loop meets its condition for no longer looping, it falls back out to the C `for` loop. This loop now increments its counter and goes to the first instruction of its loop, which is the C `while`. However, the C `while` has already completed its requirement, and thus the C `do` is not activated—and neither are its `printf` functions. Thus, you never see the results of the outer loop counting.

Conclusion

This section presented the concept of nested program loops in C. However, unlike many other programming languages, nesting of loops in C can be a bit tricky. It's important to understand how the loops behave and the action of the C ++ operator when used within the loop instruction.

Check your understanding of this section by trying the following section review.

4.4 Section Review

1. What is a nested loop?
2. Explain the structure that should be used with nested loops.
3. Which of the three C loop types may be nested?
4. State a potential problem that accompanies use of the C `do` loop.

4.5 Recursion

Discussion

A program that uses **recursion** contains a function that calls itself. As Figure 4.5 shows, a programmer may use *direct recursion* or *indirect recursion* (or both). In Figure 4.5(a), the `myself()` function calls itself directly. In Figure 4.5(b), the `first()` and `second()` functions form a recursive cycle, which repeats itself. A program that contains a recursive function works in a similar fashion to one that contains a loop. However, each time a recursive function calls itself, another complete *copy* of the functions information is placed into memory. This is not what happens during a pass through a loop.

Run Time Stack

The C programming environment supports recursion by careful use of a **run time stack,** a special data structure used to store variable and parameter values for each function in a program. As a program calls function after function, the run time stack grows. When a function terminates, the run time stack gets smaller. The run time stack is automatically managed for the programmer. What the programmer has to do is write the code in such a way that the recursion stops at some level. Otherwise, the run time stack will grow so large that it runs out of memory, and the program crashes.

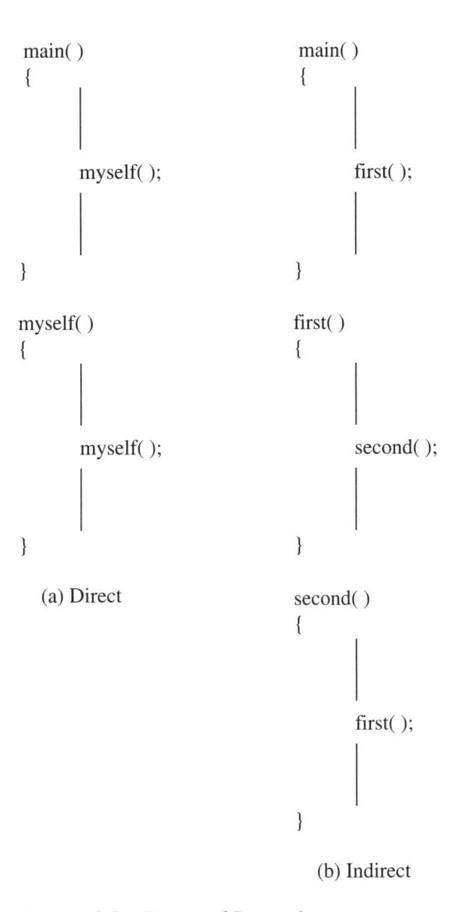

Figure 4.5 Types of Recursion

A Sample Program

Program 4.13 shows one way the level of recursion can be controlled.

Program 4.13

```
#include <stdio.h>

int fact(int num);

main()
{
        int N;

        printf("Enter the value of N => ");
        scanf("%d",&N);
        printf("\nN! equals %5d",fact(N));
}
```

```
int fact(int num)
{
        if (num == 0)
                return 1;
        else
                return num * fact(num - 1);
}
```

The `fact()` function computes the *factorial* of a non-negative integer. For example, 4 factorial (which we write in mathematical shorthand as 4!) equals 4 * 3 * 2 * 1, or 24. The `fact()` function is recursive, because of the `return` statement

```
return num * fact(num - 1);
```

which represents a direct call to `fact()`. But this is the second of two `return` statements. The first `return` statement

```
return 1;
```

does not call `fact()`. This is where the recursion will stop. Because each new call to `fact()` is done with a smaller value of `num`, eventually `fact()` will be called with `num` equal to zero, as in `fact(0)`. The test within the `if` statement

```
if (num == 0)
     return 1;
else
     return num * fact(num - 1);
```

determines when to stop recursion.

Suppose that when Program 4.13 executes, the user enters 4 for the value of N. Figure 4.6 shows the values associated with the `num` parameter and the `return` values for each level of recursion. Notice that information is passed into each new copy of `fact()` and also passed back when each one terminates.

There are four levels of recursion, one for each recursive call to `fact()`. Even though there are five complete copies of `fact()` in memory when the recursion is stopped, we do not count `main()`'s call to `fact()` as a recursive call. Only when `fact()` calls itself do we get another level of recursion.

When each recursive `fact()` function terminates, it returns a value to the previous level (from which it was called). The value 1 is returned from the last `fact()` function to force 0! to equal 1, as it is defined mathematically.

A sample execution of Program 4.13 is as follows:

```
Enter the value of N => 4
N! equals   24
```

Try other values of N as well. For example, 8! equals 40,320 but this is *not* the value returned by the program. Can you explain why?

Conclusion

Recursion is a very powerful programming technique that requires skill and patience to implement correctly. Experiment with some of the earlier programs in this chapter by con-

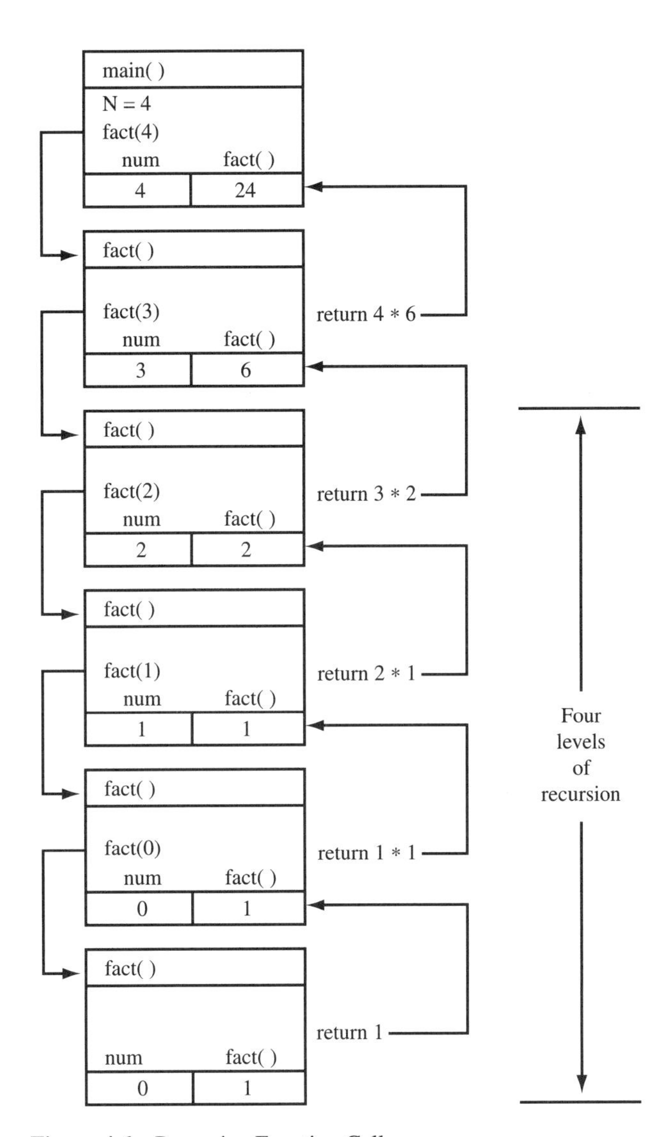

Figure 4.6 Recursive Function Calls

verting them into recursive applications. Check your understanding of this section by trying the following section review.

4.5 Section Review

1. What is a recursive function?
2. How are recursive functions implemented?
3. How is recursion stopped when two or more functions are involved in indirect recursion?
4. Why is it necessary to stop recursion at some point?
5. What `for()` statement performs the same job as `fact()`?

4.6 Example Programs

The first six programs in this section show additional uses for the loop structures covered in this chapter. A seventh program gives another example of recursion. Program 4.14 illustrates how nested loops are used to output a matrix-like list of numbers.

Program 4.14

```
#include <stdio.h>

#define rows 4              /* Number of rows.       */
#define cols 5              /* Number of columns.    */

main()
{
        int item = 0;       /* Item counter.         */
        int r,c;            /* For-loop variables.   */
        for (r = 1; r <= rows; r++)
        {
                for (c = 1; c <= cols; c++)
                {
                        printf("%2i   ",item);
                        item++;
                }
                printf("\n");
        }
}
```

Remember that a matrix consists of a group of numbers arranged in a row-column format. In the case of Program 4.14, `#define` statements are used to set the number of rows and columns to 4 and 5, respectively. The execution of Program 4.14 is as follows:

```
0    1    2    3    4
5    6    7    8    9
10   11   12   13   14
15   16   17   18   19
```

Note the use of `printf("\n")` to advance the output to the next row of numbers. Also note the `%2i` format used to display the integer item number. A great deal of time can be spent in getting a program to output data in the correct format.

Fibonacci numbers are the subject of Program 4.15. A Fibonacci number is always equal to the sum of the previous two Fibonacci numbers. Any list of Fibonacci numbers always starts the same way, with 0 and 1 as the first two terms. The third term equals 0 plus 1, or 1. The fourth term equals 1 plus 1, or 2. This concept is illustrated in Figure 4.7,

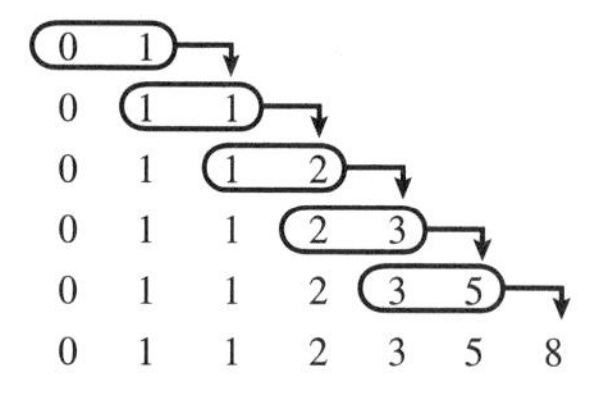

Figure 4.7 Fibonacci Series of Seven Terms

which shows a seven-term Fibonacci sequence. Program 4.15 uses a number supplied by the user to generate a Fibonacci sequence.

Program 4.15

```
#include <stdio.h>

main()
{
        int n;          /* User input number.               */
        int old,new;    /* Fibonacci sequence variables.    */
        int temp;       /* Temporary storage variable.      */
        int j;          /* For-loop counter.                */

        printf("Enter number of terms to generate => ");
        scanf("%i",&n);
        old = 0;
        new = 1;
        printf("%4i %4i ",old,new);
        for (j = 1; j <= n - 2; j++)
        {
                temp = old + new;
                old = new;
                new = temp;
                printf("%4i ",new);
        }
}
```

The last two terms in the sequence are always saved in the variables `old` and `new`. A temporary variable is used when a new term must be generated. The following is an execution of Program 4.15 showing the first 12 Fibonacci numbers:

```
Enter number of terms to generate => 12
0    1    1    2    3    5    8   13   21   34   55   89
```

What happens when the user enters a number larger than 30?

Program 4.16 shows how a simple calculating machine can be implemented in C.

Program 4.16

```
#include <stdio.h>
#include <stdlib.h>

#define FALSE 0
#define TRUE 1

main()
{
        int state;      /* Current state of machine. */
        char ch;        /* User input character.     */
        int err;        /* Error flag.               */
        int op;         /* Math operation selected.  */
```

```
int X,Y,Z;     /* Used for calculation.     */

printf("Enter an equation of the form:\n");
printf("  X + Y =   or   X - Y =\n");
printf("where X and Y are single-digit integers.\n");
printf("=> ");
state = 1;
err = FALSE;
do
{
        do      /* Skip blanks. */
        {
                ch = getch();
                printf("%c",ch);
        }
        while (ch == ' ');
        switch(state)
        {
                case 1 : if ((ch >= '0')&&(ch <= '9'))
                         {
                                X = atoi(&ch);
                                state = 2;
                         }
                         else err = TRUE;
                         break;
                case 2 : op = 0;
                         if (ch == '+') op = 1;
                         if (ch == '-') op = 2;
                         if (op == 0) err = TRUE;
                         state = 3;
                         break;
                case 3 : if ((ch >= '0')&&(ch <= '9'))
                         {
                                Y = atoi(&ch);
                                state = 4;
                         }
                         else err = TRUE;
                         break;
                case 4 : if (ch == '=') state = 5;
                         else err = TRUE;
        }
}
while ((state != 5)&&!err);
if (!err)
{
        Z = (op == 1) ? X + Y : X - Y;
        printf(" %i",Z);
```

```
        }
        else printf("\nIllegal character in input.");
}
```

The calculator performs only addition and subtraction of single-digit positive integers. A new function is used to get input from the user. This function is `getch()`, which is defined in the `<stdlib.h>` include file. `getch()` waits for a single keystroke on the keyboard and returns the associated ASCII character *without displaying the character.*

Another new function, `atoi`, is used to convert an ASCII character into a numeric integer value. This function is also part of `<stdlib.h>`.

Figure 4.8 shows a *state transition diagram* for the calculating machine. The goal of the program is to get to state 5 where the sum or difference of the two input numbers is shown. If any illegal characters are entered in *any* state, an error condition results. Multiple blanks between inputs are allowed. Sample executions are as follows:

```
Enter an equation of the form:
X + Y =   or   X - Y =
where X and Y are single-digit integers.
=> 7 + 3 = 10

Enter an equation of the form:
X + Y =   or   X - Y =
where X and Y are single-digit integers.
=> 4 - 6 = -2
```

Note that although the program does not allow negative integers to be input, they may show up in the result.

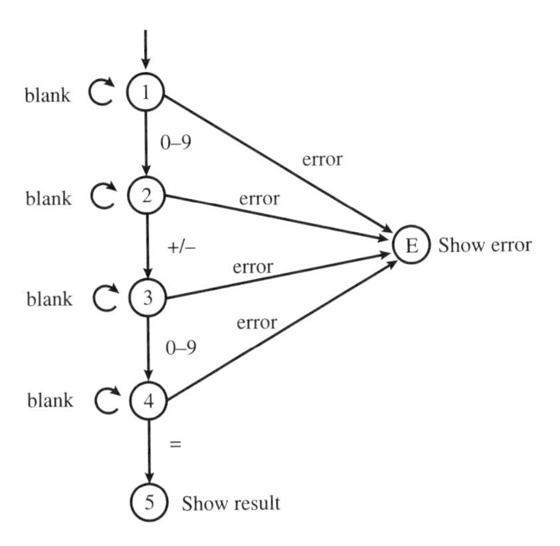

Figure 4.8 State Transition Diagram for Calculating Machine

Program 4.17 shows some of the common ways nested `for` loops are used in programs. Note the different ways the `for` variables are initialized and tested.

Program 4.17

```
#include <stdio.h>

#define N 10

main()
{
        int i,j,k;        /* For-loop counters.     */
        int num = 0;      /* Overall loop counter.  */

        for (i = 1; i <= N; i++)
                for (j = 1; j <= N; j++)
                        num++;
        printf("Group #1: The innermost loop executed %i times.\n",num);
        num = 0;
        for (i = 1; i <= N; i++)
                for (j = 1; j <= i; j++)
                        num++;
        printf("Group #2: The innermost loop executed %i times.\n",num);
        num = 0;
        for (i = 1; i <= N; i++)
                for (j = i; j <= N; j++)
                        num++;
        printf("Group #3: The innermost loop executed %i times.\n",num);
        num = 0;
        for (i = 1; i <= N; i++)
                for (j = 1; j <= N; j++)
                        for (k = 1; k <= N; k++)
                                num++;
        printf("Group #4: The innermost loop executed %i times.\n",num);
}
```

The number of times the innermost loop executes can be significantly altered by changing the parameters of the `for` statement, as shown by the following execution:

```
Group #1: The innermost loop executed 100 times.
Group #2: The innermost loop executed 55 times.
Group #3: The innermost loop executed 55 times.
Group #4: The innermost loop executed 1000 times.
```

Can you predict the results if N is changed to 11?

Program 4.18 performs a simple yet useful task, *enumeration*. To enumerate means to list all members belonging to a predefined set. For example, if the given set contains *X* and *Y*, then we can enumerate four different two-character strings: *XX*, *XY*, *YX*, and *YY*. Program 4.18 illustrates how nested loops are used to enumerate three-character strings.

Program 4.18

```
#include <stdio.h>

main()
{
	char i,j,k;		/* For-loop counters.	*/
	int num = 0;		/* Overall loop counter. */

	for (i = 'a'; i <= 'c'; i++)
		for (j = 'a'; j <= 'c'; j++)
			for (k = 'a'; k <= 'c'; k++)
			{
				num++;
				printf("%c%c%c  ",i,j,k);
			}
	printf("\nThere are %i different strings of letters.\n",num);
}
```

The execution of Program 4.18 is as follows:

```
aaa  aab  aac  aba  abb  abc  aca  acb  acc  baa  bab  bac
bba  bbb  bbc  bca  bcb  bcc  caa  cab  cac  cba  cbb  cbc
cca  ccb  ccc
There are 27 different strings of letters.
```

Program 4.18 also counts the number of strings enumerated. With simple formatting, it is possible to see a pattern emerge in the generated strings.

The next program examines the operation of a simple vending machine. The vending machine accepts nickels, dimes, and quarters. All products in the machine cost $0.50. When a product is purchased, the minimum numbers of nickels, dimes, and quarters are used to make change. For the vending machine to work properly, the user must be prevented from receiving a product until at least $0.50 has been inserted. Thus, it is necessary to keep track of the amount inserted by the user. When a product has been purchased, making minimum change may be different every time, because the user may have put much more money into the machine than required. For example, if the user inserts four nickels and two quarters, the minimum change is *two dimes*.

The steps required in the vending machine application are as follows:

1. Get money from the user one coin at a time.
2. Subtract the purchase price (always $0.50).
3. Make minimum change (if any).

Each step requires knowledge about the working amount of money in the machine. Note that change is made on the assumption that the machine will never run out of money. For example, if a very patient user enters 1000 nickels, the machine will return 198 quarters as change. It is a different matter to consider the available numbers of quarters, dimes, and nickels when making change, and this is not implemented here.

Getting the money is straightforward. The user enters 'Q', 'D', or 'N' as many times as desired. A variable is increased by 25, 10, or 5 each time, depending on the user input.

When the user wants to buy a product, 'P' must be entered. If the amount entered so far is less than $0.50, more money is needed. If the amount is sufficient, $0.50 is subtracted from the amount entered.

In order to make minimum change, three similar steps are used, in the following order:

- Subtract as many quarters as possible.
- Subtract as many dimes as possible.
- Subtract as many nickels as possible.

Examine Program 4.19.

Program 4.19

```
#include <stdio.h>

main()
{
        int amount = 0, Q = 0, D = 0, N = 0;
        char coin;

        printf("Please deposit coin...\n");
        do
        {
                coin = getche();        /* Get user response. */
                printf("\n");           /* Format display.    */
                switch(coin)            /* Make selection.    */
                {
                   case 'Q' : amount += 25; break;
                   case 'D' : amount += 10; break;
                   case 'N' : amount += 5; break;
                   case 'P' : if (amount < 50)
                                  printf("Please enter more money....\n");
                }
                if (coin != 'P')
                        printf("Amount = $%5.2f\n",amount/100.0);
        } while ((coin != 'P') || (amount < 50));

        amount -= 50;                  /* Purchase item.
        while (amount >= 25)           /* Make change in quarters. */
        {
                amount -= 25;
                Q++;
        }
        if (Q != 0)
                printf("Quarters : %2d\n",Q);

        while (amount >= 10)           /* Make change in dimes. */
        {
                amount -= 10;
                D++;
        }
```

```
	if (D != 0)
		printf("Dimes : %2d\n",D);

	while (amount >= 5)		/* Make change in nickels. */
	{
		amount -= 5;
		N++;
	}
	if (N != 0)
		printf("Nickels : %2d\n",N);
}
```

The first part of the program contains a `do` loop that accepts coins from the user. A `switch()` statement is used to determine how the `amount` variable is adjusted. If ‘P’ is entered, the `if` statement

```
case 'P' : if (amount < 50)
              printf("Please enter more money....\n");
```

is used to ensure that the user has entered enough money.

The `do` loop is repeated until the user has entered ‘P’ *and* the amount entered is at least $0.50. This condition is tested by the statement

```
} while ((coin != 'P') || (amount < 50));
```

After the purchase, the `amount` variable is decreased by $0.50. Then three similar `while` loops are used to make change. The first loop determines if any quarters are needed in the change.

```
while (amount >= 25)		/* Make change in quarters. */
{
	amount -= 25;
	Q++;
}
if (Q != 0)
	printf("Quarters : %2d\n",Q);
```

If `amount` is less than $0.25, the loop is skipped. Otherwise, `amount` is decreased by $0.25 until it is less than $0.25. The number of times this decrease is possible is indicated by the variable `Q`. If `Q` is zero, nothing is printed.

The second and third `while` loops perform the same operation with dime and nickel values.

A sample execution of Program 4.19 is as follows:

```
Please deposit coin...
Q
Amount = $0.25
N
Amount = $0.30
D
Amount = $0.40
P
Please enter more money....
```

```
D
Amount = $0.50
Q
Amount = $0.75
N
Amount = $0.80
P
Quarters :  1
Nickels :  1
```

The vending machine is a good example of a situation in which looping is used in the real world. Even though our vending machine has an unlimited supply of coins to make change with, we still need loops to make change and to accept coins from the user. To be completely useful, our vending machine should know how many coins are in each of its internal coin reservoirs. We need to pay attention to these coin reservoirs because they might be either empty or full at any point in time, forcing us to require exact change.

The last program in this section uses recursion to determine a number between 1 and 1024 that is chosen by the user. As shown in Program 4.20, direct recursion is used when `guess()` calls itself.

Program 4.20

```
#include <stdio.h>
#define MAXNUM 1024      /* Guess number between 1 and 1024 */

int guess(int lower, int upper);

main()
{
        int guesses;

        printf("Please think of a number between 1 and %4d...\n",MAXNUM);
        guesses = guess(1,MAXNUM);
        printf("\nThat took me %2d guesses.\n",guesses);
}

int guess(int lower, int upper)
{
        int newguess;
        char answer;

        newguess = lower + (upper-lower)/2;
        printf("\nIs your number %4d? ",newguess);
        answer = getche();
        switch(answer)
        {
                case 'Y' : return 1; break;
                case 'L' : return 1 + guess(lower,newguess-1); break;
                case 'H' : return 1 + guess(newguess+1,upper); break;
        }
}
```

Each new call to `guess()` reduces the range of numbers being considered by one-half. For example, suppose the chosen number is 907. A sample execution of Program 4.24 goes like this:

```
Please think of a number between 1 and 1024...

Is your number 512? H
Is your number 768? H
Is your number 896? H
Is your number 960? L
Is your number 928? L
Is your number 912? L
Is your number 904? H
Is your number 908? L
Is your number 906? H
Is your number 907? Y
That took me 10 guesses.
```

The user must correctly answer each question with an 'H' (for higher), an 'L' (lower), or a 'Y' (yes). The program always chooses the number in the middle of the guessing range. This is why the first guess is 512. Because the user answers 'H' to the first guess, the new guessing range becomes 513 through 1024. The second guess of 768 falls right in the middle of this range, and so on.

4.6 Section Review

1. Can Fibonacci numbers be generated without a loop?
2. Why is it necessary to initialize `state` to 1 in Program 4.16?
3. Show that the innermost loop counts for the sample execution of Program 4.17 are equal to N^2 (Group #1), $N*(N + 1)/2$ (Groups #2 and #3), and N^3 (Group #4).
4. How many different three-character strings are there with a choice of four letters (a through d) in each position?
5. How can recursion be used to enumerate three-character strings?
6. Explain why the maximum number of guesses in Program 4.20 is 10.

4.7 Troubleshooting Techniques

Discussion

You have probably experienced the frustration of developing programs that compile correctly and yet finding that when you execute them, you don't get the answers you expected. This kind of error is called a **run time error**. This can happen because of an incorrect formula or a mistake made entering a formula. It can also happen because you forgot to initialize certain variables or you didn't structure loops or branches correctly.

Many times, run time errors can be the most difficult bugs to analyze. The compiler is of no help because everything is correct as far as the C coding goes. The problem is embedded somewhere in the program because you are not getting the required results or what you are getting is incorrect.

There are several techniques for finding these types of errors. One way is to carefully read through your program (or have someone else read through it with you). This is a fine technique for smaller programs. However, in larger programs consisting of many functions with complex loops and branches, taking the time to read through every one may not be the most efficient method. One technique called **tracing** is used by many professional programmers. It's actually an easy addition to any program, and you may want to consider this method for your own use.

A Sample Problem

First, a program containing a problem will be presented, and the problem will be explained. Next, a debug routine will be embedded in the program. You will then see how the debug function can be activated to give you an insight into what is taking place in the program. Observe Program 4.21.

Program 4.21

```
#include <stdio.h>

main()
{
        int counter;    /* Loop counter. */

        counter = 0;

        while(counter != 9)
        {
                counter += 2;

                /* Body of loop. */
        } /* End of while. */
}
```

Program 4.21 will loop forever because the condition for stopping the loop is never met. The entire program will compile successfully because there are no syntax errors. Thus, you will experience a run time error.

What you need to do is to see what is actually going on while this part of the program is executing. A debug block usually consists of two parts, a visual section and a start/stop section.

The visual section causes the values of one or more variables to be displayed, whereas the stop/start section allows you to single step through the program. A debug block that displays the value of the loop counter is shown in Program 4.22.

Program 4.22

```
#include <stdio.h>

main()
{
        int counter;    /* Loop counter.      */
        char c;         /* Input for return. */
```

```
        counter = 0;

        while(counter != 9)
        {
                counter += 2;

                /* Debug block */ printf("Counter = %d\n",counter);
                                  scanf("%c",&c);

                /* Body of loop... */

        } /* End of while. */
}
```

When Program 4.22 is executed, the output will be

```
Counter = 0
 [Return]
Counter = 2
 [Return]
Counter = 4
 [Return]
Counter = 6
 [Return]
Counter = 8
 [Return]
Counter = 10
 [Return]
```

Notice that now you can single step though the program and observe the action of the loop counter. Its continued increase confirms what you should have suspected. However, the debug routine quickly shows you that the program never achieves the loop exit requirements.

Auto Debug

If you are in the process of developing a large program and you are testing various partially completed functions within the program, you may consider using an auto debug block. This is similar to the method discussed above, with the added advantage that you can easily turn it on or off with one simple command. This is especially helpful if you have several functions with debug blocks in them. Program 4.23 demonstrates this method.

Program 4.23

```
#include <stdio.h>
#define TRUE 1
#define FALSE 0

main()
```

```
{
        int counter;    /* Loop counter.     */
        char c;         /* Input for return. */
        int debug;      /* Debug flag.       */

        debug = TRUE;

        counter = 0;

        while(counter != 9)
        {
                counter += 2;

                /* Debug block */ if (debug)
                                  {
                                        printf("Counter = %d\n",counter);
                                        scanf("%c",&c);
                                  }

                /* Body of loop... */

        } /* End of while. */
}
```

In Program 4.23, note that TRUE and FALSE have been defined using the `#define` directive. If you set `debug = FALSE`, then the debug block will never be executed.

Auto Debug Function

The **auto debug function** is simply a C function that is included in your program. This is not feasible with all types of programs, but for certain programs where the variable types are the same, it is particularly useful. Program 4.24 illustrates an auto debug function.

Program 4.24

```
#include <stdio.h>
#define TRUE 1
#define FALSE 0

void debug(int active, int display, int startstop);

main()
{
        int counter;    /* Loop counter.     */

        counter = 0;

        while(counter != 9)
        {
                counter += 2;
```

```
                /* Debug routine */ debug(TRUE, counter, TRUE);

                /* Body of loop... */

        } /* End of while. */

        exit(0);
}

void debug(int active, int display, int startstop)
{
        char c;  /* Input for return */

        if (active == TRUE)
        {
                printf("%d",display);
                if (startstop == TRUE)
                        scanf("%c",&c);
        } /* End of if */
}
```

Note that now this function can be called by any part of the program. It will output the value of a given variable and offers a stop/start option as well as an on/off option. This can be a very useful method for debugging large programs quickly. If you use such auto debug functions, it is suggested that you highlight them so they can be removed from the final program easily. Many programmers will keep two source copies of their programs, one with the auto debug functions and the other without them. The procedure for doing this is to modify the version with the debug functions in it, make a backup copy, and then remove the debug commands; you will now have a new copy of the modified version without the debug functions. Another option is to use the conditional compilation feature presented in Section 3.7. Most C packages have built-in debuggers.

Conclusion

This section presented some of the debugging methods for large C programs. Here you saw how to include a basic debug command and then how to make it so you could easily deactivate it. You also saw how to create an automatic debugging feature that can be used for certain types of programs.

Try some of these methods in the next program you develop. For now, test your understanding of this section by trying the following section review.

4.7 Section Review

1. Explain what is meant by a run time error.
2. Does the compiler catch run time errors? Explain.
3. State what is usually contained in a debug block.
4. What is meant by an auto debug function?

4.8 Case Study: Series Resonant Circuit

Discussion

This case study shows an application using programming loops. In the previous sections of this chapter, the counting loops that you encountered had their counts determined at programming time. This case study follows the development of a C program in which the values for a counting loop are determined at run time. Neither the programmer nor the program user will know these exact values. As you will see, this type of counting loop has a wide range of applications.

Background Information

Figure 4.9 shows a series resonant RLC circuit and a corresponding plot of the impedance of the circuit over a range of frequencies. A series RLC circuit is an electrical circuit that consists of an inductor, a capacitor, and a resistor connected to a voltage source in such a way that the current from the source must trace a path through all three circuit components.

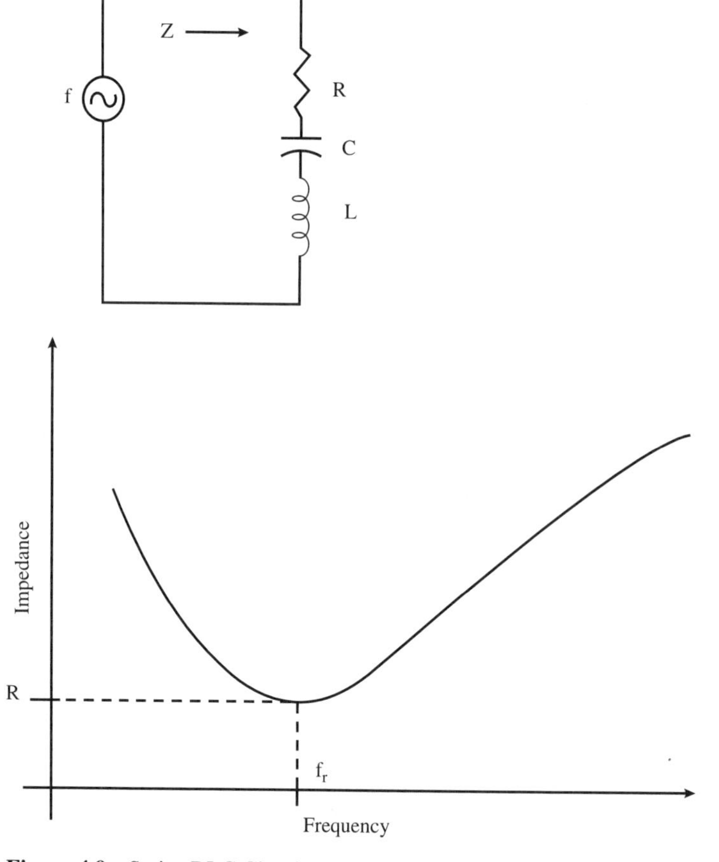

Figure 4.9 Series RLC Circuit and Frequency Plot

Impedance is defined as total opposition to current flow. For a series RLC circuit it is

$$Z = \sqrt{\left((X_L - X_C)^2 + R^2\right)}$$

Where

Z = Circuit impedance in ohms.
X_L = Inductive reactance in ohms.
X_C = Capacitive reactance in ohms.
R = Circuit resistance in ohms.

Inductive reactance (X_L) is given by

$$X_L = 2\pi fL$$

and capacitive reactance (X_C) is given by

$$X_C = 1/(2\pi fC)$$

Where

f = Applied frequency in hertz.
L = Inductance of the inductor in henrys.
C = Capacitance of the capacitor in farads.

As you can see from the impedance graph of the circuit in Figure 4.9, as the applied frequency of the source changes, so does the value of the circuit impedance. At one frequency the impedance of the circuit is a minimum and equal to the resistive value of the circuit. This is the point where

$$X_L = X_C$$

and is called *resonance*. The frequency where this happens is called the *resonant frequency* and is defined as

$$f_r = 1/(2\pi\sqrt{LC})$$

Where

f_r = The resonant frequency in hertz.
L = Inductance of the inductor in henrys.
C = Capacitance of the capacitor in farads.

This series RLC circuit is sometimes referred to as a series resonant circuit.

The Problem

Create a C program that will allow the program user to input the value of an inductor, capacitor, and resistor in a series RLC circuit. The program will compute the resonant frequency of the circuit and then display the value of the circuit impedance for a range of frequencies that range from 10% below resonance to 10% above resonance. There will be ten of these calculations below resonance and ten calculations above resonance.

The First Step—Stating the Problem

Stating the case study in writing yields

- Purpose of the Program: Compute and display the impedance of a series RLC circuit for a range of frequencies from 10% below to 10% above the resonance of the circuit.

Calculate and display 10 values of circuit impedance in the range below resonance and 10 values of circuit impedance in the range above resonance.

- Required Input Values:
  - Inductor in henrys.
  - Capacitor in farads.
  - Circuit resistance in ohms.
- Process on Input: Compute resonant frequency. Compute frequencies 10% below and 10% above resonance. Compute incremental value of frequency change. Compute circuit impedance for each of these frequency values:

$$
\begin{aligned}
f_r &= 1/(2\pi\sqrt{LC}) \\
X_L &= 2\pi \mathrm{fL} \\
X_C &= 1/(2\pi \mathrm{fC}) \\
Z &= \sqrt{((\mathrm{X_L} - \mathrm{X_C})^2 + \mathrm{R}^2)} \\
10\% \text{ below } f_r &= f_r - 0.1 f_r \\
10\% \text{ above } f_r &= f_r + 0.1 f_r \\
\text{Incremental change in } f_r &= (10\% \text{ above} - 10\% \text{ below})/20
\end{aligned}
$$

Where

f_r = Resonant frequency in hertz.
X_L = Inductive reactance in ohms.
X_C = Capacitive reactance in ohms.
L = Inductance of inductor in henrys.
C = Capacitance of capacitor in farads.
R = Resistance of circuit in ohms.
Z = Circuit impedance in ohms.

- Required Output: Values of the frequency and circuit impedance for 10 readings below resonance starting at 10% below f_r and 10 readings above resonance ending at 10% above f_r.

From this description, it should be clear exactly what the program should do. Understanding what the program should do must not depend on any knowledge of programming.

Developing the Algorithm

The steps in this program are the same as for most technology problems:

- Explain program to user.
- Get values from user.
- Do calculations.
- Display results.

For this program, a loop is required that will calculate and display the values of the circuit impedance for a range of frequencies. Since this is a counting loop that will have definite beginning and ending conditions, it was decided to use a C `for` loop.

The required programmer's block was developed as shown in Program 4.25.

Program 4.25

```
#include <stdio.h>
#include <math.h>

#define PI                    3.141592
```

```
#define square(x)             x*x
#define X_L(f,l)              2*PI*f*l
#define X_C(f,c)              1/(2*PI*f*c)
#define F_R(l,c)              1/(2*PI*sqrt(l*c))
#define Z(f,l,c,r)            sqrt(square((X_L(f,l)-X_C(f,c)))+square(r))


/************************************************************************/
/*                 Series Resonant Circuit Analyzer                    */
/************************************************************************/
/*                Developed by: A. Technology Student                  */
/*                                                                      */
/************************************************************************/
/*    This program will display the impedance of a series resonant     */
/* circuit for a range of frequencies that are 10% above and 10% below */
/* the resonant frequency of the circuit. There are ten frequency      */
/* calculations above and below resonance for this circuit.            */
/************************************************************************/
/*                       Function Prototypes                           */
/*---------------------------------------------------------------------*/

void explain_program(void);

/* This function explains the operation of the program to the program  */
/* user.                                                                */
/*---------------------------------------------------------------------*/

void get_values(void);

/* This function gets the circuit values from the program user.        */
/************************************************************************/

main()
{
       explain_program();      /* Explain program to user. */
       get_values();           /* Get circuit values.      */
       exit(0);
}

/*---------------------------------------------------------------------*/

void explain_program()         /* Explain program to user. */
{
       printf("\n\nThis program will display the impedance of a\n");
       printf("series resonant circuit for a range of frequencies\n");
       printf("that are 10%% above and 10%% below the resonant\n");
       printf("frequency of the circuit.\n");
} /* End of Explain program to user. */

/*---------------------------------------------------------------------*/

void get_values()              /* Get circuit values.                  */
```

```
{
        float resistor;          /* Value of series resistor in ohms.     */
        float inductor;          /* Value of series inductor in henrys.   */
        float capacitor;         /* Value of series capacitor in farads.  */
}
```

Program Analysis

Note in Program 4.25 that all of the formulas are defined using the `#define` directive:

```
#define PI              3.141592
#define square(x)       x*x
#define X_L(f,l)        2*PI*f*l
#define X_C(f,c)        1/(2*PI*f*c)
#define F_R(l,c)        1/(2*PI*sqrt(l*c))
#define Z(f,l,c,r)      sqrt(square((X_L(f,l)-X_C(f,c)))+square(r))
```

In order to use the `sqrt` function, the `math.h` header file had to be included:

```
#include <math.h>
```

The programmer's block then presents the necessary information (the name of the program, name of the program developer, description of the program, and preliminary function prototypes). The `main()` function is then developed to present the structure of the program:

```
main()
{
        explain_program();          /* Explain program to user.  */
        get_values();               /* Get circuit values.       */
        exit(0);
}
/*---------------------------------------------------------------*/
```

Here you can see the use of top-down design where the most general idea of the program is dealt with first, and the details are left for later. You can also see the structure of the program as shown in `main()`. What will happen is that the program will be explained to the program user and values will be received. No attempt has been made at this initial stage to go any further in the development of other program blocks.

Next, the major portion of each of the prototyped functions is described.

```
void explain_program()              /* Explain program to user. */
{
        printf("\n\nThis program will display the impedance of a\n");
        printf("series resonant circuit for a range of frequencies\n");
        printf("that are 10%% above and 10%% below the resonant\n");
        printf("frequency of the circuit.\n");

}/* End of Explain program to user.  */

/*---------------------------------------------------------------*/
```

```
void get_values()                    /* Get circuit values.              */
{
        float resistor;              /* Value of series resistor in ohms.   */
        float inductor;              /* Value of series inductor in henrys. */
        float capacitor;             /* Value of series capacitor in farads.*/
}
```

Note that all of the explanation to the program user is developed at this first stage, but only the values for the `get_values` function are declared. This is simply a matter of individual programmer choice. But it is suggested that the first coding be as simple and straightforward as possible. This is necessary in order to keep any program debugging time low and to make it so that the general structure of the program will compile and execute.

Designing the get_values Function

The `get_values()` function design is straightforward. It simply consists of a `printf` user prompt followed by a `scanf` to get the user input. As with most technology problems, the type `float` is used for the input values. This is done so that E notation can be used with the `%f` input. The developed function is shown below. Good program design would dictate the calling of another function that would calculate and display the required values.

```
void get_values()                    /* Get circuit values.              */
{
     float resistor;                 /* Value of series resistor in ohms.   */
     float inductor;                 /* Value of series inductor in henrys. */
     float capacitor;                /* Value of series capacitor in farads.*/

     printf("\n\nEnter the following values as indicated:\n");
     printf("Resistor value in ohms => ");
     scanf("%f",&resistor);
     printf("Inductor value in henrys => ");
     scanf("%f",&inductor);
     printf("Capacitor value in farads => ");
     scanf("%f",&capacitor);

     calculate_and_display(inductor, capacitor, resistor);


}    /*  End of get_values.  */
```

Designing the calculate_and_display Function

The function for calculating and displaying the required information requires several variables. It was decided that this function would be called from `get_values()`. Because of this, `calculate_and_display` requires three arguments in its parameter list so that the values of the inductor, capacitor, and resistor can be passed to it.

```
void calculate_and_display(float l, float c, float r)
```

Since this function will also contain the C `for` loop, the beginning and ending values of the frequency range as well as the frequency change are needed for the loop counter.

```
{
        float below_fr;        /* 10% below resonant frequency.   */
        float above_fr;        /* 10% above resonant frequency.   */
        float freq_change;     /* Change in frequency.            */
        float counter;         /* Loop counter.                   */
        float impedance;       /* Circuit impedance in ohms.      */
```

It was decided that the values for the C `for` loop will be computed by two other functions not yet defined. The first of these, called `calculate_below`, will calculate the frequency that is 10% below resonance and the second, called `calculate_above`, will calculate the frequency that is 10% above resonance. These values will then be used to set the counting range of the C `for` loop.

```
below_fr = calculate_below(l,c);
above_fr = calculate_above(l,c);
```

The change in frequency can then be calculated using the values of the frequency range.

```
freq_change = (above_fr - below_fr)/20.0;
```

For the convenience of the programmer, the values of the C `for` loop ranges are displayed during program execution:

```
printf("below_fr => %f hertz.\n",below_fr);
printf("above_fr => %f hertz.\n",above_fr);
printf("freq_change => %f hertz.\n",freq_change);
```

Next, the C `for` statement was developed:

```
for(counter = below_fr; counter <= above_fr; counter += freq_change)
```

The body of this loop displays the value of the frequency (done by the value of `counter`), calculates the circuit impedance (by using the `#define` for *Z*), and then uses an `if...else`. This `if...else` is used to point out to the user what frequency is the resonant frequency of the circuit. By definition, this is the frequency at which the circuit impedance is equal to the value of the circuit resistance.

```
{
        printf("Frequency => %f hertz",counter);
        impedance = Z(counter,l,c,r);
        if (impedance == r)
                printf("Impedance = %f ohms. <== Resonance!\n",impedance);
        else
                printf("Impedance = %f ohms.\n",impedance);

}  /* End of for */
```

The next step was the creation of the two functions that will calculate the frequency values 10% below and 10% above resonance. This could have been accomplished by `#define` directives at the beginning of the program, but the functions are used in this program to illustrate how the C `return()` function can be used.

Designing the calculate_below and calculate_above Functions

Since these two functions will be called, they need values passed to them. These values are the ones needed to calculate the resonant frequency of the RLC circuit. Thus, the inductor and capacitor values entered by the program user need to be formal arguments.

```
float calculate_below(float l, float c)
{
        return(F_R(l,c) - 0.1*F_R(l,c));
}

/*-----------------------------------------------------------------------*/

float calculate_above(float l, float c)
{
        return(F_R(l,c) + 0.1*F_R(l,c));
}
```

Note how the C `return` function is used in both of these. The calculation is done within the argument of this function. The resulting value of this calculation is then returned back to the calling function. Doing this saves programming space. However, keep in mind that your primary rule in developing this type of program is to keep the source code easy to read and understand.

Final Program Phase

The final completed program is shown as Program 4.26. Note that each of the functions in the program has a function prototype with comments that explain the purpose of the function as well as the meanings of all variables in the function parameter list.

Program 4.26

```
#include <stdio.h>
#include <math.h>

#define PI                          3.141592
#define square(x)                   x*x
#define X_L(f,l)                    2*PI*f*l
#define X_C(f,c)                    1/(2*PI*f*c)
#define F_R(l,c)                    1/(2*PI*sqrt(l*c))
#define Z(f,l,c,r)                  sqrt(square((X_L(f,l)-X_C(f,c)))+square(r))

/***********************************************************************/
/*                 Series Resonant Circuit Analyzer                    */
/***********************************************************************/
/*               Developed by: A. Technology Student                   */
/*                                                                     */
/***********************************************************************/
/*    This program will display the impedance of a series resonant     */
/* circuit for a range of frequencies that are 10% above and 10% below */
```

```
/* the resonant frequency of the circuit. There are ten frequency         */
/* calculations above and below resonance for this circuit.               */
/**************************************************************************/
/*                          Function Prototypes                           */
/*------------------------------------------------------------------------*/

void explain_program(void);

/* This function explains the operation of the program to the program     */
/* user.                                                                  */
/*------------------------------------------------------------------------*/

void get_values(void);

/* This function gets the circuit values from the program user.           */
/*------------------------------------------------------------------------*/

void calculate_and_display(float l, float c, float r);

/* l = Inductance in henrys.  c = Capacitance in farads.                  */
/* r = Resistance in ohms.                                                */
/* This is the function that will calculate and display the value of the  */
/* circuit impedance for the given range of frequencies.                  */
/*------------------------------------------------------------------------*/

float calculate_below(float l, float c);

/* l = Inductance in henrys. c = Capacitance in farads.                   */
/* This function calculates the frequency that is 10% below resonance.    */
/*------------------------------------------------------------------------*/

float calculate_above(float l, float c);

/* l = Inductance in henrys. c = Capacitance in farads.                   */
/* This function calculates the frequency that is 10% above resonance.    */
/**************************************************************************/

main()
{
        explain_program();      /* Explain program to user. */
        get_values();           /* Get circuit values.      */
        exit(0);
}

/*------------------------------------------------------------------------*/

void explain_program()          /* Explain program to user. */
{
        printf("\n\nThis program will display the impedance of a\n");
        printf("series resonant circuit for a range of frequencies\n");
        printf("that are 10%% above and 10%% below the resonant\n");
```

```
        printf("frequency of the circuit.\n");
} /* End of Explain program to user. */

/*-------------------------------------------------------------------------*/

void get_values()               /* Get circuit values.                     */
{
        float resistor;         /* Value of series resistor in ohms.       */
        float inductor;         /* Value of series inductor in henrys.     */
        float capacitor;        /* Value of series capacitor in farads.    */

        printf("\n\nEnter the following values as indicated:\n");
        printf("Resistor value in ohms => ");
        scanf("%f",&resistor);
        printf("Inductor value in henrys => ");
        scanf("%f",&inductor);
        printf("Capacitor value in farads => ");
        scanf("%f",&capacitor);

        calculate_and_display(inductor, capacitor, resistor);
} /* End of get_values. */

void calculate_and_display(float l, float c, float r)
{
        float below_fr;         /* 10% below resonant frequency. */
        float above_fr;         /* 10% above resonant frequency. */
        float freq_change;      /* Change in frequency.          */
        float counter;          /* Loop counter.                 */
        float impedance;        /* Circuit impedance in ohms.    */

        below_fr = calculate_below(l,c);
        above_fr = calculate_above(l,c);
        freq_change = (above_fr - below_fr)/20.0;

        printf("below_fr => %f hertz.\n",below_fr);
        printf("above_fr => %f hertz.\n",above_fr);
        printf("freq_change => %f hertz.\n",freq_change);

        for(counter = below_fr; counter <= above_fr; counter += freq_change)
        {
                printf("Frequency => %f hertz",counter);
                impedance = Z(counter,l,c,r);
                if (impedance == r)
                        printf("Impedance = %f ohms. <== Resonance!\n",
                                impedance);
                else
                        printf("Impedance = %f ohms.\n",impedance);
        } /* End of for */
} /* End of calculate_and_display */

/*-------------------------------------------------------------------------*/
```

```
float calculate_below(float l, float c)
{
        return(F_R(l,c) - 0.1*F_R(l,c));
}

/*------------------------------------------------------------------------*/

float calculate_above(float l, float c)
{
        return(F_R(l,c) + 0.1*F_R(l,c));
}

/*------------------------------------------------------------------------*/
```

Program Execution

Execution of Program 4.26 produces

```
This program will display the impedance of a
series resonant circuit for a range of frequencies
that are 10% above and 10% below the resonant
frequency of the circuit.

Enter the following values as indicated:
Resistor value in ohms => 10
Inductor value in henrys => 1e-3
Capacitor value in farads => 1e-6
below_fr => 4529.629883 hertz.
above_fr => 5536.214355 hertz.
freq_change => 50.329224 hertz.
Frequency => 4529.629883 hertz Impedance = 12.023640 ohms.
Frequency => 4579.958984 hertz Impedance = 11.648333 ohms.
Frequency => 4630.288086 hertz Impedance = 11.308162 ohms.
Frequency => 4680.617188 hertz Impedance = 11.004684 ohms.
Frequency => 4730.946289 hertz Impedance = 10.739361 ohms.
Frequency => 4781.275391 hertz Impedance = 10.513482 ohms.
Frequency => 4831.604492 hertz Impedance = 10.328093 ohms.
Frequency => 4881.933594 hertz Impedance = 10.183921 ohms.
Frequency => 4932.262695 hertz Impedance = 10.081312 ohms.
Frequency => 4982.591797 hertz Impedance = 10.020183 ohms.
Frequency => 5032.920898 hertz Impedance = 10.000000 ohms. <==Resonance!
Frequency => 5083.250000 hertz Impedance = 10.019782 ohms.
Frequency => 5133.579102 hertz Impedance = 10.078132 ohms.
Frequency => 5183.908203 hertz Impedance = 10.173290 ohms.
Frequency => 5234.237305 hertz Impedance = 10.303208 ohms.
Frequency => 5284.566406 hertz Impedance = 10.465627 ohms.
Frequency => 5334.895508 hertz Impedance = 10.658155 ohms.
Frequency => 5385.224609 hertz Impedance = 10.878351 ohms.
Frequency => 5435.553711 hertz Impedance = 11.123784 ohms.
Frequency => 5485.882813 hertz Impedance = 11.392090 ohms.
Frequency => 5536.211914 hertz Impedance = 11.681007 ohms.
```

Conclusion

This case study took you step by step through the design and development of a technology program that required the use of a program loop. More specifically, this loop has its values for counting determined at program execution time. In this case, neither the programmer nor the program user needs to know the beginning, ending, and incremental loop values. All of this information is supplied by the program. Check your understanding of this section by trying the following section review.

4.8 Section Review

1. What was the first step in the design of this case study program?
2. State why a C `for` loop was used to meet the programming requirements of this program.
3. How were the beginning and ending values of the program loop determined?
4. Explain how the C `return` was used in this program.

Interactive Exercises

DIRECTIONS

These exercises require that you have access to a computer and software that supports C. They are provided here to give you valuable experience and immediate feedback on what the concepts and commands introduced in this chapter will do.

Exercises

1. Try Program 4.27 with a character. Guess what the output will be and then give it a try for what may be a surprising result!

Program 4.27

```
#include <stdio.h>

main()
{
        char c;   /* A character. */

        for(c = 'a'; c <= 'z'; c++)
                printf(" c = %c",c);
}
```

2. Note what is different about the loop in Program 4.28. Here a number is being changed; however, it is being displayed as a character (`%c`), not as a decimal (`%d`). What do you think this loop produces? Give it a try and see!

Program 4.28

```
#include <stdio.h>

main()
{
        int i;   /* A number. */
```

```
        for(i = 97; i <= 122; i++)
                printf(" i = %c",i);
}
```

3. Your computer system probably has an extended ASCII character code set. If you tried the previous problem, you probably know what Program 4.29 will do. Don't pass up trying it!

Program 4.29

```
#include <stdio.h>

main()
{
        int i;  /* A number. */

        for(i = 128; i <= 255; i++)
                printf(" i = %c",i);
}
```

4. Program 4.30 shows all! It displays everything old ASCII is capable of doing on the monitor. Listen as well as watch on this one.

Program 4.30

```
#include <stdio.h>

main()
{
        int i;   /* A number.          */
        int j;   /* Another number.    */

        j = 0;

        for(i = 0; i <= 255; i++)
        {
                printf(" %d = %c",j,i);
                j++;
        }  /* End of for */
}
```

5. In the C `for` loop shown in Program 4.31, the loop counter has a `++i` instead of an `i++`. What difference do you suppose this will make in the outcome of the loop counter values? Try it and see. (Another good test question!)

Program 4.31

```
#include <stdio.h>

main()
{
        int i;  /* A number. */
```

```
        for(i = 0; i <= 5; ++i)
                printf(" i = %d",i);
        printf(" last i = %d",i);
}
```

6. Program 4.32 contains a "null `while`." This means that there is nothing in the argument of the C `while` statement. What do you think will happen when you try to compile this program? Check your conclusion.

Program 4.32

```
#include <stdio.h>

main()
{
        while()
                printf("Is this ever executed?");
}
```

7. Predict the output of Program 4.33. Then try it.

Program 4.33

```
#include <stdio.h>

main()
{
        int i;   /* A number. */

        for(i = 97; i <= 122; i++)
                printf(" i = %c",i);
}
```

8. How many times does Program 4.34 loop? Try it. What causes it to stop?

Program 4.34

```
#include <stdio.h>

main()
{
        int i;    /* A number. */

        i = 10;

        while(i)
        {
                printf("How many times does this loop?");
                printf(" i = %d\n",i--);
        }
}
```

9. What is printed out by Program 4.35?

Program 4.35

```
#include <stdio.h>

int raise(int exp);

main()
{
        printf("%d",raise(8));
}

int raise(int exp)
{
        if (exp == 0)
                return 1;
        else
                return 2 * raise(exp - 1);
}
```

Self-Test

DIRECTIONS

Use Program 4.26 developed in the case study section of this chapter to answer the following questions.

Questions

1. How many function prototypes are used in the program?
2. What type of a loop is used in the program?
3. In the loop used, what is the initial loop condition?
4. What is the final loop condition?
5. State what is changed each time through the loop.

End-of-Chapter Problems

General Concepts

Section 4.1

1. What is a compound statement? How is it set off in a C program?
2. Give the C notation for pre-incrementing the variable `counter`.
3. In C, what allows you to have two sequential C statements?
4. What are the major parts of a C `for` loop?

Section 4.2

5. State the construction of the C `while` loop.
6. When is the loop condition tested in a C `while` loop?
7. Give the condition for repeating a C `while` loop.
8. What is a good use of the C `while` loop?

Section 4.3

9. Give the construction of the C `do while` loop.
10. When is the loop condition of a C `do while` loop tested?
11. What are the conditions for repeating a C `do while` loop?
12. Which loop is preferred, the C `while` or the C `do while`? Explain.

Section 4.4

13. What is meant by a nested loop?
14. Which of the C loops may be nested?
15. Is there a problem with nesting a C `do` loop? Explain.

Section 4.5

16. What is necessary for recursion?
17. What is the difference between direct and indirect recursion?
18. Why must recursion be stopped at a particular level?
19. How does recursion simulate the operation of a loop?

Section 4.6

20. Explain how recursion might be used to generate a Fibonacci number.
21. How does Program 4.16 change states?
22. What happens if the same variable is used in nested `for` loops?
23. Explain how to add coin reservoirs to Program 4.19.

Section 4.7

24. Explain what is meant by a run time error.
25. What is a debug function? How is it used?
26. How is the auto debug function used?

Program Design

In developing the following C programs, use the method described in the case study. For each program you are assigned, document your design process and hand it in with your program. This process should include the design outline, process on the input, and required output as well as the program algorithm. Be sure to include all of the documentation in your final program. This should consist of, but not be limited to, the programmer's block, function prototypes, and a description of each function as well as any formal arguments you may use.

27. Create a C program that will compute the voltage drop across a resistor for a range of current values selected by the program user. The program user is to input the value of the resistor, the beginning and ending current values, and the incremental value of the current. The relationship between the resistor voltage and current is given by

$$V = IR$$

Where

V = Voltage across the resistor in volts.
I = Current in the resistor in amps.
R = Value of the resistor in ohms.

28. Develop a C program that will calculate the power dissipated in a resistor for a range of voltage values across the resistor. The program user is to select the value of the resistor and the beginning and ending resistor voltages. The relationship is given by

$$P = V^2/R$$

Where

P = Power dissipation of the resistor in watts.
V = Voltage across the resistor in volts.
R = Value of the resistor in ohms.

29. Modify the program you developed in problem 28 so that the program user will be warned when the power dissipation of the resistor exceeds a maximum power dissipation determined by the program user. An example warning could be a monitor display of

```
WARNING! Power exceeds 2 watts!!!
```

30. Modify the program in problem 28 by adding an inner loop so that the program user may observe the power dissipation for a range of resistors as well as a range of voltage values. The program user would now select the voltage range and resistance range as well as the incremental values for each.
31. Develop a C program that will display the interest compounded annually from 1 to 30 years. The program user is to input the principal and the rate of interest. The mathematical relationship is

$$Y = A(1 + N)^T$$

Where

Y = The amount.
A = The principal invested.
N = Interest rate.
T = Number of years.

32. Create a C program that will take any number to any power. The program user is to select both the base and the power.
33. What can be done to allow larger Fibonacci numbers in Program 4.15?
34. Write a C program that will output only one specific Fibonacci number. The user will specify which term in the sequence is required.
35. Modify Program 4.16 to allow multiplication and division.
36. Modify Program 4.16 to allow the use of negative integers.
37. Modify Program 4.16 to allow the use of one- and two-digit integers.
38. Write a C program that will accept only binary digits 0 and 1 from the user, until the letter 'b' or 'B' is entered. The equivalent decimal number should then be outputted. Use the `getch()` function in your program.
39. Write a C program that checks to see if a number inputted by the user is a valid real number. Examples are 1.7, –0.0354, 259.08, and –1067.25. Use the `getch()` function in your program.
40. Modify the vending machine to allow coin reservoirs. The reservoirs hold the following number of coins:

Quarters	50
Dimes	100
Nickels	70

41. Modify the vending machine to allow three choices of products, whose costs are as follows:
Product 1: $0.50
Product 2: $0.65
Product 3: $1.25
42. Write a recursive function that performs remainder division. For example, calling `remdiv(100,12)` returns a value of 4 (since 100 divided by 12 equals 8 with a remainder of 4).
43. Develop a C program that will show the change in volume of a sphere for a given change in radius. The program user selects the initial, incremental, and ending values of the radius. The mathematical relationship is

$$V = 4/3\pi r^3$$

Where

V = Volume of the sphere in cubic units.
π = The constant pi.
r = Radius of the sphere in linear units.

44. Construct a program in C that will demonstrate the change in the height of water in a cylindrical water tank as the volume of water in the tank changes (increases or decreases). The program user is to input the radius and height of the water tank as well as the incremental increase or decrease of the amount of water in the tank, expressed in gallons.
45. Construct a program in C that will compute and display the values of a range of temperatures in degrees Fahrenheit and degrees centigrade. The program user is to select the lowest and highest temperatures as well as the incremental temperature change. The mathematical relationship is

$$F = (9/5)C + 32$$

Where

F = Temperature in degrees Fahrenheit.
C = Temperature in degrees centigrade.

46. Develop a C program that will show the change in the volume of a metal cone as it is machined. Assume that the machining reduces the height of the metal cone. The program user enters the dimensions of the cone as well as the incremental changes in the cone height. The mathematical relationship is

$$V = (h/3)(A_1 + A_2 + \sqrt{A_1 A_2})$$

Where

V = Volume of the cone in cubic units.
h = Height of the cone in linear units.
A_1 = Area of the lower base in square units.
A_2 = Area of the upper base in square units.

47. Modify the program in problem 31 so that the program contains an inner loop that will display the interest for different amounts of money. The program user now selects the range for the principal and the incremental difference, and enters the rate of interest.
48. Modify the program in problem 44 so that the program user is warned when the water tank is empty or overflowing.
49. Create a C program that will display a range of lengths in feet, meters, and inches. The program user can select the beginning, ending, and incremental measurements in feet.
50. The relationship between the number of units of fertilizer and the expected crop yield is

$$Y = F/(2^F + 1)$$

Where

Y = The yield improvement factor.
F = Arbitrary units of fertilizer per acre.

Develop a C program to find what value of F produces a maximum value of Y.
51. Create a C program that will display the heartbeat of a patient in beats per minute, beats per hour, and beats per day. The program user can enter the minimum and maximum number of beats per minute and the value of the increment.
52. Develop a C program that will show a range of weights in pounds and kilograms. The program user can enter the minimum and maximum weights along with the increment value either in pounds or kilograms.
53. Write a C program that will compute the sales tax for a range of values and round the result off to the nearest cent. The program user can input the minimum and maximum dollar amounts and the percentage of the sales tax. The increment is to be determined by the program so that a printout is given for every one-cent change in the total amount.

5 Pointers

Objectives

This chapter gives you the opportunity to learn:

1. The method used by your computer to store information in its memory.
2. Different ways memory is used in your computer by the C language.
3. Methods used to represent values in your computer's memory.
4. The way negative numbers are represented in memory.
5. What pointers are in C and how they are used.
6. Methods of passing variables by using pointers.
7. Which constants and variables are known to which parts of your C program.
8. Various methods of classifying your C variables.
9. Pitfalls common to new C programmers.

Key Terms

Address
Fetch/Execute
Immediate Addressing
Direct Addressing
Byte
Word
Ones Complement
Twos Complement
Nibble
Pointer
Address Operator
Pointer Declaration
Indirection Operator
Local Variable
Variable Scope
Variable Life
Global Variable
Automatic Variable
External Variable
Static Variable
Register Variable
Initialization
`far` Pointer

Outline

Introduction

This chapter introduces the important concept of *pointers*. As you will see, this is simply another method of getting data in and out of your computer's memory. But, unlike other methods you have been using, this is a very powerful one that will give you great programming flexibility.

You should have a background in binary, octal, and hexadecimal number systems.

In this chapter, you will first be introduced to some of the internal structure of your computer. Here you will discover how the computer's memory is organized and how it is used. Having this information as a background, you will be ready to explore the features of the C pointer.

5.1 Internal Memory Organization

Discussion

This section presents important concepts about how information is stored inside your computer. Understanding memory storage will open a whole new world of programming opportunities for you. This section lays the foundation for the rest of the material presented in this chapter.

Basic Idea

Figure 5.1 presents a basic model of how a program is stored in a computer's memory.

As shown in Figure 5.1, you may visualize a computer's memory as a pile of storage locations. In order to distinguish one storage location from another, each location is assigned a unique number. This number is called an **address**. Thus, each memory location inside your computer has a unique address.

Storing a Program

Consider a program that will add two numbers and store the answer back in memory. Such a program, stored in a computer, may be visualized as shown in Figure 5.2.

Look at Figure 5.2. There are seven memory locations, each identified by an address. Observe that a memory location may contain an instruction (an action to be taken by the computer) or it may contain data (an item that is acted upon by the action of the instruction). When this program is executed, the CPU (central processing unit) starts getting

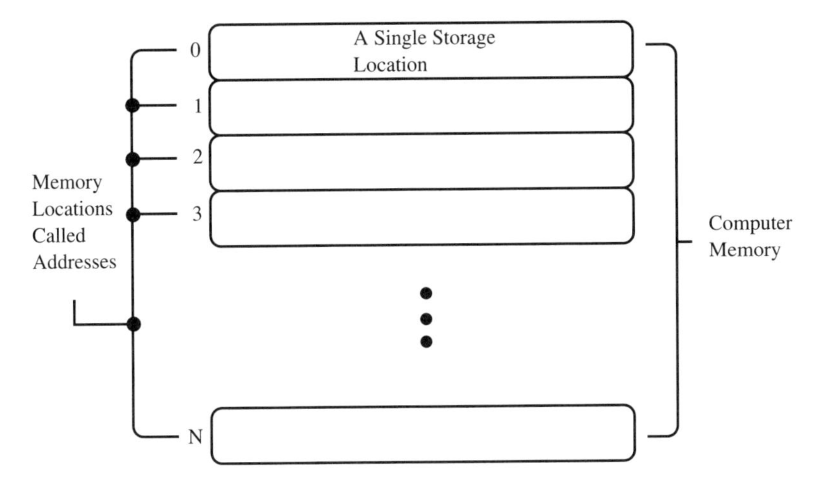

Figure 5.1 Visualization of Computer Memory

information from memory at memory location 1. As shown in Figure 5.2, the instruction in memory location 1 tells the CPU to get the number in the next memory location (location 2). After it does that, the CPU goes to the following memory location (location 3) and gets the next instruction. This process of getting an instruction, carrying out the instruction, and getting another instruction is referred to as the **fetch/execute** cycle of a computer.

The CPU will add the two numbers (3 and 2) and store the answer (5) in memory location 6. Addresses are used to distinguish one memory location from another.

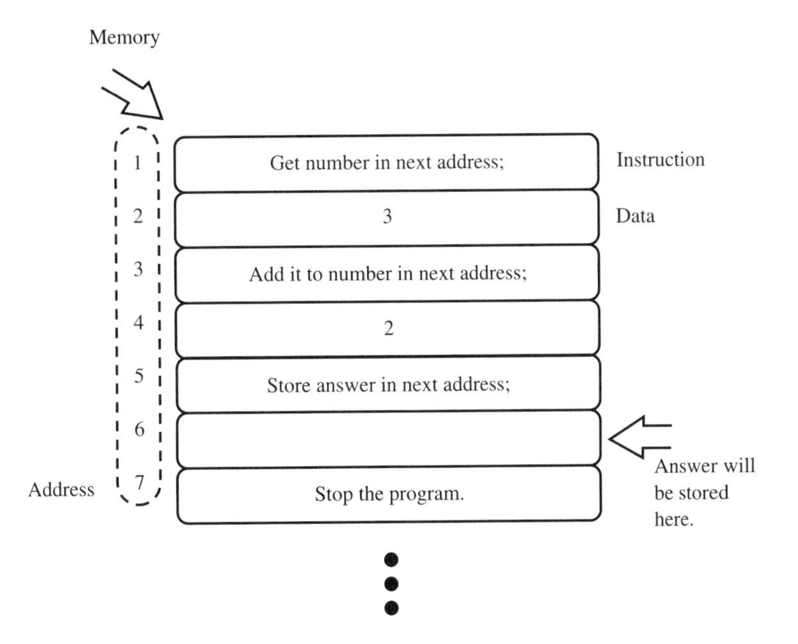

Figure 5.2 Concept of Program in Memory

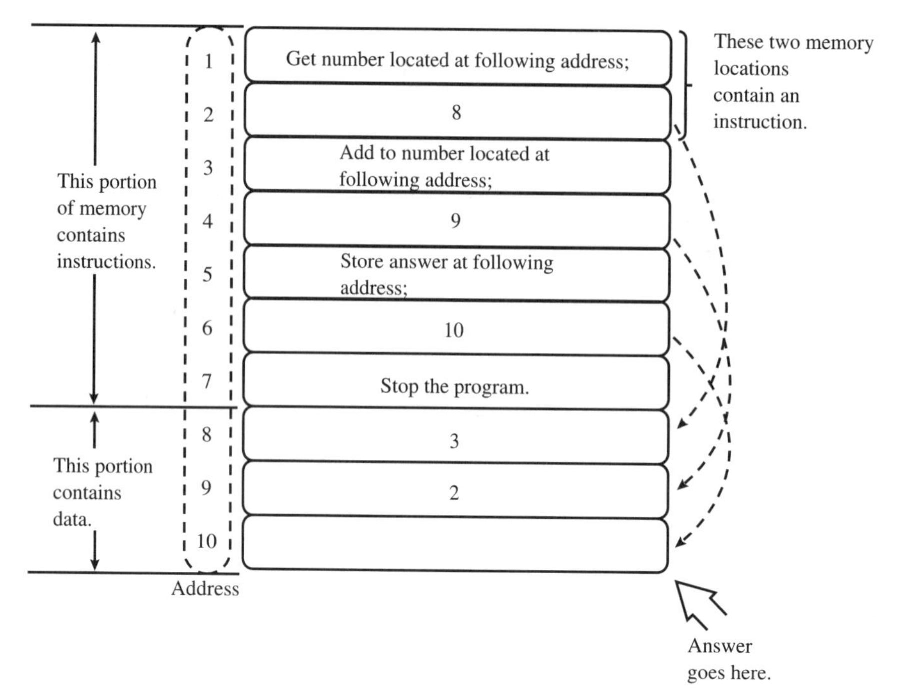

Figure 5.3 A Different Way of Doing the Same Thing

Another Way

It isn't common practice to have the data in a program in the very next memory location following an instruction. For example, consider the program shown in Figure 5.3. It does the same thing as the program shown in Figure 5.2. It adds the same two numbers (3 and 2). The difference is how this is done.

The program illustrated in Figure 5.3 executes the following process.

1. Get number located at following address: This instruction requires the CPU to go to the next memory location to get the address (new location) where the number it is to work with will be located. It gets the number 8 from the next memory location. Now the first instruction is completed and it means:
2. Get number located at address 8: So the CPU skips down to address 8 and gets the value stored there (which is the number 3). What happens here is that the first two memory locations are actually instructions to the CPU. The first piece of data for the CPU is the number 3 located at address 8.
   The CPU now returns to the address following its last instruction (to address 3, because its last instruction was at address 2) and gets its next instruction:
3. Add it to number located at following address: Again, this is only part of a complete instruction. As before, the CPU has to go down to the next address (location 4) to find the address where the next piece of data is located. Once it has done this, the completed instruction is interpreted as:

4. Add to it number located at address 9: And once again, the CPU skips down (this time to address 9) to find its next piece of data (the number 2). This is the number to be added to 3. Now the CPU returns to the address following its last instruction (location 5).
5. Store answer at following address: The CPU goes to the next memory location in order to find out where it is to store the answer. Thus it goes from address 5 to address 6 to complete this instruction. There it finds the number 10. This instruction is now completed.
6. Store answer at address 10: Thus, the CPU stores the answer (2 + 3 = 5) at memory location 10. Having done this, it goes back to the address following its last instruction (memory location 7) to get its next instruction.
7. Stop the program: And the CPU stops all further processing.

This may seem like a strange way of processing instructions and data, but both processes (Figures 5.2 and 5.3) are used by modern-day computers. In order to use many of the capabilities of the C language, it's important to understand both processes.

The first method (like the program in Figure 5.2), where the data immediately follows the instruction, is called **immediate addressing**. The second method (like the program in Figure 5.3), where the two memory locations are needed for the instruction—one to direct the program to where the data is located in memory—is called **direct addressing**.

These processes are going on every time you use your computer. The difference is that the process in Figure 5.3 uses ten memory locations. Your computer has thousands of memory locations.

Let us see how some of these memory locations can be used.

EXAMPLE 5.1

What does the program shown in Figure 5.4 do? Assume that the CPU reads the first instruction from address 3438.

Solution

This program will multiply $5 \times 12 = 60$ and store the answer in memory location 3445.

Conclusion

This section presented an important conceptual framework for understanding how instructions and data may be stored in your computer. Here you saw the important difference between immediate and direct addressing. Check your understanding of this section by trying the following section review.

5.1 Section Review

1. State how a computer's memory may be visualized.
2. Explain the difference between an instruction and data.
3. What is an address?
4. State what is meant by immediate addressing.
5. State what is meant by direct addressing.

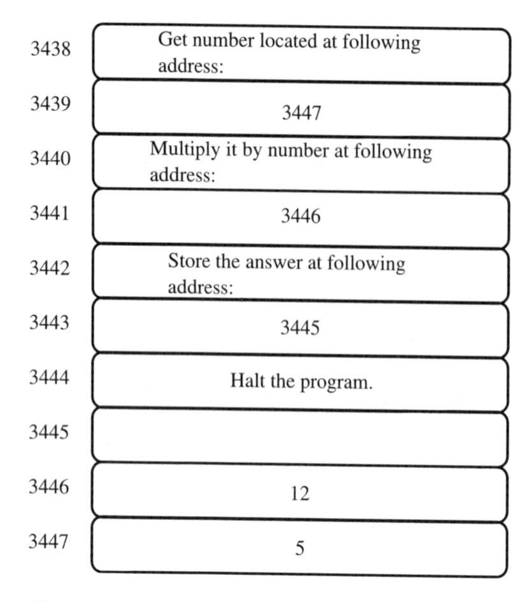

Figure 5.4 Program for Example 5.1

5.2 How Memory Is Used

Discussion

In order to benefit from the full programming power of C, you must understand how data is organized in memory. The information presented here will use material on binary, octal, and hexadecimal number systems.

Storage Size

Figure 5.5 shows how information may be classified in computer memory. Keep in mind that all instructions and data are represented by groups of electrical ONs and OFFs.

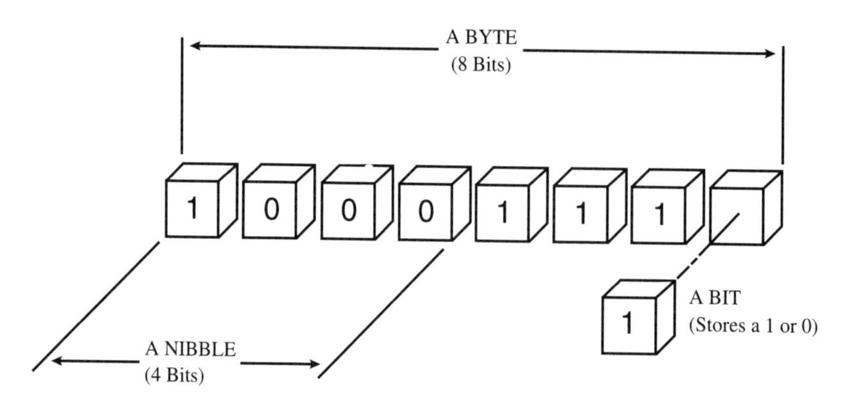

Figure 5.5 Storage Classification

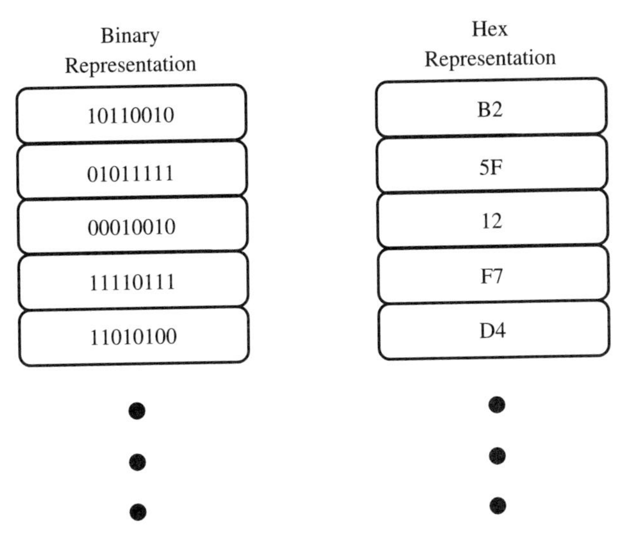

Figure 5.6 Memory Organization with 8-Bit Words

Computer memory is usually organized in groups of **bytes** called **words**. Some of the most common word sizes for computers are 8 bits (1 byte), 16 bits (2 bytes), as well as 32 and 64 bits. As an example, memory using an 8-bit word is illustrated in Figure 5.6.

char Type

In C, the type `char` occupies a single byte (8 bits) of memory. This means that the range of values that may be stored in a type `char` range from $0000\ 0000_2$ to $1111\ 1111_2$. Expressed in hex, this is 00_{16} to FF_{16}.

This is equal to 0_{10} to 255_{10}. Figure 5.7 illustrates this concept.

Program 5.1 demonstrates the limitations of 1 byte. The maximum value of a `char` is 255; if it is increased by 1 from that value, it resets to zero.

Program 5.1

```
#include <stdio.h>

main()
{
        unsigned char counter;

        counter = 0;

        while(counter != 255)
        {
                counter++;
                printf(" %d - %X ",counter,counter);
        }
}
```

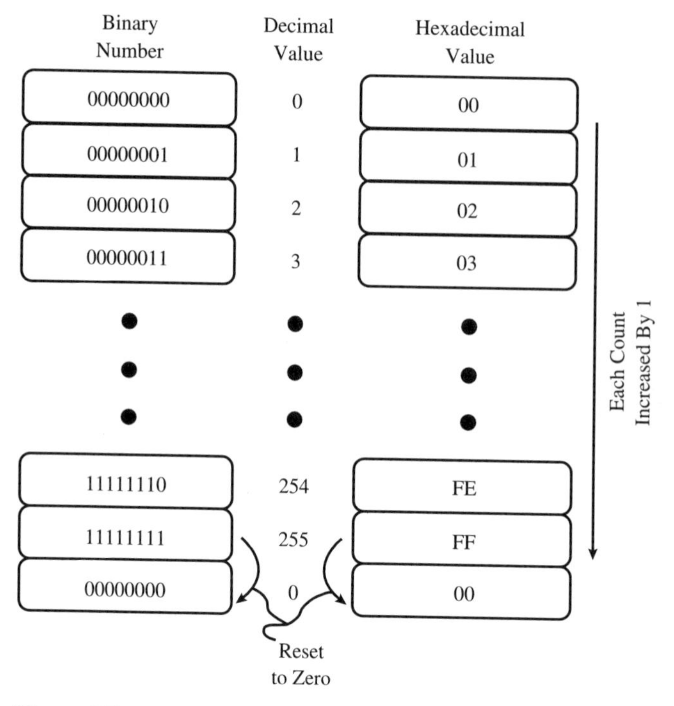

Figure 5.7 Counting in a Byte of Memory

When Program 5.1 is executed, the output will be

```
1 - 1 2 - 2 3 - 3 4 - 4 5 - 5 6 - 6 7 - 7 8 - 8 9 - 9 10 - A
11 - B 12 - C 13 - D 14 - E 15 - F 16 - 10
```

This series will continue up to

```
253 - FD 254 - FE 255 - FF
```

What has happened here is that the full count for a byte has taken place (from 0 to 255). Since there are only 8 bits, the maximum value that can be stored in binary is $1111\ 1111_2$. If one more is added to this value, it becomes $0000\ 0000_2$ (just like the odometer on your car rolling over when it reaches its maximum count). Note that in the program a type `unsigned char` is used. You will see the reason for this shortly.

The sizeof Function

The `sizeof` function in C will determine the size of a variable's storage in bytes. This is illustrated by Program 5.2.

Program 5.2

```
#include <stdio.h>

main()
{
```

```
    char character;
    int character_size;
    int integer;
    int integer_size;

    character_size = sizeof(character);
    integer_size = sizeof(integer);

    printf("Size of a character on your system is %d byte.\n"
           ,character_size);
    printf("Size of an integer on your system is %d bytes.",integer_size);
}
```

When executed, Program 5.2 will return

```
Size of a character on your system is 1 byte.
Size of an integer on your system is 2 bytes.
```

What this means is that, through the use of the `sizeof` function, C has determined the size, in bytes, of memory allocated to the storage of your particular variable.

Representing Signed Numbers

As you know by now, your computer can work with positive or negative numbers. However, all numbers stored in it must be represented by electrical ONs and OFFs (1s and 0s). This means that the sign of the number must somehow also be represented by a 1 or a 0. There is no way of representing a third symbol (such as a minus sign) in such a two-state system. Therefore negative numbers must somehow be distinguished from positive numbers through the use of the same 1s and 0s that are used to represent the values of the numbers themselves.

There are several ways of doing this. The most common method requires an understanding of binary addition.

Binary Addition

There are five rules you need to know for the addition of binary numbers. These are

$$\begin{array}{r}0\\+0\\\hline 0\end{array}\qquad\begin{array}{r}1\\+0\\\hline 1_2\end{array}\qquad\begin{array}{r}0\\+1\\\hline 1_2\end{array}\qquad\begin{array}{r}1\\+1\\\hline 10_2\end{array}\qquad\begin{array}{r}1\\1\\+1\\\hline 11_2\end{array}$$

The first three rules are self-explanatory. The fourth rule simply states that $1 + 1 = 2_{10}$ which is 10_2 in binary. The last rule is $1 + 1 + 1 = 3_{10}$ which is 11_2 in binary.

As an example, adding $2_{10} + 3_{10} = 5_{10}$ in binary is

$$\begin{array}{rl}1 & \Leftarrow\text{carry}\\ 10_2 & \\ +\,11_2 & \\\hline 101_2 = 5_{10} & \end{array}$$

As another example, the sum of $7_{10} + 3_{10} = 10_{10}$ in binary would be

$$\begin{array}{rl} 111 & \text{<=carry} \\ 111_2 & \\ +\,011_2 & \\ \hline 1010_2 = 10_{10} & \end{array}$$

Notice that a 1 is carried from the addition of the first column. Thus, the addition of the second column also produces another carry.

Test your knowledge of binary addition with Example 5.2.

EXAMPLE 5.2

Using binary addition, determine the sum of the following binary numbers:

A. $10_2 + 10_2$
B. $1010_2 + 1101_2$
C. $1111_2 + 1_2$

Solution

A.
$$\begin{array}{l} 10_2 \\ +10_2 \\ \hline 100_2 = 4_{10} \end{array}$$

B.
$$\begin{array}{l} 1010_2 \\ +1101_2 \\ \hline 10111_2 = 23_{10} \end{array}$$

C.
$$\begin{array}{ll} 1111 & \text{<= carry} \\ 1111_2 & \\ +0001_2 & \\ \hline 10000_2 = 16_{10} & \end{array}$$

It is possible to perform binary subtraction in a manner similar to binary addition. However, because this would require complex logic circuits inside the computer, binary subtraction is, as you will see, actually performed by binary addition. In order to see how this is done, you first need to know how your computer represents negative binary numbers by using just 1s and 0s.

Complementing Numbers

In computer number notation, the complements of a binary number are obtained by converting all of its 1s to 0s and all of its 0s to 1s. As an example, the complement of 1011_2 is 0100_2. This seemingly useless process is the basis for obtaining the correct result of a binary subtraction. This process may also be referred to as the **ones complement** of a binary number. This is illustrated in Example 5.3.

EXAMPLE 5.3

Determine the ones complements of the following binary numbers:

A. 101_2 B. 1111_2 C. 1000_2

Solution

A. 101_2 => 010_2
B. 1111_2 => 0000_2
C. 1000_2 => 0111_2

Negative binary numbers are actually represented inside your computer by what is called the **twos complement** notation. To find the twos complement of any binary number, simply first take the ones complement and add 1 to the result. As an example, to find the twos complement of 1011_2:

1. Find the ones complement: 0100
2. Add one to the result: 0101
3. Thus the twos complement of 1011_2 is 0101.

Example 5.4 shows the twos complements of three different numbers.

EXAMPLE 5.4

Determine the twos complement of the following binary numbers:

A. 0110_2 B. 1001_2 C. 1111_2

Solution

A. Twos complement of $0110_2 = 1001 + 1 = 1010_2$
B. Twos complement of $1001_2 = 0110 + 1 = 0111_2$
C. Twos complement of $1111_2 = 0000 + 1 = 0001_2$

Binary Subtraction

To subtract using the twos complement notation, do the following:

1. Convert the subtrahend to twos complement notation. Add the results and ignore the final carry. As an example (8 – 5 = 3) in binary is

$$\begin{array}{r} 1000_2 \\ -\ 0101_2 \\ \hline \end{array}$$

2. Convert the subtrahend to twos complement:

$$\begin{array}{r} 1000_2 \\ 1011_2 \\ \hline \end{array}$$

3. Add the results, and ignore the final carry:

$$\begin{array}{r} {}^{1} \\ 1000_2 \\ +\ 1011_2 \\ \hline 0011_2 = 3_{10} \end{array}$$

The resulting answer is 3_{10} which is what you would expect when subtracting 5 from 8.

Effectively, what was done above was to convert the value of –5 (8 – 5 was the problem) into twos complement notation. In this case it causes the MSB (most significant bit)

Table 5.1 Signed and Unsigned Nibbles

Unsigned Value	Binary	Signed Value
0	0000	0
1	0001	1
2	0010	2
3	0011	3
4	0100	4
5	0101	5
6	0110	6
7	0111	7
8	1000	–8
9	1001	–7
10	1010	–6
11	1011	–5
12	1100	–4
13	1101	–3
14	1110	–2
15	1111	–1

of the binary number to be a 1 (0101_2 in twos complement notation is 1011_2). Thus, for signed binary numbers, anytime the MSB is a 1, the number is negative and represents the twos complement of the actual number. For example, in twos complement notation, 1011_2 represents the twos complement of a negative number. To find what the value is, simply take the twos complement and prefix a – sign:

$$1011_2 => 0101_2 = -5_{10}$$

This notation of representing a signed binary nibble in twos complement notation is shown in Table 5.1.

As you can see from Table 5.1, a **nibble** in unsigned notation may represent a range of values from 0 to 15 (16 different values). However, in signed notation, it represents a range of values from –8 to +7 (still 16 different values, including 0, but now negative numbers are represented). Thus, to find the range of values used in signed binary numbers, you can use the relationship

$$\text{Range} => -(2^{N-1}) \text{ to } +(2^{N-1}-1)$$

Where

N = Bit size of the binary number.

As an example, for the 4-bit nibble, its signed range is

$$-(2^{4-1}) \text{ to } +(2^{4-1}-1) = -(2^3) \text{ to } +(2^3-1)$$
$$-8 \text{ to } (8-1) = -8 \text{ to } 7$$

EXAMPLE 5.5

Determine the range of values for a signed byte.

Solution

A byte is 8 bits. Thus, using the relationship

$$\text{Range} \Rightarrow -(2^{N-1}) \text{ to } +(2^{N-1}-1)$$

yields

$$-(2^{8-1}) \text{ to } +(2^{8-1}-1)$$
$$-2^7 \text{ to } +(2^7-1) = -128 \text{ to } +127$$

As illustrated in Example 5.5, in signed notation a byte represents a range of –128 to +127 (256 different values, including 0, or 2^8 different values), whereas in unsigned notation the same byte represents a value range of 0 to 255 (still 256 different values). This is illustrated by Program 5.3.

Program 5.3

```
#include <stdio.h>

main()
{
        char value;

        printf("Give me a two-place hex number => ");
        scanf("%X",&value);

        printf("The signed decimal value is %d",value);
}
```

Program 5.3 illustrates how negative values are stored in a single byte. For example, when this program is executed, any hex number smaller than 80_{16} that is put in by the program user (that is, a value of $7F_{16}$ or less, meaning that the MSB is not set) will produce a positive number. However, as soon as any hex value of 80_{16} or larger (1000 0000 or larger; here is where the MSB is set to 1 signifying a negative number in twos complement form), then a negative decimal value is displayed. Here's an example:

```
Give me a two-place hex number => 7F
The signed decimal value is 127
```

In this case, since the MSB was not set ($0111\ 1111_2$), 7F was interpreted as a positive value and equal to 127. However, note the following:

```
Give me a two-place hex number => 80
The signed decimal value is -128
```

Here, the MSB has been set ($80_{16} = 1000\ 0000_2$) and is interpreted as the twos complement of a negative number. Thus, taking the twos complement of 1000 0000 yields 1000 0000, which represents a –128. By the same token,

```
Give me a two-place hex number => FF
The signed decimal value is -1
```

illustrates how a negative value is stored. The hex number FF_{16} equals $1111\ 1111_2$, which represents the twos complement of a negative number. The twos complement of $1111\ 1111_2$ equals $0000\ 0001_2$, which equals -1_{10}.

This takes us back to the `unsigned char`. When this C type is used, the setting of the MSB to 1 is no longer taken to be a negative value, and the full range of numbers from 0 to 255 can now be realized for 8 bits.

Observe Program 5.4.

Program 5.4

```
#include <stdio.h>

main()
{
        unsigned char value;

        printf("Give me a two-place hex number => ");
        scanf("%X",&value);

        printf("The unsigned decimal value is %d",value);
}
```

When Program 5.4 is executed, the MSB will no longer be treated as a sign bit because of the `unsigned char` type:

```
Give me a two-place hex number = 7F
The unsigned decimal value is 127
```

This is no different from before, but now look what happens when the MSB is set to 1:

```
Give me a two-place hex number => 80
The unsigned decimal value is 128
```

and

```
Give me a two-place hex number => FF
The unsigned decimal value is 255
```

This illustrates the difference between signed and unsigned types in C. It relates directly to two important facts: how many bytes are used to store the data type and if the MSB is treated as a sign bit.

As shown in Table 5.2, the `unsigned` type modifier gives you larger positive values for the same amount of memory space because the MSB of its binary representation is no longer used to represent a negative value.

Recall that the type `char` occupies 1 byte (8 bits) of memory, thus making it capable of storing a range of signed values from –128 to +127 and unsigned values of 0 to 255. The other data types have greater ranges of values because they use more than 1 byte of memory. This is illustrated by Program 5.5.

Program 5.5

```
#include <stdio.h>

main()
{
        int integer;
        char character;
        short shortnumber;
        long longnumber;
```

```
    unsigned char unsigned_character;
    unsigned int unsigned_integer;
    unsigned long unsigned_long;
    float float_number;
    double double_number;
    long double long_double;

    printf("The size in bytes of a/an:\n");
    printf("integer => %d\n",sizeof(integer));
    printf("character => %d\n",sizeof(character));
    printf("short => %d\n",sizeof(shortnumber));
    printf("long => %d\n",sizeof(longnumber));
    printf("unsigned char => %d\n",sizeof(unsigned_character));
    printf("unsigned int => %d\n",sizeof(unsigned_integer));
    printf("unsigned long => %d\n",sizeof(unsigned_long));
    printf("float => %d\n",sizeof(float_number));
    printf("double => %d\n",sizeof(double_number));
    printf("long double => %d\n",sizeof(long_double));
}
```

When executed, Program 5.5 will give the size in bytes used by your system to store the various C data types (such as `int`, `char`, and so on.). Typically, an IBM PC uses the following memory:

```
The size in bytes of a/an:
integer => 2
character => 1
short => 2
long => 4
unsigned char => 1
unsigned int => 2
unsigned long => 4
float => 4
double => 8
long double => 8
```

Table 5.2 Range of Values for C Types

Identified by	Sometimes Called	Value Range
`char`	`signed char`	–128 to 127
`unsigned char`		0 to 255
`int`	`signed int`	–32,768 to 32,767
`unsigned int`	`unsigned`	0 to 65,535
`short`	`signed short`	–32,768 to 32,767
`unsigned short`	`unsigned short int`	0 to 65,535
`long`	`long int, signed long`	–2,147,483,648 to 2,147,483,647
`unsigned long`	`unsigned long int`	0 to 4,294,967,295
`float`		+/–3.4E +/–38 (7 digits)
`double`		+/–1.7E +/–308 (15 digits)
`long double`		+/–1.7E +/–308 (15 digits)

Conclusion

This section presented the ways in which C uses computer memory. Here you saw how values were stored in memory. Most importantly, the concept of how negative numbers are stored was presented along with programming methods that can be used to actually see how this was done. Check your understanding of this section by trying the following section review.

5.2 Section Review

1. State what is meant by a computer word.
2. What is the word size of a C `char` type?
3. How does the computer represent signed numbers?
4. Explain the difference between a signed and an unsigned data type in C.
5. What C data type uses the smallest amount of memory? The largest?

5.3 Pointers

Discussion

In this section you will discover the secrets of a very important programming concept commonly referred to as **pointers**. This is a powerful tool in the design of C programs for any area of technology. Take your time working with this section. You will find that working the short programs presented here will be helpful in establishing the concepts of pointers.

Basic Idea

Think of the memory in your computer as being constructed as shown in Figure 5.8.

As shown in the figure, each memory location can hold 1 byte of information. Since each memory location is the same size as the next, they are distinguished from each other

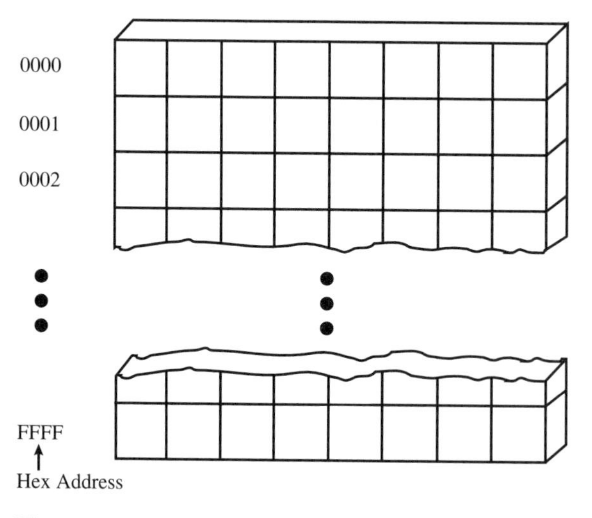

Figure 5.8 Construction of a Byte-Size Memory

by a number called an address. These addresses are sequentially numbered from 0 to the maximum size of the memory.

Program 5.6 uses a type `char` simply to place a value into a single byte of your computer's memory.

Program 5.6

```
#include <stdio.h>

main()
{
        char value;     /* A memory location to hold a value. */

        value = 97;

        printf("Value = %d",value);
}
```

When Program 5.6 is executed, its output will be

```
Value = 97
```

What the program has done is to store a number in one of the memory locations inside your computer. You know that the value of the number stored is 97. However, you do not yet know at what address (exactly where in your computer's memory) this value is stored. However, using C you can find this out by using the **address operator** `&`:

Program 5.7

```
#include <stdio.h>

main()
{
        char value;    /* A memory location to hold a value.  */

        value = 97;
        printf("%u => | %d | <= address and data of value.",&value,value);
}
```

In Program 5.7, two variables will be displayed on your monitor screen. One is the value of 97 that you put into a particular memory location. The other will be the address (the exact location inside *your* computer) at which the value of 97 is stored. For the system used in the writing of this text, the output of Program 5.7 was

```
1204 => | 97 | <= address and data of value.
```

As you can see, the address of this memory location was 1204. (This may be different on your system; it may also be different every time you run the same program—this address is assigned by the computer.) The way to find out the value of the address is by using the `&` just before the variable `&value`. This means, "Display the address of where the variable is stored in memory."

Now, look at Program 5.8. It uses a pointer. The program has a pointer declaration. A **pointer declaration** names a pointer variable and also states the type of the object to which the variable points. What happens is that a variable declared as a pointer holds a memory address. The type specifier gives the type of the object. In this program, the pointer type specifier is `char` because the object it is pointing to is of type `char`.

Program 5.8

```
#include <stdio.h>

main()
{
        char value;      /* A memory location to hold a value. */
        char *pointer;   /* A pointer.                          */

        value = 97;
        printf("%u => | %d | <= address and data of value.\n",&value,value);

        pointer = &value;
        printf("%u => | %d | <= address and data of pointer.",&pointer,pointer);
}
```

Execution of Program 5.8 yields

```
1204 => | 97 | <= address and data of value.
4562 => |1204| <= address and data of pointer.
```

The program now has two variables. One, called `value`, contains the number 97. The other, called `pointer`, contains the address where the number 97 is stored. This concept is illustrated in Figure 5.9.

Program Analysis

Program 5.8 is the first to use a pointer. The pointer is declared by using a space and the `*` symbol following the C `type`:

```
char *pointer; /* A pointer. */
```

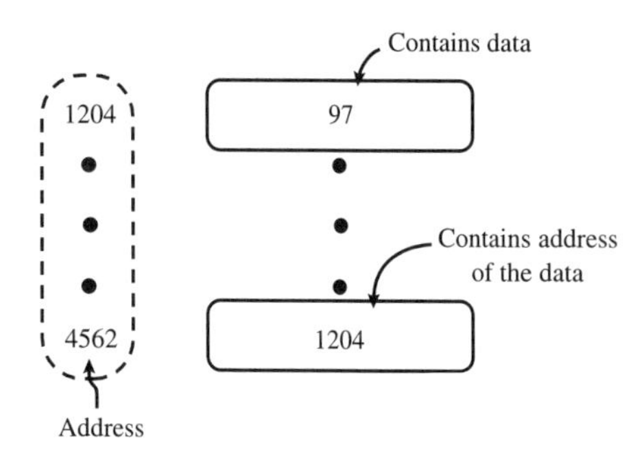

Figure 5.9 Idea of Pointer and Value

This means that the variable `pointer` will now contain the address of a given memory location. In the case of this program, the address of the memory location is placed into `pointer` by

```
pointer = &value;
```

Now the `pointer` contains the address of `value`. (*Note:* A pointer may have a type specifier of `void`. Doing this essentially delays the specification of the type to which the pointer refers. When it comes time to use the pointer, you must specify the type each time you use it. This is most easily done with a C type cast.)

Using the Pointer

Program 5.8 simply stores the address of a given variable in another variable. You don't need pointers to do this. Program 5.9 shows the significance of the `*` (called the **indirection operator**) and the `&` (address operator).

Program 5.9

```
#include <stdio.h>

main()
{
        char value;      /* A memory location to hold a character.   */
        char *pointer;   /* A pointer. */

        value = 97;
        printf("%u => | %d | <= address and data of value.\n",&value,value);

        pointer = &value;
        printf("%u => | %d | <= address and data of pointer.\n"
                         ,&pointer,pointer);

        printf("\n Value stored in pointer = %d\n",pointer);
        printf(" Address of pointer : &pointer = %u\n",&pointer);
        printf(" Value pointed to: *pointer = %d\n",*pointer);
}
```

Execution of Program 5.9 yields

```
1204 => | 97 | <= address and data of value.
4562 => |1204| <= address and data of pointer.
Value stored in pointer = 1204
Address of pointer: &pointer = 4562
Value pointed to: *pointer = 97
```

What is happening is illustrated in Figure 5.10.

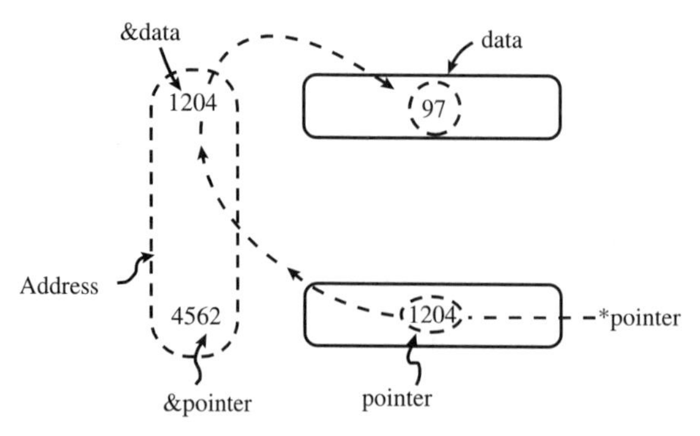

Figure 5.10 Concept of the Pointer

As you can see from the figure, the variable `*pointer` will have exactly the same value as the address of the variable to which it is pointing. What is important is to understand the meaning of the following when it comes to using pointers:

`pointer` → Contains the value stored in the variable `pointer`.
`&pointer` → Will give the address of the variable `pointer`.
`*pointer` → Will give the value stored at the memory location whose address is stored in the variable `pointer`.

Passing Variables with Pointers

You can change the value of your program variables by using pointers. This may not seem important now, but it will become very important because, as you will see, it is a way of passing more than one variable back from a called function to the calling function. Program 5.10 illustrates how this is done.

Program 5.10

```
#include <stdio.h>

main()
{
        char *pointer;   /* A pointer.                          */
        char variable;   /* A memory location to hold a value. */

        variable = 1;
        pointer = &variable;

        printf("Value stored in variable = %d\n",variable);
        printf("Value stored in pointer  = %d\n",pointer);

        *pointer = 2;

        printf("New value stored in variable = %d\n",variable);
        printf("Value stored in pointer  = %d\n",pointer);
}
```

Execution of Program 5.10 yields

```
Value stored in variable = 1
Value stored in pointer = 7732
New value stored in variable = 2
Value stored in pointer = 7732
```

Program Analysis

Program 5.10 assigns two variables:

```
char *pointer;    /* A pointer.                         */
char variable;    /* A memory location to hold a value. */
```

The first variable is a pointer because it is preceded by the * sign. The second variable is a memory location that will hold an assigned value. The pointer is of type `char` and the variable to store a given value is of type `char`, meaning that it will hold 1 byte of data.

The variable is assigned a value of 1, and the pointer is assigned the address of where the value of 1 is stored.

```
variable = 1;
pointer = &variable;
```

`variable` now contains 1, and `pointer` now contains the address of `variable`.

To demonstrate what is contained in `variable` and `pointer`, print out the values stored in `variable` and `pointer`.

```
printf("Value stored in variable = %d\n",variable);
printf("Value stored in pointer = %d\n",pointer);
```

This produces

```
Value stored in variable = 1
Value stored in pointer = 7732
```

The value of 7732 is the address assigned by the computer when the program was executed on the computer used in writing this book. This means that at address 7732, the value of 1 is stored.

A value of 2 is now assigned to `*pointer`:

```
*pointer = 2;
```

Figure 5.11 shows exactly what this instruction causes to happen inside the computer.

As you can see in Figure 5.11, the value of 2 is passed to the memory location whose address is contained in `pointer`. Thus, the value of this memory location is changed from 1 to 2. This does not affect the value stored in the variable `pointer`. This is demonstrated by the following program lines:

```
printf("New value stored in variable = %d\n",variable);
printf("Value stored in pointer = %d\n",pointer);
```

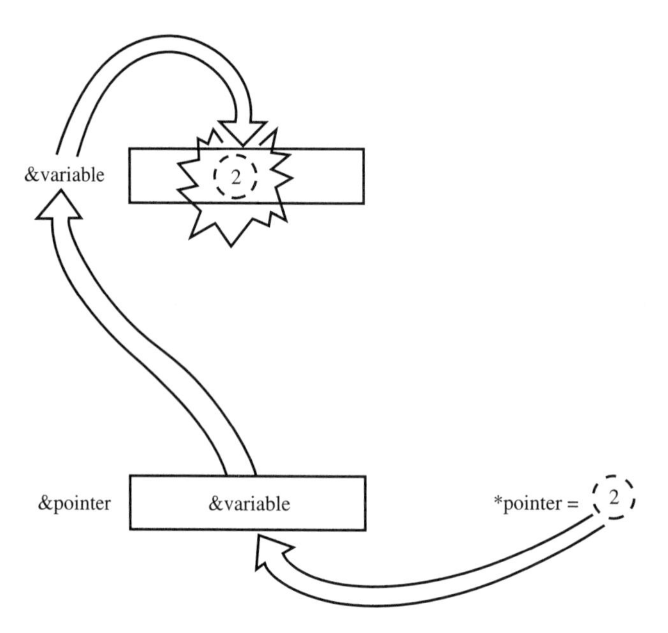

Figure 5.11 Pointer Assignment Action

The output produced by these lines is

```
New value stored in variable = 2
Value stored in pointer = 7732
```

Note that the value in `pointer` has not changed.

Some Examples

Example 5.6 offers some examples of work with pointers.

EXAMPLE 5.6

From the following program excerpt, determine the indicated values. In each case, the declared variables are

```
int number;
int *p;
```

(Assume that the address of `number` = 7735, and the address of `p` = 8364.) For each case below, determine the value of

a. `number` b. `&number` c. `p` d. `&p`

All of the results are cumulative.

A. `p` = 12; `number` = 5
B. `number` = `p`
C. `number` = `&p`
D. `p` = `&p`
E. `p` = `&number`
F. `*p` = 10

Solution

A. `number` = 5, `p` = 12, `&number` = 7735, `&p` = 8364,
`*p` = The data stored at memory location 12.

B. `number` = 12, `p` = 12, `&number` = 7735, `&p` = 8364,
`*p` = The data stored at memory location 12.

C. `number` = 8364, `p` = 12, `&number` = 7735, `&p` = 8364,
`*p` = The data stored at memory location 12.

D. `number` = 8364, `p` = 8364, `&number` = 7735, `&p` = 8364,
`*p` = The data stored at memory location 8364 which is 8364.

E. `number` = 8364, `p` = 7735, `&number` = 7735, `&p` = 8364,
`*p` = The data stored at memory location 7735 which is 8364.

F. `number` = 10, `p` = 7735, `&number` = 7735, `&p` = 8364,
`*p` = The data stored at memory location 7735 which is now 10.

All results are illustrated in Figure 5.12.

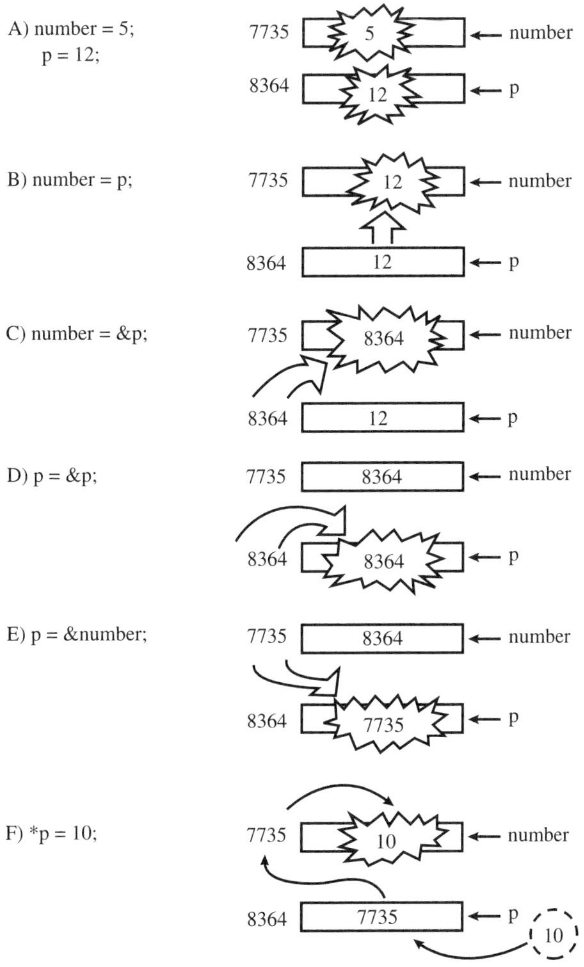

Figure 5.12 Solution for Example 5.6

Conclusion

This section presented an important introduction to the concept of pointers. Effectively, a pointer points to a memory location. Here you saw the differences among the value of a variable, its address, and a pointer that contains the address of the variable. Here you also saw how to change the value of a memory location by using a pointer. The concepts presented here will be used in the next section. For now, test your understanding of this section by trying the following section review.

5.3 Section Review

1. What is a pointer?
2. Why is a pointer called a pointer?
3. How does a pointer get the address of another memory location? Give an example.
4. How do you pass a value to a variable by using a pointer? What must you make sure of before doing this? Give an example.
5. How is a pointer declared?

5.4 Passing Variables

Discussion

This is another important section. Here you will see the application of concepts that were presented in the previous section. This section demonstrates another method of passing a value from a called function to the calling function—using pointers.

This is a very powerful tool in the design and structure of C programs. Again, devote adequate time to the study of this section. As with the previous section, you will find that entering and trying the short programs presented here will help reinforce these important concepts.

Basic Idea

Program 5.11 illustrates one way of passing a value from a called function back to the calling function. Note that if the function does not have any parameters, this is indicated by type `void`.

Program 5.11

```
#include <stdio.h>

int callme(void);    /* The function to be called by main.    */

main()
{
        int x;      /* Variable to receive a value from the called funcion. */

        x = callme();

        printf("Value of x is: %d",x);
```

```
        exit(0);
}

/*-----------------------------------------------------------------------*/

int callme(void)
{
        return(5);
}
```

Execution of Program 5.11 produces

```
Value of x is: 5
```

Essentially, all that Program 5.11 does is to get a value from a called function. This is illustrated in Figure 5.13.

As shown in Figure 5.13, the value assigned to the variable `x` in function `main()` is really received from the function `callme()`. This is done by the C `return(5)` contained in `callme()`. You have seen this method of returning a value to the calling function used before.

This procedure works fine for returning one value from a called function. It does not work for returning two or more values from a called function. As an example, suppose you were using a separate function to get the values of two or more variables from the program user and you wanted these values returned to the calling function. To return two or more values from a called function to the calling function, you need to use pointers in C.

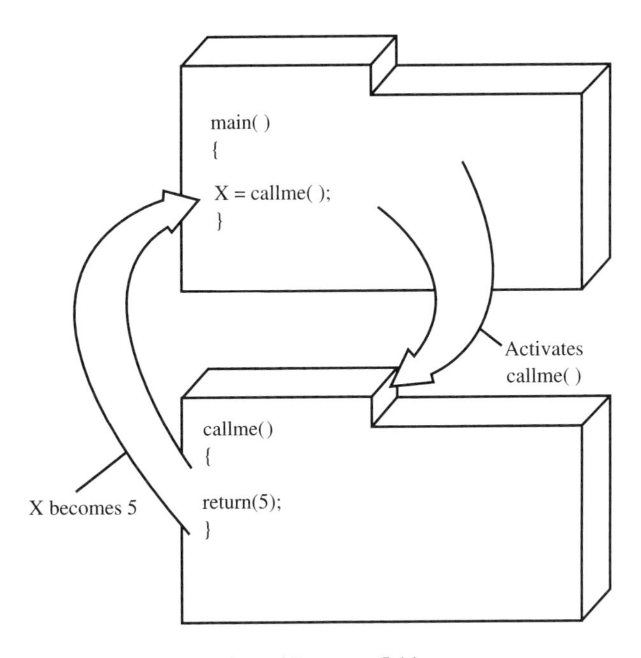

Figure 5.13 Operation of Program 5.11

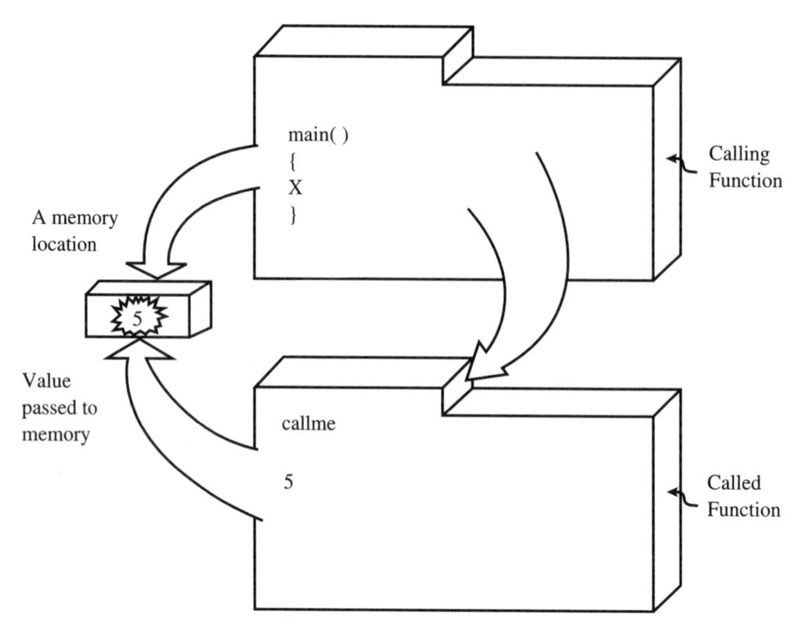

Figure 5.14 Concept of Passing a Value from a Called Function

Using Pointers

A pointer will now be used to return a single value back to the calling function. What will happen here is that a value of 5 will be assigned to a variable of the calling function by the called function, as shown in Figure 5.14. The final result is no different from what was done in Program 5.11. What happens now is that the same thing will be done using a pointer instead of the C `return()`.

To do this, the function `callme()` will use a pointer in its formal argument:

```
void callme(int *p);
```

It is helpful to think of this as if it sets up a memory location to act as a pointer—meaning that what will be stored there will be the address (memory location) of another variable. This concept is shown in Figure 5.15.

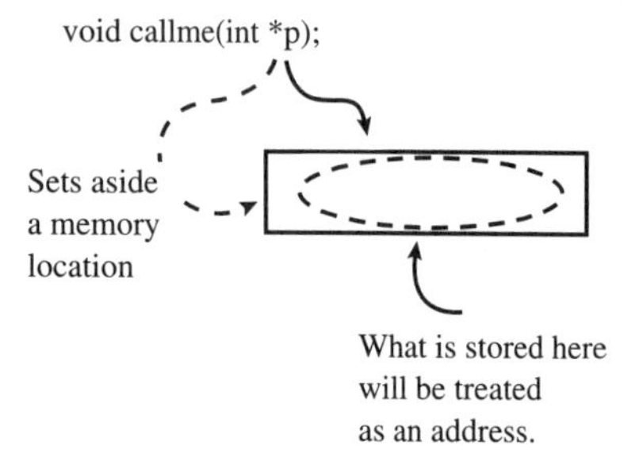

Figure 5.15 Setting a Pointer Argument

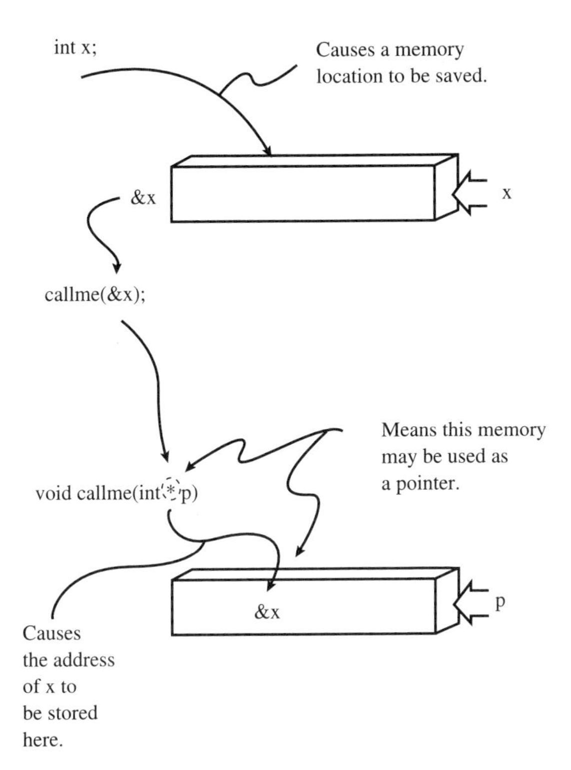

Figure 5.16 Action of Calling the Function

When the function `callme` is called, the address of the variable to be changed will be passed to it:

```
main()
{
      int x;
      callme(&x);
```

What has happened is shown in Figure 5.16.

When the function is called, it places the value 5 at the address pointed to by the pointer:

```
void callme(int *p)
{
      *p = 5;
}
```

This is shown in Figure 5.17.

The program that makes all of this happen is Program 5.12. Since the function `callme` does not return a value, it is declared a type `void`.

Execution of Program 5.12 yields

```
The value of x is 0
The new value of x is 5
```

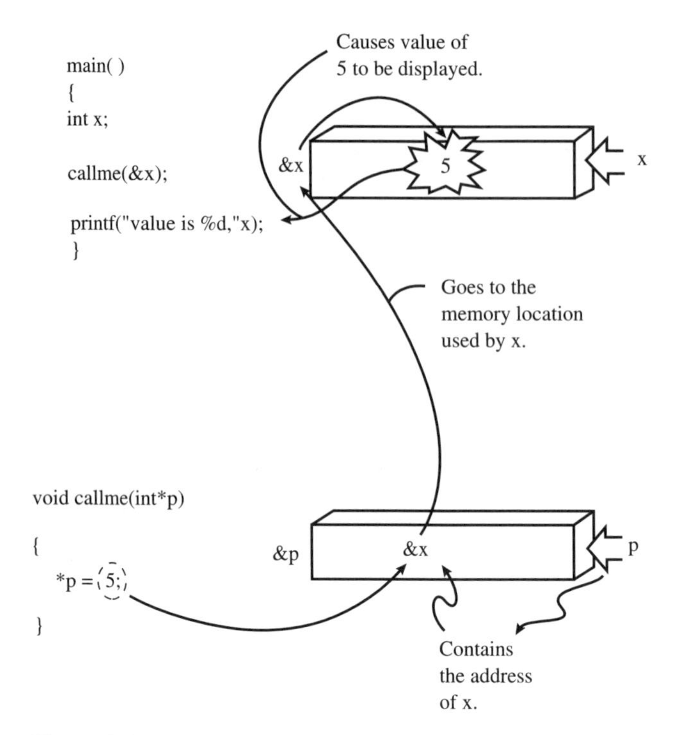

Figure 5.17 Action of Called Function

Program 5.12

```
#include <stdio.h>

void callme(int *p);

main()
{
        int x;

        x = 0;
        printf("The value of x is %d\n",x);

        callme(&x);
        printf("The new value of x is %d",x);
}

void callme(int *p)
{
        *p = 5;
}
```

This seems like a lot of effort just to assign a value of 5 to the variable x. However, this example demonstrates the passing of a value from one function to another using a pointer. What follows is an application of this principle.

Pointer Application

Consider Program 5.13. It calculates the current in a resistor given the value of the resistor and the voltage across it. The program uses `main()` to simply call functions. It is these called functions that do the work in the program. The function `main()` simply shows the structure of the program.

Program 5.13

```
#include <stdio.h>

void explain_program(void);
void get_values(float *r, float *v);
float do_calculations(float resistance, float voltage);
void display_answer(float current);

main()
{
        float resistor;
        float volts;
        float current;

        explain_program();
        get_values(&resistor, &volts);
        current = do_calculations(resistor,volts);
        display_answer(current);
}

/*-----------------------------------------------------------------------*/

void explain_program()
{
        printf("This program calculates the value of the current\n");
        printf("in amps. You need to enter the value of the resistor\n");
        printf("and the voltage in volts.\n");
}  /* end of explain_program() */

/*-----------------------------------------------------------------------*/

void get_values(float *r, float *v)
{
        float resistance;
        float voltage;

        printf("\n\nInput the resistance in ohms = ");
        scanf("%f",&resistance);
        printf("Input the voltage in volts = ");
        scanf("%f",&voltage);

        *r = resistance;
```

```
        *v = voltage;
}   /* end of get_values() */

/*------------------------------------------------------------------------*/

float do_calculations(float resistance, float voltage)
{
        float current;

        current = voltage/resistance;
        return(current);
}   /* end of do_calculations() */

/*------------------------------------------------------------------------*/

void display_answer(float current)
{
        printf("The value of the current is %f amps.",current);
}
```

Assuming that the program user enters a value of 10 for the voltage and 5 for the resistor, execution of Program 5.13 yields

```
This program calculates the value of the current
in amps. You need to enter the value of the resistor
and the voltage in volts.

Input the resistance in ohms = 5
Input the voltage in volts = 10
The value of the current is 2 amps.
```

Program Analysis

The first thing to note about Program 5.13 is how the functions are prototyped. The design of the program follows the typical sequence of an interactive technical program:

1. Explain program to user.
2. Get values from user.
3. Do the calculations.
4. Display the answer(s).

This is reflected in the names of the function prototypes:

```
void explain_program(void);
void get_values(float *r, float *v);
float do_calculations(float resistance, float voltage);
void display_answer(float current);
```

Observe that the function `get_values` uses two pointers in its formal argument. It does this because these values will have to be passed back to the calling function. As demonstrated at the beginning of this section, this can be done with pointers. When two or more variables are to be passed back, it *must* be done with pointers.* Note that only one of the function prototypes is not `void`: this is `float do_calculations`. The reason for this is that the function itself will return a single value—the value of the current. Since only one value needs to be returned, the C `return()` function can be used, and no pointers are necessary (though one could have been used instead in the formal argument).

The function `main()` declares three variables. These contain the values passed between the called functions:

```
main()
{
        float resistor;
        float volts;
        float current;
```

The body of `main()` now contains calls to the various functions.

```
explain_program();
get_values(&resistor, &volts);
current = do_calculations(resistor, volts);
display_answer(current);
```

Note in these calls that `get_values` contains the address operators in its actual arguments. This is necessary because the formal argument of this function identifies each of these to be a pointer. The addresses of the two variables must be passed to this function so that the values entered by the program user can be returned back to `main()`.

The first function called simply explains the purpose of the program to the program user:

```
void explain_program()
{
        printf("This program calculates the value of the current\n");
        printf("in amps. You need to enter the value of the resistor\n");
        printf("and the voltage in volts.\n");
}/* end of explain_program() */
```

The next function called, `get_values`, is the one that has the pointers in its formal argument. It also defines two variables that will be used to store the values entered by the program user:

```
void get_values(float *r, float *v)
{
        float resistance;
        float voltage;
```

The body of this function gets the values from the program user using the C `scanf`:

```
printf("\n\nInput the resistance in ohms = ");
scanf("%f",&resistance);
printf("Input the voltage in volts = ");
scanf("%f",&voltage);
```

*This refers to information presented up to this point in the text.

Now these values are assigned to the pointers. Since these pointers contain the addresses of actual arguments used in `main()`, the values entered by the program user are now actually being passed back to the two variables in the calling function `main()`.

```
*r = resistance;
*v = voltage;
```

Now that `main()` has the values entered by the program user, these are passed to the next function, where the actual calculation is performed:

```
float do_calculations(float resistance, float voltage)
{
        float current;

        current = voltage/resistance;
        return(current);
}/* end of do_calculations() */
```

Since there is only one value to return back to the calling function, this is done with the C `return()`.

The last called function simply has the actual value of the current passed to it so it may be displayed:

```
void display_answer(float current)
{

        printf("The value of the current is %f amps.",current);
}
```

Key Points

Even though Program 5.13 does only a simple division calculation that could have easily been done in your head, it illustrates an ideal structure of a C program. First, the function `main()` does nothing but call other functions. This is important because, for more complex programs, it allows you to see exactly what the structure of the program looks like.

The next point is that each called function does one specific task. If it needed to do more than one task, it would in turn call another function. Doing this makes a program easy to understand, debug, modify, and develop. You only need to develop one function at a time, test it, and then continue with the development of another function.

Conclusion

This section illustrated the passing of values through the use of pointers. You will use the concept of pointers throughout the remaining chapters of this text. Test your understanding of this section by trying the following section review.

5.4 Section Review

1. Name two ways of passing a value from a called function to the calling function.
2. What is the limitation of using the C `return` to pass a value back to the calling function?
3. State how more than one value can be returned to the calling function from the called function.

4. Describe the mechanism for passing values to the called function that uses pointers in its formal argument.
5. State why it is necessary to use separate functions.

5.5 Scope of Variables

Discussion

This section presents information about your C variables. Here you will be introduced to the concept of the relationships between your C functions and C variables. This information will help you understand what you should not do as well as what you should do when developing technical programs in C.

Local Variables

All of the variables declared in the programs up to this point have been local. The concept of a **local variable** is illustrated in Program 5.14. This program is attempting to use a variable in a second function that was declared only in the calling function.

Program 5.14

```
#include <stdio.h>

void other_function(void);

main()
{
        int a_variable;

        a_variable = 5;

        printf("The value of a_variable is %d",a_variable);
        other_function();
}

void other_function()
{
        printf("The value of a_variable in this function is %d",a_variable);
}
```

Program 5.14 will not compile because the second function does not know the meaning of the variable `a_variable` declared in the calling function `main()`. The reason for this is that when a variable is declared by you, within a function, it is known only to that function and none of the others. This means that the variable is local to the function in which it is declared. Another way of saying this is that the **scope** of a local variable is only within the function in which it is declared and the **life** of that variable lasts only while the function in which it is declared is active.

Global Variables

In order to make a variable known to all the functions in your C program, it must be declared ahead of `main()`. This is illustrated in Program 5.15. This program will compile because the variable is now known to both functions.

Program 5.15

```
#include <stdio.h>

void other_function(void);
int a_variable;

main()
{
        a_variable = 5;

        printf("The value of a_variable is %d\n",a_variable);
        other_function();
}

void other_function()
{
        printf("The value of a_variable in this function is %d",a_variable);
}
```

For Program 5.15, since the variable `a_variable` is declared outside the first function, it is now known to all the functions within the program. Execution of Program 5.15 will yield

```
The value of a_variable is 5
The value of a_variable in this function is 5
```

When a variable is declared in this fashion it is called a **global variable**. The scope of a global variable is every function within the program. The life of a global variable lasts as long as any part of the program is active. Figure 5.18 illustrates.

Caution with Global Variables

Using global variables can sometimes lead to unexpected results because any function can change the value of a global variable. Thus, when another function uses a global variable it may now have a value different from what you might expect. This is especially true when a recursive function uses a global variable.

It is considered good programming practice to keep your variables as local as possible. In this manner you are protecting these variables from being changed by other functions. This is why variables are passed between functions as arguments. When this is done, the value of each of these variables is protected.

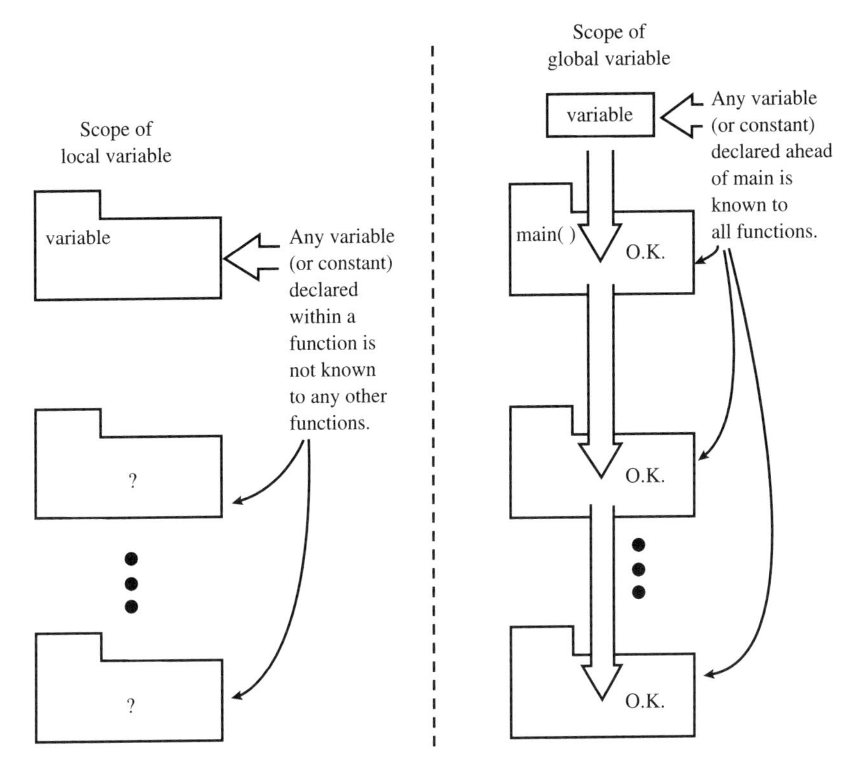

Figure 5.18 Concept of Scope

Again, it's a good rule to avoid global variables in your programs. They were introduced here so you would know what they look like. Program 5.16 illustrates one reason why you should be very cautious about using them.

Program 5.16

```
#include <stdio.h>

void other_function(void);
int counter;

main()
{
        counter = 0;

        while (counter < 5)
        {
                printf("The count is %d\n",counter);
                counter++;
        }

        other_function();
}
```

```
void other_function()
{
        while (counter < 6)
        {
                printf("The count in this function is %d\n",counter);
                counter++;
        }
}
```

When Program 5.16 is executed, the results are

```
The count is 0
The count is 1
The count is 2
The count is 3
The count is 4
The count in this function is 5
```

Observe that the output of the second function could be a surprise to the programmer (or someone else trying to read and understand the program). Since a global variable is used, it retains its counting value from the function `main()`. (Recall that the life of a global variable lasts as long as any part of the program is active.) Thus the count for the second function starts off with the value of the variable `counter` equal to 5. This could have a disastrous effect in larger programs where different functions are written by different programmers.

Conclusion

This section introduced the concept of the scope and life of program variables. What applies to variables also applies to program constants. You also saw the difference between local and global variables. Here you learned that it is considered good programming practice to keep your program variables as local as possible. Not doing so could produce undesirable results during program run time.

Check your understanding of this section by trying the following section review.

5.5 Section Review

1. What is meant by a local variable?
2. State the meaning of the term "scope" as applied to variables in C.
3. How is a global variable declared? A local variable?
4. Is it considered good programming practice to use global variables? Explain.
5. If global variables are not used, state how values may be passed between functions.

5.6 Variable Class

Discussion

This section presents various ways of classifying your data in C. Being able to do this gives you great flexibility in how you treat your variables. The material presented in this section makes use of the concepts introduced in the previous section.

Constants

You can think of one class of data as a constant. In C a constant is a value that never changes anywhere during program execution. It has a particular value assigned to it, and from then on, its value is never allowed to change.

C allows you to declare this type of data by using the keyword

```
const
```

This is illustrated in Program 5.17.

Program 5.17

```
#include <stdio.h>

const float TWO_PI = 6.28;

main()
{
        printf("The value of 2 times pi is %f",TWO_PI);
}
```

In Program 5.17, if you attempt to change the value of the global constant (`TWO_PI`) anywhere in the program, you will get a warning message during compile time. It is OK to use global constants because their values can never be changed. As a matter of fact, the use of global constants is considered good programming practice.

Automatic Variables

In C, the local variables that you have been using are called **automatic variables.** By default, any variable you have been specifying inside a function is treated as an automatic variable. If you wish to emphasize this (to show, for example, that you are overriding an external function definition) you may use the keyword

```
auto
```

This is demonstrated in Program 5.18.

Program 5.18

```
#include <stdio.h>

main()
{
        auto int value;

        value = 5;

        printf("The value is %d",value);
}
```

In Program 5.18, you could have left off the keyword `auto`, and nothing in the program would have changed as far as its operation is concerned.

External Variables

Variables of another class allowed by C are external. An **external variable** is similar to a global variable except that it may also have a life between different C files. (C file operations are presented in Chapter 9 of this text.) For now, you may consider an external variable to be one that has the entire program as its scope. The keyword of this class of variables is:

```
extern
```

Static Variables

In C, a **static variable** is a variable that is local. But, unlike a local, its life lasts as long as the program is active—even though its scope is only within the function that declared it. What this means is that a "normal" (automatic) local variable is completely forgotten by the computer once the function in which it was declared is no longer active. When the function is called again, the automatic variable no longer has the value it was left with when the function terminated.

This is not the case with a static variable. Its value is retained by the computer so that when the function that declared it is called again, its previous value is used. This concept is illustrated by Program 5.19.

Program 5.19

```
#include <stdio.h>

void second_function(void);

main()
{
        second_function();
        second_function();
        second_function();
}

void second_function()
{

        static int number;

        number++;
        printf("This function has been called %d times.\n",number);
}
```

Execution of Program 5.19 yields

```
This function has been called 1 times.
This function has been called 2 times.
This function has been called 3 times.
```

The important point about Program 5.19 is that the value of the `static` variable is remembered between calls to the function. It makes no difference from how many different functions this is done within the same program; its value will still be retained as long as the program is active. However, it is still a local variable known only to the function in which it was defined.

Register Variables

Another class of variable used in C is the **register variable**. This type of variable tells C that you want the quickest possible speed in the use of this variable. This can be accomplished by having the variable stored inside one of the registers in your computer's microprocessor rather than inside one of your computer's memory locations. Doing this saves time because the value of the variable does not have to be exchanged between your computer's microprocessor and its memory. You can only do this with a few variables during any one program. The reason for this is that your computer's microprocessor has a limited number of internal registers, and some of these will be saved for other needed types of processing. However, when you request a register variable, your computer will try to honor the request if possible; otherwise, the variable will be stored in a memory location.

The keyword for this is

```
register
```

Program 5.20 illustrates the use of a register variable.

Program 5.20

```
#include <stdio.h>

main()
{
        register counter;

        for (counter = 0; counter > 5; counter++);
}
```

Program 5.20 will execute faster with a `register` variable because there is no need to access memory each time to increment the value of the counter.

Overview

Table 5.3 presents the various classes of data available in C.

Conclusion

This section presented the various classes of data available to you in C. Here you also saw the effects of different C variable types. The section presented various programs to illustrate the effects of the different classes as well as the variable types. Check your understanding of this section by trying the following section review.

Table 5.3 Classes of Data

Type	Keyword	Where Declared	Scope	Life
Global	None	Ahead of all functions.	Entire program.	Until program ends.
Global	`extern`	Ahead of all functions within a file.	All files including other files where declared extern.	While any of these files are active.
Local	None or `auto`	Within the function.	Only within the function where it is declared.	Until function is no longer active.
Local	`register`	Within the function.	Only within the function where it is declared.	Until function is no longer active.
Local	`static`	Within the function.	Only within the function where it is declared.	Until program ends.

5.6 Section Review

1. What effect does the C keyword `const` have when used to identify a data type?
2. State what is meant by an automatic variable.
3. State what is meant by a static variable.
4. What is a register variable?

5.7 Example Programs

In this section we will look at a few more examples of how pointers can be used to access data.

In Program 5.21, the bitwise AND operation is used in combination with the shift-right operation to convert a decimal integer into binary. A binary *mask* pattern is used to select one bit at a time from the binary representation of the input number. The value of the selected bit (0 or 1) is displayed. The bit mask begins at 80 hexadecimal (0x80 in C notation), which is 10000000 binary. This pattern is used to check the most significant bit. The mask pattern is shifted right one bit position each time through the `while` loop. Thus, after eight passes, each bit in the input number has been selected by the bit mask. Note that the address of the number is passed to `tobin()`.

Program 5.21

```
#include <stdio.h>

void tobin(char *value);

main()
{
        unsigned char number;    /* Storage for user input number.   */
```

```
        printf("Enter a number from 0 to 255 => ");
        scanf("%d",&number);
        printf("The number %d in binary is ",number);
        tobin(&number);
}

void tobin(char *value)
{
        unsigned char position = 0x80;
        char temp;

        while (position != 0)
        {
                temp = *value & position;
                (temp == 0) ? printf("0 ") : printf("1 ");
                position = position >> 1;
        }
}
```

A sample execution of Program 5.21 is as follows:

```
Enter a number from 0 to 255 => 143
The number 143 in binary is 1 0 0 0 1 1 1 1
```

This is yet another technique for performing this type of conversion. Others we have already seen involved the use of recursion and remainder division.

Program 5.22 shows how a simple sequence of binary patterns can be generated and displayed.

Program 5.22

```
#include <stdio.h>

void tobin(char *value);

main()
{
        unsigned char pattern = 0xe0;    /* Initial pattern.  */
        int counter,loops = 0;

        while (20 > loops++)
        {
                for (counter = 1; counter <= 5; counter++)
                {
                        tobin(&pattern);
                        printf("\n");
                        pattern = pattern >> 1;
                }
                for (counter = 1; counter <= 5; counter++)
                {
```

```
                        tobin(&pattern);
                        printf("\n");
                        pattern = pattern << 1;
                }
        }
}

void tobin(char *value)
{
        unsigned char position = 0x80;
        char temp;

        while (position != 0)
        {
                temp = *value & position;
                (temp == 0) ? printf("0 ") : printf("1 ");
                position = position >> 1;
        }
}
```

Imagine that a suitable output circuit exists for your computer that allows eight LEDs to be controlled by the bits displayed on the output screen. For example, if the program has just outputted the pattern

```
0 0 1 1 1 0 0 0
```

then we have the following LED activity:

```
off off on on on off off off
```

Program 5.22 uses two loops together with the left- and right-shift operations to produce an interesting sequence of output patterns. A sample from the sequence looks like this:

```
1 1 1 0 0 0 0 0
0 1 1 1 0 0 0 0
0 0 1 1 1 0 0 0
0 0 0 1 1 1 0 0
0 0 0 0 1 1 1 0
0 0 0 0 0 1 1 1
0 0 0 0 1 1 1 0
0 0 0 1 1 1 0 0
0 0 1 1 1 0 0 0
0 1 1 1 0 0 0 0
1 1 1 0 0 0 0 0
```

Although Program 5.27 gives a simplified example of how bit operations can be used to turn lights on and off, it should be easy to extend the idea to that of a traffic control system, where each bit controls one of the red, yellow, or green lights on a traffic signal.

The third example, Program 5.23, shows how a *pseudo*-random sequence of numbers can be generated.

Program 5.23

```
#include <stdio.h>

void rand(unsigned char *random);

main()
{
        unsigned char random;        /* Random number storage.  */
        int counter;

        printf("Enter random number seed (0 to 255) => ");
        scanf("%d",&random);
        for (counter = 1; counter <= 10; counter++)
        {
                rand(&random);
                printf("The next random number is %d\n",random);
        }
}

void rand(unsigned char *random)
{
        unsigned char temp1,temp2;  /* Used in calculations.   */

        temp1 = *random & 0x80;
        temp2 = *random & 0x04;
        *random <<= 1;
        if ((0 == temp1 | temp2) || (0x84 == temp1 | temp2))
                *random = *random | 0x01;
}
```

Remember that all integers have a unique binary representation. If we fiddle with the bits in the representation we get a new number. For example, begin with any 8-bit binary number. Clear one of the bits, set one of the bits, and complement a third bit in the number. You will always get a different number in the end. Program 5.23 attempts to generate a random sequence of integers using a similar technique. A random number *seed* is supplied by the user. Then ten random numbers are generated using the following technique:

1. Examine the most significant bit using pattern 0x80.
2. Examine bit 2 (the LSB is bit 0) using pattern 0x04.
3. Shift number one bit position to the left.
4. If the two bits examined are equal, set the LSB to 1.

An execution with a starting seed of 34 (00100010 binary) gives the following results:

```
Enter random number seed (0 to 255) => 34
The next random number is 69
The next random number is 139
The next random number is 22
The next random number is 45
The next random number is 91
The next random number is 183
The next random number is 111
```

```
The next random number is 223
The next random number is 191
The next random number is 127
```

The sequence of numbers displayed *looks* random, although it was actually generated in a systematic way. Try starting seeds of 73, 118, and 137 yourself and observe the overall results. Are there limitations to this technique?

Conclusion

In this section we saw additional examples of how data is passed between functions by the use of pointers. Check your understanding of this material with the following section review.

5.7 Section Review

1. Why do you think `scanf()` requires an address for the number scanned?
2. Is the value of the `pattern` variable in Program 5.22 changed inside the `tobin()` function?
3. Are pointers really necessary in Program 5.23?

5.8 Troubleshooting Techniques

Problems in C Programming

This section presents some of the most common errors encountered by C programmers when using pointers. Bringing these to your attention now may save you hours of frustration later. As you will discover, the unique characteristics of C open the door for some specific programming problems that may, at first, be difficult to spot.

Pointers

Understandably, many of those who are new to C have some difficulty with the concept and application of pointers. As a first step in understanding pointers, you should make sure you understand the differences among the various forms of the pointer variable. Table 5.4 summarizes these important differences.

Table 5.4 Pointer Nomenclature

Given: `int *ptr;`
`int data;`
[It then follows that:]

Operation	Operation	Meaning
Address Operator	`&data`	Address of the variable `data`
Assignment	`ptr = &data;`	Places the address of `data` into the variable `ptr`
Variable	`ptr`	Value of the variable `ptr` which now contains the address of `data`
Indirection Operator	`*ptr`	Value in the variable `data`

The indirection operator (*) accesses a value indirectly, through a pointer. The operand must be a pointer value.

It is often helpful to try short programs that work with the concepts presented in Table 5.4. The interactive exercise section for this chapter contains several of these programs.

Not Initializing Pointers

A common error made by those new to C is the failure to initialize pointers. **Initialization** means that you must know what value is in the pointer before you use it because this value will be the actual memory location that the pointer will be pointing to. As an example, the following program excerpt is not correct:

```
main()
{
int *ptr;

        *ptr = 5;
}
```

Here the programmer is attempting to assign a value of 5 to the memory location pointed to by the pointer. However, no one (including the programmer) knows where in the computer memory the value of 5 will be stored. Doing this in small programs may have no noticeable effect. However, in larger programs, this could have a disastrous effect because the number 5 could be placed in a memory location that is being used by your program for other necessary items. The point to remember here is that you should always know where the pointer is pointing before using it as a pointer.

Forgetting to Use the Address Operator

Another common error is forgetting to pass values by address. This is illustrated in the following program excerpt:

```
main()
{
        int input;

        printf("Give me a whole number => ");
        scanf("%d",input);

        printf("The value you entered was %d",input);

}
```

In this excerpt, the value entered by the program user will not do what is expected. The `scanf` function requires that you use the address operator `&` in order to get the entered value returned to the calling function. For the program excerpt above, the `scanf` function must be changed to

```
scanf("%d",&input);
```

Now you will get the expected results.

The same problem occurs when you define your own functions that must return values back to the calling function. As with the `scanf` function, you must use the address operator `&` as the actual parameter.

Conclusion

This section presented some of the most common programming pitfalls encountered by beginning C programmers. A table was presented to emphasize the meanings of the different aspects of using pointers. A reminder of how to pass by address was also presented. Check your understanding of this section by trying the following section review.

5.8 Section Review

1. What is the address operator in C? What does it do?
2. What is the indirection operator in C? What does it do?
3. What happens if you forget to use the address operator?

5.9 Case Study: Electronic Sketchpad

Discussion

This case study illustrates a simple pointer application. Through the use of a pointer into the video screen memory area of the personal computer, it is possible to place characters and symbols of various colors onto the display screen. The pointer is used to write data into the screen memory. The data is then displayed by the video hardware. By manipulating the pointer, we are able to control where the data is placed into memory. This allows us to create simple sketches using red, green, or blue asterisks. This is the essence of the Electronic Sketchpad.

Background Information

The personal computer contains a reserved memory area that is accessed by special video hardware in such a way that each character on the video screen is represented in memory by a pair of bytes. The first byte stores the ASCII character that will be displayed on the screen. The second byte stores the *attribute* of the character. The attribute byte controls the colors used to display the character. Figure 5.19 shows an example of how a red capital A on a black background is stored in memory.

The video screen is composed of many of these two-byte pairs. For example, a display screen containing 25 rows of characters and 80 characters on each line requires 2000 pairs, or a total of 4000 bytes. The first two locations (0 and 1) control the character in the upper left corner of the display. The last two bytes (locations 3998 and 3999) control the character in the lower right corner. We can use a pointer to access the memory locations where these bytes are stored and write new character and attribute data. The pointer value, which controls where the new data is written, will be updated by the user in the Electronic

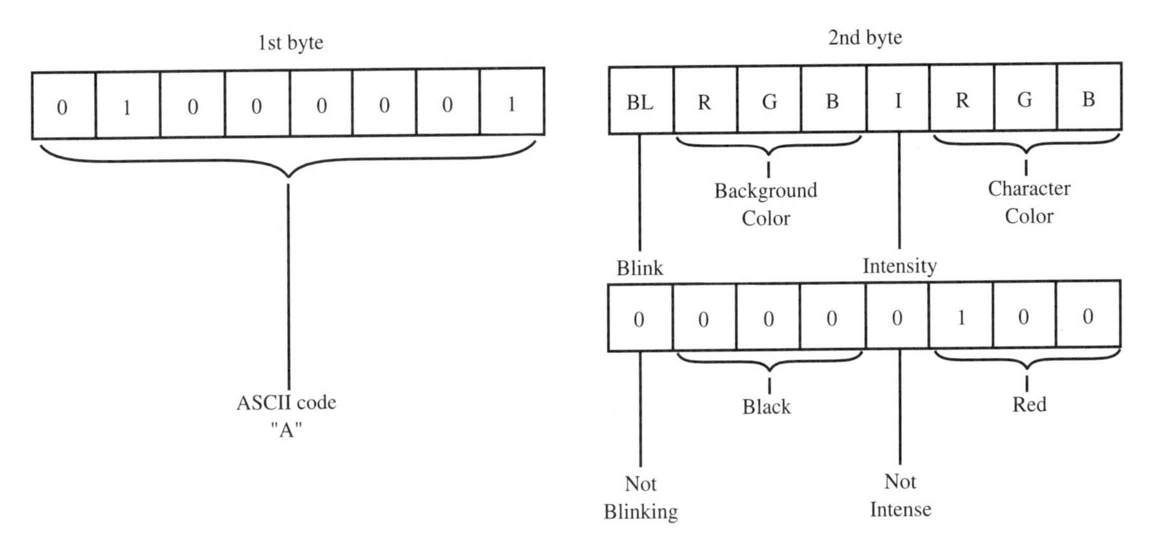

Figure 5.19 Character/Attribute Bytes

Sketchpad. This will be accomplished by assigning commands (such as move or change color) to specific keyboard keys. Pressing the up arrow, for example, causes the pointer to be changed so that it can access memory in the previous row of characters.

One important point to keep in mind is that we must be careful in the way the pointer is moved around in memory. For example, what happens when we try to move the pointer to line 26 on a 25 line display? Writing to an improper pointer location may cause unpredictable results in the personal computer. So, it is best to "keep inside the lines," and perform some type of *boundary checking* on the pointer to prevent it from ever reaching an illegal value.

The Problem

Before we can use the pointer to update video memory, it must be initialized to some location inside the block of locations making up the display memory. The problem is that the personal computer uses over one million locations when running DOS. Where in these one million locations does the 4000 byte block of screen memory reside? Fortunately for us, this information is available in most DOS references: the starting address of the video display memory is 0xB8000. To use this type of address we will need to employ a **`far` pointer,** a type of pointer in C that is tailored to the adressing scheme used in the personal computer. The statement used to declare our video pointer looks like this:

```
unsigned char far *vbase = (char far *)0xB80000001;
```

The `(char far *)` specifier requires the address of the video memory to be written as a pair of four-digit hexadecimal values. The first value, called the *segment address,* is 0xB800. The second value is the *offset* into the segment. The offset is 0x0000, which places the pointer `*vbase` at the first location in the video memory segment.

Developing the Algorithm

Sketching a drawing on the video screen will require a good deal of interaction from the user. By pressing certain keys on the keyboard, the user will be able to control the sketchpad. The user will be able to do all of the following:

- change the color to red, blue, or green
- switch from drawing mode to skip mode and back
- move the drawing element (an asterisk)
- quit

Single letter commands from the keyboard can be used to represent these operations. A function will be needed to read the keyboard and check for a valid command. This command key will be processed by another function to perform the action of the command. A loop allowing unlimited commands to be entered will only be exited when the player quits.

Furthermore, it is nice to start a sketch with a "blank sheet," or in the case of the personal computer, a blank display screen. This will require all pairs to be loaded with the ASCII code for a blank (0x20). Program 5.24 shows the initial coding of the Electronic Sketchpad.

Program 5.24

```
#include <stdio.h>
#include <conio.h>

/*******************************************************************/
/*                    Electronic Sketchpad                         */
/*******************************************************************/
/*                  Developed by Ann Artist                        */
/*                                                                 */
/*******************************************************************/
/*      This program allows the operator to draw a pattern of      */
/*      asterisks on the screen in red, green, and blue. The       */
/*      screen is updated using a far pointer to the video display */
/*      buffer.                                                    */
/*******************************************************************/
/*                    Function Prototypes                          */
/*-----------------------------------------------------------------*/

/* This function explains the operation of the program to the user. */
/*-----------------------------------------------------------------*/

void directions(void);

/* This function clears the video screen. */
/*-----------------------------------------------------------------*/

void clear(void);

/* This function gets a valid key from the user. */
/*-----------------------------------------------------------------*/
```

```
char getkey(void);

/* This function processes the key pressed by the user. */
/*-----------------------------------------------------------------*/

void process_key(char command);
/* This function moves the asterisk on the screen. */
/*-----------------------------------------------------------------*/

void xymove(void);

/*****************************************************************/
/*                         Global Variables                      */
/*-----------------------------------------------------------------*/

        /* Pointer to video screen memory. */
        unsigned char far *vbase = (char far *)0xB8000000l;

/*-----------------------------------------------------------------*/

main()
{
        char key;                         /* User selection. */

        directions();                     /* Explain program to user. */
        clear();                          /* Clear the screen. */
        do
        {
                key = getkey();
                if (key != 'Q')
                process_key(key);
        } while (key != 'Q');
}

/*-----------------------------------------------------------------*/

void directions()
{
        printf("\n\nKey\t\tFunction\n");
        printf("R\t\tDraw with red *\n");
        printf("G\t\tDraw with green *\n");
        printf("B\t\tDraw with blue *\n");
        printf("S\t\tEnter skip mode\n");
        printf("D\t\tEnter draw mode\n");
        printf("arrow\t\tChange position\n");
        printf("Q\t\tQuit\n");
        printf("\nPress any key to begin...\n");
        getch();
}

/*-----------------------------------------------------------------*/
```

```
void clear()
{
}

/*------------------------------------------------------------------*/

char getkey()
{
}

/*------------------------------------------------------------------*/

void process_key(char command)
{
}

/*------------------------------------------------------------------*/

void xymove()
{
}
```

Note the simplicity of the do loop used in main():

```
main()
{
     char key;                       /* User selection. */

     directions();                   /* Explain program to user. */
     clear();                        /* Clear the screen. */
     do
     {
          key = getkey();
          if (key != 'Q')
               process_key(key);
     } while (key != 'Q');
}
```

The user is given directions, the screen is cleared, and the do loop processes each keypress until 'Q' is encountered. Let us see how each function does its job.

directions()

The user is provided with directions that explain what keys to use on the keyboard. The user is allowed to view the directions indefinitely.

```
void directions()
{
     printf("\n\nKey\t\tFunction\n");
     printf("R\t\tDraw with red *\n");
     printf("G\t\tDraw with green *\n");
```

```
    printf("B\t\tDraw with blue *\n");
    printf("S\t\tEnter skip mode\n");
    printf("D\t\tEnter draw mode\n");
    printf("arrow\t\tChange position\n");
    printf("Q\t\tQuit\n");
    printf("\nPress any key to begin...\n");
    getch();
}
```

The `getch()` function waits for any key to be pressed.

`clear()`

The screen is cleared by writing the ASCII code for a blank character (0x20) into each character byte position in the video memory. This is easily accomplished by using a nested loop.

```
void clear()
{
    int r,c;

    /* Write the ASCII code for blank into each character */
    /* position on the screen. */

    for(r = 0; r < 25; r++)
        for(c = 0; c < 80; c++)
        {
            vbuff(r,c,0) = BLANK;
            vbuff(r,c,1) = WHITE;
        }
}
```

The `r` (row) and `c` (column) variables are used to control the position accessed within video memory. The actual address is generated by `vbuff()`, which is defined as follows:

```
/* Screen position calculation. */
#define  vbuff(r,c,n)   *(vbase + (r * 160) + (c * 2) + n)
```

An entire row of 80 characters requires 160 bytes, and each character requires two bytes (ASCII code and attribute). The offset into the video memory is found by multiplying `r` and `c` by 160 and 2, respectively, and adding the products plus the offset `n` to the base video address. For example, to access the character in column 25 of row 10, we use `vbuff(10,25,0)`. It is convenient to use these parameters to refer to a location in screen memory, instead of its actual address (0xB8000000 + (10 * 160) + (25 * 2) + 0). This (r,c,n) representation has the added advantage of working on different screen sizes (25 by 40, for example) without having to change any of the function code. Just the `#define` for `vbuff()` would have to be modified.

`getkey()`

The user presses keyboard keys to control the Electronic Sketchpad. If the key pressed by the user is a lowercase character, it is converted into uppercase. This allows the user to enter g or G for example, and not worry about having to hit their CAPS-LOCK key. The getkey() function takes care of this step.

```
char getkey()
{
     char kyb;

     kyb = getch();            /* Get a key from the user. */
     if ((kyb >= 'a') && (kyb <= 'z'))        /* Is key lowercase? */
          kyb &= 0xdf;                        /* Convert to uppercase. */
     switch(kyb)
     {
          case 'R' : return('R');
          case 'G' : return('G');
          case 'B' : return('B');
          case 'S' : return('S');
          case 'D' : return('D');
          case 'Q' : return('Q');
          case  SPECIAL : dir = getch();      /* Read extended key code. */
                  return('M');                /* Check direction. */
          default  : return('X');             /* Not a valid key. */
     }
}
```

The if statement checks to see if kyb contains a lowercase character. If so, the sixth bit of the ASCII code stored in kyb is cleared, which converts a lowercase character code into the respective uppercase code.

If the user presses one of the four arrow keys, getch() returns a zero instead of an ASCII code. In this case it is necessary to call getch() a second time to read the *extended key code* of the arrow key that was pressed. A number of #defines are used to represent these special values:

```
#define SPECIAL 0     /* Extended key code. */
#define UP      72    /* Key code for up arrow. */
#define DOWN    80    /* Key code for down arrow. */
#define LEFT    75    /* Key code for left arrow. */
#define RIGHT   77    /* Key code for right arrow. */
```

The directional #defines are the values returned by getch() and stored in dir when an arrow key is pressed. dir is a global variable used to control how the video pointer is updated.

If the user does not enter a valid key, getkey() returns an 'X'. Note that only uppercase characters are returned by getkey().

process_key()

The key codes returned by getkey(), which represent sketchpad commands, are processed by process_key().

```
void process_key(char command)
{
     switch(command)              /* Command is users key code. */
     {
```

```
            case 'R' : color = RED;
                       vbuff(r,c,1) = color;
                       break;
            case 'G' : color = GREEN;
                       vbuff(r,c,1) = color;
                       break;
            case 'B' : color = BLUE;
                       vbuff(r,c,1) = color;
                       break;
            case 'S' : skipflag = TRUE; break;
            case 'D' : skipflag = FALSE; break;
            case 'M' : xymove(); break;
      }
}
```

Two new global variables are used to help control how the asterisk is placed and moved. `color` keeps track of the color attribute stored with the asterisk. `skipflag` is a Boolean variable used by `xymove()` that determines if the asterisk is temporarily or permanently written as it is moved.

xymove()

When `getkey()` returns an `'M'` as the result of the user pressing an arrow key, `process_key()` calls `xymove()` to move the asterisk. Moving the asterisk requires a number of steps:

1. If `skipflag` is true, restore the previous contents of the screen position. The character and attribute of the previous contents are saved in the global variables `oldchar` and `oldcolor`.
2. Perform boundary checking. Limit row values to the range 0...24. Limit column values to the range 0...79. The row and column of the asterisk's position are stored in the globals `r` and `c`.
3. Read the character and attribute at the new screen position. Save them in `oldchar` and `oldcolor`.
4. Write a colored asterisk into the new screen position.

These steps are accomplished as follows:

```
void xymove()
{
      if (skipflag)              /* In skip mode? */
      {
            /* Restore previous character and color. */
            vbuff(r,c,0) = oldchar;
            vbuff(r,c,1) = oldcolor;
      }
      switch(dir)
      {
            /* Check for screen edges. */
            case UP    : r = (r > 0) ? r - 1 : 0; break;
            case DOWN  : r = (r < 24) ? r + 1 : 24; break;
            case LEFT  : c = (c > 0) ? c - 1 : 0; break;
```

```
                case RIGHT : c = (c < 79) ? c + 1 : 79; break;
            }

            /* Save a copy of current character and color. */
            oldchar = vbuff(r,c,0);
            oldcolor = vbuff(r,c,1);

            /* Display colored asterisk. */
            vbuff(r,c,0) = '*';
            vbuff(r,c,1) = color;
        }
```

Note that `vbuff()` is used to both read and write the screen.

Final Program

Program 5.25 shows how the `#defines`, global variables, and function definitions are all combined to make the Electronic Sketchpad.

Program 5.25

```
#include <stdio.h>
#include <conio.h>

#define FALSE   0
#define TRUE    1
#define SPECIAL 0     /* Extended key code. */
#define UP      72    /* Key code for up arrow. */
#define DOWN    80    /* Key code for down arrow. */
#define LEFT    75    /* Key code for left arrow. */
#define RIGHT   77    /* Key code for right arrow. */
#define BLANK   32    /* ASCII code for blank. */
#define BLUE    1     /* Blue attribute value. */
#define GREEN   2     /* Green attribute value. */
#define RED     4     /* Red attribute value. */
#define WHITE   7     /* White attribute value. */

/* Screen position calculation. */
#define  vbuff(r,c,n)  *(vbase + (r * 160) + (c * 2) + n)

/*******************************************************************/
/*                       Electronic Sketchpad                      */
/*******************************************************************/
/*                     Developed by Ann Artist                     */
/*                                                                 */
/*******************************************************************/
/*     This program allows the operator to draw a pattern of       */
/*     asterisks on the screen in red, green, and blue. The        */
/*     screen is updated using a far pointer to the video display  */
/*     buffer.                                                     */
/*******************************************************************/
```

```
/*                        Function Prototypes                         */
/*------------------------------------------------------------------*/

/* This function explains the operation of the program to the user. */
/*------------------------------------------------------------------*/

void directions(void);

/* This function clears the video screen. */
/*------------------------------------------------------------------*/

void clear(void);

/* This function gets a valid key from the user. */
/*------------------------------------------------------------------*/

char getkey(void);

/* This function processes the key pressed by the user. */
/*------------------------------------------------------------------*/

void process_key(char command);

/* This function moves the asterisk on the screen. */
/*------------------------------------------------------------------*/

void xymove(void);

/******************************************************************/
/*                        Global Variables                          */
/*------------------------------------------------------------------*/

        /* Pointer to video screen memory. */
        unsigned char far *vbase = (char far *)0xB8000000l;

        int r,c;                 /* Current position of asterisk. */
        unsigned char color;     /* Color attribute for asterisk. */
        int skipflag;            /* False means draw, true means skip. */
        char dir;                /* Direction to move asterisk. */
        unsigned char oldchar;   /* Original character at current position. */
        unsigned char oldcolor;  /* Original color attribute. */

/*------------------------------------------------------------------*/

main()
{
        char key;                         /* User selection. */

        directions();                     /* Explain program to user. */
        clear();                          /* Clear the screen. */
        skipflag = TRUE;                  /* Initial mode is skip. */
```

```
        color = GREEN;                  /* Initial color is green. */
        r = 12;                         /* Initial row is 12. */
        c = 40;                         /* Initial column is 40. */
        vbuff(r,c,0) = '*';             /* Display asterisk and color. */
        vbuff(r,c,1) = color;
        oldchar = BLANK;                /* Initial update character. */
        oldcolor = WHITE;
        do
        {
                key = getkey();
                if (key != 'Q')
                        process_key(key);
        } while (key != 'Q');
}

/*------------------------------------------------------------------*/

void directions()
{
        printf("\n\nKey\t\tFunction\n");
        printf("R\t\tDraw with red *\n");
        printf("G\t\tDraw with green *\n");
        printf("B\t\tDraw with blue *\n");
        printf("S\t\tEnter skip mode\n");
        printf("D\t\tEnter draw mode\n");
        printf("arrow\t\tChange position\n");
        printf("Q\t\tQuit\n");
        printf("\nPress any key to begin...\n");
        getch();
}

/*------------------------------------------------------------------*/

void clear()
{
        int r,c;

        /* Write the ASCII code for blank into each character */
        /* position on the screen, and set color to white. */

        for(r = 0; r < 25; r++)
                for(c = 0; c < 80; c++)
                {
                        vbuff(r,c,0) = BLANK;
                        vbuff(r,c,1) = WHITE;
                }
}

/*------------------------------------------------------------------*/

char getkey()
{
```

```
        char kyb;

        kyb = getch();          /* Get a key from the user. */
        if ((kyb >= 'a') && (kyb <= 'z'))       /* Is key lowercase? */
                kyb &= 0xdf;                    /* Convert to uppercase. */
        switch(kyb)
        {
                case 'R' : return('R');
                case 'G' : return('G');
                case 'B' : return('B');
                case 'S' : return('S');
                case 'D' : return('D');
                case 'Q' : return('Q');
                case  SPECIAL : dir = getch();  /* Read extended key code. */
                           return('M');         /* Check direction. */
                default   : return('X');        /* Not a valid key. */
        }
}

/*------------------------------------------------------------------*/

void process_key(char command)
{
        switch(command)             /* Command is user's key code. */
        {
                case 'R' : color = RED;
                           vbuff(r,c,1) = color;
                           break;
                case 'G' : color = GREEN;
                           vbuff(r,c,1) = color;
                           break;
                case 'B' : color = BLUE;
                           vbuff(r,c,1) = color;
                           break;
                case 'S' : skipflag = TRUE; break;
                case 'D' : skipflag = FALSE; break;
                case 'M' : xymove(); break;
        }
}

/*------------------------------------------------------------------*/

void xymove()
{
        if (skipflag)           /* In skip mode? */
        {
                /* Restore previous character and color. */
                vbuff(r,c,0) = oldchar;
                vbuff(r,c,1) = oldcolor;
        }
        switch(dir)
        {
```

```
            /* Check for screen edges. */
            case UP    : r = (r > 0) ? r - 1 : 0; break;
            case DOWN  : r = (r < 24) ? r + 1 : 24; break;
            case LEFT  : c = (c > 0) ? c - 1 : 0; break;
            case RIGHT : c = (c < 79) ? c + 1 : 79; break;
    }

    /* Save a copy of current character and color. */
    oldchar = vbuff(r,c,0);
    oldcolor = vbuff(r,c,1);

    /* Display colored asterisk. */
    vbuff(r,c,0) = '*';
    vbuff(r,c,1) = color;
}
```

The `main()` function has additional statements that are needed to initialize the global variables and the initial screen settings. A green asterisk is placed at position (12,40), which is the center of the 25 by 80 display screen. The character/attribute stored in that position is saved in the `oldchar` and `oldcolor` variables.

Figure 5.20 is a screen dump of a cube drawn using the Electronic Sketchpad. Even with only one character (the asterisk) available for drawing, it is still possible to create interesting and meaningful displays.

Conclusion

In this section we saw how a pointer to video memory could be manipulated and used to place colored characters onto the display screen. Check your understanding with the following section review.

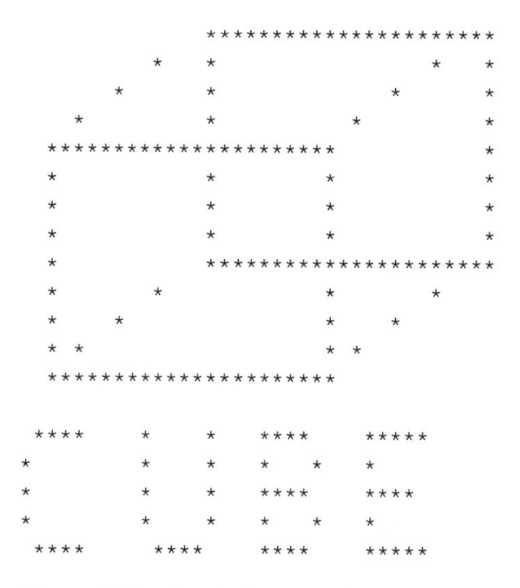

Figure 5.20 Simple Cube Drawn with the Electronic Sketchpad

5.9 Section Review

1. Why are global variables used to keep track of eveything? How else might all the information be handled?
2. What two methods are used to send information back from `getkey()`?
3. Is the video memory pointer `*vbuff` ever changed once it is initialized?
4. How are edge boundaries enforced in `xymove()`?
5. What changes are necessary to utilize a 25 by 40 display?

Interactive Exercises

DIRECTIONS

These exercises require that you have access to a computer and software that supports C. They are provided here to give you valuable experience and immediate feedback on what the concepts and commands introduced in this chapter will do.

Exercises

1. In Program 5.26, which of the displayed values are predictable? Try the program to check your answer.

Program 5.26

```
#include <stdio.h>

main()
{
        int data = 15;
        int *point_to;

        point_to = &data;

        printf("%d\n", point_to);
        printf("%d\n", &point_to);
        printf("%d\n", *point_to);
}
```

2. Program 5.27 has a global constant. Predict what the output will be, then try the program to test your prediction.

Program 5.27

```
#include <stdio.h>

const int a_value = 15;
int function(void);

main()
{
        int a_value = 12;
```

```
        printf("%d\n",a_value);
        printf("%d\n",function());
}

int function(void)
{
        return(a_value);
}
```

3. Will Program 5.28 compile? If not, what needs to be changed?

Program 5.28

```
#include <stdio.h>

        int a = 5;

void abc(int b);

main()
{
        int c = 10;

        abc(c);
}

void abc(int b)
{
        printf("a equals %d",a);
        printf("b equals %d",b);
        printf("c equals %d",c);
}
```

Self-Test

DIRECTIONS

Program 5.29 was developed to illustrate many of the key concepts presented in this chapter. The questions for this Self-Test pertain to this program.

Program 5.29

```
#include <stdio.h>

void function_1(void);
void function_2(char *letter1, char *letter2);
int AND_bits(int this_byte, int that_byte);

const unsigned int number_1 = 57532;
```

```
int *look_at;

main()
{
        int memory_location_1;
        char this_value;
        char that_value;
        int result;

        look_at = &memory_location_1;
        function_1();

        this_value = 'a';
        that_value = 'b';
        function_2(&this_value, &that_value);

        result = AND_bits(15,15);

        printf("The constant is %u.\n",number_1);
        printf("The value of memory_location_1 is %d\n",memory_location_1);
        printf("The contents of this_value = %c\n",this_value);
        printf("The result is %X.\n",result);
}

void function_1()
{
        *look_at = 375;
}

void function_2(char *letter1, char *letter2)
{
        *letter1 = 'b';
        *letter2 = 'a';
}

int AND_bits(int this_byte, int that_byte)
{
        return(this_byte & that_byte);
}
```

Questions

1. State what the output of each of the `printf` functions in `main()` will be when the program is executed.
2. Which data value(s) is (are) global for the entire program?
3. Which data value(s) is (are) global for a part of the program? What is its scope?
4. Explain how the variable `memory_location_1` receives the value of 375.
5. What statement in `main()` causes the value of `memory_location_1` to become 375?
6. Explain why the output of the last `printf` function in `main()` is F.
7. State why `function_1` does not require pointers in its argument and `function_2` does require them.

8. Does `function_1` need to be of type `double`? Explain.
9. State how `function_1` "knows" the pointer `*look_at` since it is not declared within the function.
10. Why does `function_1` cause the value of the variable `memory_location_1` to change?

End-of-Chapter Problems

General Concepts

Section 5.1

1. State a way of visualizing a computer's memory as suggested in this chapter.
2. How is one memory location distinguished from another in computer memory? What is this called?
3. What is the process by which the CPU gets an instruction from memory and then executes the instruction called?
4. State the difference between immediate addressing and direct addressing.

Section 5.2

5. State some of the most common word sizes used by computers.
6. What is the purpose of the `sizeof` function in C?
7. Determine the binary sums of the following binary numbers:
   A. $0101_2 + 0001_2$ B. $0111_2 + 0001_2$
8. Find the ones complements of the following binary numbers:
   A. 0001_2 B. 1111_2 C. 1010_2
9. Find the twos complements of the following binary numbers:
   A. 0001_2 B. 1111_2 C. 1010_2
10. Using the MSB as the sign bit in twos complement notation, determine the values (including the signs) of the decimal values of the following binary numbers:
   A. 0111_2 B. 1000_2 C. 1101_2
11. Using signed twos complement notation, convert the following hex values to their signed decimal equivalents:
   A. A_{16} B. 8_{16} C. $9C_{16}$

Section 5.3

12. What is the address operator in C? How is it used?
13. What is the indirection operator in C? How is it used?
14. What is a pointer? How is it used?
15. Assuming the following declarations, answer the questions below:

```
int data;
int *pointer;

    pointer = &data;
    *pointer = 5;
```

   A. What is the value of `pointer`?
   B. What is the value of `data`?
   C. What is the value of `*pointer`?
16. State how more than one variable is passed from a called function to the calling function.

Section 5.4

17. State why separate functions are used in a C program.
18. What is the limitation of the C `return()` used in a called function?
19. Explain the use of the `&` operator in returning values in the arguments of a called function.
20. What must be in the formal argument of a function definition that is to return values to the calling function through its argument?

Section 5.5

21. What is the name of a variable declared within a function whose life lasts only as long as the function is active?
22. When is the life of an automatic variable the duration of the program?
23. What is the name of a variable that is declared at the beginning of the program before `main()`? What is the scope of such a variable?
24. State what is considered good programming practice concerning the use of variables.

Section 5.6

25. What is the C keyword used to ensure that an assigned value cannot be changed during program execution?
26. Give the name of a variable that is local to the function but whose value persists during the life of the program.
27. State the name of the variable class that requests the compiler to keep the variable in one of the internal registers of the microprocessor.
28. What is the name for a variable class that has a scope of more than one function?

Section 5.7

29. Show how a pointer can be used with a recursive function to convert a decimal value into 8-bit binary.
30. What changes must be made to Program 5.23 to generate floating point random numbers?

Section 5.8

31. State how you would obtain the address of the variable `data` in a C program.
32. Show how you would pass the address of `data` to a pointer called `ptr` in C.
33. What is the C symbol used for equality? For assignment?
34. What terminology is used when an address operator is used to assign values to function arguments?

Section 5.9

35. What is the hexadecimal address produced by `vbuff(10,50,0)`?
36. Explain how to add a command that allows the user to specify the drawing character in Program 5.25.
37. Explain how a command to turn the screen upside down would work.

Program Design

In developing the following C programs, use the method developed in Program 5.13. This means that you will use top-down design and block structure with no global variables. The function `main()` is to do little more than call other functions. When more than one variable is to be passed back to the calling function, then pointers are to be used. As before, be sure to include all of the documentation in your final program. This should consist of, but not be limited to, the programmer's block, function prototypes, and a description of each function as well as any formal arguments you may use.

38. Create a C program that will compute the voltage across each component of a series resonant circuit consisting of a resistor, capacitor, inductor, and AC voltage source. The program user

is to input the value of each component along with the applied frequency and source voltage. The relationships are

$$X_L = 2\pi f L$$

Where

X_L = Inductive reactance in ohms.
f = Frequency in hertz.
L = Inductance of inductor in henrys.

$$X_C = 1/(2\pi f C)$$

Where

X_C = Capacitive reactance in ohms.
f = Frequency in hertz.
C = Capacitance of capacitor in ohms.

$$Z = \sqrt{\left((X_L - X_C)^2 + R^2\right)}$$

Where

Z = Circuit impedance in ohms.
X_L = Inductive reactance in ohms.
X_C = Capacitive reactance in ohms.
R = Resistance in ohms.

$$I_t = V_t / Z$$

Where

I_t = Total circuit current in amps.
V_t = Applied source voltage in volts.
Z = Circuit impedance in ohms.

$$V_L = I_t X_L$$
$$V_C = I_t X_C$$
$$V_R = I_t R$$

Where

V_L = Inductor voltage in volts.
V_C = Capacitor voltage in volts.
V_R = Resistor voltage in volts.

39. Create a C program that represents the input of a cash register. The program user will see a display of ten items as follows:
A. Hamburger—Regular
B. Hamburger—Double
C. Hamburger—Super
D. Fries—Small
E. Fries—Medium
F. Fries—Large
G. Shake—Small
H. Shake—Medium
I. Shake—Large
T. Total:

The program user selects as many items as desired by letter along with the quantity of each. When done, the letter T is entered for a total. The program then totals the order and includes a 6% sales tax. As the programmer, you enter the price of each item in your source code.

40. Create a C program that allows the program user to select any of the following six bitwise operations:
    A. COMPLEMENT
    B. AND
    C. OR
    D. XOR
    E. Shift Left
    F. Shift Right

    Once the user selects the operation, then the values for the bitwise operation may be entered.

41. Write a C program that will convert an input number into two-digit hexadecimal notation. For example, 100 becomes '64' and 255 becomes 'FF'. Use bitwise and shift operations to do the conversion.

42. Write a C program that will control the lights at an imaginary intersection of two streets. The two streets, North St. and East St., are two-way streets. The bit pattern that controls the traffic light is defined as follows:

7	6	5	4	3	2	1	0
?	?	Red	Yellow	Green	Red	Yellow	Green
		|-----------	North	-----------|	|------------	East	------------|

A 1 in any position turns the associated light on. Use bitwise operations to generate the sequence of control patterns for this traffic pattern:

North	*East*
Red	Green
Red	Yellow (flash twice)
Red	Red
Green	Red
Yellow (flash once)	Red
Red	Red

43. Devise your own random number generator. Write a C program to generate 40 terms from a starting seed. Display eight terms on a single line. How well does your algorithm work?

44. Write a C program that will make an 8-bit *palindrome* out of any 4-bit number. A palindrome is a number whose digits read the same forwards and backwards. For example, if supplied with 0011, your program should produce the palindrome 11000011. The user enters a number from 0 to 15. Use bitwise operations to test for and duplicate the 1s.

6 Strings

Objectives

This chapter provides you the opportunity to learn:

1. The relationship between characters and pointers.
2. Methods of putting characters in memory to form strings.
3. The use of strings in C.
4. How string functions manipulate strings.
5. Methods of sorting string data.

Key Terms

String
Array
Character
NULL
Element
String Initialization
Rectangular Array
Ragged Array

Outline

Introduction

This chapter will give you an opportunity to apply what you have learned about C pointers. Here you will see the relationships among pointers and strings. Understanding how these are related will help you in developing C programs that are capable of many powerful applications, including sorting.

With the information in this chapter, you will have a solid foundation for the development of future C programs that will solve a wide variety of complex technology programs.

6.1 Characters and Strings

Discussion

This section presents the relationships among characters, pointers, and strings. Here you will see how pointers are used to store characters in memory. You will receive your first introduction to arrays.

The use of strings is an important part of any technology program. Understanding how to use strings in your programs opens another dimension of technical programming. The ability to use the names of objects, people, and data is a powerful addition to your programming skills. This section introduces this important tool.

Storing Strings

Recall from Chapter 1 that a character can be thought of as a single memory location that contains an ASCII code. A **string** is nothing more than an arrangement of characters. The word **array** means arrangement; therefore, a string can be thought of as an array of characters. A **character** is any individual ASCII byte of data. How a string is stored in memory is illustrated in Figure 6.1.

Such a sequential arrangement of data within memory is referred to as an array. Note that the last character of the string array in Figure 6.1 consists of the C **NULL** character (represented by a `'\0'`). All of C's character strings require the null character to let C know where the string ends. The null character is automatically placed at the end of the string by C when the string is defined.

To tell C you want a string (an array of characters), you must use the brackets `[]` immediately following your string identifier. This is demonstrated in Program 6.1.

Program 6.1

```
#include <stdio.h>

char string[] = "Hello";

main()
{
        printf("The string is %s.",string);
}
```

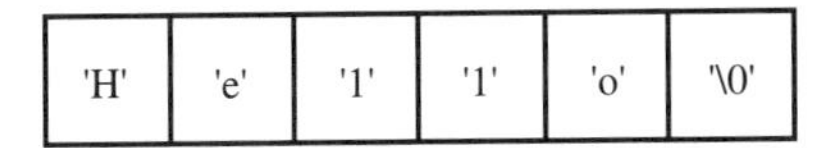

Figure 6.1 String Storage in Memory

When Program 6.1 is executed, the output will be

```
The string is Hello.
```

Note that the format specifier is `%s` for a string variable.

Program Analysis

The statement `char string[] = "Hello";` tells C to reserve enough consecutive memory spaces to hold the string of characters: `H e l l o`. The variable `string` is now an array variable because it represents an arrangement of information in memory, not just one memory location.

The `%s` as the format specifier lets C know to display the array variable as a string of characters.

An Inside Look

An array is said to be made up of **elements**. As an example, in the string array of "Hello," each character is an element of the array. Elements in a C string array are numbered beginning with zero, so the zero element of this array is the letter H. To represent any single element of an array, simply place the element number inside the square brackets of the array variable; `string[0]` in this case is the letter H. This is illustrated by Program 6.2.

Program 6.2

```
#include <stdio.h>

        char string[] = "Hello";

main()
{
        printf("The string is %s.\n",string);
        printf("The characters are:\n");

        printf("%c\n",string[0]);
        printf("%c\n",string[1]);
        printf("%c\n",string[2]);
        printf("%c\n",string[3]);
        printf("%c\n",string[4]);
        printf("%c\n",string[5]);
}
```

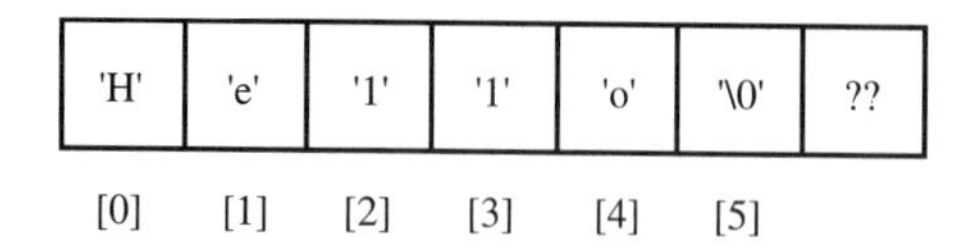

Figure 6.2 String Storage in Memory with Element Numbers

Execution of Program 6.2 yields

```
The string is Hello.
The characters are:
H
e
l
l
o
```

This shows some interesting facts about string arrays in C. First, they all start with the 0 element of the array. Second, all string arrays in C require a null terminator as the last element of the array. This terminator is necessary so that the program knows when the string array ends. This concept is illustrated in Figure 6.2.

Program Analysis

Program 6.2 again uses the statement

```
char string[] = "Hello";
```

to define the string array `string`. The individual characters are then displayed by the `printf` function using the character specifier `%c` along with a single array element:

```
printf("%c\n",string[0]);
```

This statement, for example, causes the first element (the 0 element) of the string array to display its contents as a character, and hence the capital letter H is displayed. In a like manner, the other `printf` statements cause each of the other individual array elements to be displayed.

Where Are the Elements?

In a string array, each element is a character and represents a single memory location of one byte. As with other data, a pointer can also be used to access any element of the array. Consider Program 6.3. It defines the same string array and also a pointer (of type `char`). It then uses the `*ptr` to access each individual element of the array (each memory location).

Program 6.3

```
#include <stdio.h>

char string[] = "Hello";

main()
```

```
{
        char *ptr;

        ptr = string;

        printf("The string is %s.\n",string);
        printf("The characters are:\n");
        printf("%c\n",*ptr);
        printf("%c\n",*(ptr + 1));
        printf("%c\n",*(ptr + 2));
        printf("%c\n",*(ptr + 3));
        printf("%c\n",*(ptr + 4));
        printf("%c\n",*(ptr + 5));
}
```

Execution of Program 6.3 yields

```
The string is Hello.
The characters are:
H
e
l
l
o
```

As you can see, the output of Program 6.3 is identical to that of Program 6.2. This means that the following are exactly equal:

```
*ptr = string[0]
*(ptr + 1) = string[1]
*(ptr + 2) = string[2]
*(ptr + 3) = string[3]
*(ptr + 4) = string[4]
*(ptr + 5) = string[5]
```

The reason for this equality is because the variable `string` is declared as an arrayed variable (it was followed by the square brackets `[]`). When the statement

```
ptr = string;
```

is used, this automatically places the address of the first string element (`string[0]`) in `ptr`. Thus, if you add 1 to the value contained in `ptr` (as with `*(ptr + 1)`), you are adding 1 to the address contained there, which is the memory location of the next character. This concept is illustrated in Figure 6.3. Note the use of sample memory addresses (5321 through 5326).

It's important to observe the difference between

```
*(ptr + 1)
```

and

```
ptr + 1
```

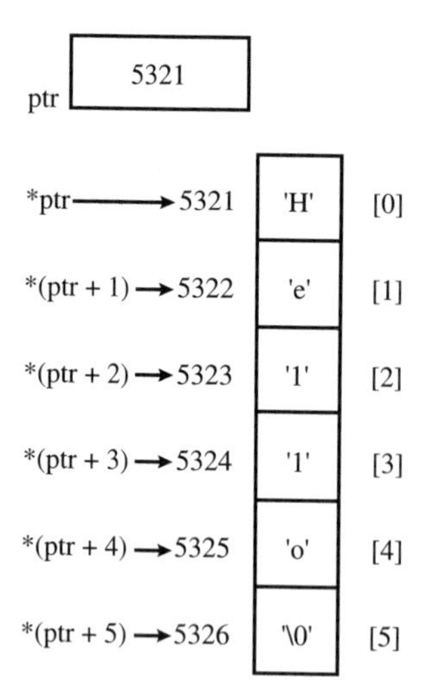

Figure 6.3 Using a Pointer to Access String Characters

In the first case, the program will be directed to the address that follows the address stored in `ptr`. In the second case, the value stored in `ptr` will have the number 1 added to it. This is illustrated by Program 6.4.

Program 6.4

```
#include <stdio.h>

        char string[] = "Hello";

main()
{
        char *ptr;

        ptr = string;

        printf("The string is %s.\n",string);
        printf("The characters are:\n");

        printf("Address => %d | %c |\n",ptr, *ptr);
        printf("Address => %d | %c |\n",ptr + 1, *(ptr + 1));
        printf("Address => %d | %c |\n",ptr + 2, *(ptr + 2));
        printf("Address => %d | %c |\n",ptr + 3, *(ptr + 3));
        printf("Address => %d | %c |\n",ptr + 4, *(ptr + 4));
        printf("Address => %d | %c |\n",ptr + 5, *(ptr + 5));
}
```

When Program 6.4 is executed, the output is

```
Address => 5321  | H |
Address => 5322  | e |
Address => 5323  | l |
```

```
Address => 5324   | l |
Address => 5325   | o |
Address => 5326   |   |
```

This demonstrates three things. First, the addresses of the character array are contiguous. Second, `ptr + 1` is a value that is 1 larger than the value contained in `ptr`. Third, `*(ptr + 1)` represents the character stored in the memory location whose address is 1 larger than the address stored in `ptr`.

Conclusion

This section introduced you to the use of strings in C. You were also introduced to the concept of an array as a contiguous set of data elements in memory. Here you saw that the first element number of an array in C is 0. You also saw that a string array must be terminated with the C null character `'\0'`.

The relationship between arrays and pointers was also demonstrated. Check your understanding of this section by trying the following section review.

6.1 Section Review

1. State what is meant by a string.
2. Explain how you indicate a `char` string in C.
3. For a string consisting of five characters, how many array elements are required? Explain.
4. What is the element number of the first character in a C string array?
5. Explain the relationship between pointers and string array elements.

6.2 Initializing Strings

Discussion

In this section we will examine a number of ways a C string of characters may be **initialized.** Recall from Section 6.1 that the statement

```
char string[] = "Hello";
```

automatically reserves the correct number of memory locations to store each element of the string, including the final null character `'\0'`. The same thing is accomplished with this new statement

```
char string[] = {'H', 'e', 'l', 'l', 'o', '\0'};
```

which you might agree is more cumbersome than a simple matching pair of double quotes.

When it is necessary to specify the length of a string, an integer length value must be placed between the `brackets`. For example, in the statement

```
char string[5] = "Hello";
```

the actual length of the string (5) is specified between the square brackets. Note that the null character is not counted as part of the string's length. It is thus possible to create a string of zero length by writing the null character into element 0's position, as in

```
string[0] = '\0';
```

Note that specifying more characters for the string than are required is common and relieves the programmer of the burden of counting the string characters while writing the program. So the statement

```
char string[80] = "Hello";
```

is also acceptable and allows more string characters to be added at a later time, if desired. Keep in mind that the extra character locations may contain random numbers that may not be correctly interpreted as valid characters without initialization.

In some cases we only wish to allocate memory for the string at the beginning of a program and stuff characters into the locations at a later time, as shown in the following code:

```
char string[5];

string[0] = 'H';
string[1] = 'e';
string[2] = 'l';
string[3] = 'l';
string[4] = 'o';
string[5] = '\0';
```

Note that it is necessary for us to place the null character into memory as well. This is required so that C knows where the string ends. If we were to forget to do this, our string would not merely be `"Hello"` but `"Hello"` followed by all characters in memory following the `'o'` until a null character is found. Also note the difference between using a single quote (as in `'x'`) and a double quote (as in `"x"`). Single quotes define a *single character*, whereas double quotes define a *string*. Thus, `'x'` represents the single ASCII character `'x'`, and `"x"` represents a string containing the two characters `'x'` and `'\0'`.

Another way to initialize a string is to reserve storage for it and then use `scanf` to read the string in. Consider the following code:

```
char string[20];

printf("Enter a string => ");
scanf("%s",string);
```

Notice that the `&` character is not used before the variable name (e.g., `&string`) as it usually is when `scanf` is used to read in a number. This is because the string's variable name automatically represents the starting address of the string, and that is what `scanf` requires.

During execution, suppose that the user enters

```
Enter a string => Hello
```

The characters stored into memory by `scanf` will be `'H'`, `'e'`, `'l'`, `'l'`, and `'o'` followed by the null character. However, `scanf` has its limitations. For instance, if the user had instead entered

```
Enter a string => Hello there.
```

`scanf` would have only loaded the same five characters, because it terminates scanning when it encounters the blank character between `'Hello'` and `'there.'`. This is unfortunate, because we would like to be able to enter strings containing blank characters (as in a word-processing application). The solution to this situation lies in the `gets()` function.

`Gets()` will read the entire line of text entered by the user, up to the carriage return. So, a better technique for reading a string into memory is as follows:

```
char string[20];

printf("Enter a string => ");
gets(string);
```

The `gets()` function only requires the name of the string to be supplied, and is included in `stdio.h` as a standard input function. If we enter

```
Enter a string => Oh, I get it!
```

the `gets()` function will store all 13 characters entered, plus a null character to terminate the string.

So there are many options available to the C programmer when it comes to initializing strings. One last point deserves mention. It is often necessary to fill an entire string with blanks (or some other character). This is easily accomplished as follows:

```
char string[80];
int i;

for(i = 0; i < 80; i++)
     string[i] = ' ';
string[80] = '\0';
```

Once again, note the importance of writing the null character into memory to complete the string definition.

Check your understanding of this section by trying the following section review.

6.2 Section Review

1. How many characters are placed into memory by the following statement?

   ```
   char string[] = "abcdefghijklmnopqrstuvwxyz";
   ```

2. What would be stored in memory by `scanf` if the following text were entered?

   ```
   "Oh, I get it!"
   ```

3. What is the difference between `"\0"` and `'\0'`?
4. Why use the statement `char string[80] = "Hello"` instead of `char string[] = "Hello"`?

6.3 Passing Strings Between Functions

Since strings are nothing more than character arrays, you can pass strings between functions by passing only a pointer to the first element of the string. This is demonstrated by Programs 6.5 and 6.6.

Program 6.5

```
#include <stdio.h>

void function1(char name[]);
```

```
main()
{
        char string[20];

        printf("What is your name => ");
        gets(string);

        function1(string);
}

void function1(char name[])
{
        printf("Hello there %s!",name);
}
```

Program 6.5 illustrates the passing of a string to a called function. Execution of the program produces

```
What is your name => Joe Smith
Hello there Joe Smith!
```

Program 6.6 shows the string being passed back to the calling function.

Program 6.6

```
#include <stdio.h>

        void function1(char name[]);

main()
{
        char string[20];

        function1(string);
        printf("Hello there %s!\n\n",string);
}

void function1(char name[])
{
        printf("Enter your name => ");
        gets(name);
}
```

The output of Program 6.6 is identical to that of Program 6.5. Note that only the starting address of the string is passed between `function1` and `main`.

To pass multiple strings between functions you need only to include all string names in the function header. For example, Program 6.7 contains a function called `equal_size()` that compares the lengths of two strings passed into it. If both strings have the same number of characters before the null character, they have the same length. The starting addresses of both strings are passed into `equal_size`, which returns a value of 1 if the strings do have an equal number of characters, and 0 if they do not.

Program 6.7

```
#include <stdio.h>

int equal_size(char s1[], char s2[]);

main()
{
        char str1[] = "One";
        char str2[] = "Two";
        char str3[] = "Three";

        if(1 == equal_size(str1,str2))
                printf("\"%s\" and \"%s\" are equal in length.\n",str1,str2);
        else
                printf("\"%s\" and \"%s\" are not equal in length.\n"
                        ,str1,str2);
        if(1 == equal_size(str1,str3))
                printf("\"%s\" and \"%s\" are equal in length.\n",str1,str3);
        else
                printf("\"%s\" and \"%s\" are not equal in length.\n"
                        ,str1,str3);
}

int equal_size(char s1[], char s2[])
{
        int i = 0, j = 0;
        while('\0' != s1[i])
                i++;
        while('\0' != s2[j])
                j++;
        if(i == j)
                return(1);
        else
                return(0);
}
```

An important point to remember concerns the passing of a string *back* from a function. It is necessary for the storage space for the passed string to be contained within the *calling* function. Otherwise, the memory contents of the string may be lost when the function terminates and its local storage is returned to the storage pool. For this reason, many programmers use global string definitions when string passing is required.

Check your understanding of this section by trying the following section review.

6.3 Section Review

1. Would Programs 6.5 and 6.6 work the same way if `gets()` were replaced by `scanf()`?
2. Why is it not necessary to specify the string size when passing a string to a function?
3. What would happen during the execution of `equal_size` if one or both of the strings were missing a terminating null character?

6.4 Working with String Elements

Discussion

In this section we examine a number of built-in functions that allow us to input string elements, examine them, and even convert them from one form to another. Each operation is useful in a program that accepts text from a user.

A Character at a Time

One useful built-in C function that gets one character at a time from the input is called `getchar()`. As you will see, the advantage of doing this is that each individual character may be tested. Doing this will determine if a character is a letter of the alphabet, a number, or some other kind of input such as a punctuation mark. The use of this function is illustrated in Program 6.8.

Program 6.8

```
#include <stdio.h>

main()
{
        char ch;

        printf("Give me a single input => ");
        ch = getchar();

}
```

The function `getchar()` reads a single character from input. For example, a string of characters may be entered by the program user until a newline marker is encountered (meaning the program user has pressed the carriage return). This is demonstrated in Program 6.9.

Program 6.9

```
#include <stdio.h>

main()
{
        char ch;

        printf("Give me a number => ");

        while((ch = getchar()) != '\n');
}
```

In Program 6.9, the program user will be able to continually input characters from the keyboard until the return key is depressed. When this happens, the C `while` will terminate.

Checking Characters

There are several different types of built-in C functions that will check the type of character that is entered into a C program. Table 6.1 lists them.

Program 6.10 illustrates an application of one of the built-in C character checking functions.

Program 6.10

```
#include <stdio.h>
#include <ctype.h>

main()
{
        char ch;

        printf("Give me a sentence => \n");

        while ((ch = getchar()) != '\r')
                printf("%d",isalpha(ch));
}
```

Table 6.1 C Character Classifications in <ctype.h>

Function	**Meaning (Returns a nonzero value if character meets the test, otherwise zero.)**	**Example** (`ch` **is character being tested)**
`isalnum()`	Alphanumeric	`if(isalnum(ch) != 0)` `printf(%c is alphanumeric.\n",ch);`
`isalpha()`	Alphabetic	`if(isalpha(ch) != 0)` `printf(%c is alphabetical.\n",ch);`
`iscntrl()`	Control character	`if(iscntrl(ch) != 0)` `printf(%c is a control ch.\n",ch);`
`isdigit()`	Digit	`if(isdigit(ch) != 0)` `printf(%c is a digit.\n",ch);`
`isgraph()`	Checks if character is printable Excludes the space character	`if(isgraph(ch) != 0)` `printf(%c is printable.\n",ch);`
`islower()`	Checks if character is lowercase	`if(islower(ch) != 0)` `printf(%c is lowercase.\n",ch);`
`isprint()`	Checks if character is printable Includes the space character	`if(isprint(ch) != 0)` `printf(%c is printable.\n",ch);`
`ispunct()`	Checks if character is punctuation	`if(ispunct(ch) != 0)` `printf(%c is punctuation.\n",ch);`
`isspace()`	Space	`if(isspace(ch) != 0)` `printf(%c is a space.\n",ch);`
`isupper()`	Checks if character is uppercase	`if(isupper(ch) != 0)` `printf(%c is uppercase.\n",ch);`
`isxdigit()`	Hexadecimal digit	`if(isxdigit(ch) != 0)` `printf(%c is hex digit.\n",ch);`

Execution of Program 6.10 produces

```
Give me a sentence =>
<Learn C>  — entered by user but not printed
 1222201   —displayed as user enters characters
```

As you can see from the above output, the `isalpha()` function returns a 0 when the character is not a letter of the alphabet (such as the blank space). It returns a 1 for any uppercase character and 2 for any lowercase character.

More Character Checking

Just as Program 6.10 checks for alphabetical characters, you can also check for numerical characters or any other type of character presented in Table 6.1.

Program 6.11 asks for a string of text from the user and prints out the string minus any punctuation or numerical digits. Lowercase alphabetic characters are converted to uppercase as well. This technique is useful in a word-processing application where spell checking is required and it is necessary to eliminate all nonalphabetic characters. Getting all characters into uppercase provides for easier searching and comparing with the spell check dictionary.

Program 6.11

```
#include <stdio.h>
#include <ctype.h>

main()
{
        char string[80];
        int i = 0;

        printf("Enter a string of text including punctuation and numbers: \n");
        gets(string);

        while(string[i] != '\0')
        {
                if ((0 == ispunct(string[i])) && (0 == isdigit(string[i])))
                {
                        if(0 != islower(string[i]))
                                string[i] = string[i] & 0xdf;
                        printf("%c", string[i]);
                }
                i++;
        }
}
```

A sample execution of Program 6.11 is as follows:

```
Enter a string of text including punctuation and numbers:
Hello Agent 99, this is Max! Where's the chief?
HELLO AGENT  THIS IS MAX WHERES THE CHIEF
```

Looking for Numbers

When the user enters a numeric string from the keyboard, the digits of the number are stored as ASCII character codes, not actual numeric digits. This makes it necessary to scan the string and convert each numeric ASCII code into its associated decimal value and combine all digits into a single number. Program 6.12 does just this for signed integers. The format of the signed integer is as follows:

```
[+ or -][digits]
```

where the + or - sign is optional (with no sign indicating a positive integer). There must be at least 1 digit in the `[digits]` portion. Other types of numbers have similar, though more detailed, formats, as in

```
[+ or -][digits][.][digits]     (real numbers)
[+ or -][digits][.][digits][e or E][+ or -][digits]   (scientific numbers)
```

Program 6.12 asks the user to enter a signed integer and then proceeds to examine the input string and build a resulting integer value based on the ASCII characters entered. Note that no error checking is provided to enforce the required format.

Program 6.12

```
#include <stdio.h>
#include <ctype.h>

main()
{
        char string[10];
        int i = 0;
        int sign = 1,number = 0;

        printf("Enter a signed integer => ");
        gets(string);
        if('-' == string[0])
        {
                sign = -1;
                i++;
        }
        if('+' == string[0])
                i++;
        while(string[i] != '\0')
        {
                number *= 10;
                number += string[i] - 0x30;
                i++;
        }
        number *= sign;
        printf("You entered %d.",number);
}
```

An alternative way to convert a string-based integer is through the use of the `atoi()` (ASCII to Integer) function found in `<stdlib.h>`. This function provides an easy way to perform ASCII-to-integer conversion. Program 6.13 demonstrates the use of `atoi()`.

Program 6.13

```
#include <stdio.h>
#include <stdlib.h>

main()
{
        char string[10];
        int number;

        printf("Enter a signed integer => ");
        gets(string);
        number = atoi(string);
        printf("You entered %d.",number);
}
```

Other related functions in `<stdlib.h>` are:

- `atof()` ASCII to Float
- `atol()` ASCII to Long

Although these built-in functions make it easy to perform necessary conversions, it is still a good programming challenge to write the conversion code ourselves (with the added benefit of additional error checking).

Check your understanding of this section by trying the following section review.

6.4 Section Review

1. Explain the operation of `getchar()`.
2. What is meant by the C character classifications? Give examples.
3. Will Program 6.11 output any characters besides alphabetic ones?
4. Can Program 6.11 be rewritten with a different `<ctype.h>` function and still output only alphabetic characters?
5. What happens if the user enters 40000 during execution of Program 6.12?

6.5 String Handling Functions

The programming examples covered so far in this chapter have dealt with operations on character strings on a character-by-character basis. From a different standpoint, what are the operations we might need to perform on an *entire* string? A few examples might be:

- Find the length of a string.
- Combine two strings together.
- Compare two strings.
- Search a string for a character (or substring).

Table 6.2 String Functions Available Through `<string.h>`

Function	Meaning
`strlen()`	String length
`strcat()`	String concatenation
`strncat()`	String concatenation of n characters
`strcmp()`	String comparison
`strncmp()`	String comparison of n characters
`strchr()`	String has character
`strrchr()`	String has character (search from end)
`strstr()`	String has substring
`strpbrk()`	String pointer break
`strtod()`	String to double conversion
`strtol()`	String to long conversion
`strtoul()`	String to unsigned long conversion

C provides these operations, and many more, through the functions found in the `include` file `string.h`. A subset of the available string handling functions is shown in Table 6.2. In this section, we will examine the operation and use of all functions shown in Table 6.2. In many cases, two example programs will be used to illustrate a function. The first program will show how the string function is used, and the second program will show how the string function can be implemented with C statements.

strlen()

The `strlen()` function determines the length of a string. The length of a string is an integer value indicating the number of characters in the string up to, but not including, the null character. For example, a string defined as

```
char alpha[] = "abcdefghijklmnopqrstuvwxyz";
```

will cause

```
strlen(alpha)
```

to return a value of 26. Program 6.14 illustrates how `strlen()` is used.

Program 6.14

```
#include <stdio.h>
#include <string.h>

main()
{
        char alpha[] = "abcdefghijklmnopqrstuvwxyz";

        printf("The string \"%s\" contains %d characters.",alpha
                ,strlen(alpha));
}
```

It is important to use the `#include <string.h>` statement to make the `strlen()` function available during compilation. Program 6.14 outputs the following message:

```
The string "abcdefghijklmnopqrstuvwxyz" contains 26 characters.
```

Keep in mind that the double quotes (`"`) are output because of the `\"` used in the `printf` statement, and not by `%s` itself.

To implement the `strlen()` function it is necessary to count the number of string characters, beginning with element `[0]`, until the `'\0'` is found. If element `[0]` contains the null character, the string length is zero. Program 6.15 shows how a function can be written to perform the same operation as `strlen()`.

Program 6.15

```
#include <stdio.h>

int lenstr(char text[]);

main()
{
        char alpha[] = "abcdefghijklmnopqrstuvwxyz";

        printf("The string \"%s\" contains %d characters."
                ,alpha,lenstr(alpha));
}

int lenstr(char text[])
{
        int ccount = 0;

        while (text[ccount] != '\0')
                ccount++;
        return(ccount);
}
```

Notice that `lenstr()` is used as the name of the function to avoid confusion with the name of the string handling function.

strcat() and strncat()

The `strcat()` and `strncat()` functions are used to *concatenate* two strings. When two strings are concatenated, the contents of the second string are copied to the end of the first string, as shown in Figure 6.4. In this case, `strcat()` is used to combine the two strings together. It is important for the first string to have a predefined length that is long enough to hold the characters of the concatenated string. As Figure 6.4 shows, the `strcat()` function requires two arguments—the names of the strings to concatenate. The first string named will be the destination for the new string.

When it is not necessary to concatenate the entire second string, the `strncat()` function should be used. This function only concatenates the first *n* characters of the second string. Because the value of *n* must be supplied, `strncat()` requires three arguments, as follows:

```
strncat(str1,str2,n)
```

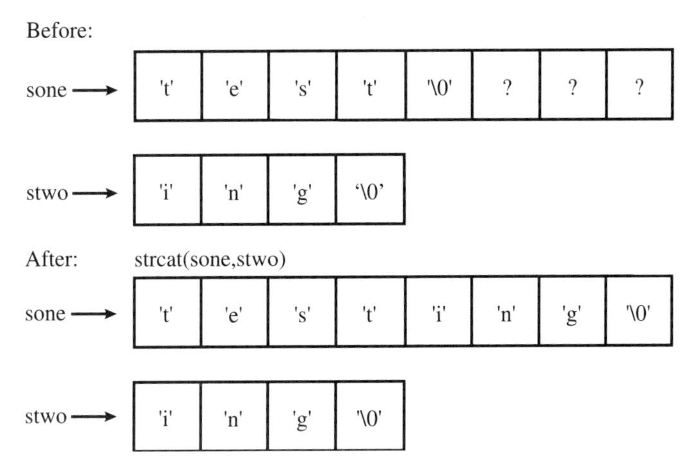

Figure 6.4 Concatenating Two Strings

Suppose that `str1` and `str2` are defined as follows:

```
char str1[20] = "Where is ";
char str2[] = "Kenneth?";
```

What does `str1` look like after `strncat(str1,str2,3)` is executed? Since the value of *n* is 3, only the first 3 characters of `str2` will be concatenated, giving `"Where is Ken"` as the resulting string.

Program 6.16 shows how the `strcat()` function is used.

Program 6.16

```
#include <stdio.h>
#include <string.h>

main()
{
        char astring[80] = "C programming";
        char bstring[80] = " is fun";

        strcat(astring,bstring);
        printf("The new string is \"%s\".",astring);
}
```

Notice that the string size in Program 6.16 is defined as 80 characters, even though each string actually contains a much smaller number of characters. This advance planning helps to avoid problems later on, when multiple concatenations might possibly lead to large string lengths. Program 6.16 outputs the following message:

```
The new string is "C programming is fun".
```

Program 6.17 shows how the `strcat()` function is implemented.

Program 6.17

```
#include <stdio.h>
#include <string.h>

int catstr(char str1[], char str2[]);

main()
{
        char astring[80] = "C programming";
        char bstring[80] = " is fun";

        catstr(astring,bstring);
        printf("The new string is \"%s\".",astring);
}

int catstr(char str1[], char str2[])
{
        int aptr,length,j;

        aptr = strlen(str1);
        length = strlen(str2);
        for(j = 0; j < length; j++)
        {
                str1[aptr] = str2[j];
                aptr++;
        }
        str1[aptr] = '\0';
}
```

Once again, the name has been changed (to `catstr`) to avoid confusion. The `catstr()` function uses `strlen()` to determine the ending position of the first string. It then uses `strlen()` again to get the length of the second string and proceeds to copy characters from the second string to the end of the first string. It is necessary to place the null character at the end of the first string to guarantee its ending position.

strcmp() and strncmp()

The `strcmp()` and `strncmp()` functions are used to *compare* two strings. The comparison is performed on a character-by-character basis. Each function returns a value based on the result of the comparison. The integer value returned is:

- 0 if the strings are identical (same characters *and* length)
- <0 if the first string alphabetically precedes the second
- >0 if the second string alphabetically precedes the first

For example, consider this short list of last names:

"Doe"
"Morris"
"Morrison"
"Smith"

The four names are listed in alphabetical order. Thus, “D” comes before “M,” and “M” comes before “S.” Negative values will be returned by `strcmp("Doe","Morris")` and `strcmp("Morris","Smith")`. In the case of “Morris” and “Morrison,” both names match for the first six characters. “Morrison” comes second because it contains more letters than “Morris.” A negative value is returned by `strcmp("Morris","Morrison")` and a positive value by `strcmp("Morrison","Morris")`. When strings are listed in this fashion, they are said to be in *lexicographic order*. Thus, `strcmp()` returns a value related to the lexicographic ordering of its two arguments.

The `strncmp()` function requires a third argument, *n*, which specifies the number of characters to compare. Thus, “Morris” and “Morrison” look identical when `strncmp("Morris","Morrison",6)` is used.

Program 6.18 shows how the `strcmp()` function is used. Four strings are defined and compared in three different ways to illustrate what `strcmp()` does.

Program 6.18

```
#include <stdio.h>
#include <string.h>

void compare(char str1[], char str2[]);

main()
{
        char astring[] = "shopper";
        char bstring[] = "shopping";
        char cstring[] = "shopper";
        char dstring[] = "howdy";

        compare(astring,bstring);
        compare(astring,cstring);
        compare(astring,dstring);
}

void compare(char str1[], char str2[])
{
        int value;

        value = strcmp(str1,str2);
        if (value == 0)
                printf("\"%s\" is 'equal to' \"%s\".\n",str1,str2);
        else
        if (value < 0)
                printf("\"%s\" is 'less than' \"%s\".\n",str1,str2);
        else
                printf("\"%s\" is 'greater than' \"%s\".\n",str1,str2);
}
```

The output of Program 6.18 is as follows:

```
"shopper" is 'less than' "shopping".
"shopper" is 'equal to' "shopper".
"shopper" is 'greater than' "howdy".
```

Program 6.19 uses the function `cmpstr()` to emulate the operation of `strcmp()`. If the two strings have unequal length, the longer string length is used to drive the comparison loop. Comparisons continue as long as the two strings have identical characters in each position.

Program 6.19

```
#include <stdio.h>
#include <string.h>

int cmpstr(char str1[], char str2[]);
void compare(char str1[], char str2[]);

main()
{
        char astring[] = "shopper";
        char bstring[] = "shopping";
        char cstring[] = "shopper";
        char dstring[] = "howdy";

        compare(astring,bstring);
        compare(astring,cstring);
        compare(astring,dstring);
}

int cmpstr(char str1[], char str2[])
{
        int a,b,length;
        int j = 0;

        a = strlen(str1);
        b = strlen(str2);
        length = (a > b) ? a : b;
        while((str1[j] == str2[j]) && (j < length))
                j++;
        if ((str1[j] == '\0') && (str2[j] == '\0'))
                return(0);
        else
        if (str1[j] < str2[j])
                return(-1);
        else
                return(1);
}

void compare(char str1[], char str2[])
{
        int value;

        value = cmpstr(str1,str2);
        if (value == 0)
                printf("\"%s\" is 'equal to' \"%s\".\n",str1,str2);
        else
```

```
        if (value < 0)
                printf("\"%s\" is 'less than' \"%s\".\n",str1,str2);
        else
                printf("\"%s\" is 'greater than' \"%s\".\n",str1,str2);
}
```

strchr() and strrchr()

Both the `strchr()` and `strrchr()` functions search a string for a specified character. Two arguments are required. The first is the string to be searched. The second is the character to search for. Both functions return a pointer to the character's position within the string, if found, or null if not found. The position of the first occurrence of the search character is returned by `strchr()`. The position of the last occurrence of the search character is returned by `strrchr()`. Program 6.20 shows how `strchr()` is used. The `search()` function uses `strchr()` to determine the position of the search character and then tests the position for a null value to see if the search was successful.

Program 6.20

```
#include <stdio.h>
#include <string.h>
#include <ctype.h>

void search(char text[], char letter);

main()
{
        char alpha[] = "abcdefghijklmnopqrstuvwxyz";

        search(alpha,'e');
        search(alpha,'E');
}

void search(char text[], char letter)
{
        int position;

        position = strchr(text,letter);
        if (position == '\0')
                printf("The letter %c is not in the string.\n",letter);
        else
                printf("The letter %c is located at address %X in memory.\n"
                        ,letter,position);
}
```

Execution of Program 6.20 results in the output

```
The letter e is located at address E0E in memory.
The letter E is not in the string.
```

This indicates that `strchr()` is *case sensitive*. Thus, lowercase characters and uppercase characters are not equal to each other. This makes sense, because lowercase and uppercase characters have unique ASCII codes assigned to them.

strstr()

The `strstr()` function searches for the first occurrence of a substring within a string. If the substring is found, `strstr()` returns a pointer to the beginning of the substring. If the substring is not found, `strstr()` returns null. For example, `strstr("yes indeed","in")` returns a pointer to the `"in"` substring within `"yes indeed"`, whereas `strstr("yes indeed","huh?")` returns null, because the substring `"huh?"` is not found anywhere.

Program 6.21 uses `strstr()` to find all occurrences of the word `"the"` in a sample of text.

Program 6.21

```
#include <stdio.h>
#include <string.h>

main()
{
        char text[] = "It is important for this block of text to "
                      "contain the word 'the' as many times as "
                      "possible, since the function strstr() will "
                      "count the number of times 'the' is seen. "
                      "This is a simple mathematical operation. "
                      "How many times does 'the' appear?";
        char *ptr;
        int thecount = 0;

        ptr = text;
        do
        {
                ptr = strstr(ptr,"the");
                if(ptr != '\0')
                {
                        thecount++;
                        ptr++;
                }
        } while(ptr != '\0');
        printf("The word 'the' appears %d times.",thecount);
}
```

Each time `strstr()` is called and a substring is found, the pointer returned is used to indicate where the next call to `strstr()` should begin searching. This process continues until the null value is returned.

Can you think of how `strstr()` might be useful in a word-processing spell checker?

strpbrk()

The `strpbrk()` function searches for the first occurrence of any character of a substring within a string. For example, `strpbrk("good morning","time out")` will return a pointer to the first `'o'` in `"good morning"`, because that is the first character in `"good morning"` that is also in `"time out"`. This function is useful when it is necessary to determine quickly if a string contains a specific character. For instance, if the supplied string represents a mathematical expression such as `"5*(2 - 3)"` it may be necessary to find the positions of all numbers in the formula. The statement

```
ptr = strpbrk(expr,"0123456789")
```

is a simple way to do this. Performing the same chore with `strchr()` would require multiple statements.

Program 6.22 shows how `strpbrk()` is used to remove all standard punctuation from an input string.

Program 6.22

```
#include <stdio.h>
#include <string.h>

main()
{
        char text[80];
        char *ptr;

        printf("Enter a string containing punctuation:\n");
        gets(text);
        ptr = text;
        while(ptr != '\0')
        {
                ptr = strpbrk(ptr,".,!;'\"?-");
                if(ptr != '\0')
                        *ptr = ' ';
        }
        printf("\n%s",text);
}
```

The punctuation searched for in Program 6.22 is `".,!;'\"?-"`. When it is found, it is replaced by a blank character. A sample execution is as follows:

```
Enter a string containing punctuation:
It's really sunny today! Don't you think so?

It s really sunny today  Don t you think so
```

This technique is useful when we wish to strip off meaningless or unimportant information, or data that we do not want to process at a particular time. If we were encrypting a text string, the punctuation would most likely be removed, because it might give clues about the lengths of words or sentences.

strtod(), strtol(), and strtoul()

The string functions `strdtod()`, `strtol()`, and `strtoul()` provide the means for converting strings into double- and long-value numbers. This use is similar to that of the `atoi()` and `atof()` functions covered in the previous section. These functions simply allow larger numbers to be converted.

Check your understanding of this section by trying the following section review.

6.5 Section Review

1. What `include` statement is needed to make string functions available?
2. List the common string operations.
3. Why is it necessary for `strlen()` to encounter the null character?
4. What would a string *substitution* function require? For example, `strsub("abcde","bc", "howdy")` would result in the string `"ahowdyde"`.
5. Why is the length requirement of the first string argument supplied to `strcat()`?
6. What does it mean that `strchr()` is *case sensitive*?
7. Which string functions are case sensitive? You may need to try examples to answer this question.

6.6 String Sorting

In this section we will examine one way a collection of strings can be sorted into alphabetical order. The collection of strings is stored in memory as a *two-dimensional* array of characters. A two-dimensional array of five names with each name containing up to eight characters (the ninth position is reserved for the null character) is defined as follows:

```
char names[5][9];
```

This type of array is also referred to as a **rectangular array**.

Initializing the two-dimensional array is straightforward and can be accomplished in the following way:

```
char names[5][9] = { {'k', 'r', 'i', 's', 't', 'e', 'n'},
                     {'t', 'u', 'r', 'n', 'e', 'r'},
                     {'k', 'e', 'n', 'n', 'y'},
                     {'v', 'i', 'c', 't', 'o', 'r', 'i', 'a'},
                     {'k', 'i', 'm', 'b', 'e', 'r', 'l', 'y'} };
```

Notice that each individual string in the array is surrounded by a matching pair of braces `{}`, and that the entire array is also enclosed in braces. When a name has less than seven characters (as in `"kenny"` and `"turner"`), this method of intialization automatically fills the remaining character positions with null characters, as shown in Figure 6.5.

As Program 6.23 shows, a single name from the list of names is referenced by using a single subscript, as in `names[j]` or `names[k]`). For example, the `strcpy()` function requires only a single subscript in this fashion to access the entire name.

Program 6.23

```
#include <stdio.h>
#include <string.h>
#define NUM 5
```

```
main()
{
        char names[5][9] = { {'k', 'r', 'i', 's', 't', 'e', 'n'},
                             {'t', 'u', 'r', 'n', 'e', 'r'},
                             {'k', 'e', 'n', 'n', 'y'},
                             {'v', 'i', 'c', 't', 'o', 'r', 'i', 'a'},
                             {'k', 'i', 'm', 'b', 'e', 'r', 'l', 'y'} };
        char swapname[9];
        int j,k;

        printf("The original list is:\n");
        for(j = 0; j < NUM; j++)
                printf("%s\n",names[j]);
        for(j = 0; j < NUM - 1; j++)
                for(k = j + 1; k < NUM; k++)
                        if(strcmp(names[j],names[k]) > 0)
                        {
                                strcpy(swapname,names[j]);
                                strcpy(names[j],names[k]);
                                strcpy(names[k],swapname);
                        }
        printf("\nThe sorted list is:\n");
        for(j = 0; j < NUM; j++)
                printf("%s\n",names[j]);
}
```

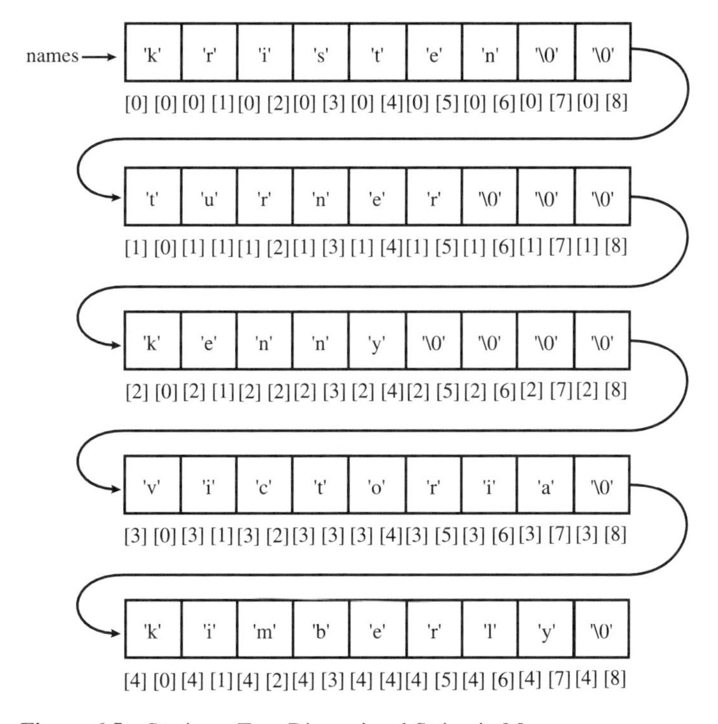

Figure 6.5 Storing a Two-Dimensional String in Memory

The technique used to sort the list of names in Program 6.23 is called *bubble sorting*. A bubble sort is actually a nested loop that performs $N - 1$ passes over a list of N items, swapping item X with item $X + 1$ during each pass if item X is 'greater' than item $X + 1$. For example, the list of names defined in Program 6.23 begins with `"kristen"`. Since `"kristen"` is lexicographically greater than two other names in the list (`"kenny"` anmd `"kimberly"`), `"kristen"` will be swapped twice to end up in the correct position. Sample execution of Program 6.23 is as follows:

```
The original list is:
kristen
turner
kenny
victoria
kimberly

The sorted list is:
kenny
kimberly
kristen
turner
victoria
```

The bubble sort is one of the most simple sorting techniques, and also one of the most inefficient. Consider a list of 100 names. A bubble sort of the list will require, unfortunately, almost 10,000 comparisons. In Chapter 7 we will examine other, more efficient, sorting techniques.

An alternative way of defining the two-dimensional array of characters for the sorting program is as follows:

```
char *names[5] = { {"kristen "},
                   {"turner  "},
                   {"kenny   "},
                   {"victoria"},
                   {"kimberly"} };
```

This method of intialization also defines a rectangular array but requires the user to insert the correct number of blanks after each name to guarantee that each string in the array occupies the same amount of storage space. If this were not the case, the built-in string functions (`strcpy`, `strcmp`) would give unpredictable results. Program 6.24 shows how this new array is sorted. There is no change in the sorting code, only in the way the array is defined.

Program 6.24

```
#include <stdio.h>
#include <string.h>
#define NUM 5

main()
{
        char *names[5] = { {"kristen "},
                           {"turner  "},
```

```
                                {"kenny   "},
                                {"victoria"},
                                {"kimberly"} };
        char swapname[9];
        int j,k;

        printf("The original list is:\n");
        for(j = 0; j < NUM; j++)
                printf("%s\n",names[j]);
        for(j = 0; j < NUM - 1; j++)
                for(k = j + 1; k < NUM; k++)
                        if(strcmp(names[j],names[k]) > 0)
                        {
                                strcpy(swapname,names[j]);
                                strcpy(names[j],names[k]);
                                strcpy(names[k],swapname);
                        }
        printf("\nThe sorted list is:\n");
        for(j = 0; j < NUM; j++)
                printf("%s\n",names[j]);
}
```

When the trailing blanks are not used, as in

```
char *names[5] = { {"kristen"},
                   {"turner"},
                   {"kenny"},
                   {"victoria"},
                   {"kimberly"} };
```

the result is called a **ragged array**. This is illustrated in Figure 6.6. As you can see, the ragged array is not rectangular. It requires only 39 storage locations, whereas the rectangular array defined with `names[5][9]` requires 45. Thus, a ragged array is a more efficient storage method than a rectangular array, but it is also harder to work with. Consider the swap code from Program 6.23:

```
strcpy(swapname,names[j]);
strcpy(names[j],names[k]);
strcpy(names[k],swapname);
```

During the swap of `"turner"` with `"kenny"`, we get this unfortunate result:

```
names[0] = "kristen"        names[0] = "kristen"
names[1] = "turner"         names[1] = "kenny"
names[2] = "kenny"          names[2] = "turner"
names[3] = "victoria"       names[3] = ""
names[4] = "kimberly"       names[4] = "kimberly"
```

This results from the fact that the internal pointers for `names[1]`, `names[2]`, etc. are defined when the array is initialized. Thus, the individual lengths of each string within the ragged array are fixed to specific values (6 for `"kenny"` including the null character, 7 for

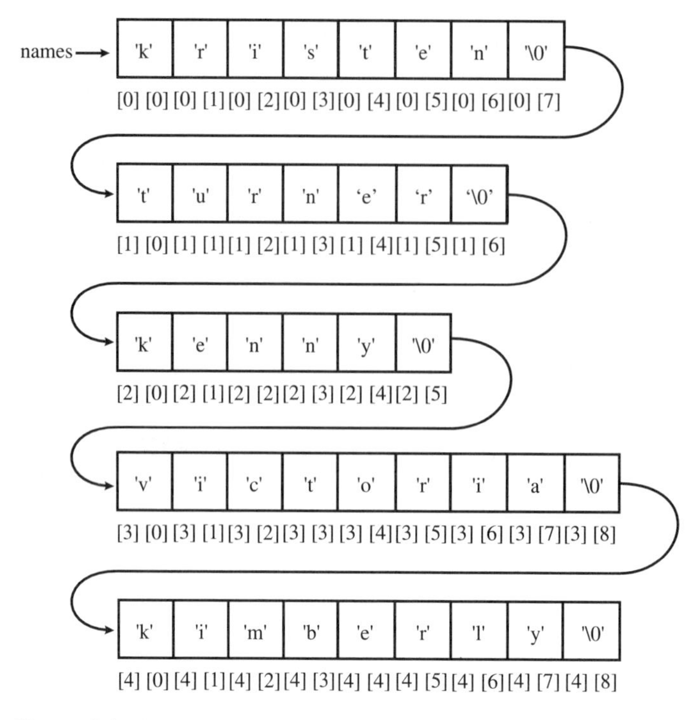

Figure 6.6 A Ragged Array of Characters

`"turner"`, and so on). Overwriting any of these strings may result in one string running over into another string's location, as we can see with `names[3]` becoming `""`. A good programmer should remember the strengths and weaknesses of using one array definition over another.

Check your understanding of this section by trying the following section review.

6.6 Section Review

1. How many storage locations are required by a rectangular array defined as follows: `char array[7][10]`?
2. How many comparisons are required by the bubble sort in Program 6.23?
3. Show the entire names array after each complete pass of the outer `for` loop in Program 6.23.
4. Why does the bubble sort make approximately N^2 comparisons for N items?
5. What are the strengths and weaknesses of rectangular and ragged arrays?
6. Explain how the `strcpy()` function produced `""` in `names[3]` during the `"turner"` and `"kenny"` swap.

6.7 Example Programs

The example programs presented in this section show how to do many interesting and useful things with character strings. You are encouraged to study them to gain a full understanding of how they work.

ISBN Checker

Program 6.25 checks a user-supplied ISBN (International Standard Book Number) code to determine if it is a valid sequence.

Program 6.25

```
#include <stdio.h>

main()
{
        char input[10];
        int total = 0;
        int i,rem;

        printf("Enter the first 9 digits of an ISBN code => ");
        scanf("%s", input);

        for(i = 0; i < 9; i++)
                total += (input[i] - 0x30) * (i + 1);

        rem = total % 11;
        printf("The last character should be ");
        if(rem != 10)
                printf("%c",(rem + 0x30));
        else
                printf("X");
}
```

The format of an ISBN code is as follows:

- Group code (1 digit)
- Publisher code (4 digits)
- Book code (4 digits)
- Check character/digit (1 character/digit)

As an example, the ISBN code for a typical engineering book is

0 6 7 5 2 0 9 9 3 5

The check character/digit is found in two steps. First, each of the first nine digits are multiplied by the first nine integers, and their products are added together:

$$\begin{array}{ccccccccc} 0 & 6 & 7 & 5 & 2 & 0 & 9 & 9 & 3 \\ \times 1 & 2 & 3 & 4 & 5 & 6 & 7 & 8 & 9 \\ \hline \end{array}$$

$$0 + 12 + 21 + 20 + 10 + 0 + 63 + 72 + 27 = 225$$

The sum is then divided by 11, and the integer remainder saved. So, 225/11 gives a remainder of 5. This is the computed check digit, which matches the check digit in the original ISBN code. If the remainder turns out to be 10, the character X is used instead. Thus,

0 6 7 5 2 0 7 7 2 X

is also a valid ISBN code.

Sample execution of Program 6.25 produces

```
Enter the first 9 digits of an ISBN code => 067520772
The last character should be X
```

Note that Program 6.25 converts each ASCII input digit into its associated numeric value by subtracting 0x30 (30 hexadecimal) from it. This is possible because the ASCII codes for '0' through '9' are 0x30 through 0x39.

Try Program 6.25 with the ISBN codes from your other textbooks.

Postorder Expression Generator

In dealing with mathematical expressions, it is important to preserve the order of mathematical operations. For example, what is the value of the following expression?

$$5 + 6 * 3$$

If your answer is 23, you are correct. If your answer is 33, you violated a basic rule of mathematical operations: *multiplication/division is performed before addition/subtraction!* Some people who have trouble remembering this rule require parentheses in the expression to guarantee the right answer, as in:

$$5 + (6 * 3)$$

In this case, the multiplication *must* be performed before the addition, because it is contained within parentheses.

One common form an expression may take is called *postorder notation.* A postorder expression does not contain parentheses, but instead contains numbers (or variables) and math operation symbols in a particular order. The postorder expression:

$$5\ 6\ 3 * +$$

is interpreted as follows. When a math operation symbol is encountered, use the previous two numbers in the list to perform the desired operation, then place the new result back into the list. So, the * symbol requires us to multiply the 6 and the 3, giving 18. The new expression becomes

$$5\ 18 +$$

which then reduces to 23, the correct answer.

By a similar method, the postorder expression

$$ABC - D*+$$

is evaluated as follows:

$$A + ((B - C) * D)$$

The purpose of Program 6.26 is to take any mathematical expression in standard form and create the postorder notation that represents it. A number of simple rules are used to accomplish the conversion. These rules are based on the fact that * and / precede + and – and that anything in () is evaluated first.

Program 6.26

```
#include <stdio.h>
#include <ctype.h>

main()
{
        char expr[80];              /* Input expression */
        int expidx = 0;             /* Input expression index */
        char pfexpr[80];            /* Postfix expression */
        int pfeidx = 0;             /* Postfix expression index */
        char pfstk[80];             /* Postfix operator stack */
        int pfsidx = 0;             /* Postfix operator stack index */
        char ch, tos;

        printf("Enter expression => ");
        scanf("%s",expr);
        while(expr[expidx] != '\0')         /* While more input exists... */
        {
                ch = expr[expidx];          /* Read input character */
                expidx++;
                if (ch == '(')              /* Check for '(' operator */
                {                           /* Save on operator stack */
                        pfsidx++;
                        pfstk[pfsidx] = '(';
                }
                if (ch == ')')              /* Check for ')' operator */
                {
                        /* Pop operator stack until '(' is found */
                        while(pfstk[pfsidx] != '(')
                        {
                                /* Pop operator off stack and */
                                /* add it to the postfix expression */
                                pfexpr[pfeidx] = pfstk[pfsidx];
                                pfeidx++;
                                pfsidx--;
                        }
                        pfsidx--;           /* Eat the '(' operator */
                }
                /* Do + or - now? */
                if ((ch == '+') || (ch == '-'))
                {
                        /* Read top of operator stack */
                        if (pfsidx > 0)
                                tos = pfstk[pfsidx];
                        else
                                tos = '\0';

                        /* Push operator if priority allows */
                        if ((tos == '\0') || (tos == '('))
                        {
```

```
                    pfsidx++;
                    pfstk[pfsidx] = ch;
                }
                else
                {
                    /* Add operator to postfix expression */
                    pfexpr[pfeidx] = pfstk[pfsidx];
                    pfeidx++;
                    pfsidx--;
                    expidx--;
                }
        }
        /* Do * or / now? */
        if ((ch == '*') || (ch == '/'))
        {
                /* Read top of operator stack */
                if (pfsidx > 0)
                    tos = pfstk[pfsidx];
                else
                    tos = '\0';

                /* Push operator if priority allows */
                if ((tos == '\0') || (tos == '+')
                    || (tos == '-') || (tos == '('))
                {
                    pfsidx++;
                    pfstk[pfsidx] = ch;
                }
                else
                {
                    /* Add operator to postfix expression */
                    pfexpr[pfeidx] = pfstk[pfsidx];
                    pfeidx++;
                    pfsidx--;
                    expidx--;
                }
        }

        /* Add symbol/digit to postfix expression? */
        if (isalpha(ch) != 0)
        {
                pfexpr[pfeidx] = ch;
                pfeidx++;
        }
}

/* End of input expression reached. Pop any remaining operators */
/* off operator stack and add them to postfix expression*/
while(pfsidx > 0)
{
        pfexpr[pfeidx] = pfstk[pfsidx];
```

```
                pfeidx++;
                pfsidx--;
        }
        pfexpr[pfeidx] = '\0';     /* Fix string end */
        printf("Postorder notation => %s",pfexpr);
}
```

A few sample executions will verify the examples previously given.

```
Enter expression => A+B*C
Postorder notation => ABC*+
```

This execution used the variables *A, B,* and *C* to represent the numbers 5, 6, and 3.

```
Enter expression => A+((B-C)*D)
Postorder notation => ABC-D*+
```

This execution gives the same result seen before.

A more complex expression is converted in this example:

```
Enter expression => A+B*(C-D)/E-F
Postorder notation => ABCD-*E/+F-
```

which uses parentheses to force subtraction to be performed on *C* and *D* before multiplication by *B*.

Postorder expressions are easily evaluated on *stack-based* machines. Hewlett-Packard calculators are one example of stack-based machines. Postorder notation is also referred to as *Reverse-Polish* notation.

Vowel Counter

Program 6.27 counts vowels (a, e, i, o, or u) found in a text string. Since uppercase and lowercase vowels are identical, they must both be counted. Five integer variables are used to keep track of the vowel counts. To reduce the number of comparisons required, each character from the input string is converted into uppercase before the comparison is made. Lowercase ASCII characters in the range ‘a’ to ‘z’ are easily converted into uppercase by ANDing their codes with `0xdf`.

Program 6.27

```
#include <stdio.h>
#include <string.h>
#include <ctype.h>

main()
{
        char text[] = "OUR instructor SPEAKS clearly in EACH class.";
        char vstr[] = "AEIOU";
        static int vowels;
        static int acount,ecount,icount,ocount,ucount;
        int j;
```

```
        printf("The input string is => \"%s\"",text);
        for(j = 0; j < strlen(text); j++)
        {
                if (isalpha(text[j]) != 0)
                        text[j] = text[j] & 0xdf;
                if (strchr(vstr,text[j]) != '\0')
                        vowels++;
                switch(text[j])
                {
                        case 'A' : acount++; break;
                        case 'E' : ecount++; break;
                        case 'I' : icount++; break;
                        case 'O' : ocount++; break;
                        case 'U' : ucount++; break;
                }
        }
        printf("\nThe input string contains %d vowels.\n",vowels);
        printf("There are %d A's,  %d E's,  %d I's,  %d O's, and  %d U's",
                acount,ecount,icount,ocount,ucount);
}
```

Execution of Program 6.27 produces

```
The input string is => "OUR instructor SPEAKS clearly in EACH class."
The input string contains 13 vowels.
There are 4 A's,  3 E's,  2 I's,  2 O's, and  2 U's
```

Count the vowels for yourself to verify correct program operation.

Palindrome Checker

A *palindrome* is a string of symbols that reads the same forwards and backwards. Palindromes play an important role in the study of languages. Some sample palindromes are:

mom   radar   otto   11011011

Even entire expressions can be palindromes. Ignoring punctuation,

a man, a plan, a canal, panama

is also a palindrome.

Program 6.28 allows the user to enter a string of symbols. The program then determines if the string is a valid palindrome by checking symbol equality beginning at each end of the string.

Program 6.28

```
#include <stdio.h>

main()
{
        char palstr[80];
        int lchar,rchar,stopped;
```

```
        printf("Enter a string => ");
        gets(palstr);
        lchar = 0;
        rchar = strlen(palstr) - 1;
        stopped = 0;
        while((lchar <= rchar) && !stopped)
        {
                if(palstr[lchar] != palstr[rchar])
                        stopped = 1;
                lchar++;
                rchar--;
        }
        if(!stopped)
                printf("\"%s\" is a palindrome.",palstr);
        else
                printf("\"%s\" is not a palindrome.",palstr);
}
```

A few sample executions are shown to illustrate the operation of Program 6.28.

```
Enter a string => radar
"radar" is a palindrome.

Enter a string => 11011011
"11011011" is a palindrome.

Enter a string => howdy
"howdy" is not a palindrome.
```

Note that Program 6.28 does *not* ignore punctuation!

A Tokenizer

The structure of a *compiler* program is shown in Figure 6.7. A compiler takes a source file (any .c text file, for example), breaks each line of the source file into individual *tokens*, passes the tokens to a *parser* which checks and enforces correct syntax, and finally builds a file of object code representative of the original source program. Because the study of compilers is an advanced topic, we will not go into any detail on their operation. Instead, we will examine the basic operation of the first part of a compiler, the tokenizer.

Consider this sample C statement:

```
char string[] = "howdy";
```

This single statement is composed of the following tokens:

1. `char`
2. `string`
3. `[`
4. `]`
5. `=`
6. `"howdy"`
7. `;`

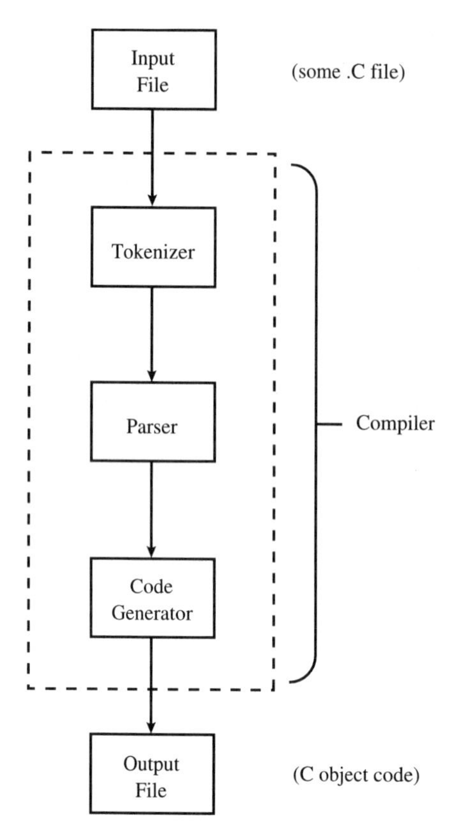

Figure 6.7 Structure of a Compiler

You can see that the purpose of a tokenizer is to break the input statement down into the smallest parts of a language (in this case, of the C language).

Program 6.29 is a limited *tokenizer*. It will not correctly tokenize any C statement that you enter, only those statements that contain the defined token symbols shown in the *singles* and *doubles* arrays.

Program 6.29

```
#include <stdio.h>
#include <string.h>
#include <ctype.h>

main()
{
        char singles[] = ";{},=:()[]*?'\"\\";
        char doubles[] = "*= /= %= += -= || && == != <= >= << >> ++ --";
        char ch,input[80],ds[3];
        int pos = 0;

        printf("Enter a C statement => ");
```

```
gets(input);
while (pos < strlen(input))
{
        printf("Token: ");
        do
        {
                ch = input[pos];
                pos++;
        } while(ch == ' ');      /* Skip blanks */

        /* Do we have a symbol name? */
        if(isalpha(ch) != 0)
        {
                printf("%c",ch);        /* Print first letter */

                /* Print remaining letters/digits */
                while(isalnum(input[pos]))
                {
                        printf("%c",input[pos]);
                        pos++;
                }
        }
        else

        /* Do we have a number? */
        if(isdigit(ch))
        {
                printf("%c",ch);        /* Print first digit */

                /* Print remaining digits */
                while(isdigit(input[pos]))
                {
                        printf("%c",input[pos]);
                        pos++;
                }
        }
        else

        /* Check for a single token */
        if(strchr(singles,ch) != '\0')
        {
                printf("%c",ch);        /* Print token */

                /* String or character token? */
                if((ch == '\"') || (ch == '\''))
                {
                        do      /* Print remaining token text */
                        {
                                ch = input[pos];
                                printf("%c",ch);
                                pos++;
                        } while((ch != '\"') && (ch != '\''));
```

```
                }
                else

                /* Check for == token */
                if((ch == '=') && (input[pos] == '='))
                {
                        printf("=");
                        pos++;
                }
        }
        else

        /* Check for double token */
        {
                ds[0] = ch;             /* Load double token */
                ds[1] = input[pos];
                ds[2] = '\0';
                if(strstr(doubles,ds) != '\0')
                {
                        printf("%s",ds);        /* Print it */
                        pos++;
                }
                else
                        printf("%c",ch);
        }
        printf("\n");
    }
}
```

The basic operation of Program 6.29 is as follows: If an input character is alphabetic, assume the beginning of a variable or function name. Otherwise, scan the singles and doubles arrays for a match. If no match is found, the unmatched input text is simply outputted as an undefined token.

A sample execution is as follows:

```
Enter a C statement => a = max(values,n)*25;
Token: a
Token: =
Token: max
Token: (
Token: values
Token: ,
Token: n
Token: )
Token: *
Token: 25
Token: ;
```

As you can see, even a simple assignment statement is composed of many individual tokens. You are encouraged to think of ways to improve the tokenizer. For example, how would hexadecimal or float numbers be tokenized?

Encoding Text

In the interest of security, many organizations now encode their computer data to keep prying eyes from gathering information. Many techniques exist for encoding (or encrypting) text, some of which have been around for hundreds of years.

One of the simplest encoding techniques is called *transposition* encoding. In this technique, the input text is written as a two-dimensional array of characters. Then the array is transposed (rows and columns reversed) and the characters read out. Consider this sample input:

c programming is fun

Not including the blanks between words, we have 17 characters. Writing the input string down in matrix form, we have:

```
c p r o g
r a m m i
n g i s f
u n
```

Now, reading the characters out of the matrix a column at a time (ignoring blanks), we have:

crnupagnrmiomsgif

which is the transposition-encoded string. Blanks are ignored, because their positions might give away information about where words begin or end.

Program 6.30 performs transposition encoding by creating a square matrix whose dimensions are based on the length of the input string. Unused matrix elements are filled with blanks to eliminate the possibility of stray characters appearing in the output.

Program 6.30

```
#include <stdio.h>
#include <string.h>
#include <math.h>

main()
{
        char input[80];
        char encoder[9][9];
        int r,c,p,n,i;

        printf("Enter a text string to encode => ");
        gets(input);
        i = 0;
        for(n = 0; n < strlen(input); n++)
                if(input[n] == ' ')
                        i++;
        n = 1+sqrt(strlen(input) - i);
        for(r = 0; r < 9; r++)
                for(c = 0; c < 9; c++)
                        encoder[r][c] = ' ';
        p = 0;
        r = 0;
        c = 0;
```

```
        while(input[p] != '\0')
        {
                if(input[p] != ' ')
                {
                        encoder[r][c] = input[p];
                        r++;
                        if(r == n)
                        {
                                r = 0;
                                c++;
                        }
                }
                p++;

        }
        printf("Transposition encoding: ");
        for(r = 0; r < 9; r++)
                for(c = 0; c < 9; c++)
                        if(encoder[r][c] != ' ')
                                printf("%c",encoder[r][c]);
}
```

Running Program 6.30 with the sample string previously used gives the following results:

```
Enter a text string to encode => c programming is fun
Transposition encoding: crnupagnrmiomsgif
```

Reverse transposition encoding gives the original string back. It is up to you to figure out how this is done.

Conclusion

In this section we examined some useful string applications. Test your understanding of this material with the following section review.

6.7 Section Review

1. Why is it necessary to subtract 0x30 from each digit in the ISBN checker?
2. Determine the postorder expression for this input expression:

   20-(2/(6-4)+3*(1+7))/4

   Check your answer with the one given by Program 6.26.
3. How is the postfix stack manipulated in Program 6.26?
4. How are lowercase string characters converted into uppercase?
5. What technique is used to check for palindromes in Program 6.28?
6. What does a tokenizer do? Give an example.
7. Use transposition encoding to encode this string: `ken is here`

6.8 Troubleshooting Techniques

Discussion

In this section we will review a number of techniques designed to help prevent compile and run time errors from occurring when working with strings. Keep these techniques in mind when you write your own string functions.

Reserve the Right Amount of Storage

There are two methods of defining the length of a string. In the first method, the compiler counts the number of characters between quotes, and reserves the exact number of locations required. For example, in:

```
char string[] = "Hello.";
```

the compiler reserves 7 locations (6 for characters and the seventh for the NULL character that terminates the string). This string cannot be made larger (no characters can be added onto it) without unpredictable results at runtime.

In the second method, the number of characters in the string is specified by the programmer, as in:

```
char string[80];
```

In this case, the string may contain from 0 to 80 characters. This method is preferred when the string represents interactive input from a user and the length of the input data is unknown. This type of string can be made larger safely during runtime, as long as the maximum of 80 characters is not exceeded.

Initialize the String

When defining a string like this:

```
char string[80];
```

it may be necessary to initialize the string so that its contents will be correctly interpreted by functions that use the string. For a local string variable, it may be necessary to write the NULL character into the first position so that the string looks empty:

```
string[0] = '\0';
```

This is not needed when the string is defined globally or as a static type, as the string locations are preloaded with zeros.

The string may also be initialized by input from the user, via `scanf()` or some other string function. In this case, it is important to reserve enough string space to contain all the user input.

Know Your Dimensions

In Section 6.6 we examined the operation of a bubble-sorting application for strings. Sorting was accomplished by selectively swapping elements of the two-dimensional input string. The swapping was performed by the `strcpy()` function. `strcpy()` worked correctly on two-dimensional strings that were defined like this:

```
char names[5][9] = { {"kristen "},
                  {"turner  "},
                  {"kenny   "},
                  {"victoria"},
                  {"kimberly"} };
```

Note that both array dimensions are specified in the variable definition for names. This is a rectangular string array.

`strcpy()` gives unpredictable results when the array is defined as a ragged array, as in:

```
char *names[5] = { {"kristen"},
                  {"turner"},
                  {"kenny"},
                  {"victoria"},
                  {"kimberly"} };
```

where the length of each string element is not fixed. Recall that the swap of the `"turner"` and `"kenny"` elements resulted in `""` (an empty string) where `"victoria"` used to be, which is not a correct result. The ragged array has the advantage in regards to the amount of storage space required for the entire two-dimensional string, but is not accessed the same way as a rectangular array. As the programmer, you must know which type of string to use in a particular application. For two-dimensional string arrays that will not be modified, the ragged array will provide an efficient storage method. If the string elements are going to be modified (swapped or increased in size), a rectangular string array would be the better choice.

Conclusion

This section pointed out a number of details that must be considered when handling strings of one or more dimensions. Check your understanding of the material with the following section review.

6.8 Section Review

1. Why are the number of locations reserved for a string important?
2. Explain how an uninitialized string may cause the string function `strlen()` to give an incorrect result.
3. Is it possible to correctly swap two elements of a ragged string array? If so, show the associated string definition.

6.9 Case Study: Text Formatter

Discussion

String handling is such an important part of many programming applications that we now choose to devote the case study for this chapter to a string-based application. The application presented here takes the form of a text formatter. Text formatters are commonly available with most word-processing programs. The purpose of the text formatter is to adjust the way a block of text is displayed or printed by inserting the right number of blanks between words on any given line in such a way that all lines exactly fit between the predefined left and right margins. For example, consider this block of input text:

```
The microcomputer is an important part of any engineering
curriculum. The machine itself can be used to teach many
hardware-based course topics, such as interfacing, memory
design, and microprocessor fundamentals. The software packages
now available offer the student a wide variety of
applications, including graphics, word processing,
spreadsheets, programming, and analog/digital design. All in
all, the microcomputer represents a valuable teaching and
productivity aid.
```

It may be necessary to reformat the text so that each line of text contains no more than 45 characters—possibly because of a printer requirement (the width of special paper loaded into the printer) or some other personal preference. When the block of text is reformatted to 45-character lines, we get:

```
The microcomputer is an important part of any
engineering  curriculum.  The  machine itself
can  be  used  to  teach  many hardware-based
course  topics,  such  as interfacing, memory
design,  and microprocessor fundamentals. The
software  packages  now  available  offer the
student   a  wide  variety  of  applications,
including      graphics,     word processing,
spreadsheets, programming, and analog/digital
design.   All   in   all,  the  microcomputer
represents    a    valuable    teaching   and
productivity aid.
```

Notice how lines (such as the third line from the bottom) that contain fewer than 45 characters are expanded with multiple blanks inserted between every two words. This block of formatted text certainly looks neater than the previous one, where each line ended at a different position.

The application program developed here creates this type of output format.

The Problem

The problem with this type of formatting is twofold. First, how do we determine how many words will fit on a single line of text, and second, each line of displayed text may require a different number of blanks to be inserted to completely fill the margins. For example, consider the third line from the bottom. Its original form, before expansion, looked like this:

```
-->design. All in all, the microcomputer        <--
```

Why was the next word in the block of text being formatted ("represents") not included? The answer is simple. The --> and <-- are used to signify the desired 45-character margins. The line of text contains 37 characters, which leaves 45 – 37 = 8 blanks needed for expansion. The next word ("represents") requires ten character positions and there are only eight remaining. Thus it is necessary to expand the line as it currently looks (to place "represents" at the beginning of the *next* line). The problem then becomes this: How do we insert an equal number of blanks between every two words in the line of text? One solution is to advance through the line of text until a blank is found and then insert a new blank, then advance again until the end of the next word is found and insert a blank there. This process is repeated until the desired number of blanks have been inserted. The third line of text is expanded in the following way:

```
-->design.  All in all, the microcomputer       <--
-->design.  All  in all, the microcomputer      <--
-->design.  All  in  all, the microcomputer     <--
-->design.  All  in  all,  the microcomputer    <--
-->design.  All  in  all,  the  microcomputer   <--
-->design.   All  in  all,  the  microcomputer  <--
-->design.   All   in  all,  the  microcomputer <--
-->design.   All   in   all,  the  microcomputer<--
```

Although there may be other techniques for performing this expansion, this is the one used here.

Developing the Algorithm

The steps in the text formatting application are as follows:

1. Load new words into output buffer until no more words fit.
2. Expand the output buffer to fill the margins.
3. Display the output buffer.
4. Repeat steps 1 through 3 until no more words are left.
5. Do not expand the last line of text stored in the output buffer.

Each step requires examination of the input text and output text buffer. The built-in string handling functions will be used as necessary.

The Overall Process

The text formatter works in the following way: For each line of text that will be formatted, we must:

1. Start with an empty output buffer (buffer filled with nulls).
2. Determine the length of the next word of text.
3. If there is room in the output buffer for the next word, copy it in and go back to step 2.
4. If there is not enough room for the next word, expand the output buffer, display it, and start the new output buffer with the next word.
5. Repeat steps 1 through 4 until there are no more words available.
6. If the output buffer is not empty when the end of the input text is reached, simply display the output buffer without expansion.

This process is implemented by a number of functions called from `main()`, which are:

`initbuff()`	Initialize empty output buffer.
`get()`	Get next word from input text.
`loadword()`	Load next word into output buffer.
`expand_line()`	Expand output buffer.

The code for `main()` is as follows:

```
main()
{
     spaceleft = WIDTH;
     where = 0;
     initbuff(buffer);
     do
     {
          next = get(text, where, nextword);
          if(spaceleft >= strlen(nextword))
          {
               loadword(buffer,nextword);
               where = next;
          }
          else
          {
                expand_line(buffer,EXP);
                initbuff(buffer);
                spaceleft = WIDTH;
          }
      } while(0 != strlen(nextword));
      if(0 != strlen(buffer))
          expand_line(buffer,NOEXP);
}
```

The value of `WIDTH` is predefined to be the size of the margins. The `where` variable points to the current position in the input text, which is saved in the string `text`. The output buffer is called `buffer` and is utilized by `initbuff()`, `get()`, `loadword()`, and `expand_line()`. `EXP` and `NOEXP` stand for *expand* and *no-expand* and are used to control what `expand_line()` does with the output buffer.

Program 6.31 shows how the text formatter is implemented.

Program 6.31

```
#include <stdio.h>
#include <string.h>
#define EXP 1
#define NOEXP 0
#define WIDTH 45

void initbuff(char buff[]);
int get(char data[], int ptr, char word[]);
void loadword(char buff[], char word[]);
```

```
void expand_line(char buff[], int how);

        char text[] = "The microcomputer is an important part of any"
                      " engineering curriculum. The machine itself"
                      " can be used to teach many hardware-based"
                      " course topics, such as interfacing, memory"
                      " design, and microprocessor fundamentals."
                      " The software packages now available offer"
                      " the student a wide variety of applications,"
                      " including graphics, word-processing,"
                      " spreadsheets, programming, and analog/digital"
                      " design. All in all, the microcomputer represents"
                      " a valuable teaching and productivity aid.";
        char buffer[80];
        char nextword[80];
        int spaceleft,where,next;

main()
{
        spaceleft = WIDTH;
        where = 0;
        initbuff(buffer);
        do
        {
                next = get(text, where, nextword);
                if(spaceleft >= strlen(nextword))
                {
                        loadword(buffer,nextword);
                        where = next;
                }
                else
                {
                        expand_line(buffer,EXP);
                        initbuff(buffer);
                        spaceleft = WIDTH;
                }
        } while(0 != strlen(nextword));
        if(0 != strlen(buffer))
                expand_line(buffer,NOEXP);
}

void initbuff(char buff[])
{
        int z;

        for(z = 0; z <= WIDTH; z++)
                buff[z] = '\0';
}
```

```
int get(char data[], int ptr, char word[])
{
        int i = 0;

        /* Skip blanks between words */
        while((data[ptr] == ' ') && (data[ptr] != '\0'))
                ptr++;

        /* Copy characters until blank or NULL found */
        while((data[ptr] != ' ') && (data[ptr] != '\0'))
        {
                word[i] = data[ptr];
                i++;
                ptr++;
        }
        word[i] = '\0';       /* Fix string end */
        return ptr;
}

void loadword(char buff[], char word[])
{
        strcat(buff,word);               /* Copy new word to buffer */
        spaceleft -= strlen(word);       /* Adjust remaining buffer space */
        if(WIDTH > strlen(buff))         /* Is buffer full? */
        {
                strcat(buff," ");        /* Add blank after word if not */
                spaceleft--;
        }
}

void expand_line(char buff[], int how)
{
        int n,k,exp;

        if(how == 0)                     /* Do not expand buffer? */
                printf("%s\n",buff);
        else
        {
                /* Fill end of buffer with NULL's */
                n = WIDTH;
                while((buff[n] == '\0') || (buff[n] == ' '))
                {
                        buff[n] = '\0';
                        n -= 1;
                }

                /* Determine number of blanks to expand */
                exp = WIDTH - strlen(buff);
```

```
        n = 0;
        while(exp > 0)      /* While there are blanks to insert... */
        {
                do                  /* Skip current word */
                {
                        n++;
                } while((buff[n] != ' ') && (buff[n] != '\0'));

                /* Reset buffer index if end is reached */
                if(buff[n] == '\0')
                        n = 0;
                else
                {                   /* Otherwise, insert one */
                        do          /* new blank into buffer */
                        {
                                n++;
                        } while(buff[n] == ' ');
                        for(k = WIDTH - 1; k > n; k--)
                                buff[k] = buff[k-1];
                        buff[n] = ' ';
                        n++;
                        exp--;  /* One less blank to insert */
                }
        }
        printf("%s\n",buff);    /* Print expanded buffer */
    }
}
```

Conclusion

The text formatting application requires us to really understand how to use and manipulate strings. You are encouraged to think of a different way to achieve the same goal. For example, instead of loading one word at a time, you may choose to scan the input text until you know that you have reached the format width and then load the entire block of text at that point. You may also perform the expansion in a different way, by keeping track of the number of words on any given line and using the word count to mathematically compute the required number of spaces between words.

Check your understanding of this section by trying the following section review.

6.9 Section Review

1. Why is text formatting desired?
2. How is the length of the next word in the input text determined?
3. What happens if the output buffer is completely filled with text?
4. What happens when the `WIDTH` value is changed?
5. What happens if the `WIDTH` value is changed to 10?
6. Should restrictions be placed on the value of `WIDTH`?
7. How are blanks inserted in `expand_line()`?
8. Can ragged arrays be utilized to make the formatting work more easily?

Interactive Exercises

DIRECTIONS

Execute each program on your computer and note the results of its execution. In many cases you are asked to predict what the program will do. Compare your predictions with the actual program execution.

Exercises

1. Program 6.32 contains an uninitialized string. What will be displayed when the program is executed?

Program 6.32

```
#include <stdio.h>

main()
{
        char string[10];

        printf("%s",string);
}
```

2. What does Program 6.33 display when executed?

Program 6.33

```
#include <stdio.h>

main()
{
        char string[10];
        int i;

        for(i = 0; i <= 10; i++)
                string[i] = i + 65;
        printf("%s",string);
}
```

3. What does Program 6.34 display when executed?

Program 6.34

```
#include <stdio.h>

void xyz(char abc[]);

main()
{
        char *string;

        xyz(string);
        printf("%s",string);
```

```
}

void xyz(char abc[])
{
        char def[] = "Does this work?";

        abc = def;
}
```

4. What does Program 6.35 display when executed?

Program 6.35

```
#include <stdio.h>
#include <string.h>

main()
{
        char abc[] = "Is this printed?";
        int i;

        abc[7] = abc[strlen(abc)];
        printf("%s",abc);
}
```

5. What does Program 6.36 display when executed?

Program 6.36

```
#include <stdio.h>
#include <string.h>

main()
{
        char sa[80] = "The dark side ";
        char sb[20] = "of midnight.";
        char sc[] = "the moon.";

        strncat(sa,sb,3);
        strcat(sa,sc);
        printf("%s",sa);
}
```

6. Is the integer output of Program 6.37 positive or negative?

Program 6.37

```
#include <stdio.h>
#include <string.h>

main()
```

```
{
        char sa[] = "cpu";
        char sb[] = "CPU";

        printf("%d",strcmp(sa,sb));
}
```

7. What does Program 6.38 display when executed?

Program 6.38

```
#include <stdio.h>
#include <string.h>

main()
{
        char string[] = "This is only a test...";

        printf("%s",strchr(string,'t'));
}
```

8. What does Program 6.39 display when executed?

Program 6.39

```
#include <stdio.h>
#include <string.h>

main()
{
        char string[] = "This is only a test...";

        printf("%s",strrchr(string,'t'));
}
```

9. What does Program 6.40 display when executed?

Program 6.40

```
#include <stdio.h>
#include <string.h>

main()
{
        char sa[] = "C programming is fun!";
        char sb[] = "program";
        char sc[] = "fun ";

        printf("%s\n",strstr(sa,sb));
        printf("%s\n",strstr(sa,sc));
}
```

Self-Test

DIRECTIONS

Answer the following questions by referring to the programs in Section 6.7 (Example Programs).

Questions

1. When is the X used in an ISBN number? How does Program 6.25 generate the X?
2. How are priorities determined before accessing the postfix stack in Program 6.26?
3. What happens when a ')' is encountered in the input expression when building a postfix expression?
4. Why are the `vowels` and `vcount` declared as `static int`, and not simply `int` in Program 6.27?
5. How does Program 6.28 know when it has completely tested a palindrome string?
6. Why is it more difficult to test for tokens in the doubles array than in the singles array in Program 6.29?
7. How is a string extracted as a single token in Program 6.29?
8. How is the encoder matrix filled in Program 6.30?

End-of-Chapter Problems

General Concepts

Section 6.1

1. What is the index of the first character in a C string?
2. State how many array elements are required in a C string consisting of eight characters. Explain.
3. Explain the format of a C character string.
4. How are character string variables declared in C?

Section 6.2

5. What are three ways of initializing the string `"Data"`?
6. Why dimension a string variable with a number larger than that required? For instance, why use `char string[80] = "Hello";` instead of `char string[] = "Hello";`?
7. What are the differences between `scanf()` and `gets()` in relation to string input?

Section 6.3

8. What is actually passed between functions when one is dealing with strings?
9. Where must storage space for a string be reserved when strings are passed between functions?

Section 6.4

10. Why make the function `isprint()` available? What value does it have?
11. Does `getchar()` echo characters to the display?
12. Why write our own conversion functions when `atoi()` and `atof()` exist?

Section 6.5

13. If the null character is missing from the end of a character string, what does `strlen()` do?
14. What happens to each string used in a `strcat()` operation?

15. Why does it make sense that “Harris” has a smaller value than “Harrison”? How would you determine this?
16. How could `strchr()` be used to scan a string for math symbols +, -, *, and /?
17. What other string operations might be useful? Explain how they might be implemented.

Section 6.6

18. What is a rectangular array of characters? How is it stored in memory?
19. What are braces `{}` used for in a string declaration?
20. What are the different ways of defining a two-dimensional character string?
21. What is a ragged array?

Section 6.7

22. How might an expression be evaluated using recursion?
23. Why are `static ints` used in the vowel counter (Program 6.27)?
24. What are the values of `lchar` and `rchar` when Program 6.28 completes execution with an input string of `"100101"`?

Section 6.8

25. What three details are important to any string usage?
26. A series of 7-digit phone numbers must be sorted. Should a rectangular string array or a ragged string array be used?

Section 6.9

27. How is a very long character string defined?
28. What is the technique used to expand a line of text?
29. Why is the last `if` statement needed in the `main()` routine of Program 6.31?

Program Design

In developing the following C programs, use the method described in the case study. For each program you are assigned, document your design process and hand it in with your program. This process should include the design outline, process on the input, and required output as well as the program algorithm. Be sure to include all of the documentation in your final program. This should consist of, but not be limited to, the programmer's block, function prototypes, and a description of each function as well as any formal arguments you may use.

30. Write a C program that will ask the user to enter a number in scientific notation (see Section 6.4) and check the number for validity. If valid, convert the number into the appropriate floating value.
31. Write a C program that will generate and display a random 10-by-10 array of letters from the alphabet.
32. Modify the program of problem 31 so that the array is scanned row by row and column by column for any of the following words:

one	two	three	four	five	six
seven	eight	nine	ten	help	see
saw	boy	girl	fast	slow	up
down	left	right	top	go	stop

33. Write your own `expand_line()` function for the text formatter presented in the case study section.
34. Write a C function that scans the two-dimensional tic tac toe matrix TICTAC[3][3] for three ‘X’s in any row, column, or diagonal. Return 1 if three ‘X’s are found and 0 otherwise.

35. Write a C program that checks a user-supplied binary string to determine if it is of the form WW, where W is any binary string. For example, 110110 is accepted, because W is 110. 10111011 is also accepted, because W must be 1011. However, 100001 is not accepted, because 100 does not match 001.
36. Improve the tokenizer developed in Program 6.29 so that more standard C operators are identified.
37. Write a C program that decodes a transposition-encoded string.
38. Create a C program that will give the program user the color of an area of the grid system shown in Figure 6.8. The program user must enter the row and column number.

Column

Row	1	2	3	4
1	Red	Green	Blue	White
2	Violet	Amber	Brown	Black
3	Orange	Pink	Magenta	Yellow
4	Silver	Gold	Slate	Pink

Figure 6.8 Grid System for Problem 38

7 Arrays

Objectives

This chapter provides you the opportunity to learn:

1. What an array is.
2. How to create and initialize an array.
3. Applications of arrays.
4. How to pass arrays between C functions.
5. The basic concepts of sorting with numeric arrays.
6. How to merge two sorted arrays.
7. Matrix algebra.

Key Terms

Dimension
Array Initialization
Array Passing
Array Index
Bubble Sort
Bucket Sort
Merge Sort
Quick Sort
Pivot
Multidimensional Array
Matrix
Row
Column
Field Width Specifier
Bounds Checking
Backtracking
Row-Major Order

Outline

Introduction

You will note some similarities between the operation of string variables (covered in Chapter 6) and the operation of numeric array variables. Both utilize multiple elements accessed with an index, and must be dimensioned or initialized beforehand to work properly. In this chapter, we will concentrate on the use of numeric arrays in new applications, and examine the methods used to access, partition, sort, merge, and search them.

7.1 Numeric Arrays

Discussion

In this section you will learn about numeric arrays. You will discover how to dimension a numeric array and learn more about the relationship between array elements and pointers.

Basic Idea

To **dimension** a numeric array, you simply place a number inside its brackets. As an example, `int array[3];` means that an array of three elements has been declared. These elements will consist of `array[0]`, `array[1]`, and `array[2]`.

Program 7.1 shows the relationship between `array` and `&array[0]`.

Program 7.1

```
#include <stdio.h>

main()
{
        int array[3];

        printf("array = %X\n",array);
        printf("&array[0] = %X\n",&array[0]);
}
```

Execution of Program 7.1 produces the same value for each of the two variables:

```
array = 7325
&array[0] = 7325
```

What this means is that the name of the array variable and the address of the first element of the array have the same value. This is the memory location of the first element of the array. This is illustrated in Figure 7.1.

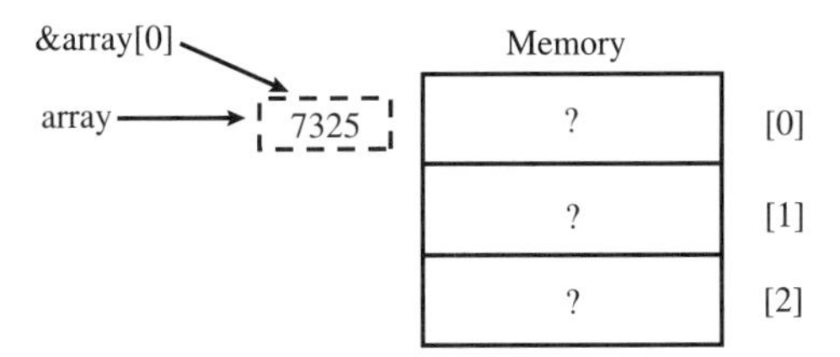

Figure 7.1 First Address of the Array

The Type int Array

Note that Program 7.1 uses an `int` as an array. Recall that on a personal computer an `int` uses *two* 8-bit memory locations (a `char` uses one). To illustrate this, the same array will be looked at again in Program 7.2. The difference this time is that the address of each array element is displayed.

Program 7.2

```
#include <stdio.h>

main()
{
        int array[3];

        printf("address array[0] => %d\n",&array[0]);
        printf("address array[1] => %d\n",&array[1]);
        printf("address array[2] => %d\n",&array[2]);
}
```

Execution of Program 7.2 yields

```
address array[0] => 7325
address array[1] => 7327
address array[2] => 7329
```

Note from the output that each address is 2 larger than the previous one. This is because a type `int` is the arrayed variable which allocates two bytes per location. What is going on here is illustrated in Figure 7.2.

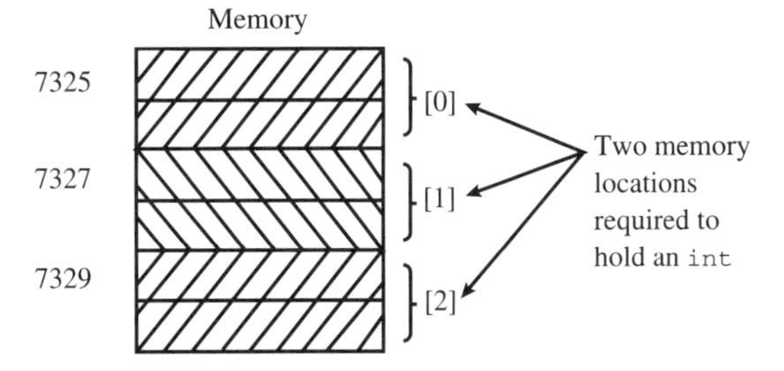

Figure 7.2 Address with Type `int` Arrays

What this means is that C will automatically designate the required memory space depending on the data type you are using in the array. However, the fact remains that the elements of the array are contiguous in memory.

Inside Arrays

In Program 7.2, you saw the address of each array element. What is inside each of these elements (what is stored at these addresses)? Program 7.3 gives you an idea of what is there.

Program 7.3

```
#include <stdio.h>

main()
{
        int array[3];

        printf("contents of array[0] => %d\n",array[0]);
        printf("contents of array[1] => %d\n",array[1]);
        printf("contents of array[2] => %d\n",array[2]);
}
```

Execution of Program 7.3 produces

```
contents of array[0] => 0
contents of array[1] => -5672
contents of array[2] => 58
```

Where did these values come from? They are just what happened to be in those memory locations at the time the program was executed. What has happened is that the program has set aside enough memory space for the three-element array, but it hasn't put anything in these memory locations. Now consider a slight variation of the same program. This time, however, the array is being declared as a global variable (Program 7.4).

Program 7.4

```
#include <stdio.h>

        int array[3];

main()
{
        printf("contents of array[0] => %d\n",array[0]);
        printf("contents of array[1] => %d\n",array[1]);
        printf("contents of array[2] => %d\n",array[2]);
}
```

Look at what happens when Program 7.4 is executed:

```
contents of array[0] => 0
contents of array[1] => 0
contents of array[2] => 0
```

Now all of the elements of the array have been set to zero. Setting the elements of an array to a known quantity is called **array initialization**.

The only difference between the last two programs is that in Program 7.3 the array was declared automatic (local), whereas in Program 7.4 the array was declared global (external). This illustrates an important point in C. Arrays declared global are initialized to zero by default, whereas local arrays are not.

When it is necessary to initialize a local numeric array to zero, an alternative technique that may be used is as follows:

```
main()
{
     static int array[3];
     .
     .
     .

}
```

Notice the use of the word *static* in the array declaration. A static integer array is automatically initialized to all zeros the first time the program is executed. However, since *static* has other features, such as control over the variables' lifetime, it is best for the beginning programmer to use the intialization technique described in the next section.

Putting in Your Own Values

You can place your own values into an array. One method of doing this is shown in Program 7.5.

Program 7.5

```
#include <stdio.h>

main()
{
        int array[3];

        array[0] = 10;
        array[1] = 20;
        array[2] = 30;

        printf("contents of array[0] => %d\n",array[0]);
        printf("contents of array[1] => %d\n",array[1]);
        printf("contents of array[2] => %d\n",array[2]);
}
```

Execution of Program 7.5 produces

```
contents of array[0] => 10
contents of array[1] => 20
contents of array[2] => 30
```

As you can see from Program 7.5, each element of the array now contains a value entered by you. You could also have used pointers to get the contents of each of your array elements in the program. Program 7.6 shows how to do this.

Program 7.6

```
#include <stdio.h>

main()
{
        int array[3];
        int *ptr;

        array[0] = 10;
        array[1] = 20;
        array[2] = 30;

        ptr = array;

        printf("contents of array[0] => %d\n",*ptr);
        printf("contents of array[1] => %d\n",*(ptr + 1));
        printf("contents of array[2] => %d\n",*(ptr + 2));
}
```

Execution of Program 7.6 yields exactly the same results as did Program 7.5:

```
contents of array[0] => 10
contents of array[1] => 20
contents of array[2] => 30
```

This demonstrates the following equalities:

```
array[0] = *ptr;
array[1] = *(ptr + 1);
array[2] = *(ptr + 2);
```

Note that `*ptr` is declared as type `int`. This lets C know that every increment of `*ptr` is to be two bytes and not one (because type `int` uses two memory locations on a personal computer).

Passing Numeric Arrays

You can pass numeric arrays from one function to another. **Array passing** is shown in Program 7.7.

Program 7.7

```
#include <stdio.h>

void function1(int this[]);

main()
{
```

```
        int array[3];

        array[0] = 10;
        array[1] = 20;
        array[2] = 30;

        function1(array);
}

void function1(int this[])
{
        printf("contents of array[0] => %d\n",this[0]);
        printf("contents of array[1] => %d\n",this[1]);
        printf("contents of array[2] => %d\n",this[2]);
}
```

Execution of Program 7.7 produces

```
contents of array[0] => 10
contents of array[1] => 20
contents of array[2] => 30
```

Program Analysis

Program 7.7 declares a function prototype of type `void` that contains an array argument:

```
void function1(int this[]);
```

Notice that the array is called `this[]`. The indirection operator `*` is not used (it could have been) because `this[]` is a pointer that will point to the first element of the array.

The function `main()` also defines an array type consisting of three elements; then it initializes each element:

```
int array[3];

array[0] = 10;
array[1] = 20;
array[2] = 30;
```

After this initialization, the function `function1` is then called:

```
function1(array);
```

Note that the actual argument did not use the `&` operator. (It could have used `&array[0]`.) However, `array` is the beginning address of the array. This is what is passed to `function1`.

Next, `function1` receives the value of the address of the first array element and then proceeds to display the values of the first three elements.

```
printf("contents of array[0] => %d\n",this[0]);
printf("contents of array[1] => %d\n",this[1]);
printf("contents of array[2] => %d\n",this[2]);
```

What is happening is shown in Figure 7.3.

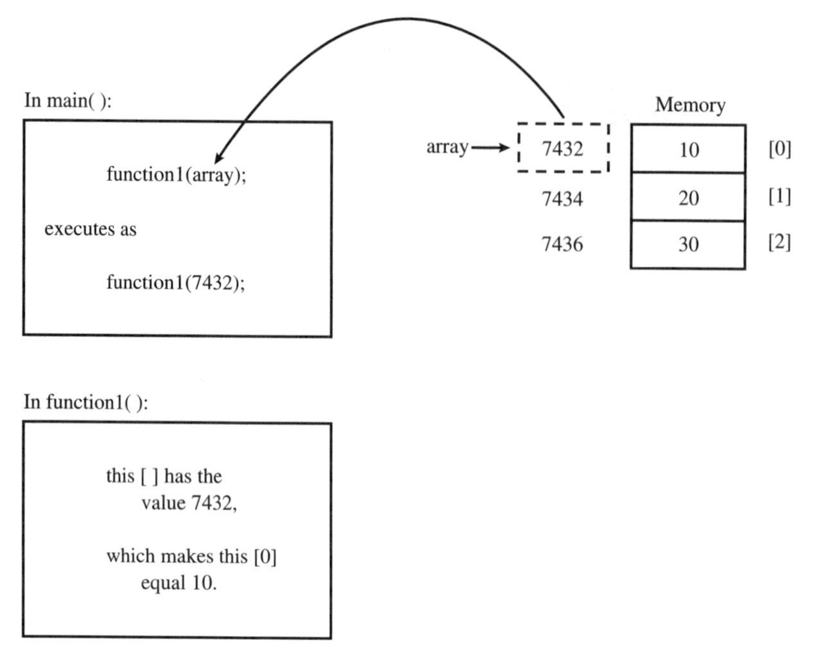

Figure 7.3 Passing an Array Address to a Called Function

The next program (Program 7.8) illustrates the passing of an array from a called function.

Program 7.8

```
#include <stdio.h>

void functionl(int this[]);

main()
{
        int array[3];

        functionl(array);

        printf("contents of array[0] => %d\n",array[0]);
        printf("contents of array[1] => %d\n",array[1]);
        printf("contents of array[2] => %d\n",array[2]);
}

void functionl(int this[])
{
        this[0] = 20;
        this[1] = 40;
        this[2] = 60;
}
```

Passing Numeric Arrays Back

Passing a numeric array back from a called function implies that the entire array is passed, although in actuality nothing is passed back. What really happens is that the called function uses the starting address of the array passed to it by the calling function to write new data into the memory locations reserved for the numeric array. Program 7.8 demonstrates this principle by having `main()` pass the starting address of the array to `function1()`. During `function1()`'s execution, the array elements are initialized (*passed back*) by the fact that `function1` knows the address of the array elements and can access the actual memory location for each element.

The output of Program 7.8 is

```
contents of array[0] => 20
contents of array[1] => 40
contents of array[2] => 60
```

Again, the starting address of the array is passed to the called function by `function1(array);` in `main()`. Since the starting address is known by `function1`, each element of the array can be initialized. Figure 7.4 illustrates the process.

Conclusion

As you can see from the preceding programs, arrayed variables can be easily passed between functions in a C program.

Check your understanding of this section by trying the following section review.

7.1 Section Review

1. How can you let C know how many elements an array will have?
2. What is always the index of the first element of a C array?

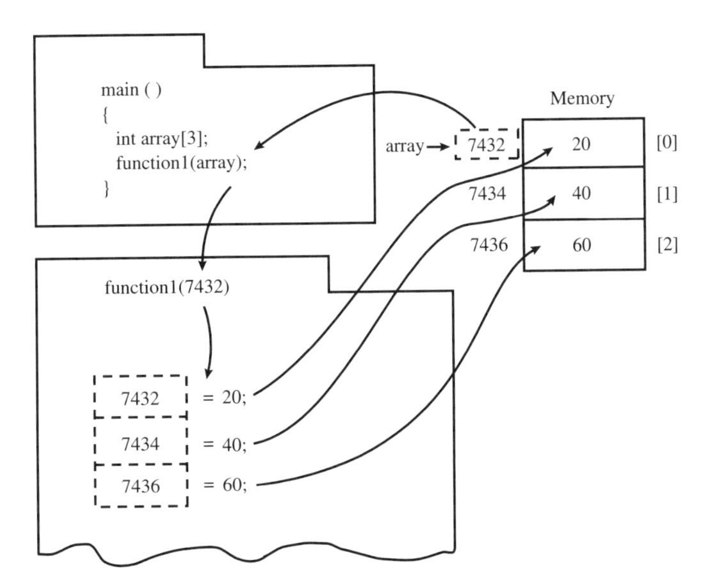

Figure 7.4 Process of Initialization in a Called Function

3. If the variable `int value[5];` is declared in a function, what is the relationship between `value`, `&value`, and `&value[0]`?
4. What is the difference between declaring a local array and declaring a global array?
5. Why do `char abc[3];` and `int def[3];` reserve different amounts of memory?

7.2 Introduction to Numeric Array Applications

Discussion

This section introduces you to the fundamental concepts of applying arrays in technical programs. Here you will see how to use arrays to rearrange a list of numbers. This will be developed into a program that will then be able to find the smallest value from a list of user input values. The material presented here lays the foundation for the information in the next section of this chapter.

Working with the Array Index

The one idea behind array applications is the ability to work with the index of the array. It is important not to think of the value of the **array index** as a fixed number. These values can be manipulated in any manner you choose. Understanding this concept can produce many powerful technical programs. Program 7.9 illustrates this concept.

Program 7.9

```
#include <stdio.h>

main()
{
        int number_array[7];
        int index;

        /*   Place values in the array.   */

        number_array[0] = 0;
        number_array[1] = 2;
        number_array[2] = 4;
        number_array[3] = 8;
        number_array[4] = 16;
        number_array[5] = 32;
        number_array[6] = 64;

        for(index = 1; index <= 5; index++)
        {
                printf("number_array[%d] = %d",index,number_array[index]);
                printf(" number_array[%d + 1] = %d",index,number_array[index+1]);
                printf(" number_array[%d - 1] = %d\n",index,
                        number_array[index-1]);
        }
}
```

Execution of Program 7.9 produces

```
number_array[1] = 2 number_array[1 + 1] = 4 number_array[1 - 1] = 0
number_array[2] = 4 number_array[2 + 1] = 8 number_array[2 - 1] = 2
number_array[3] = 8 number_array[3 + 1] = 16 number_array[3 - 1] = 4
number_array[4] = 16 number_array[4 + 1] = 32 number_array[4 - 1] = 8
number_array[5] = 32 number_array[5 + 1] = 64 number_array[5 - 1] = 16
```

As you can see from this output, changing the array index value causes a corresponding change in the program output value. In Program 7.9, values are assigned to each element of the array. Then, the output of the array elements are displayed. However, to demonstrate the effect on the output of changing the index values, each index value is increased by 1 and then decreased by 1. This is done in order to demonstrate the resulting output. Thus, `number_array[3]` contains the same value as `number_array[2 + 1]` and `number_array[4 - 1]`.

Changing the Sequence

Program 7.10 further illustrates the manipulation of the array index. This program allows the program user to enter nine numbers. The program will then display the numbers in the order opposite to that in which they were entered.

Program 7.10

```
#include <stdio.h>

#define maxnumber 9      /* Maximum number of array elements. */

void run_backwards(int user_array[]);

main()
{
        int number[maxnumber];
        int index;

        printf("Give me nine numbers and I'll print them backwards.\n");

        for(index = 0; index < maxnumber; index++)
        {
                printf("Number[%d] = ",index);
                scanf("%d",&number[index]);
        }

        printf("Thank you...\n");

        run_backwards(number);
}

void run_backwards(int user_array[])
{
        int index;
```

```
    printf("\n\nHere are the numbers you entered displayed\n");
    printf("in the reverse order of entry:\n");

    for(index = maxnumber - 1; index >= 0; index--)
        printf("number[%d] = %d\n",index, user_array[index]);
}
```

Assume that the program user enters the following numbers. Then execution of Program 7.10 will produce

```
Give me nine numbers and I'll print them backwards.
Number[0] = 1
Number[1] = 2
Number[2] = 3
Number[3] = 4
Number[4] = 5
Number[5] = 6
Number[6] = 7
Number[7] = 8
Number[8] = 9
Thank you...

Here are the numbers you entered displayed
in the reverse order of entry:
Number[8] = 9
Number[7] = 8
Number[6] = 7
Number[5] = 6
Number[4] = 5
Number[3] = 4
Number[2] = 3
Number[1] = 2
Number[0] = 1
```

Program Analysis

The program first defines the maximum size of the array with a `#define`:

```
#define maxnumber 9  /* Maximum number of array elements.  */
```

A function prototype that uses an array in its formal argument is then declared:

```
void run_backwards(int user_array[]);
```

Next a C `for` loop gets input from the program user. Note that in this loop, the array index is being incremented, and it goes from 0 (first array element) to 1 less than `maxnumber` (last array element). This is an efficient method of getting array element values from the program user:

```
printf("Give me nine numbers and I'll print them backwards.\n");
for(index = 0; index < maxnumber; index++)
{
```

```
   printf("Number[%d] = ",index);
   scanf("%d",&number[index]);
}
```

The program then calls on the function that will now display the array in reverse order. Note that the pointer of the starting address of the array is used in the actual parameter.

```
run_backwards(number);
```

The called function uses another C `for` loop, but this time the loop starts the array index at `maxnumber - 1` (the maximum index value for the array) and ends at 0, doing a C `--` on the variable `index`.

```
for(index = maxnumber - 1; index >= 0; index--)
   printf("number[%d] = %d\n",index, user_array[index]);
```

In each case, the value of the variable is printed out.

Finding a Minimum Value

Program 7.10 illustrates the changing of the order in which an array is displayed by reordering the array index. This concept will be developed further in Program 7.11, where the smallest value from a list of numbers is extracted. The program user enters the numbers, and the program searches through the list and extracts the smallest value and displays it on the screen.

Program 7.11

```
#include <stdio.h>

#define maxnumber 9   /*  Maximum number of array elements.   */

int minimum_value(int user_array[]);

main()
{
        int number[maxnumber];
        int index;

        printf("Give me nine numbers and I'll find the minimum value:\n");
        for(index = 0; index < maxnumber; index++)
        {
                printf("Number[%d] = ",index);
                scanf("%d",&number[index]);
        }
        printf("Thank you...\n");

        printf("The minimum value is %d\n",minimum_value(number));
}

int minimum_value(int user_array[])
{
```

```
        register int index;
        int minimum;

        minimum = user_array[0];
        for(index = 1; index < maxnumber; index++)
                if(user_array[index] < minimum)
                        minimum = user_array[index];
        return(minimum);
}
```

Assuming the program user enters the following series of numbers, execution of Program 7.11 yields

```
Give me nine numbers and I'll find the minimum value:
Number[0] = 12
Number[1] = 21
Number[2] = 58
Number[3] = 3
Number[4] = 5
Number[5] = 8
Number[6] = 19
Number[7] = 91
Number[8] = 105
Thank you...
The minimum value is 3
```

Program Analysis

Program 7.11 is different from Program 7.10 in the called function. First, Program 7.11 does a function prototype for the new function:

```
int minimum_value(int user_array[]);
```

The function `main()` still gets nine values from the program user in the same manner as before. The difference is function `minimum_value`. This function defines a variable called `minimum`. This will hold the minimum value of the array. First `minimum` is initialized to the value of the first element of the array:

```
minimum = user_array[0];
```

This is done so that the variable `minimum` has a value from the array with which to compare the other values. This comparison is done in a C `for` loop.

```
for(index = 1; index < maxnumber; index++)
    if(user_array[index] < minimum)
        minimum = user_array[index];
```

What happens is that the next element of the array is compared with `minimum`. If this element is less than `mininum`, then `minimum` is given this value. If this is not the case, `minimum` retains its previous value. It is in this manner that the smallest value of the array is selected. This value is then returned to the calling function:

```
return(minimum);
```

Conclusion

In this section you were introduced to the concept of accessing arrayed data. Here you saw how to modify the index of an array in order to accomplish this. This section presented a method of rearranging how an array is displayed to show a minimum value from a list of values entered by the program user. Test your understanding of this section by trying the following section review.

7.2 Section Review

1. State the basic idea behind array applications.
2. Explain what programming method was used in order to get arrayed values from the program user.
3. State the method used to cause a series of entered values to be displayed in the opposite order from which they were entered.
4. What method was used in order to extract a minimum value from a list of entered values?
5. How can the maximum value be extracted from a list of values?

7.3 Sorting with Numeric Arrays

Several different sorting techniques will be examined in this section. These techniques go by the interesting names *bubble sort*, *bucket sort*, *merge sort*, and *quick sort*.

One measure of a sorting algorithm's efficiency is the number of comparisons it makes while sorting. Since the methods used in each technique are fundamentally different, some sorting techniques are more efficient than others (with bubble sort generally regarded as the least efficient). We will examine each technique and see how an array of integers is sorted by each one. The number of comparisons made will be computed in each case for efficiency evaluation.

Bubble Sort

The **bubble sort** technique gets its name from the fact that individual numbers in the array being sorted are *bubbled* to the top of the list as the sort progresses.

As an example, consider the following list of numbers.

6
3
6
8

Suppose it is your job to arrange this list in ascending order (smallest number first, largest last). Assume that you are to use the bubble sorting technique to accomplish this. Here are the rules you would use:

Bubble Sorting Rules (to sort in ascending order):

1. Test only two numbers at a time, starting with the first two numbers.
2. If the top number is smaller, leave as is. If the top number is larger, switch the two numbers.

3. Go down one number and compare that number to the number that follows it. These two will be a new pair.
4. Continue this process until no switch has been made in an entire pass through the list.

To sort in descending order, simply change rule 2 as follows:

2. If the top number is larger, leave as is. If the top number is smaller, switch the two numbers.

To sort the example list of data, start with the first rule. Test only two numbers at a time, starting with the first two numbers. So, using the example list, start with the top two numbers:

6
3

The top number is larger, so using rule 2, switch the two numbers:

3
6

The list now looks like this:

3
6
6
8

Go down one number and compare it with the number that follows (a new number pair):

6
6

These are both the same. Since they are equal, it makes no difference what you do to them!

Go to the next new number pair:

6
8

Since the smaller number is already on top, leave them as is. You have completed the list, but a switch was made on this pass through the list; therefore, you must make another pass through the list:

Testing the first two numbers

3
6

no switch is necessary. Now the next two:

6
6

Again, no switch is necessary. The next two:

6
8

Still no required switch. Because there were no switches in this pass through the list, the sorting is completed and the resulting list is

3
6
6
8

Sample Program

A sample program that implements a bubble sort is shown in Program 7.12.

Program 7.12

```
#include <stdio.h>

#define maxnumber 9     /* Maximum number of array elements.  */

int bubble_sort(int user_array[]);
void display_array(int sorted_array[]);

main()
{
        int number[maxnumber];
        int index,compares;

        printf("Give me nine numbers and I'll sort them:\n");
        for(index = 0; index < maxnumber; index++)
        {
                printf("Number[%d] = ",index);
                scanf("%d",&number[index]);
        }

        compares = bubble_sort(number);

        display_array(number);
        printf("\nThe number of comparisons is %d",compares);

}

int bubble_sort(int user_array[])
{
        int index;
        int switch_flag;
        int temp_value;
        int valtest = 0;

        do
        {

                switch_flag = 0;
```

```
            for (index = 0; index < maxnumber; index++)
            {
                    valtest++;
                    if((user_array[index] > user_array[index + 1])
                            &&(index != maxnumber - 1))
                    {
                            temp_value = user_array[index];
                            user_array[index] = user_array[index + 1];
                            user_array[index + 1] = temp_value;
                            switch_flag = 1;
                    }
            }
        } while (switch_flag);
        return(valtest);
}

void display_array(int sorted_array[])
{
        int index;

        printf("\n\nThe sorted values are:\n");

        for (index = 0; index < maxnumber; index++)
            printf("Number[%d] = %d\n",index,sorted_array[index]);
}
```

Assuming that the program user enters the numbers as shown, execution of Program 7.12 produces

```
Give me nine numbers and I'll sort them:
Number[0] = 6
Number[1] = 7
Number[2] = 5
Number[3] = 8
Number[4] = 4
Number[5] = 9
Number[6] = 3
Number[7] = 0
Number[8] = 2

The sorted values are:
Number[0] = 0
Number[1] = 2
Number[2] = 3
Number[3] = 4
Number[4] = 5
Number[5] = 6
Number[6] = 7
Number[7] = 8
Number[8] = 9

The number of comparisons is 72
```

Program Analysis

After the `#define maxnumber 9`, which is used to set the maximum size of the array, the program defines two function prototypes:

```
int bubble_sort(int user_array[]);
void display_array(int sorted_array[]);
```

The formal parameters of both functions declare `int` arrays. The declaration `user_array[]` is a pointer just as `*user_array` is a pointer. It will be used to store the address of the first element (`[0]`) of the array. The first function `bubble_sort` will be used by `main()` to do the actual sorting. The next function `display_array` is used to display the sorted array.

Function `main()` simply gets nine numbers from the program user:

```
int number[maxnumber];
int index,compares;

printf("Give me nine numbers and I'll sort them:\n");
for(index = 0; index < maxnumber; index++)
{
     printf("Number[%d] = ",index);
     scanf("%d",&number[index]);
}
```

The program excerpt above defines an array `number` consisting of `maxnumber` elements. This means that the first element of the array will be `number[0]` and the last element `number[8]`. When the program user has entered all nine numbers, the program then calls the two functions—one to sort the given numbers, and the other to display them. Note that the actual parameter passed is the address of the first element. Recall that this contains the address of the first element of the array. It is the same as using `&number[0]` as the parameter.

```
compares = bubble_sort(number);

display_array(number);
```

The function `bubble_sort` declares three variables: `index`, `switch_flag`, and `temp_value`. The variable `index` will be used as the index for each of the array elements. `switch_flag` will let the program know when the sorting of the array is completed, and `temp_value` will temporarily store the value of one array element while it is being switched with another array element. You can think of a flag as a variable that is either up (ON) or down (OFF). Thus, for this example, `switch_flag` has just one of two values.

```
int index;
int switch_flag;
int temp_value;
```

In the body of the `bubble_sort` function, a C `do` loop is used to repeatedly go through the array to check if a switch is needed between array elements. The loop starts by setting the `switch_flag` to 0 (thus making `switch_flag` false).

```
do
{
       switch_flag = 0;
```

Next there is a nested `for` loop that causes a scan through each element of the array.

```
for (index = 0; index < maxnumber; index++)
```

Inside the C `for` loop, a comparison of each array element is made, and a check is made to ensure that no more than the maximum number of array elements is tested (`user_array[8]` is the largest array element).

```
if((user_array[index] > user_array[index + 1])&&(index! = maxnumber - 1))
```

If a switch is needed, the following then takes place:

```
{
   temp_value = user_array[index];
   user_array[index] = user_array[index + 1];
   user_array[index + 1] = temp_value;
   switch_flag = 1;
}
```

Notice how the variable `temp_value` is used. This is similar to having a cup of milk and a glass of orange juice and having the task of putting the milk in the glass and the orange juice in the cup. One method of doing this is to get a third container (call it `temp_container`). Pour the milk into the temporary container, pour the juice into the cup, then pour the milk from the temporary container into the glass. `temp_value` is used in this manner. If a switch is made, then `switch_flag` is set to 1 (making it true). This means that the outer `do` loop will have to be repeated:

```
}while (switch_flag);
```

Recall that the way of ensuring that the sorting of the loop is completed is by not having a switch on a comparison through the loop.

The display function simply uses an index counter in a C `for` loop to cause the now sorted values to be displayed on the monitor:

```
for (index = 0; index < maxnumber; index++)
   printf("Number[%d] = %d\n",index,sorted_array[index]);
```

Notice that the number of comparisons made (72) is almost equal to the number of items squared ($9^2 = 81$). Thus, bubble sorting of 50 numbers might require almost 2500 comparisons.

Bucket Sort

Unlike the bubble sort, which can sort numbers of any size, a **bucket sort** requires the numbers to be within a predetermined range. For example, consider the same numbers used in the execution of Program 7.12:

6 7 5 8 4 9 3 0 2

None of these numbers is larger than 9. In addition, none of the numbers is duplicated. These are both important requirements for the bucket sort. Knowing in advance that we will be sorting nonduplicated numbers less than or equal to 9 allows us to initialize a 10-element *bucket array*, as shown in Figure 7.5. As you can see, each element of the array is initialized to 0, which is used to represent an *empty bucket*.

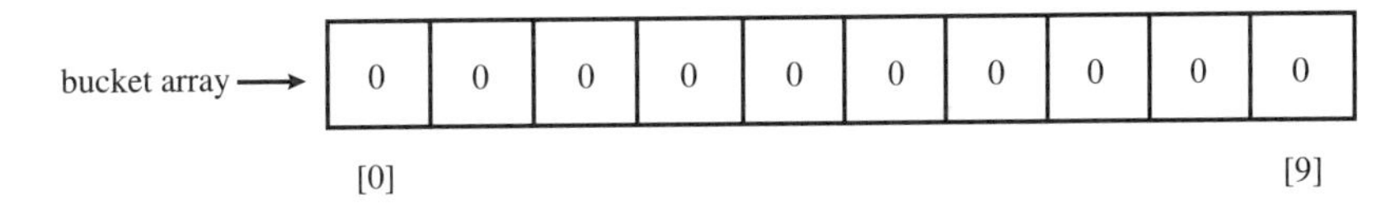

Figure 7.5 Initializing the Bucket Array

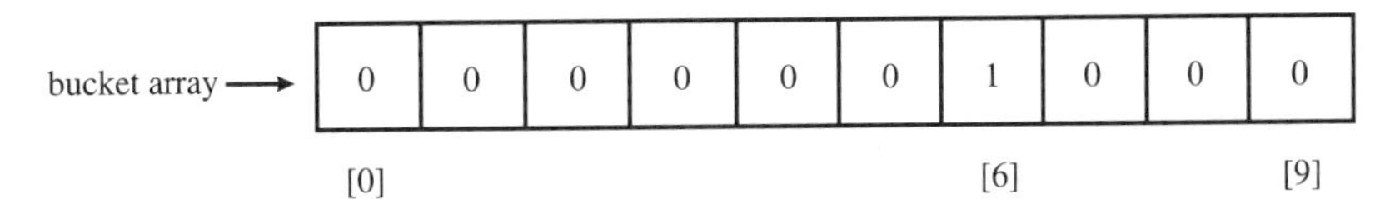

Figure 7.6 Bucket Array After Processing First Element

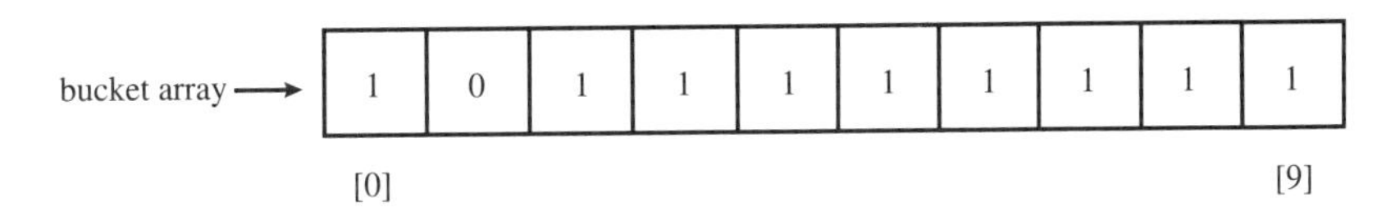

Figure 7.7 Final Bucket Array

The process of bucket sorting is now very straightforward: For each element in the input, we set the associated element in the bucket array equal to 1 (or any nonzero integer), signifying a *full* bucket. For example, Figure 7.6 shows the bucket array after the first input element (6) has been processed. Notice the 1-value in element position [6].

After all nine input elements have been processed, the bucket array contains the element values shown in Figure 7.7. The lone zero remaining indicates that we did not see the value '1' in our input list.

To display the sorted array, it is only necessary to start at the beginning of the bucket array and display each index number that contains a value of 1.

Program 7.13 shows how a bucket sort can be implemented. Because the input numbers are the same as those used in Program 7.12, the user is saved the trouble of having to enter them, since the program initializes the input array to these numbers.

Program 7.13

```
#include <stdio.h>

int bucket_sort(int in[], int k, int out[]);


main()
{
        int inarray[20] = {6, 7, 5, 8, 4, 9, 3, 0, 2};
        int n = 9;
        int outarray[20];
        int i,compares;

        printf("Unsorted array => ");
        for(i = 0; i < n; i++)
```

```
                printf("%4d",inarray[i]);
        printf("\n");
        compares = bucket_sort(inarray,n,outarray);
        printf("Sorted array   => ");
        for(i = 0; i < 20; i++)
                if(outarray[i] != 0)
                        printf("%4d",i);
        printf("\nThe number of comparisons is %d",compares);
}

int bucket_sort(int in[], int k, int out[])
{
        int j;

        for(j = 0; j < 20; j++)
                out[j] = 0;
        for(j = 0; j < k; j++)
                out[in[j]] = 1;
        return(k);
}
```

The execution of Program 7.13 is as follows:

```
Unsorted array =>   6   7   5   8   4   9   3   0   2
Sorted array   =>   0   2   3   4   5   6   7   8   9
The number of comparisons is 9
```

The number of comparisons is misleading, because no comparisons are actually made. Instead, we should think of "comparisons" as "steps" in the bucket sort application. Recall from the bubble sort that 72 comparisons were required. If you think of these comparisons as 72 "steps," then the bucket sort is clearly more efficient. Do not forget, however, that the bucket sort can be used only in limited sorting situations.

Merge Sort

The **merge sort** technique has an efficiency that falls somewhere between those of the bucket and bubble sorts. Unlike the bucket sort, merge sort can work with numbers of unspecified size and unlimited duplication. The savings in the number of comparisons comes from the fact that the entire list of input numbers is divided into smaller lists, which require fewer comparisons to sort and merge back together. Figure 7.8 shows the first step in the process. The array of input numbers is identical to that of Program 7.13.

In step 1, the middle of the initial array is determined to be the element at position [4]. The array is broken into two subarrays containing elements 0 through 4 and elements 5 through 8.

In step 2, each subarray is split into two parts. Notice now that three of the subarrays contain only two elements. At this point a single comparison is performed on each of the three two-element subarrays to determine the order of each element.

In step 3, the subarray consisting of elements 0 through 3 is split in half again, resulting in the arrays shown in step 4. The resulting subarrays have two and one elements each.

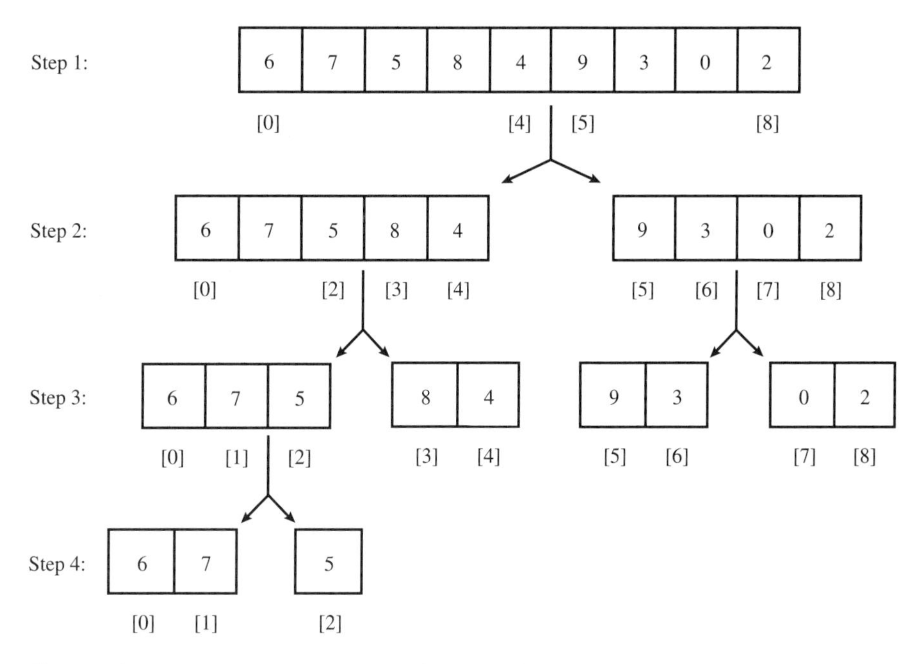

Figure 7.8 Breaking Down Input Array in Merge Sort

A single comparison is performed on the two-element array, and no comparison is necessary on the one-element array. So, even though we started with nine numbers, we have made only four comparisons so far to determine relative ordering.

Figure 7.9 shows all five subarrays prior to the beginning of the merge process. The numbers in the third and fourth subarrays have been switched as a result of their comparisons. So, each of the five subarrays contains elements that are in numeric order.

Merging two ordered arrays is a straightforward task. Elements in each array are compared one at a time. The smaller element is written to the result array and its associated index pointer is advanced. If the end of either array is reached, the remaining numbers in the other array are simply copied to the result array. For example, suppose the initial ordered arrays (and their index pointers) are as follows:

```
17  22  25    and    15  19  21
^                    ^
```

Because 17 is larger than 15, 15 is written to the result and its array index is advanced, giving

```
17  22  25    and    15  19  21
^                        ^
```

6	7		5		4	8		3	9		0	2
[0]	[1]		[0]		[0]	[1]		[0]	[1]		[0]	[1]

Figure 7.9 Subarrays Prior to Merging

Now, 19 is larger than 17, so 17 is written to the output and its index pointer is advanced. Now we have

```
17  22  25     and     15  19  21
     ^                      ^
```

Comparison of 19 and 22 causes 19 to be written to the result and its index pointer to be adjusted. This gives us

```
17  22  25     and     15  19  21
     ^                          ^
```

Comparing 22 and 21 sends 21 to the result. Since the pointer in the second array is now at the end, the remaining elements of the first array (22 and 25) are simply copied to the result.

When this merge technique is applied to our sample set of numbers, we get the result shown in Figure 7.10. The subarrays are merged back into the original arrays from which they were split, except that the elements are now in order. The final array contains all nine elements sorted into numeric order. Program 7.14 shows how the merge sort is implemented.

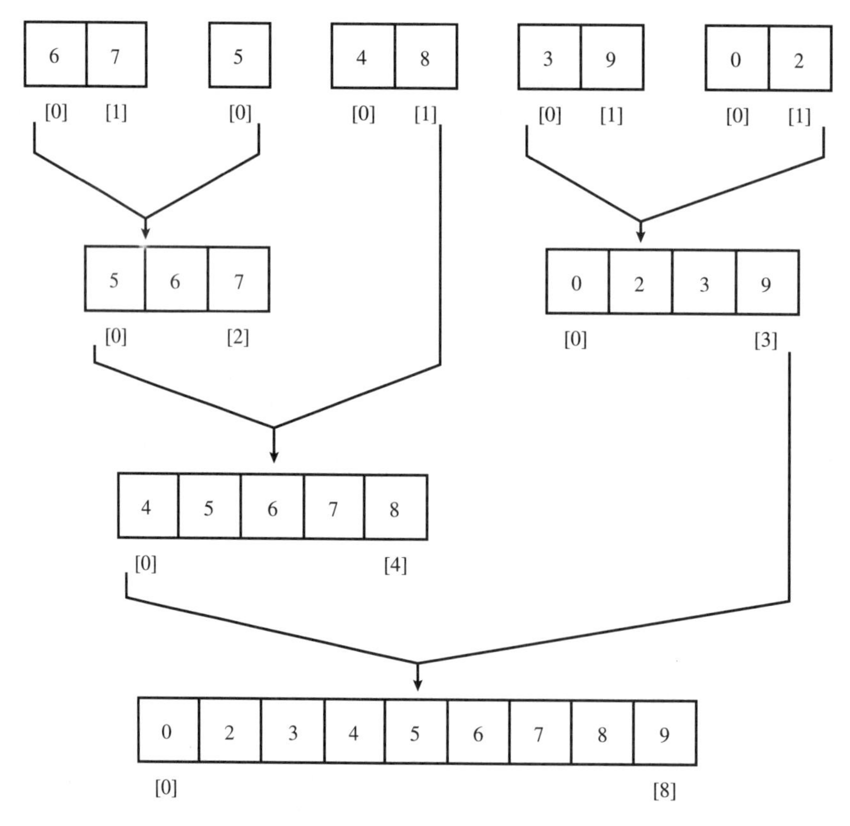

Figure 7.10 Merging Subarrays

Program 7.14

```
#include <stdio.h>

void merge_sort(int in[], int a, int b, int out[]);
void showarray(int in[],int k);
void merge_array(int in1[], int in2[], int n1, int n2, int out[]);
int compares = 0;

main()
{
        int inarray[20] = {6, 7, 5, 8, 4, 9, 3, 0, 2};
        int n = 9;
        int outarray[20];

        printf("Unsorted array => ");
        showarray(inarray,n);
        merge_sort(inarray,0,n-1,outarray);
        printf("Sorted array   => ");
        showarray(outarray,n);
        printf("\nThe number of comparisons is %d",compares);
}

void showarray(int in[], int k)
{
        int i;

        for(i = 0; i < k; i++)
                printf("%4d",in[i]);
        printf("\n");
}


void merge_sort(int in[], int a, int b, int out[])
{
        int m;
        int out1[20], out2[20];

        /* Does array contain a single element? */
        if(a == b)
                out[0] = in[a];          /* Return single element */
        else

        /* Does array contain only two elements? */
        if(1 == (b - a))
        {
                if(in[a] <= in[b])       /* Do not swap elements */
                {
                        out[0] = in[a];
                        out[1] = in[b];
                }
```

```
                else                        /* Ok, swap the two elements */
                {
                        out[0] = in[b];
                        out[1] = in[a];
                }
                compares++;
        }
        else
        {
                /* Must partition array of 3-or-more elements */
                m = a + (b - a)/2;                  /* Find the middle */
                merge_sort(in,a,m,out1);            /* Sort first half */
                merge_sort(in,m+1,b,out2);          /* Sort second half */

                /* Now merge the sorted halfs together */
                merge_array(out1,out2,1+m-a,b-m,out);
        }
}

void merge_array(int in1[], int in2[], int n1, int n2, int out[])
{
        int i = 0,j = 0,k = 0;

        while((i < n1) && (j < n2))
        {
                /* Is first array element the smallest? */
                if(in1[i] <= in2[j])
                {
                        out[k] = in1[i];            /* Write to output */
                        i++;                        /* Adjust in1 pointer */
                }
                else            /* Second array element is smaller */
                {
                        out[k] = in2[j];            /* Write to output */
                        j++;                        /* Adjust in2 pointer */
                }
                k++;            /* Adjust output pointer */
                compares++;
        }

        /* Are any elements left in first array? */
        if(i != n1)
        {
                do      /* Write remaining in1 elements to output */
                {
                        out[k] = in1[i];
                        i++;
                        k++;
                } while(i < n1);
        }
        else            /* Write remaining elements from in2 to output */
```

```
        {
                do
                {
                        out[k] = in2[j];
                        j++;
                        k++;
                } while(j < n2);
        }
}
```

The execution of Program 7.14 gives the following result:

```
Unsorted array =>   6   7   5   4   8   3   9   0   2
Sorted array   =>   0   2   3   4   5   6   7   8   9

The number of comparisons is 19
```

The number of comparisons is composed of two portions: the comparisons made when the arrays are subdivided, and the comparisons made when they are merged. The 19 comparisons made here are significantly smaller than the 72 of bubble sort.

Where are all the subarrays stored? This is a good question. The merge sort shown in Program 7.14 relies on *recursion* to keep track of the subarrays. When `merge_sort` is initially called, we have

```
merge_sort(inarray,0,8,outarray)
```

This causes C to create a memory image (containing data and pointers) to execute `merge_sort`, which computes the middle of the input array and splits it into two parts. Each part is passed to its *own* copy of `merge_sort`, as in

```
merge_sort(in,0,4,out1)
merge_sort(in,5,8,out2)
```

There are now *three* copies of `merge_sort` stored in memory. Copy 2 splits its input array into the two parts

```
merge_sort(in,0,2,out1)
merge_sort(in,3,4,out2)
```

and copy 3 splits its input into the two parts

```
merge_sort(in,5,6,out1)
merge_sort(in,7,8,out2)
```

There are now *seven* copies of `merge_sort` in memory.

Finally, copy 4 splits its input into

```
merge_sort(in,0,1,out1)
merge_sort(in,2,2,out2)
```

Thus there are a total of *nine* copies of `merge_sort` in memory simultaneously. This is the one disadvantage of merge sort. It requires a substantial amount of memory to perform its job. This indicates that there is a need to balance efficiency (the number of comparisons) against the amount of storage required. Small arrays are not a problem. But consider an array containing 10,000 elements. There could easily be over 8000 separate copies of `merge_sort` running (two with 5000 numbers, four with 2500, eight with 1250, 16 with 625, etc.). An array this large would surely eat up all available memory before sorting could be completed.

However, for small arrays without numeric restrictions on the elements, merge sort is a good solution.

Quick Sort

The **quick sort** technique—the last of our sorting techniques—is generally regarded as the most efficient when used in its ideal form. The form used here is not ideal, as you will see, but works very well and is comparable to the efficiency of merge sort.

Figure 7.11 shows an array of numbers that will be sorted using quick sort. The idea behind quick sort is to choose an array *partitioning* element, which is called a **pivot.** The array is then divided into two parts. One part contains elements that are less than or equal to the pivot. The other part contains elements that are greater than the pivot. As Figure 7.11 shows, the first element in the initial array (3) is chosen as the pivot. The pivot is then compared with the other elements, resulting in the new arrays shown in step 2. Notice that the pivot element is in the correct position between the two sets of elements. The pivot has already been sorted into its correct position!

In step 2, new pivot values are chosen for both arrays and the arrays are partitioned again, resulting in the arrays shown in step 3. When an array contains a single element, we are done partitioning. When an array contains two elements, we need to perform one comparison to get the two elements into their correct order. For example, single comparisons are needed in step 3 on the 1,0 and 7,6 arrays.

At the completion of step 3 (in this example), all numbers are in their correct positions. The individual arrays need to be combined back into a single result array. This does not require any additional comparisons (as merge sort does), since the elements are already completely sorted.

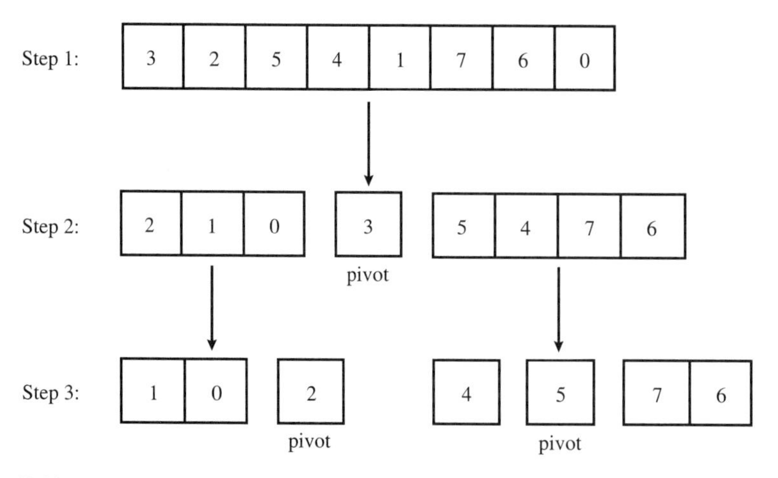

7.11 Partitioning an Array During Quick Sort

Program 7.15 shows how quick sort is implemented. Once again you will see that *recursion* is used to partition subarrays at each step. The original input array is partitioned into two subarrays with each new call to `quick_sort()`. Thus we are required to provide storage space for the subarrays (`in1` and `in2`) within `quick_sort()`.

Program 7.15

```
#include <stdio.h>

void quick_sort(int in[], int a, int b, int out[]);
void showarray(int in[],int k);
int compares = 0;

main()
{
        int inarray[20] = {6, 7, 5, 8, 4, 9, 3, 0, 2};
        int n = 9;
        int outarray[20];

        printf("Unsorted array => ");
        showarray(inarray,n);
        quick_sort(inarray,0,n-1,outarray);
        printf("Sorted array   => ");
        showarray(outarray,n);
        printf("\nThe number of comparisons is %d",compares);
}

void showarray(int in[], int k)
{
        int i;

        for(i = 0; i < k; i++)
                printf("%4d",in[i]);
        printf("\n");
}


void quick_sort(int in[], int a, int b, int out[])
{
        int pivot, i = 0, j = 0, k = 1, z = 0;
        int in1[20], in2[20];
        int out1[20], out2[20];

        if(b != -1)                /* Just one element? */
                if(a == b)
                        out[0] = in[a];
                else

                /* Only two elements? */
                if(1 == (b - a))
```

```
{
        if(in[a] <= in[b])                  /* Do not swap */
        {
                out[0] = in[a];
                out[1] = in[b];
        }
        else                                /* Swap them */
        {
                out[0] = in[b];
                out[1] = in[a];
        }
        compares++;
}
else            /* Handle 3-or-more elements */
{
        pivot = in[0];                      /* Pick pivot */
        while(k <= b)
        {
                /* Choose an output array to write to */
                if(pivot > in[k])           /* Compare pivot */
                {
                        /* Write smaller output array */
                        in1[i] = in[k];
                        i++;
                }
                else    /* Pivot is smaller */
                {
                        /* Write larger output array */
                        in2[j] = in[k];
                        j++;
                }
                k++;            /* Adjust input pointer */
                compares++;
        }

        /* Sort smaller partition */
        quick_sort(in1,0,i-1,out1);

        /* Sort larger partition */
        quick_sort(in2,0,j-1,out2);

        /* Write smaller array to output */
        for(k = 0; k < i; k++)
        {
                out[z] = out1[k];
                z++;
        }
        out[z] = pivot;         /* Write pivot to output */
        z++;
```

```
            /* Write larger array to output */
            for(k = 0; k < j; k++)
            {
                    out[z] = out2[k];
                    z++;
            }
    }
}
```

Executing `quick_sort` on our sample array gives the following result:

```
Unsorted array =>   6   7   5   8   4   9   3   0   2
Sorted array   =>   0   2   3   4   5   6   7   8   9

The number of comparisons is 21
```

This is only slightly less efficient than merge sort (which required only 19 comparisons). What is interesting is this: if the ordering of the original data is changed, we get a different result. Consider this second execution of quick sort with element values 5 and 3 interchanged:

```
Unsorted array =>   6   7   3   8   4   9   5   0   2
Sorted array   =>   0   2   3   4   5   6   7   8   9

The number of comparisons is 17
```

We have saved four comparisons by changing the original order of the input array. If this is done with merge sort, 19 comparisons are still needed. So, in quick sort, the order of the array elements has an effect on the efficiency. Thus, in an ideal quick sort application, the pivot element is not simply chosen as the first element in each subarray. Instead, the *median* element is chosen. For instance, given the numbers

6 7 5 8 4 9 3 0 2

the first pivot element would be 5, since there are as many element values below 5 as there are above it. Using the median element value as the pivot helps keep the size of the partitioned arrays equal, resulting in fewer levels of recursion (and fewer comparisons).

C provides a built-in quick sort function in `<stdlib.h>`. The function name is `qsort()` and works with *any* numeric data type. The `qsort()` function requires four parameters: (1) a pointer to the array to be sorted, (2) the size of the array, (3) the size of elements in the array, and (4) a compare function.

The size of each element is specified because `qsort()` is able to sort int-, float-, and double-valued elements. The compare function is written for a specific data type and is used by `qsort()` to perform element comparisons. The compare function must return the following when comparing elements a and b:

- a negative integer if $a < b$
- 0 if $a = b$
- a positive integer if $a > b$

Program 7.16

```
#include <stdio.h>
#include <stdlib.h>

void showarray(int in[],int k);
int cmp(const void *a, const void *b);
int compares = 0;

main()
{
        int inarray[20] = {6, 7, 5, 8, 4, 9, 3, 0, 2};
        int n = 9;

        printf("Unsorted array => ");
        showarray(inarray,n);
        qsort(inarray,n,sizeof(inarray[0]),cmp);
        printf("Sorted array   => ");
        showarray(inarray,n);
        printf("\nThe number of comparisons is %d",compares);
}

int cmp(const void *a, const void *b)
{
        compares++;
        return(*((int *) a) - *((int *) b));
}

void showarray(int in[], int k)
{
        int i;

        for(i = 0; i < k; i++)
                printf("%4d",in[i]);
        printf("\n");
}
```

Program 7.16 shows how the `qsort()` function is used to sort our sample array. The compare function, called `cmp()`, has a rather cryptic definition:

```
int cmp(const void *a, const void *b);
```

which indicates that pointers are passed to the `cmp()` function, but the data type of the pointer is unspecified. Within the actual function, we have

```
return(*((int *) a) - *((int *) b));
```

which uses a technique called *type casting* to force the pointers to access integer values. Thus,

```
*((int *) a)
```

actually represents an integer value stored at location `a`. So, the `return()` statement is actually subtracting two integers in this case.

If you need to sort floating-point numbers with `qsort()`, the `cmp()` function must be changed to

```
return(*((float *) a) - *((float *) b));
```

In addition, you will notice that the call to `qsort` in Program 7.16 contains the `sizeof()` function:

```
qsort(inarray,n,sizeof(inarray[0]),cmp);
```

The `sizeof()` function returns a value to `qsort()` indicating what type of elements are being sorted.

Execution of Program 7.16 produces the output

```
Unsorted array =>   6   7   5   8   4   9   3   0   2
Sorted array   =>   0   2   3   4   5   6   7   8   9

The number of comparisons is 23
```

Conclusion

As you can see, the `qsort()` routine is comparable in efficiency to the quick sort and merge sort routines covered in this section.

Test your knowledge of the ideas presented in this section by trying the following section review.

7.3 Section Review

1. Why does bubble sort require more comparisons than merge sort?
2. For any list of data, what is the minimum number of times a sorting program must go through all the numbers before the sort is considered complete? Under what circumstances would this happen?
3. What determines if a list will be sorted in ascending or descending order?
4. In order to sort into ascending order the list

   8 7 3 1

   how many passes through the list are required by:
   (a) bubble sort?
   (b) bucket sort?
   (c) merge sort?
5. What makes the bucket sort so efficient? Are there any limitations to a bucket sort?
6. How many levels of recursion are needed by merge sort to decompose the following list of numbers?

   0 3 8 1 9 2 6 4 7 5

7. What are the differences between merge sort and quick sort?
8. Is there a difference between sorting numbers and sorting character strings?

7.4 Multidimensional Numeric Arrays

Discussion

In the last few sections, you worked with arrays that had a single dimension. This means that there was only one index. In technology applications, it is common to find arrays with more than one dimension. These are called **multidimensional arrays**. This section introduces two-dimensional arrays and shows an application with them.

Basic Idea

The only kinds of arrays you can initialize are static and external. In working with strings, you could initialize a string array as shown in Program 7.17.

Program 7.17

```
#include <stdio.h>

main()
{
        static char array[7] = {'H', 'e', 'l', 'l', 'o'};

        printf("%s",array);
}
```

Execution of Program 7.17 will produce

```
Hello
```

The array initialization is performed using the `{}`. Inside these, separated by commas, are the individual characters to be used in the array. All arrays may be initialized this way. This is not the most efficient method for strings, but it is quite easy for numbers. Consider the one-dimensional array in Program 7.18.

Program 7.18

```
#include <stdio.h>

main()
{
        static int array[3] = {10,20,30};
        int index;

        for(index = 0; index < 3; index++)
        {
                printf("array[%d] = %d\n",index,array[index]);
        }
}
```

Execution of Program 7.18 produces

```
array[0] = 10
array[1] = 20
array[2] = 30
```

To initialize this array, the `{}` are again used. Values assigned to each element of the array are again separated by commas: `{10,20,30}`.

Arrays of more than one dimension may be initialized as shown in the next program. This program has a two-dimensional array. You can visualize a two-dimensional array as a checkerboard pattern where each square is identified by a unique row and column number. Each square contains data. This concept is shown in Figure 7.12. Such an arrangement is usually referred to as a **matrix**.

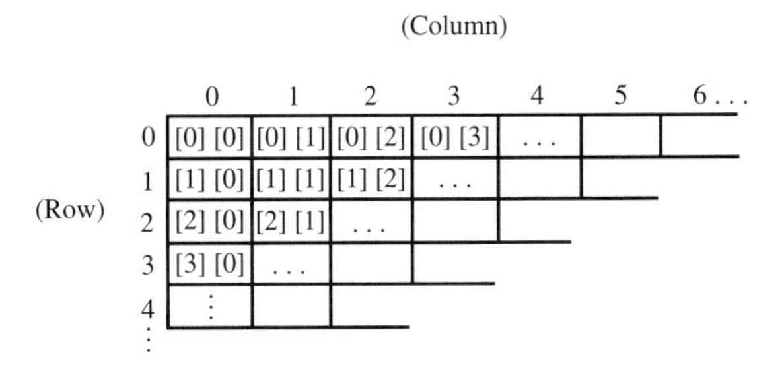

Figure 7.12 Concept of a Matrix

Program 7.19 shows how a two-dimensional array can be utilized.

Program 7.19

```
#include <stdio.h>

main()
{
        static int array[2][3] = {
                                   {10,20,30},
                                   {11,21,31}
                                 };
        int row;
        int column;

        for(row = 0; row < 2; row++)
        {
                for(column = 0; column < 3; column++)
                        printf("%5d",array[row][column]);
                printf("\n\n");
        }
}
```

Execution of Program 7.19 produces

```
10   20   30

11   21   31
```

Program Analysis

First a two-dimensional array is declared and initialized by

```
static int array[2][3] = {
                          {10,20,30},
                          {11,21,31}
                         };
```

The indexes `[2][3]` indicate that the array is a two-dimensional array containing two **rows** (numbered 0 and 1) and **columns** (numbered 0 through 2). Note that each dimension

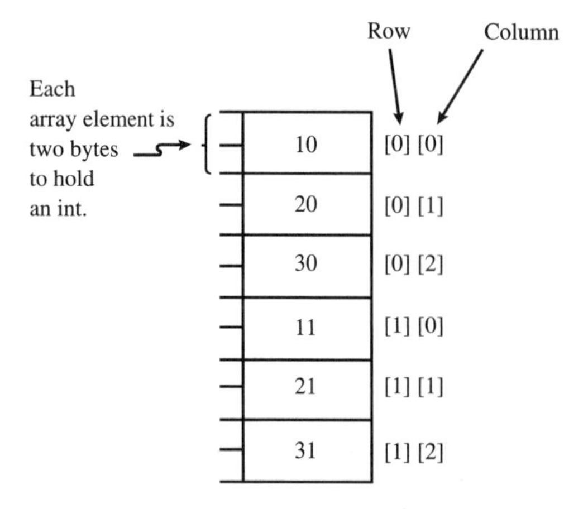

Figure 7.13 Memory Allocation with Rows and Columns for a Two-Dimensional Array

of the array uses a new set of `[]`. The way in which C lays out memory is to lay out one row at a time with a given number of columns. This arrangement in memory is illustrated in Figure 7.13, and is called *row-major* order.

The method of initializing each element is done by specifying each element in the first row, then each element in the second row. This could have been done in the declaration by

```
{{10,20,30},{11,21,31}}
```

However, it is more descriptive of an actual two-dimensional matrix to lay it out as shown.

Next, two variables are declared:

```
int row;
int column;
```

These two variables will be used to step through each element of the array so its value can be displayed. This is accomplished by the nested C `for` loops:

```
for(row = 0; row < 2; row++)
{
   for(column = 0; column < 3; column++)
      printf("%5d",array[row][column]);
   printf("\n\n");
}
```

Remembering that the index of `row` starts with 0 and ends with 1, whereas the index for `column` starts with 0 and ends with 2, the two loops cause each element value in the array to be displayed through the variable reference

```
array[row][column]
```

To see how this is done, note in Figure 7.13 that, the first time through the loop, `row = 0` and `column = 0`; thus, the first element to be displayed by the `printf` function is `array[0][0]`, which has the value of 10. The next element to be displayed is `array[0][1]`, which has a value of 20. This process continues until the nested `for` loop reaches the count of 3 and then the `printf("\n\n");` function causes two spaces, and the outer loop increments and the process begins again, this time displaying the values of each column in the second row.

There is a **field width specifier** used with the format specifier in the `printf` function of this program:

```
printf("%5d",array[row][column]);
```

Note the `%5d`. The 5 means to use five characters when printing the output.

Array Applications

Arrays have many applications in the solution of technical problems. Many times the use of arrays requires manipulation of the array elements. Hence it's important to develop programming skills that allow calculations with arrays.

One of the simplest processes to perform with array arithmetic is to determine the sum of any column or any row. Program 7.20 determines the sum of the first column of the two-dimensional array of Program 7.19.

Program 7.20

```
#include <stdio.h>

int add_column(int arrayin[][3]);

main()
{
        static int array[2][3] = {
                                    {10,20,30},
                                    {11,21,31}
                                 };

        int row;
        int column;
        int first_column_sum;

        for(row = 0; row < 2; row++)
        {
                for(column = 0; column < 3; column++)
                        printf("%5d", array[row][column]);

                printf("\n\n");
        }

        first_column_sum = add_column(array);
        printf("The sum of the first column is %d",first_column_sum);
}

int add_column(int arrayin[][3])
{
        int row;
```

```
        int column_sum;

        column_sum = 0;

        for(row = 0; row < 2; row++)
                column_sum += arrayin[row][0];

        return(column_sum);
}
```

Execution of Program 7.20 produces

```
        10    20    30

        11    21    31
The sum of the first column is 21
```

This seems like a lot of work just to add two numbers. However, this is just a simple example to demonstrate a powerful concept—arithmetic operations using arrays.

Program Analysis

The program starts by first declaring a function prototype:

```
int add_column(int arrayin[][3]);
```

Since an array of more than one dimension will be used, C requires that the limits of other dimensions be specified within the formal parameter list. The reason for this is the way arrays are stored in C. In this case, for the two-dimensional array, C needs to know how many columns will be required for each of the rows. In this manner it can properly set its array pointer.

The array elements are declared as in Program 7.19:

```
static int array[2][3] = {
                           {10,20,30},
                           {11,21,31}
                         };
```

And the variables for `main()` are declared:

```
int row;
int column;
int first_column_sum;
```

Note the addition of the new variable `first_column_sum`, which will be used to store the value of the sum for the first column.

Next, the array is displayed as in Program 7.19:

```
for(row = 0; row < 2; row++)
{
   for(column = 0; column < 3; column++)
```

```
        printf("%5d",array[row][column]);
    printf("\n\n");
}
```

Next, the function that will take the sum of the first column and return the value is called:

```
first_column_sum = add_column(array);
```

Note, as before, that when passing the address of the array to the called function it is only necessary to pass the identifier for the array variable. This, of course, is the starting address of the array.

This is a good time to present the details of the called function. The function is of type `int` and defines two variables:

```
int add_column(int arrayin[][3])
{
    int row;
    int column_sum;
```

The variable `column_sum` will be used to store the value of the required sum. It is first initialized to `0`.

```
column_sum = 0;
```

Then it is used in a `for` loop that uses the += operation to generate the sum of `arrayin[0] [0] + arrayin[1][0]`. This is accomplished by the following loop:

```
for(row = 0; row < 2; row++)
    column_sum += arrayin[row][0];
```

The final value is returned to the calling function:

```
return(column_sum);
```

After this, the final answer is displayed:

```
printf("The sum of the first column is %d",first_column_sum);
```

The next program illustrates a method of adding both columns of the matrix.

Adding More Columns

Consider Program 7.21. It expands on Program 7.20 and adds all columns of the given matrix.

Program 7.21

```
#include <stdio.h>

void add_columns(int arrayin[][3], int column_value[]);

main()
{
        static int array[2][3] = {
                                    {10,20,30},
```

```
                                    {11,21,31}
                                  };
      int row;
      int column;
      int column_value[3];

      for(row = 0; row < 2; row++)
      {
              for(column = 0; column < 3; column++)
                      printf("%5d",array[row][column]);
              printf("\n\n");
      }

      add_columns(array, column_value);

      for(column = 0; column < 3; column++)
              printf("The sum of column %d is %d\n",column
                      ,column_value[column]);
}

void add_columns(int arrayin[][3], int column_value[])
{
      int row;
      int column;

      for(column = 0; column < 3; column++)
      {
              column_value[column] = 0;
              for(row = 0; row < 2; row++)
                      column_value[column] += arrayin[row][column];
      }
}
```

When Program 7.21 is executed, the output will be

```
   10   20   30

   11   21   31


The sum of column 0 is 21
The sum of column 1 is 41
The sum of column 2 is 61
```

Note now that each column in the matrix is totaled. Being able to do this has applications in the operation of business spreadsheets.

Program Analysis

Program 7.21 incorporates a few additions to Program 7.20. First, the function prototype is changed to a type `void`, and its argument is expanded to include an array that will contain the sum of each column:

```
void add_columns(int arrayin[][3], int column_value[]);
```

The function definition is also expanded:

```
void add_columns(int arrayin[][3], int column_value[])
{
   int row;
   int column;

   for(column = 0; column < 3; column++)
   {
      column_value[column] = 0;
      for(row = 0; row < 2; row++)
            column_value[column] = arrayin[row][column];
   }
}
```

Note now that there is a C `for` loop that sums each column of the array using the += operator. Also note that within the loop, the array variable `column_value[column]` is initialized to zero each time. This is to ensure that the += operation starts with a zero value in this variable. The answers for the sum of each column are stored in the arrayed variable `column_value[column]`. It is this array that is returned to the calling function (`main()`):

```
add_columns(array, column_value);
```

Observe that `main()` has defined a new array variable:

```
int column_value[3];
```

This is used as the actual parameter of the called function.

What is important to realize is that two things are happening when this function is being called. First, the starting address of the arrayed variable `array` that contains the values of the array is being passed to the called function. Second, the arrayed variable `column_value` is being used to return the starting address of a second array that contains the values of the required column sums.

When `add_columns()` returns, the output values are displayed as follows:

```
for(column = 0; column < 3; column++)
   printf("The sum of column %d is %d\n",column,column_value[column]);
```

The last program in this section computes the sum of each column and each row of the matrix. Program 7.22 uses two separate functions to do this. One function calculates the column sum; the other calculates the row sum.

Program 7.22

```
#include <stdio.h>

int add_columns(int arrayin[][3], int column_value[]);
int add_rows(int arrayin[][3], int row_value[]);

main()
{
        static int array[2][3] = {
```

```
                                  {10,20,30},
                                  {11,21,31}
                                };

        int row;
        int column;
        int column_value[3];
        int row_value[2];

        for(row = 0; row < 2; row++)
        {
                for(column = 0; column < 3; column++)
                        printf("%5d",array[row][column]);

                printf("\n\n");
        }

        add_columns(array,column_value);

        for(column = 0; column < 3; column++)
                printf("The sum of column %d is %d\n",
                        column,column_value[column]);

        add_rows(array, row_value);

        for(row = 0; row < 2; row++)
                printf("The sum of row %d is %d\n",row,row_value[row]);
}

int add_columns(int arrayin[][3], int column_value[])
{
        int row;
        int column;

        for(column = 0; column < 3; column++)
        {
                column_value[column] = 0;

                for(row = 0; row < 2; row++)
                        column_value[column] += arrayin[row][column];
        }
}

int add_rows(int arrayin[][3], int row_value[])
{
        int row;
        int column;

        for(row = 0; row < 2; row++)
        {
                row_value[row] = 0;

                for(column = 0; column < 3; column++)
```

```
                row_value[row] += arrayin[row][column];
        }
}
```

Execution of Program 7.22 produces

```
        10   20   30

        11   21   31


The sum of column 0 is 21
The sum of column 1 is 41
The sum of column 2 is 61
The sum of row 0 is 60
The sum of row 1 is 63
```

Matrix multiplication is another useful matrix operation, since we are often required to multiply matrices together to get a desired result. The general rule for multiplying two matrices together requires that the number of columns in the first matrix equal the number of rows in the second matrix. For example, any *square* matrix (a matrix having the same number of rows and columns) may be multiplied by itself, because it has an equal number of rows and columns. Matrices with different dimensions must conform to the general rule or they cannot be multiplied. Consider the following list of matrices:

- A[2][3]*B[3][4]
- C[1][4]*D[4][2]
- E[2][3]*F[4][3]

Only the A*B and C*D matrix products are allowed, because they conform to the general rule. Product E*F is not possible, since E's three columns do not equal F's four rows.

Figure 7.14 shows why the general rule is required. Two matrices (MATA and MATB) are being multiplied. To get any new element in the result matrix (MATC), each individual

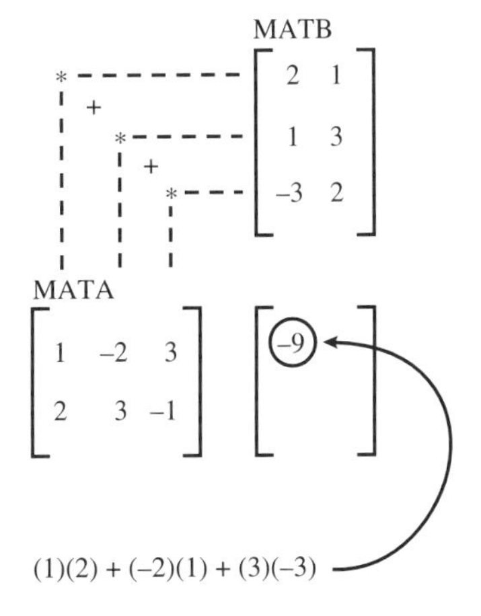

Figure 7.14 Finding One Element of a Matrix Product

MATB

$$\begin{bmatrix} 2 & 1 \\ 1 & 3 \\ -3 & 2 \end{bmatrix}$$

MATA MATC

$$\begin{bmatrix} 1 & -2 & 3 \\ 2 & 3 & -1 \end{bmatrix} \quad \begin{bmatrix} -9 & 1 \\ 10 & 9 \end{bmatrix}$$

Figure 7.15 Product of Two Matrices

row element in the first matrix (MATA) is multiplied by a corresponding column element in the second matrix (MATB). The individual products are added together to create the result element (in MATC). The general rule guarantees that there are matched pairs of numbers in each row and column.

To multiply the entire pair of matrices, each row in the first matrix is multiplied by each column in the second matrix. The result of this process is shown in Figure 7.15. Study the figure and make sure you see how each element of the result matrix (MATC) is found.

Program 7.23 implements this process and multiplies a [2][3] matrix by a [3][2] matrix, which results in a [2][2] matrix. A set of nested `for` loops is used to generate the row and column indices in computing the individual elements of the result matrix. The dimensions of the result matrix are based on the dimensions of the two matrices being multiplied. The number of rows of the first matrix and the number of columns of the second matrix are the dimensions of the result matrix.

Program 7.23

```
#include <stdio.h>

main()
{
        int mata[2][3] = {{1, -2,  3},
                          {2,  3, -1}};
        int matb[3][2] = {{ 2, 1},
                          { 1, 3},
                          {-3, 2}};
        int matc[2][2];
        int r,c,v,psum;

        for(r = 0; r < 2; r++)
                for(c = 0; c < 2; c++)
                {
                        psum = 0;
                        for(v = 0; v < 3; v++)
                                psum += mata[r][v] * matb[v][c];
                        matc[r][c] = psum;
                }
        printf("The array product is:\n");
```

```
        for(r = 0; r < 2; r++)
        {
                for(c = 0; c < 2; c++)
                        printf("%5d",matc[r][c]);
                printf("\n");
        }
}
```

Conclusion

This section presented the concept of arrays with more than one dimension. Specifically, a two-dimensional array was presented. You learned how to initialize the array and to pass values between functions. You saw an application that did arithmetic operations on the array. Check your understanding of this section by trying the following section review.

7.4 Section Review

1. State a way of initializing an array.
2. How do you declare an array with more than one dimension in C?
3. Explain how to declare a formal parameter in C that uses more than one dimension.
4. State the method used in the programs of this section that allows arrayed values to be added.
5. What must you do before using the += operator with an uninitialized array?
6. What are the resulting matrix dimensions for these products?
   (a) [1][2] * [2][3]
   (b) [3][3] * [3][1]
   (c) [2][1] * [1][4]
   (d) [3][1] * [2][4]

7.5 Example Programs

The example programs presented in this section show how to do many interesting and useful things with numeric arrays. You are encouraged to study them to gain a full understanding of how they work.

Dynamic Linking

Dynamic linking is a technique used for efficient storage and retrieval of disk files. It is used in conjunction with a special area on the disk called the *File Allocation Table (FAT)*. The FAT is a block of numbers describing how sectors (or allocation units) on the disk have been allocated. Each entry in the FAT is interpreted as one of the following:

- Free sector
- Allocated sector
- Bad sector
- Final sector in a file

One of the most interesting benefits of using a FAT is that files need not be stored in consecutive sectors (as many people imagine they are). For example, suppose that the entire disk does not contain a contiguous group of sectors numbering more than six. A disk operating system that does not use a FAT (together with dynamic linking) will not be able to store a file containing ten sectors of data. However, with dynamic linking, the ten sectors can reside *anywhere* on the disk.

Dynamic linking works as follows. A sector's entry in the FAT contains a number that indicates one of the following:

- this is the last sector in the file
- the next sector is sector___

Thus, to string a bunch of sectors together, we need only place the next sector number in the previous FAT entry. To illustrate this concept, consider this simple 16-sector FAT:

```
0     0   5   4     2   13   12   9
0    99   0   0    99    6    0   0
```

Each nonzero element in the FAT indicates a used sector, except for element value 99, which indicates the last sector in a file. Each zero element indicates a free sector. Sector 0 may not be assigned, because it is where we are storing the FAT.

The following directory information associated with this FAT indicates that two files are stored:

File	*Starting Sector*
ABC	3
DEF	7

What sectors are assigned to each file? File ABC begins at sector 3. Sector 3's entry in the FAT contains 4. This indicates that the next sector in file ABC is sector 4. Sector 4's entry is 2. This is the third sector in file ABC. Continuing with this process we get the *chain* of sector numbers for file ABC:

3 4 2 5 13 6 12 99

A similar method gives these sector numbers for file DEF:

7 9 99

which is a smaller file.

The chain of sector numbers associated with each file is the dynamic chain for that file.

Program 7.24 uses this technique to access a FAT capable of holding 64 sectors (of which only 63 are usable, because one sector is reserved for the FAT). A function called `dylink()` is used to search through the FAT until the 99 is seen. A second function, `free_sectors()`, is used to determine how many sectors are available for use.

Program 7.24

```
#include <stdio.h>

void dylink(int cp);
void free_sectors(void);
```

```
        int fat[64] = {0,  0,  99, 54, 5,  6,  38, 24,
                       0,  10, 49, 15, 59, 0,  12, 0,
                       0,  99, 0,  0,  2,  0,  44, 0,
                       29, 0,  11, 0,  0,  3,  0,  0,
                       4,  17, 0,  99, 0,  0,  39, 9,
                       0,  0,  6,  0,  45, 46, 26, 0,
                       0,  33, 0,  0,  0,  0,  55, 56,
                       57, 99, 0,  0,  61, 20, 0,  0};

main()
{
        int chains[] = {7, 32, 60, 14};
        int i;

        for(i = 0; i < 4; i++)
        {
                printf("File #%d dynamic chain: ",i+1);
                dylink(chains[i]);
        }
        printf("\n");
        free_sectors();
}

void dylink(int cp)
{
        do
        {
                printf("%d\t",cp);
                if((cp != 0) && (cp != 99))
                        cp = fat[cp];
        } while ((cp != 0) && (cp != 99));
        if(cp == 0)
                printf("?\nError! Lost chain!");
        printf("\n");
}

void free_sectors(void)
{
        int j,free = 0;

        for(j = 1; j < 64; j++)
                if(fat[j] == 0)
                        free++;
        printf("There are %d free allocation units.",free);
}
```

Execution of Program 7.24 yields

```
File #1 dynamic chain: 7        24      29      3       54
55      56      57
```

```
File #2 dynamic chain: 32     4     5     6     38
39      9      10      49     33    17
File #3 dynamic chain: 60     61    20    2
File #4 dynamic chain: 14     12    59    ?
Error! Lost chain!

There are 30 free allocation units.
```

Notice that there is a problem with file #4. The fourth sector in the file does not exist. This is one of the problems inherent with the use of dynamic file allocation. Sometimes, owing to incorrect use of the file system, a file's dynamic chain is corrupted. In the case of file #4, sector 59's slot contains a zero, which indicates that the chain for file #4 was broken. There is another lost chain beginning at sector 44.

The FAT in Program 7.24 contains other problems as well. For instance, the slot for sector 12 contains the value 59, but no other entries in the FAT contain 12. Thus, there is no way to get to sector 12's information. This type of problem is referred to as a *lost allocation unit.* Try to find the other lost allocation units. Lost allocation units reduce the usable storage space on the disk. It is possible to identify lost allocation units and reclaim them. You should figure out a way to do this yourself.

A Histogram Generator

When large amounts of data are examined, one useful technique for evaluation of the entire block of data is the histogram. A histogram contains the frequency of occurrence for each different data value in the block of data. For example, consider this short list of numbers:

3 7 4 2 3 6 7 8 2 3 9

This list contains two 2's, three 3's, one 4, one 6, two 7's, one 8, and one 9. Thus, the histogram might look like this:

```
0: 0
1: 0
2: 2
3: 3
4: 1
5: 0
6: 1
7: 2
8: 1
9: 1
```

The purpose of Program 7.25 is to generate a histogram for a block of 100 data values. The range of the data values is from 1 to 10.

Program 7.25

```
#include <stdio.h>
#include <stdlib.h>

main()
```

```
{
        int values[100];
        static int hist[10];
        int i;

        for(i = 0; i < 100; i++)
                values[i] = 1 + rand() % 10;
        printf("The initial set of numbers is:\n");
        for(i = 0; i < 100; i++)
        {
                if(0 == i % 10)
                        printf("\n");
                printf("%4d",values[i]);
        }
        for(i = 0; i < 100; i++)
                hist[values[i] - 1]++;
        printf("\n\nThe histogram for the given set of numbers is:\n\n");
        for(i = 0; i < 10; i++)
                printf("%4d:\t%d\n",i + 1,hist[i]);
}
```

The data values are generated by the `rand()` function found in `<stdlib.h>`. The `rand()` function returns an integer in the range 0 to `RAND_MAX` (the value of which is predefined in `<stdlib.h>`). The 1 to 10 data value is determined by the following formula:

```
values[i] = 1 + rand() % 10;
```

The `%` operator calculates the integer *remainder* of the `rand()` value divided by 10. The remainder must be between 0 and 9. Thus, `values[i]` will be between 1 and 10.

Execution of Program 7.25 produces

```
The initial set of numbers is:

   2   8   5   1  10   5   9   9   3   5
   6   6   2   8   2   2   6   3   8   7
   2   5   3   4   3   3   2   7   9   6
   8   7   2   9  10   3   8  10   6   5
   4   2   3   4   4   5   2   2   4   9
   8   5   3   8   8  10   4   2  10   9
   7   6   1   3   9   7   1   3   5   9
   7   6   1  10   1   1   7   2   4   9
  10   4   5   5   7   1   7   7   2   9
   5  10   7   4   8   9   9   3  10   2

The histogram for the given set of numbers is:

   1:   7
   2:   14
   3:   11
   4:   9
   5:   11
   6:   7
```

```
    7:   11
    8:    9
    9:   12
   10:    9
```

Verify for yourself that the histogram is correct.

Finding Standard Deviation

Economists and statisticians love finding the standard deviations of groups of numbers. For the student, standard deviation has always been simply a tough programming assignment. Program 7.26 uses the formula shown in Figure 7.16 to determine the standard deviation of a set of numbers. In the formula, N equals the number of data values, A equals the average of the data values, and X_i indicates data value i. So, to find the standard deviation of a group of numbers, we must first find the average A. This average is then subtracted from each data value. The difference is squared and added to a running total. This total is then divided by $N - 1$ (for theoretical reasons that make the deviation more accurate). Then the square root is taken to find the standard deviation.

Program 7.26

```
#include <stdio.h>
#include <stdlib.h>
#include <math.h>

main()
{
        int values[24];
        int i, sum = 0;
        float ave, temp, squares = 0.0, stdev;

        for(i = 0; i < 24; i++)
                values[i] = 50 + rand() % 45;
        printf("The set of test scores is:\n");
        for(i = 0; i < 24; i++)
        {
                if(0 == i % 8)
                        printf("\n");
                printf("%4d",values[i]);
        }
        for (i = 0; i < 24; i++)
                sum += values[i];
        ave = sum / 24.0;
        printf("\n\nThe average is %6.2f\n",ave);
        for(i = 0; i < 24; i++)
        {
                temp = values[i] - ave;
                squares += temp * temp;
        }
        stdev = sqrt(squares/23.0);
        printf("The standard deviation is %6.2f",stdev);
}
```

$$\text{S.D.} = \sqrt{\frac{1}{N-1}\left(\sum_{i=1}^{N}(X_i - A)^2\right)}$$

Figure 7.16 Standard Deviation Formula

Sample execution of Program 7.26 produces

```
The set of test scores is:

  91  67  84  90  94  69  53  68
  57  79  85  70  66  92  66  91
  75  67  62  86  86  74  82  68

The average is  75.92
The standard deviation is  11.95
```

Here we see the standard deviation computed for a typical set of test scores. Note that the square-root function used to compute the standard deviation is called `sqrt()` and is found in `<math.h>`.

Hash Tables

The term "hash table" is a strange name for a data object that stores information at locations whose index is computed through the use of a *hash function*. A hash table can be composed of a one-dimensional array, or some other, more complicated data structure. In this example we will use a simple hash function to generate an index for a hash table that may contain 32 entries.

Why is a hash table needed? Consider the process of compiling a C program. Each new variable or user-defined function name that the compiler encounters must be stored in a special data structure called a *symbol table*. The symbol table is used to store the value of the variable, if defined, or the parameters of a function definition. Through the use of the symbol table, the compiler can determine if a variable is illegally redefined or never defined at all.

When many variables and function names are used in a program, the process of searching the symbol table can become very time consuming. Imagine that a lengthy C program contains 750 symbol table entries during compilation. This is a lot of entries to search through.

The purpose of a hash table and its associated hash function is to practically eliminate the need to search. The variable or function name itself is used to generate an index in the hash table where the information is stored. For example, consider the two variable names *ABC* and *DEF*. If we find the sum of the ASCII code values for each letter in the variable name and then divide by three, we get:

$$ABC = 66$$
$$DEF = 69$$

Clearly, these two variables have different hash values. The 66 and the 69 can be used as indexes into a hash table containing the value of each variable.

Program 7.27 shows how a simple hash function is implemented and used. The hash table is a 32-element array of integers initialized to all zeros. Thus, an empty array element is represented by the value 0. The user enters a variable name, which gets passed to the hash function `hasher()`. This function finds the sum of the ASCII codes for the first, middle, and last characters in the variable name. The sum is then shifted two bits to the right and then ANDed with 31 to put it into the range 0 to 31. We are guaranteed to get a valid array index for any variable name passed to `hasher()`.

Program 7.27

```
#include <stdio.h>
#include <string.h>

int hasher(char name[]);

main()
{
        static int hashtable[32], done;
        int hashaddr;
        char var[10];
        do
        {
                printf("Enter variable name => ");
                gets(var);
                if(0 != strcmp(var,"stop"))
                {
                        hashaddr = hasher(var);
                        if(0 == hashtable[hashaddr])
                             printf("New variable, hash value = %d\n"
                                     ,hashaddr);
                        else
                        {
                             printf("Collision with %d other variables!\n"
                                     ,hashtable[hashaddr]);
                             printf("Hash value = %d\n",hashaddr);
                        }
                        hashtable[hashaddr]++;
                }
                else
                        done++;
        } while(!done);
}

int hasher(char name[])
{
        int hashval, mid;

        mid = strlen(name) / 2;
        hashval = name[0] + name[mid] + name[strlen(name) - 1];
        hashval >>= 2;
```

```
    hashval &= 31;
    return(hashval);
}
```

Sample execution of Program 7.27 produces

```
Enter variable name => index
New variable, hash value = 17
Enter variable name => var1
New variable, hash value = 6
Enter variable name => var7
New variable, hash value = 7
Enter variable name => row
New variable, hash value = 22
Enter variable name => num
New variable, hash value = 20
Enter variable name => var2
Collision with 1 other variables! Hash value = 6
Enter variable name => count
New variable, hash value = 19
Enter variable name => sum
New variable, hash value = 21
Enter variable name => value
Collision with 1 other variables! Hash value = 17
Enter variable name => col
New variable, hash value = 15
Enter variable name => stop
```

Notice that some of the variable names hash to the same value. When this happens, we have a *collision*. A good hash function tries to minimize the number of collisions. A collision is a problem because the hash table can store only one value in each table position. Because it may not be possible to eliminate all collisions (and it is almost impossible to predict when they will occur), simple one-dimensional arrays are not usually used for hash tables. Instead, a two-dimensional structure is often used (an array or some other kind of data structure) with the variable name *and* value stored at each position. Each position also has the ability to store more than one variable. So, when a collision occurs, the new variable name is compared with the variable names stored at the hashed position. The collision is resolved if a match is found. Otherwise, a new entry is made to the hash table at the current position.

Magic Squares

We finish this section with an interesting matrix application. A *magic square* is a square matrix having an odd number of rows and columns, whose rows and columns (and even both diagonals) add up to the same value. For example, examine this 3-by-3 magic square:

$$\begin{matrix} 6 & 1 & 8 \\ 7 & 5 & 3 \\ 2 & 9 & 4 \end{matrix}$$

The numbers in each row and column add up to 15. Even both diagonals add up to 15.

The technique used to generate a magic square is very simple. Begin with the value 1 in the middle element of the first row in the matrix. Then always move up and to the left one position when writing the next element value. The rows wrap around in such a way that the top row and bottom row are *next to* each other. The same is true for the left and right columns. If the new position is already occupied, return to the previous position and go down one row and resume.

Program 7.28 shows how this technique is implemented. Because even-sized matrices are not allowed, the program tests the input size to guarantee it is odd.

Program 7.28

```
#include <stdio.h>

main()
{
        static int magic[9][9];
        int row,col,k,n,x,y;

        printf("Enter size of magic square => ");
        scanf("%d",&n);
        if(0 == n % 2)
        {
                printf("Sorry, you must enter an odd number.");
                exit(0);
        }
        k = 2;
        row = 0;
        col = (n - 1)/2;
        magic[row][col] = 1;
        while(k <= n*n)
        {
                x = (row - 1 < 0) ? n - 1 : row - 1;
                y = (col - 1 < 0) ? n - 1 : col - 1;
                if(magic[x][y] != 0)
                {
                        x = (row + 1 < n) ? row + 1 : 0;
                        y = col;
                }
                magic[x][y] = k;
                row = x;
                col = y;
                k++;
        }
        for(row = 0; row < n; row++)
        {
                for(col = 0; col < n; col++)
                        printf("\t%d", magic[row][col]);
                printf("\n");
        }
}
```

A few sample executions of Program 7.28 follow:

```
Enter size of magic square => 3
         6       1       8
         7       5       3
         2       9       4

Enter size of magic square => 4
Sorry, you must enter an odd number.

Enter size of magic square => 5
         15      8       1       24      17
         16      14      7       5       23
         22      20      13      6       4
         3       21      19      12      10
         9       2       25      18      11
```

Can you determine a formula that predicts the sum of any row/column for a given size N?

Conclusion

The example programs presented in this section covered a variety of topics. Check your understanding of this material with the following section review.

7.5 Section Review

1. Explain how the `rand()` function is used in Program 7.25.
2. Why is a hash table useful?
3. What is the order in which the elements are accessed in Program 7.28?

7.6 Troubleshooting Techniques

One of the most common problems encountered when working with arrays is the use of improper index values. The compiler is not able to guarantee that an index is valid, since the index value may not be known until runtime. For example, the statement:

```
sum += mat[r][c];
```

contains two index variables `r` and `c` that may take on any value at runtime. If the value of `r` or `c` is larger than the associated matrix dimension, the data pointed to by the index variables will not be valid. It is therefore necessary for the programmer to enforce the proper index values at runtime. This is referred to as **bounds checking**.

For sequential access to an array, the index values can be bounded by the parameters of a loop, as in:

```
for(r = 0; r < 3; r++)
{
        sum += mat[r][0];
        .
        .
        .
}
```

In this case the values of the index variable r are limited to 0, 1, and 2.

When the index variable is based on data provided by the user, a different method must be used to perform the bounds checking. For example, the program might ask the user to enter an item number in a database. The item number entered by the user might be greater than the number of items stored in the database.

Consider an array defined as follows:

```
int speeds[10];
```

Legal index values for speeds are 0 through 9. Bounds checking can be applied to any access to speeds using this technique:

```
if (inbounds(k))
{
        .
        .
        access to speeds[k]
        .
        .
}
```

where inbounds() is defined as follows:

```
#define inbounds(x) ((x >= 0) && (x < 10))
```

The true/false condition evaluated by inbounds() is used by the if statement to restrict access to speeds to those values of k that are legal.

A similar technique can be used for multidimensional arrays. You can determine this yourself as you take the section review.

7.6 Section Review

1. What is bounds checking?
2. What happens if an illegal index value is used during runtime?
3. How can bounds checking be performed on a two-dimensional array defined like this:

```
int points[4][7];
```

7.7 Case Study: Eight-Queens Puzzle

Discussion

This case study solves an interesting chess puzzle. Two-dimensional arrays are used to represent the chessboard and generate index values. Bounds checking is performed to restrict array access.

The Problem

The problem being solved is called the *Eight-Queens Puzzle*, which asks "Is it possible to place eight Queens on a chessboard, so that none of the Queens occupy the same row, column, or diagonal?"

One complication of this problem is that there are so many different ways to place the eight Queens. There are, in fact, 8^8 different Queen patterns, which is just over 16.7 *million.* Only a fraction of these 16 million patterns are legal solutions to the Eight-Queens puzzle. Is it possible to find a solution without having to check all 16 million patterns?

The Algorithm

The algorithm used to solve the Eight-Queens puzzle employs recursion to implement a search technique called **backtracking.** The recursive search function finds a legal Queen position on a specific row on the chessboard. Backtracking refers to taking a step backwards when you make a wrong turn. When the algorithm discovers an illegal Queen position, it backtracks to a position where it can make a different choice. This may involve backing up to a previous row, and not just to a different position on the current row. The basic algorithm is as follows:

1. Pick a position for the Queen.
2. If legal, go to next row.
3. If illegal, pick the next position.
4. If no legal position is found, back up one row.

If legal positions are found for all eight rows, the problem is solved.

The statements that make up the `main()` function for the Eight-Queens puzzle look like this:

```
main()
{
     if (solve_row(0))              /* If puzzle has a solution */
          show_board();             /* display the chessboard. */
     else
          printf("No solution!\n");
}
```

The `solve_row()` function is called to start the solution at row 0. If `solve_row()` is able to recursively find a solution, it returns TRUE. This in turn allows the `if` statement to display the results by calling `show_board()`.

If no solution is possible, `solve_row()` returns FALSE. This can happen at any level of the solution process, and is the means of implementing backtracking by returning to a previous recursive level in the solution process.

The recursive `solve_row()` function uses a `while` loop to check all possible Queen positions in row r:

```
int solve_row(int r)
{

     int c;                  /* Column counter. */
     int done;               /* TRUE when a legal position is found. */

     if (r == 8)             /* Break recursive loop? */
          return(TRUE);
```

```
        c = 0;                  /* Initial Queen position within row. */
        done = FALSE;
        while((c < 8) && !done)         /* Try all 8 positions if necessary. */
        {
             if (legal(r,c))            /* Is the Queen safe? */
             {
                  board[r][c] = TRUE;         /* Mark position. */
                  done = solve_row(r+1);      /* Solve next row based */
                                              /* on this position. */
                  if (!done)          /* Was a solution possible? */
                       board[r][c] = FALSE;     /* No, try again. */
             }
             c++;            /* Next Queen position. */
        }
        return(done)         /* Return FALSE if no positions */
                             /* were legal. */
}
```

The `legal()` function pretends that a new Queen is placed at location `[r][c]`. All rows, columns, and diagonals from this position are scanned for previously placed Queens. Figure 7.17 shows the eight search directions. If any Queens are found, `legal()` returns FALSE.

The code for `legal()` is as follows:

```
int legal(int r, int c)
{
     /* This array contains the row and column deltas */
     /* used to adjust the row and column counters. */
     int dirs[8][2] = { {-1,0}, {-1,1}, {0,1}, {1,1}
                    {1,0}, {1, -1}, {0,-1}, {-1,-1}};
     int i;           /* Direction counter. */
     int cr,cc;       /* Check row and column. */

     for(i = 0; i < 8; i++)          /* Check all 8 directions. */
     {
          cr = r + dirs[i][0];       /* Get first position to check. */
          cc = c + dirs[i][1];
          while(inbounds(cr) && inbounds(cc))
               if (board[cr][cc])        /* TRUE if occupied. */
                    return(FALSE);       /* Move is illegal. */
               else
               {
                    cr += dirs[i][0];    /* Get next position. */
                    cc += dirs[i][1];
               }
     }
     return(TRUE);          /* Never saw a Queen, move is legal. */
}
```

The `cr` and `cc` variables contain the position of the next `board` element to check. `cr` and `cc` are updated by adding the corresponding delta from the `dirs` array. Positions are scanned until a Queen is found or the edge of the board is encountered.

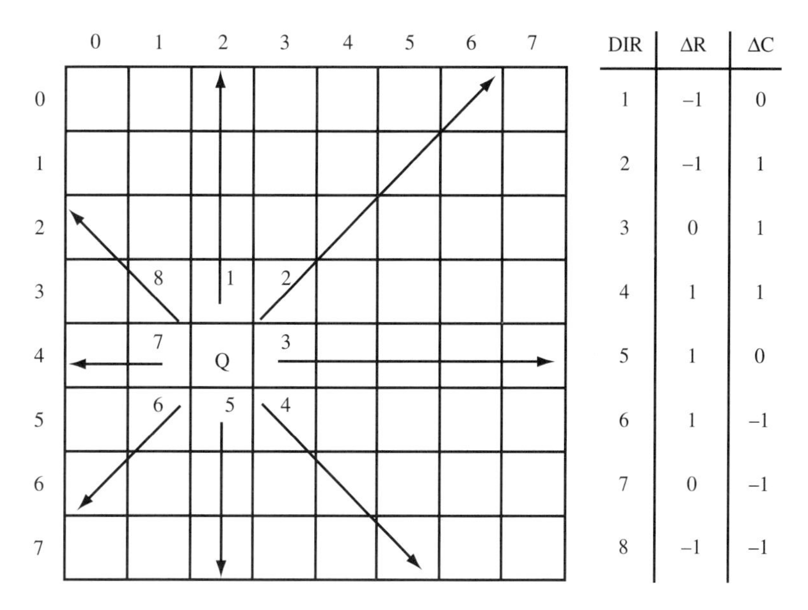

DIR	ΔR	ΔC
1	–1	0
2	–1	1
3	0	1
4	1	1
5	1	0
6	1	–1
7	0	–1
8	–1	–1

Figure 7.17 Search Directions for Each Queen and Associated Deltas

The Complete Program

Program 7.29 shows the complete Eight-Queens Puzzle solution. The `show_board()` function uses the contents of the global `board` array to display the Eight-Queens pattern if a solution exists.

Program 7.29

```
#include <stdio.h>

#define FALSE 0
#define TRUE  1
#define inbounds(x) ((x >= 0) && (x < 8))

/********************************************************************/
/*                        Eight - Queens Puzzle                     */
/********************************************************************/
/*                        Developed by A. King                      */
/*                                                                  */
/********************************************************************/
/*     This program uses recursion to find the positions where      */
/*     eight Queens may be placed on a chessboard so that:          */
/*          * No two Queens are on the same row.                    */
/*          * No two Queens are in the same column.                 */
/*          * No two Queens are on the same diagonal.               */
/********************************************************************/
/*                        Function Prototypes                       */
/*------------------------------------------------------------------*/
```

```
/* This function displays an 8 by 8 chess board with Queens on it.  */
/*------------------------------------------------------------------*/

void show_board(void);

/* This function recursively solves the puzzle beginning at row r.  */
/* Returns TRUE if a solution exists beginning at row r.            */
/*------------------------------------------------------------------*/

int solve_row(int r);

/* This function returns TRUE if a Queen can be placed at position  */
/* (r,c) on the board.                                              */
/*------------------------------------------------------------------*/

int legal(int r, int c);

/********************************************************************/
/*                          Global Variables                        */
/*------------------------------------------------------------------*/

        int board[8][8];                /* 8 by 8 chessboard. */

/*------------------------------------------------------------------*/

main()
{
        if (solve_row(0))               /* If puzzle has a solution */
                show_board();           /* display the chessboard. */
        else
                printf("No soltuion!\n");
}

/*------------------------------------------------------------------*/

void show_board()
{
        int r,c;

        printf("The position of each Queen is:\n  ");
        for(c = 0; c < 8; c++)
                printf(" %1d",c);
        printf("\n");
        for(r = 0; r < 8; r++)
        {
                printf("%1d ",r);
                for(c = 0; c < 8; c++)
                        if (board[r][c])
                                printf(" Q");
                        else
                                printf("  ");
```

```
                printf("\n");
        }
}

/*--------------------------------------------------------------------*/

int solve_row(int r)
{

        int c;                  /* Column counter. */
        int done;               /* TRUE when a legal position is found. */

        if (r == 8)             /* Break recursive loop? */
                return(TRUE);
        c = 0;                  /* Initial Queen position within row. */
        done = FALSE;
        while((c < 8) && !done)         /* Try all 8 positions if necessary. */
        {
                if (legal(r,c))                 /* Is the Queen safe? */
                {
                        board[r][c] = TRUE;     /* Mark position. */
                        done = solve_row(r+1);  /* Solve next row based */
                                                /* on this position. */
                        if (!done)              /* Was a solution possible? */
                                board[r][c] = FALSE;    /* No, try again. */
                }
                c++;                    /* Next Queen position. */
        }
        return(done);                   /* Return FALSE if no positions */
                                        /* were legal. */
}

/*--------------------------------------------------------------------*/

int legal(int r, int c)
{
        /* This array contains the row and column deltas */
        /* used to adjust the row and column counters. */
        int dirs[8][2] = { {-1,0}, {-1,1}, {0,1}, {1,1},
                           {1,0}, {1, -1}, {0,-1}, {-1,-1}};
        int i;          /* Direction counter. */
        int cr,cc;      /* Check row and column. */

        for(i = 0; i < 8; i++)          /* Check all 8 directions. */
        {
                cr = r + dirs[i][0];    /* Get first position to check. */
                cc = c + dirs[i][1];
                while(inbounds(cr) && inbounds(cc))
                        if (board[cr][cc])              /* TRUE if occupied. */
                                return(FALSE);          /* Move is illegal. */
                        else
```

```
                {
                        cr += dirs[i][0];       /* Get next position. */
                        cc += dirs[i][1];
                }
        }
        return(TRUE);               /* Never saw a Queen, move is legal. */
}
```

The execution of Program 7.29 results in the following output:

```
The position of each Queen is:
   0 1 2 3 4 5 6 7
0  Q
1          Q
2                Q
3            Q
4      Q
5              Q
6    Q
7        Q
```

Note that no two Queens occupy the same row, column, or diagonal.

Figure 7.18 represents the same solution in an 8-by-8 grid that allows the diagonals to be examined easier.

It would be very interesting to count the number of times the `legal()` function is called. This number will surely be significantly smaller than the 16 million different Queen patterns required when not using backtracking.

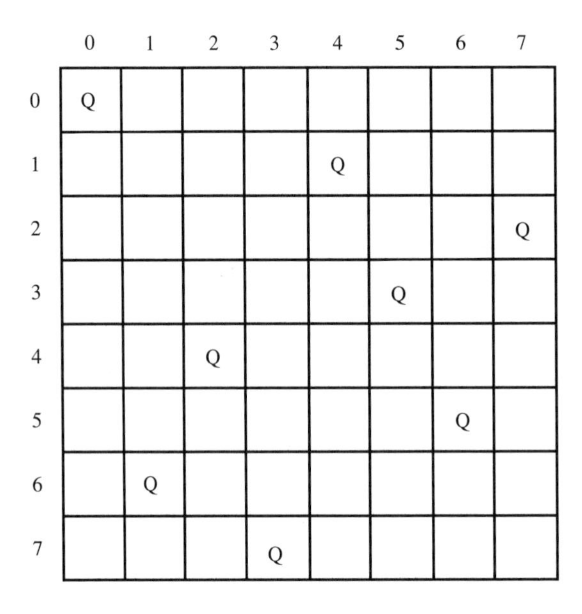

Figure 7.18 Eight-Queens Solution

Conclusion

In this section we examined how arrays can be combined with recursion to provide a very powerful solution-space search technique called backtracking. Test your understanding of this material with the following section review.

7.7 Section Review

1. How are all 64 board locations initialized to empty (FALSE)?
2. How is recursion stopped in `solve_row()`?
3. What is backtracking?
4. What does `inbounds()` do?
5. What board positions are examined when `legal(3,5)` is called?

Interactive Exercises

DIRECTIONS

Execute each program on your computer and note the results of its execution. In many cases you are asked to predict what the program will do. Compare your predictions with the actual program execution.

Exercises

1. Program 7.30 displays the elements of an uninitialized array. What is displayed?

Program 7.30

```
#include <stdio.h>

main()
{
        int array[5];
        int index;

        for(index = 0; index < 5; index++)
                printf("%d\n",array[index]);
}
```

2. Examine the array indexes of Program 7.31. Do you think this program will compile and execute? What happens when you try?

Program 7.31

```
#include <stdio.h>

#define MAXVALUE 1000

main()
{
        int array[MAXVALUE];
        int index;
```

```
        for (index = 0; index > -3; index--)
                printf("%d\n",array[index]);

}
```

3. What number is displayed by Program 7.32?

Program 7.32

```
#include <stdio.h>

main()
{
        int v1[4] = {1, 2, 4, 8};
        int v2[4] = {1, 0, 1, 1};
        int i,j = 0;

        for(i = 0; i < 4; i++)
                j += v1[i]*v2[i];
        printf("%d",j);
}
```

4. What number is displayed by Program 7.33?

Program 7.33

```
#include <stdio.h>

        int v1[4] = {3, 40, 100};

int mbt(int index);

main()
{
        printf("%d",mbt(2));
}

int mbt(int index)
{
        if(0 == index)

                return(v1[index]);
        else
                return(v1[index] + mbt(index - 1));
}
```

Self-Test

DIRECTIONS

Answer the following questions by referring to the programs in Section 7.3 (Sorting with Numeric Arrays).

Questions

1. Why are more comparisons performed in Program 7.12 if the switch-detection statement is removed?
2. Why is `temp_value` needed to perform a swap?
3. How does the bucket sort in Program 7.13 determine what numbers to print out in sorted order?
4. How could the bucket sort be used to sort positive and negative numbers?
5. Why does merge sort require more overall storage space to sort a 20-element array than bubble sort does?
6. How is the midpoint of the array chosen in merge sort?
7. What function does the pivot perform in quick sort?
8. How can the median element of an array be determined?
9. Why is the comparison function defined as it is in quick sort?
10. Given the following array:

    10 9 8 7 6 5 4 3 2 1

    which technique is more efficient—merge sort or quick sort? Explain.

End-of-Chapter Problems

General Concepts

Section 7.1

1. What is meant by the term *numeric array*?
2. State how you would declare a numeric array of ten elements in a C program.
3. What is the index of the first element of the array in question 2? What is the index of the last element?
4. What is the difference between a local array and a global array?
5. What is accomplished when a `static` array is declared?

Section 7.2

6. Explain the basic idea behind applications of arrays in a C program.
7. Explain how a C program could be developed to find the largest element in an array.
8. What is an easy method of getting user input values into an array?

Section 7.3

9. Explain the basic differences among bubble sort, bucket sort, merge sort, and quick sort.
10. How does element order affect sorting efficiency?
11. Explain how to merge two ordered arrays.
12. How does recursion assist the sorting process?
13. Why is the choice of a pivot so important in quick sort?

Section 7.4

14. What is meant by row-major order in regard to the storage of matrix elements?
15. How is a matrix passed between functions?
16. How many math operations (`+` and `*`) are required to multiply a 3-by-4 matrix by a 4-by-2 matrix?

Section 7.5

17. What is the technique behind dynamic linking?
18. How is the `hist` array indexed in Program 7.25?

19. Why is an array needed to store the numbers in Program 7.26?
20. What is the purpose of a hash function?

Section 7.6

21. Why is it impossible for the compiler to perform bounds checking?
22. Why use a `#define` for `inbounds()` instead of a function?

Section 7.7

23. How can a counter variable be added to `legal()` to keep track of how many times `legal()` is called?
24. Describe the actions of `solve_row(7)` when Program 7.29 executes.

Program Design

All of the following C programs will require the use of a one- or two-dimensional array. In order to test each program you will have to make up your own set of test data, and properly initialize any arrays that require data from the user.

25. The circuit in Figure 7.19 is a series parallel circuit. Develop a C program that will compute the total resistance of any branch selected by the program user. The total resistance of any one branch is:

$$R_T = R_1 + R_2 + R_3$$

Where

R_T = Total branch resistance in ohms.
R_1, R_2, R_3 = Value of each resistor in ohms.

26. Modify the program you developed in problem 25 so that the total resistance of any combination of branches may be found by the program user. The total resistance of any parallel branch is found by first determining the total resistance of that branch, and then using the parallel resistance formula for finding the total resistance. The parallel resistance formula is

$$R_T = 1/(1/R_{T1} + 1/R_{T2} \ldots + 1/R_{TN})$$

Where

R_T = Total resistance in ohms.
$R_{T1}, R_{T2}, \ldots R_{TN}$ = Total branch resistance in ohms.

27. Expand the modified program from problem 26 so that the program user can enter the value of each resistor and the resistors will be displayed in numerical order.
28. Develop a C program that computes the power dissipation of each resistor in a series circuit. The program user may select how many resistors there will be in the circuit. The program will

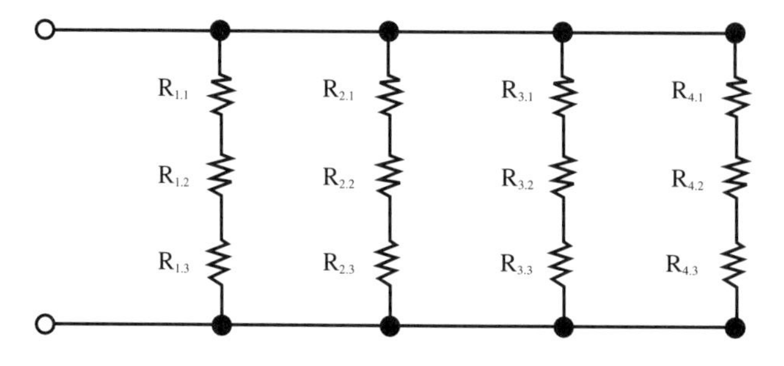

Figure 7.19 Circuit for Problems 25, 26, and 27

sort the resistors by their power dissipation and display their values and subscript numbers (the subscript numbers represent the order in which they appear in the circuit). The program user enters the value of the voltage source in volts. Power dissipation in a resistor may be determined by

$$P = I^2R$$

Where

P = Power dissipation in watts.
I = Current in the resistor in amps.
R = Value of resistor in ohms.

The current in each resistor in a series circuit is the same. It may be determined from

$$I = V_S/R_T$$

Where

I = Circuit current in amps.
V_S = Source voltage in volts.
R_T = Total circuit resistance in ohms.

29. Write a C program that will normalize the 20 floating numbers saved in the array STATS. To normalize an array of numbers, first find the largest number in the array, then divide each number in the array by the largest number. Each resulting value will then be between 0 and 1.
30. Figure 7.20 shows the schematic of a low-pass filter. The corner frequency of the filter is found by the formula $Fc = 1/(2\pi RC)$. The operation of the low-pass filter is as follows: Input signals whose frequencies are lower than *Fc* pass through the filter with a small loss in amplitude. Signals whose frequencies are larger than *Fc* pass through with a large loss in amplitude. The output voltage of the low-pass filter is found by the formula:

    Vo= *Vi*/(sqrt(1 + (*F*/*Fc*)^2))

    where *F* is the applied frequency. Write a C program that will display the frequency and output voltage for a low-pass filter at the following frequencies:

    0.1**Fc* 0.25**Fc* 0.5**Fc* 0.75**Fc* 0.9**Fc*
    Fc 2**Fc* 5**Fc* 7.5**Fc* 10**Fc*

    Let $Vi = 100$ volts, $R = 1000$ ohms, and $C = 0.1$ microfarads. Save each frequency/voltage pair in an array and scan the array for the frequency at which *Vo* is closest to 70.7 volts. Display the resulting frequency value also.
31. Write a C program that will match each resistor from list A with each resistor from list B (totaling 20 pairs), and compute the total resistance, the current supplied by a 50-volt battery to

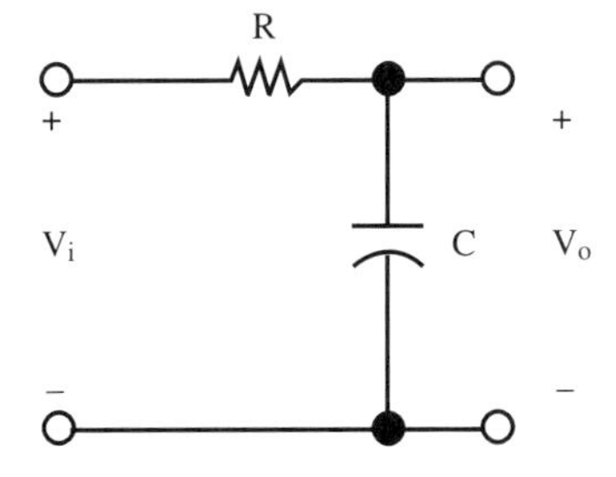

Figure 7.20 Circuit for Problem 30

the total resistance, and the voltage across each resistor. For example, the first pair of resistors is 100 ohms, 270 ohms. The resulting display should be:

```
R1     R2     Rt     I          V1       V2

100    270    370    0.135      13.5     36.5
```

The lists of resistor values are as follows:

List A	*List B*
100	270
150	330
220	470
270	560
330	

32. Develop a C program that will display the amount of money in any safety deposit box that is contained in a wall which has 10 rows by 8 columns of these boxes.
33. Write a C program that will generate and display a random 10-by-10 array of numbers between 50 and 200.
34. A five-card poker hand is represented by the two-dimensional integer array CARDS[2][5]. The first row in CARDS contains the card numbers, which are:

2 through 10	: 2 through 10
Jack	: 11
Queen	: 12
King	: 13
Ace	: 14

    The second row contains the suit number for each card. The suit numbers are defined as:

Clubs	: 1
Diamonds	: 2
Hearts	: 3
Spades	: 4

    Write a C function that will analyze the poker hand and determine if any of the following are present:

    Royal flush (10, J, Q, K, A of same suit)
    Straight flush (any consecutive group of five cards in same suit)
    Four of a kind
    Full house (three of a kind and two of a kind)
    Flush (all five cards in same suit)
    Straight (any consecutive group of five cards from any suit)
    Three of a kind
    Two pair
    One pair

    Assign any values you wish to the poker hands shown.
35. The knight on a chess board may move from its current position to any of the positions shown in Figure 7.21. Assuming row and column numbers are from 1 to 8, write a function that determines if a knight is being legally moved from OldRow, OldColumn to NewRow, NewColumn. Return 1 if the move is valid and 0 otherwise.

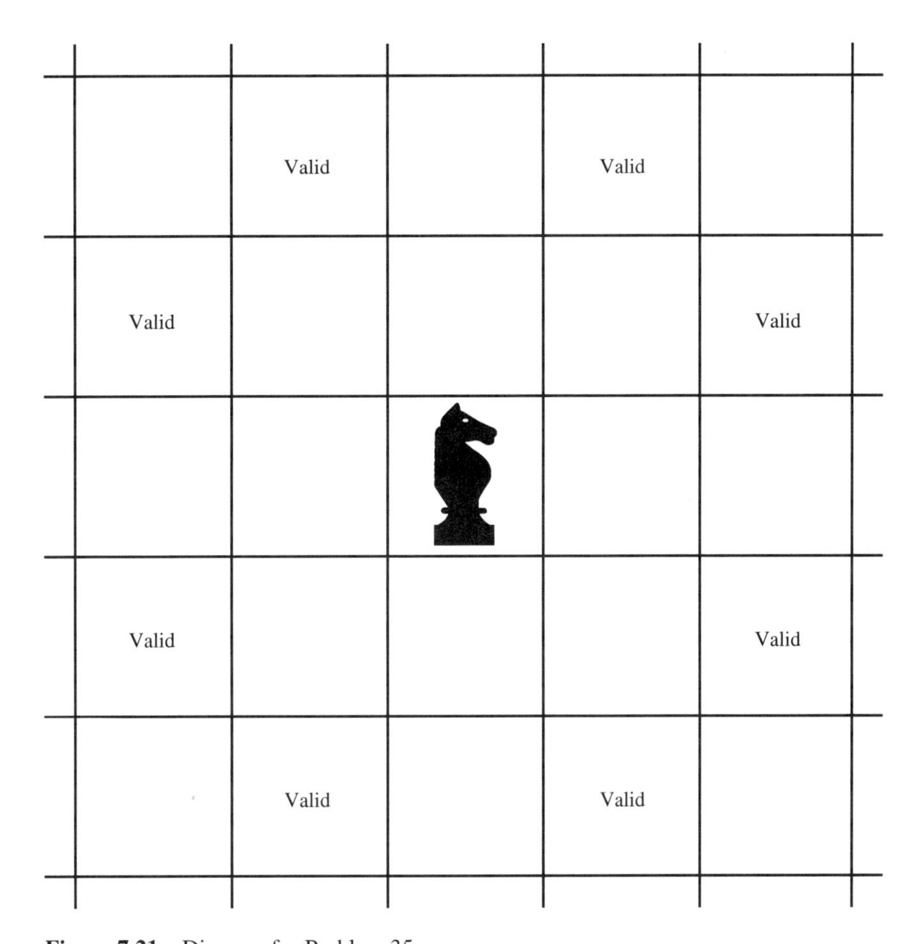

Figure 7.21 Diagram for Problem 35

36. Write a C program that computes all factors of a user-supplied integer. Store the factors in an array called FACTORS.
37. Modify the program in problem 36 to determine if the input number is a *perfect* number. A perfect number is a number whose factors add up to itself. For example, 6 is a perfect number, because its factors (1, 2, and 3) add up to 6. The number itself is not considered a factor in this case.
38. Another technique used to sort numbers is called *insertion* sort. In this technique, each new number is inserted into the correct position of an in-order array. For example, given the array

    5 6 9 12 24 39 52

    the new value 18 will be inserted between the 12 and the 24 by moving elements 24, 39, and 52 up one position in the array. Write a C program that performs an insertion sort. Use the example array as the initial array. Allow the user to enter three numbers, one at a time. Show the resulting array after each insertion.
39. Write a C function that determines the minimum value and maximum value of an array of K integers using only one pass through the numbers.
40. A two-dimensional array LINES[3][4] contains a set of three endpoint coordinates. For example, if the first row of LINES contains the numbers 10, 20, 100, and 20, the endpoints of the

first line are (10,20) and (100,20). Write a C function that determines if the three lines form a triangle. Return 1 if a valid triangle exists and 0 otherwise.

41. Write a C function that transposes a square matrix. For example, given this matrix:

```
1 2 3
4 5 6
7 8 9
```

your function should create the following transposed matrix:

```
1 4 7
2 5 8
3 6 9
```

42. Add a function to Program 7.24 that identifies lost allocation units.
43. Write a C program that finds the histogram of a two-dimensional matrix of integers DATA[8][8]. Then find the histogram element that has the highest frequency count and use this value to do the following: if a DATA element is smaller than the histogram element, replace the DATA element with 0, otherwise replace it with 1.
44. Write your own `hasher()` function for Program 7.27. Compare its number of collisions with the number of collisions shown in the sample execution of Program 7.27 in Section 7.5.
45. Determine a formula to predict the sum of any row or column in a magic square of size N.

8 Data Structures

Objectives

This chapter provides you the opportunity to learn:

1. The meaning of enumerated data types in C.
2. How to define your own data types.
3. The meaning of data structures.
4. Why data structures are useful.
5. Arranging data structures as arrays.
6. How to develop complex data structures.
7. How to use the C `union`.
8. How to use advanced data structures.

Key Terms

Enumerated
Type Definition
Structure Member
Structure
Member of Operator
Template
Structure Tag
Member Declaration List
`struct`
`union`
Structure Array
Node
Linked-List
`malloc`
Stack
LIFO
Queue
FIFO

Binary Tree
Traversal
Depth-First Search
Breadth-First Search
Emulator

Outline

Introduction

Up to this point, you have been concerned primarily with working with only one data type at a time. For example, when you worked with arrays, the array was of a single type (such as `int` or `char`). A single data type has a simple data structure. In complex applications, you may use more than one data type at a time. This chapter will show you how to do this.

This is a very useful chapter. Essentially this chapter will show you how to develop C programs that will handle more complex technology problems.

8.1 Enumerating Types

Discussion

This section presents another way of expressing data in C. Up to this point you have had the `#define` available. Now you will have the opportunity to learn another method for expressing data that will give your programs even greater readability. The material you learn here will help set the stage for the remaining sections of this chapter.

Expressing Data

There is another way of expressing data in C called the `enum` (for **enumerated**). What this does is to hold one integer value from a fixed set of identified integer constants.

The form of `enum` is as follows:

```
enum tag { enumeration-list }
```

The above enumeration declaration presents the name of an enumeration variable and then defines the names within the enumeration list. The declaration begins with the keyword `enum`. The resulting `enum` variable can then be used anywhere an `int` type is used. An example is illustrated in Program 8.1.

Program 8.1

```
#include <stdio.h>

        enum color_code {black, brown, red, orange, yellow};

main()
{
        enum color_code color;

        char value;

        printf("Input an integer from 0 to 4 => ");
        value = getchar();

        switch(value)
        {
                case '0' : color = black;
                           break;
                case '1' : color = brown;
                           break;
                case '2' : color = red;
                           break;
                case '3' : color = orange;
                           break;
                case '4' : color = yellow;
                           break;
        }

        switch(color)
        {
                case black  : printf("Color is black.");
                              break;
                case brown  : printf("Color is brown.");
                              break;
                case red    : printf("Color is red.");
                              break;
                case orange : printf("Color is orange.");
                              break;
                case yellow : printf("Color is yellow.");
                              break;
        }
}
```

Program 8.1 asks the program user to input an integer from 0 to 4. The program will then give the program user the equivalent resistor color code for that number. As an example

```
Input an integer from 0 to 4 => 3
Color is orange.
```

The important point of this program is that it uses the `enum` data type to make the source code more readable.

Program Analysis

First, an enumerated data type is declared outside of `main()`:

```
enum color_code {black, brown, red, orange, yellow}
```

What this does is declare `color_code` to be an enumerated type. The list within the braces shows the constant names that are valid values of the `enum color_code` variable.

Next, inside `main()` is the declaration of an enumerated variable of type `color_code`:

```
main()
{
        enum color_code color;
```

This means that the variable `color` may have any of these values: black, brown, red, orange, or yellow.

Look at what the first C `switch` does. From the user input

```
switch (value)
{
        case '0' : color = black;
                   break;
        case '1' : color = brown;
                   break;
        case '2' : color = red;
                   break;
        case '3' : color = orange;
                   break;
        case '4' : color = yellow;
                   break;
}
```

it assigns color value to the variable `color`. Hence, if the user selects 3 for an input, the variable `color` will be assigned the value of `orange`. This data is then used in the next C `switch`:

```
switch (color)
{
        case black  : printf("Color is black.");
                      break;
        case brown  : printf("Color is brown.");
                      break;
        case red    : printf("Color is red.");
                      break;
        case orange : printf("Color is orange.");
                      break;
        case yellow : printf("Color is yellow.");
                      break;
}
```

This results in a very descriptive C `case`. You should note that `enum` does not introduce a new basic data type. Variables of the `enum` type are treated as if they were of type `int`. What `enum` does is improve the readability of your programs.

Another Form

Another method of displaying an enum declaration is in the vertical form, as follows:

```
enum color_code
{
     black,
     brown,
     red,
     orange,
     yellow
}
```

Enumeration Example

To demonstrate exactly what is going on in an enum data type, look at Program 8.2:

Program 8.2

```
#include <stdio.h>

enum color_code {black, brown, red, orange, yellow}

main()
{
        enum color_code color;

        for(color = black; color <= yellow; color++)
                printf("Digital value of enum type color => %d\n", color);
}
```

Execution of Program 8.2 yields

```
Digital value of enum type color => 0
Digital value of enum type color => 1
Digital value of enum type color => 2
Digital value of enum type color => 3
Digital value of enum type color => 4
```

Note that the enum type is used in the C for loop just as if it were of type int.

Assigning enum Values

The C enum type may also be assigned values—as long as they are integer values. This is demonstrated in Program 8.3. This program solves for the total current for a given voltage source when the Thevenin* equivalent voltage and resistance are known. In this program,

*The Thevenin equivalent form of any resistive circuit consists of an equivalent voltage source and an equivalent resistance. These values depend on the values of the original circuit. This method is used to simplify resistive circuits to a single voltage source and single resistor.

the Thevenin equivalent voltage is 12 volts and the Thevenin equivalent resistance is 150 ohms. Note how these values are set by the C enum type.

Program 8.3

```
#include <stdio.h>

        enum thevenin {source = 12, resistance = 150};

main()
{
        float load;
        float current;

        printf("Input value of load resistor => ");
        scanf("%f",&load);

        current = source/(resistance + load);

        printf("The circuit current is %f amps.", current);
}
```

In Program 8.3, the enum data type thevenin is assigned values of source (which is set equal to 12) and resistance (set equal to 150). These constant names are then used inside the program to do the necessary calculations. Again, the purpose here is to make the program more readable and allow easy changes in the constant assignments used in the program.

Conclusion

This section introduced the C enum. Here you discovered another method of using data in your C program to improve its readability. You also saw that the use of the C enum is compatible with the C int. Check your understanding of this section by trying the following section review.

8.1 Section Review

1. In C, what does the keyword enum mean?
2. State what an enum data type does in C.
3. For the following code, what is the integer value of first?

```
enum numbers {first, second, third}
```

4. What is the main purpose of using a C enum?
5. Can a C enum constant be set to a given integer value? Give an example.

8.2 Naming Your Own Data Types

Discussion

In Section 8.1, you were introduced to a method of making your program easier to read. This section elaborates on that theme. However, unlike the previous section, this section will show you how to name your own data types.

Inside typedef

The use of the `typedef` (**type definition**) in C allows you to define names for a C data type. A `typedef` declaration is similar to a variable declaration except that the keyword `typedef` is used. The form is

```
typedef type-specifier declarator
```

For example, you could have:

```
typedef char LETTER;
```

This now declares `LETTER` as a synonym for `char`. This means that `LETTER` can now be used as a variable declaration such as

```
LETTER character;
```

instead of

```
char character;
```

As an example, suppose you have a program that is part of an automated testing system. In this program the user inputs the status of an indicator light and the program gives a response depending on the given status. The program could be written as shown in Program 8.4.

Program 8.4

```
#include <stdio.h>

        void test_it(void);
        int check_lights(char condition);

main()
{
        test_it();
}


void test_it()
{
        char input;

        printf("\n\n1] Red  2] Green  3] Off \n\n");
        printf("Select light condition by number => ");
        input = getchar();

        check_lights(input);
}

int check_lights(char condition)
{
        switch(condition)
        {
                case '1': printf("Check system pressure.");
                          break;
```

```
                case '2': printf("System OK.");
                          break;

                case '3': printf("Check system fuse.");
                          break;
        }
}
```

Execution of Program 8.4 produces

```
1] Red  2] Green  3] Off

Select light condition by number => 2
System OK.
```

However, the program could be made more descriptive by defining a data type called `light_status`. The same program using this new type definition is illustrated by Program 8.5.

Program 8.5

```
#include <stdio.h>

        typedef enum light_status {Red, Green, Off};
        void test_it(void);
        int check_lights(enum light_status condition);

main()
{
        test_it();
}


void test_it()
{
        char input;
        enum light_status reading;

        printf("\n\n1] Red  2] Green  3] Off \n\n");
        printf("Select light condition by number => ");
        input = getchar();

        switch (input)
        {
                case '1' : reading = Red;
                           break;
                case '2' : reading = Green;
                           break;
                case '3' : reading = Off;
                           break;
        } /* End of switch */

        check_lights(reading);
```

```
}

int check_lights(enum light_status condition)
{
        switch(condition)
        {
                case Red   : printf("Check system pressure.");
                             break;

                case Green: printf("System OK.");
                             break;

                case Off   : printf("Check system fuse.");
                             break;
        } /* End of switch */
}
```

The output of Program 8.5 is the same as that of Program 8.4. The difference is that a new C type name is created for an existing data type. This allows new C identifiers to be defined as this data type. This can be done in the body of functions as well as in the function parameters. It's important to note that the `typedef` declaration does not create types. It simply creates synonyms for existing types.

Program Analysis

The C keyword `typedef` is used to indicate that a new definition is to be used for a data type.

```
typedef enum light_status {Red, Green, Off};
```

In this case, the data type is an `enum` type called `light_status`. It consists of the enumerated constants `Red`, `Green`, and `Off`. Now other variables may be defined in terms of this new data type. Look at the formal parameter of the function prototype:

```
int check_lights(enum light_status condition);
```

It declares a new variable `condition` as a type `light_status`. It is used again in the definition of function `test_it`.

```
void test_it()
{
        char input;
        enum light_status reading;
```

This means that the variable `reading` is the enumerated data type `light_status` and can assume the values `Red`, `Green`, and `Off`.

After the variable `reading` is assigned one of these three values by the C `switch`, it is then passed to the called function `check_lights()`.

```
switch (input)
{
        case '1' : reading = Red;
                   break;
```

```
        case '2' : reading = Green;
                   break;
        case '3' : reading = Off;
                   break;
} /* End of switch  */

check_lights(reading);
```

Next it is used in the called function as part of the C `switch` to make the code more readable:

```
int check_lights(enum light_status condition)
{

        switch (condition)
        {
                case Red   : printf("Check system pressure.");
                             break;
                case Green : printf("System OK.");
                             break;
                case Off   : printf("Check system fuse.");
                             break;
        }   /* End of switch */

}
```

Other Applications

Some other applications of the C `typedef` are:

```
typedef char *STRING;
```

This allows you to use the new word `STRING` as a pointer to `char`:

```
STRING inputname;
```

Here, `inputname` is defined as a type `STRING` which is really a pointer to `char` `(char *)`.

You will see even more powerful applications of the C `typedef` in the next section of this chapter.

Conclusion

This section introduced you to a method for naming data types using C. Here you saw how to implement this in a program where you could create new identifiers in terms of this new data type. Check your understanding of this section by trying the following section review.

8.2 Section Review

1. State what the C `typedef` does.
2. For the following code, what is the resulting new name of the data type?

   ```
   typedef char name[20];
   ```

3. What is the main purpose of using the C `typedef`?

8.3 Introduction to Data Structures

Discussion

Up to now, you have been using arrays and pointers to create items of the same data type. You have never mixed a type `char` and a type `int` within a single data type. Understandably, you may not have even considered the possibility. However, most of the everyday "keeping track of things" requires the use of more than one data type. Consider your checking account. Each check you write must have your name on it (a type `char`), the amount of the check (a type `float`), and the check number (a type `int`). All of this information (plus a lot more) is contained on this one item called a check. Since this is a very natural way of arranging information, C has provided a method for you to accomplish the same thing. Doing this makes your program more readable and makes it easier to handle complex data consisting of many different types that are logically related.

The C Structure

A structure declaration names a structure variable and then states a sequence of variable names—called **structure members**—which may have different types. The basic form is

```
struct
{
        type   member_identifier1;
        type   member_identifier2;
        .
        .
        .
        type    member_identifierN;
}  structure_identifier;
```

Note that structure declarations begin with the keyword `struct`.

As an example of the use of a C structure, consider a box of parts as shown in Figure 8.1.

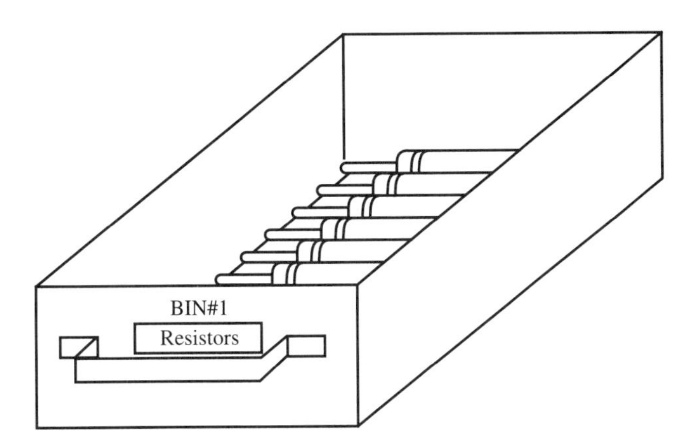

Figure 8.1 Box of Resistors

Assume that this box of parts consists of one type of resistor. Assume for the moment that there are three things you wish to keep track of concerning these resistors: the name of the manufacturer, the quantity of resistors in the box, and the price of each resistor. You could develop a C program that would easily do this without using the concept of a structure, but to keep things simple for now, the C **structure** will be used to keep this inventory. Program 8.6 illustrates the construction of a C structure.

Program 8.6

```
#include <stdio.h>

main()
{
        struct
        {
                char  manufacturer[20]; /* Resistor manufacturer.    */
                int   quantity;         /* Number of resistors left. */
                float price_each;       /* Cost of each resistor.    */
        } resistors;
}
```

Figure 8.2 illustrates the key points for the design of a C structure.

Note that the structure in Program 8.6 consists of a collection of data elements, called members, which may consist of the same type or different types that are logically related. The data types are `char` for the variable `manufacturer`, `int` for the variable `quantity`, and `float` for the variable `price_each`. The general form is

```
struct
{
        type    member_identifier1;
        type    member_identifier2;
        .
        .
        .
        type    member_identifierN;
}  structure_identifier;
```

Where

`struct`	=	Keyword indicating that a structure follows.
`{`	=	Necessary opening brace to indicate that a list of structure elements is to follow.
`type`	=	The C type for each element.
`member_identifier`	=	The variable identifier for the structure member.
`}`	=	Necessary closing brace.
`structure_identifier`		Defines the structure variable.

Putting Data into a Structure

Now that you've seen what a C structure looks like, you need to know how to get values into the structure members. Program 8.7 shows you how this is done. The program gets

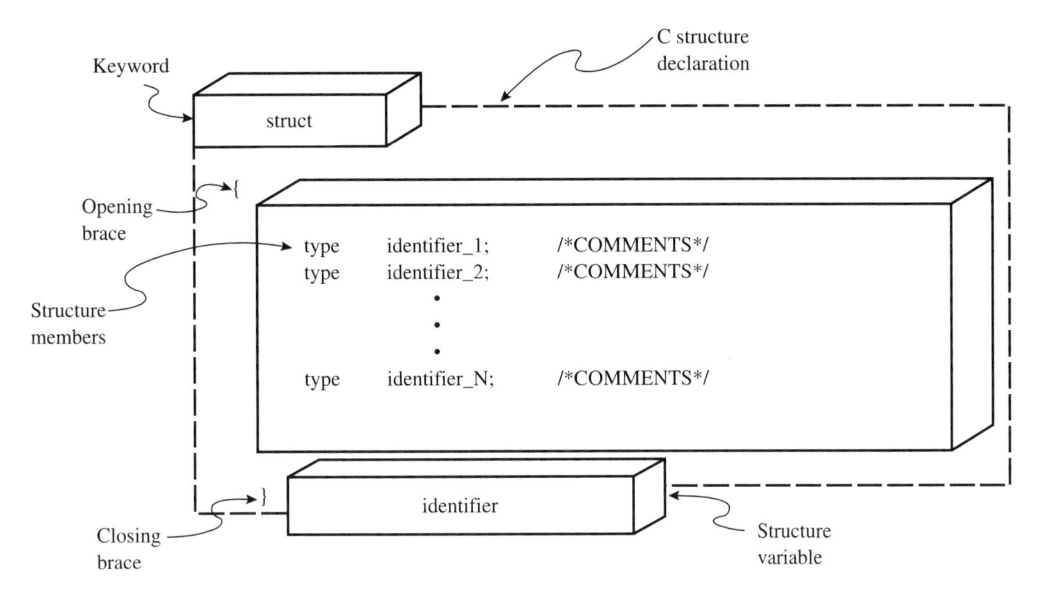

Figure 8.2 Key Parts of a C Structure

information from the program user concerning the manufacturer of the resistors, the quantity of resistors in the bin, and the unit price of the resistors. It then calculates the total value of all the resistors, displaying this along with the information put in by the program user.

The method of getting data into and out of a structure member is by using the **member of operator**. The member of operator specifies the name of the structure member and the structure of which it is a member. For example, in Program 8.7

```
resistors.manufacturer
```

represents the variable `manufacturer`, which is a member of the structure `resistors`. Note that the member of operator is represented by the period (.).

Program 8.7

```
#include <stdio.h>

main()
{
        struct
        {
                char  manufacturer[20];      /* Resistor manufacturer.      */
                int   quantity;              /* Number of resistors left.   */
                float price_each;            /* Cost of each resistor.      */
        } resistors;

        float total_value;                   /* Total value of parts.       */

        /* Display variables:                 */
```

```
        /* Get name of manufacturer:          */

        printf("Name of manufacturer => ");
        gets(resistors.manufacturer);

        /* Get number of parts left:          */

        printf("Number of parts left => ");
        scanf("%d",&resistors.quantity);

        /* Get cost of each part:             */

        printf("Cost of each part => ");
        scanf("%f",&resistors.price_each);

        /* Calculate total value:             */

        total_value = resistors.quantity * resistors.price_each;

        /* Display variables:                 */

        printf("\n\n");
        printf("Item:            Resistors\n\n");
        printf("Manufacturer:    %s\n",resistors.manufacturer);
        printf("Cost each:       $%f\n",resistors.price_each);
        printf("Quantity:        %d\n",resistors.quantity);
        printf("Total value:     $%f\n",total_value);

}
```

Assuming the user inputs the following, execution of Program 8.7 results in

```
Name of manufacturer => Ohmite
Number of parts left => 10
Cost of each part => 0.05

Item:            Resistors

Manufacturer:    Ohmite
Cost each:       $0.050000
Quantity:        10.000000
Total value:     $0.500000
```

Program Analysis

Program 8.7 starts with the same C structure as before and also declares another variable, `total_value`.

```
main()
{

        struct
```

```
{
        char   manufacturer[20];    /* Resistor manufacturer.     */
        int    quantity;            /* Number of resistors left. */
        float  price_each;          /* Cost of each resistor.    */
} resistors;

float total_value;                  /* Total value of parts.     */
```

The variable `total_value` will be used to hold the value that represents the total value of all the resistors in the parts bin.

Next the program gets the name of the resistor manufacturer from the program user.

```
/* Get name of manufacturer:          */

printf("Name of manufacturer => ");
gets(resistors.manufacturer);
```

Note how this is done. The name of the resistor manufacturer must be placed in the member `char manufacturer[20]`. Since the variable `manufacturer[20]` is contained in the structure referred to by the variable `resistors`, this must somehow be shown in the program. This is done by the member of operator ".".

```
resistors.manufacturer
```

The member of operator specifies the name of the structure member and the structure of which it is a member. This C member of operator is used again to get the data for the remaining two structure member variables:

```
/* Get number of parts left:          */

printf("Number of parts left => ");
scanf("%d",&resistors.quantity);

/* Get cost of each part:             */

printf("Cost of each part => ");
scanf("%f",&resistors.price_each);
```

Note that this time the address of `&` operator is needed because of the use of the `scanf()` function. But the member of operator is still the same:

```
structure_name.member_name
```

Next, a calculation is performed using the member of operator:

```
/* Calculate total value:             */

total_value = resistors.quantity * resistors.price_each;
```

Again, the name of the structure and the name of the member are used to identify the structure variables. The results are then displayed. Note that the variable `total_value` does not require a member of operator because it is not a member of any structure.

```
/* Display variables:                        */

printf("\n\n");
printf("Item:              Resistors\n\n");
printf("Manufacturer:      %s\n",resistors.manufacturer);
printf("Cost each:         $%f\n",resistors.price_each);
printf("Quantity:          %d\n",resistors.quantity);
printf("Total value:       $%f\n",total_value);
```

Conclusion

This section introduced you to the concept of a C structure. In the next section you will see other more powerful ways of applying a C structure. For now, test your understanding of this section by trying the following section review.

8.3 Section Review

1. State how the information on a check from a checking account could be treated as a structure.
2. What is a C structure?
3. Describe the structure presented in this section.
4. State how a structure member variable is programmed for getting data and displaying data.
5. Give an example of problem 4 above.

8.4 More Data Structure Details

Discussion

In the last section you were introduced to the concept and form of a C structure. In this section, you will see different ways of letting your C program know that you are constructing a structure. You will also discover how to name function types, declare functions as structures, and pass structures between functions.

The Structure Tag

In the last section, the structure for an inventory of resistors of the same type contained in a storage box was demonstrated. Take a closer look at where the structure was defined:

```
main()
{

    struct
    {
        char   manufacturer[20];   /* Resistor manufacturer.     */
        int    quantity;           /* Number of resistors left. */
        float  price_each;         /* Cost of each resistor.     */
    } resistors;
```

It was defined inside `main()` and as a consequence will only be known to `main()`. Another method of developing a structure is to define the structure before `main()`. This method creates a global variable.

C has a method of announcing the construction of a structure that will serve as a **template** that can then be used by any function within the program. This is illustrated in Program 8.8. The output of this program is exactly the same as that of Program 8.7. The difference is the way the C structure is placed in the program.

Program 8.8

```
#include <stdio.h>

struct parts_record
{
        char  manufacturer[20];          /* Resistor manufacturer.      */
        int   quantity;                  /* Number of resistors left.   */
        float price_each;                /* Cost of each resistor.      */
};

main()
{
        struct parts_record resistors;  /* Structure variable.         */
        float total_value;              /* Total value of parts.       */

        /* Get name of manufacturer:    */

        /* Get number of parts left:    */

        /* Get cost of each part:       */

        /* Calculate total value:       */

        /* Display variables:           */
}
```

Program Analysis

Program 8.8 declares a structure type called `parts_record`. There is no variable identifier; `parts_record` is called a **structure tag**. The structure tag is an identifier that names the structure type defined in the **member declaration list**. Note that this was done after the keyword `struct`.

```
struct parts_record
{
        char    manufacturer[20];    /* Resistor manufacturer.    */
        int     quantity;            /* Number of resistors left. */
        float   price_each;          /* Cost of each resistor.    */
};
```

Now any function within the program can use this structure by making reference to the structure tag. This was done in `main()`.

```
main()
{
        struct parts_record resistors; /* Structure variable. */
```

Note the format of the declaration above. It uses the keyword `struct` and the tag identifier `parts_record`, which defines the variable `resistors` as the structure tagged `parts_record`. The general form of the declaration when a structure tag is used is

```
struct tag variable_identifier;
```

The same method is used to access the individual members of the structure as before, using the member of operator (.). For example, `resistors.quantity` identifies the structure member `quantity`. Using the tag method and declaring the structure outside of `main` sets a template that may now be used by any function within the program.

Naming a Structure

Another method of identifying a structure is to use the C `typedef`. This can be accomplished as shown in Program 8.9.

Program 8.9

```
#include <stdio.h>

typedef struct
{
        char  manufacturer[20];         /* Resistor manufacturer.       */
        int   quantity;                 /* Number of resistors left.    */
        float price_each;               /* Cost of each resistor.       */
} parts_record;                         /* Name for this structure.     */

main()
{
        parts_record resistors;         /* Structure variable.          */
        float total_value;              /* Total value of parts.        */

        /* Get name of manufacturer:   */

        /* Get number of parts left:   */

        /* Get cost of each part:      */

        /* Calculate total value:      */

        /* Display variables:          */
}
```

Program 8.9 does exactly the same as Program 8.8. The difference is in how the C structure is declared. This time, the C `typedef` is used to name the structure type `parts_record`.

```
typedef struct
{
        char   manufacturer[20];    /* Resistor manufacturer.    */
        int    quantity;            /* Number of resistors left. */
        float  price_each;          /* Cost of each resistor.    */
} parts_record;                     /* Name for this structure.  */
```

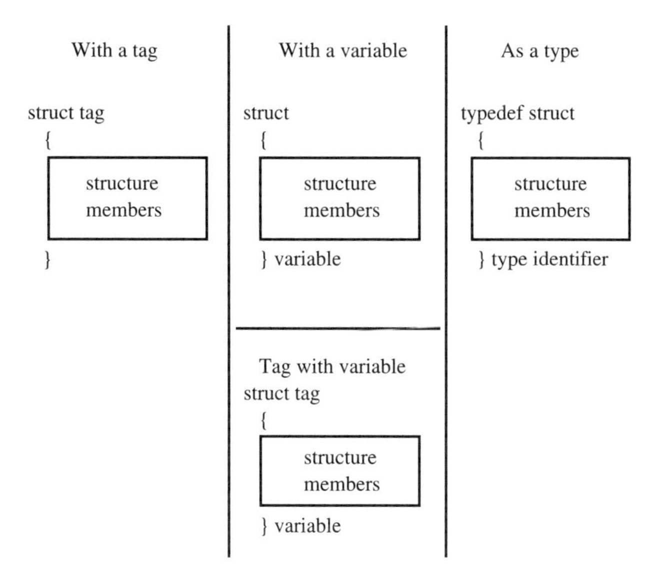

Figure 8.3 Methods of Declaring Structures

Note that a structure tag is not used. Instead, the structure variable `parts_record` is now being named as a data type by the keyword `typedef`. The advantage of doing this is that (as before) you can use the defined structure in any function. However, you only need to declare local structure variables in terms of this new type definition. This is done inside `main()` to keep the record variable `resistors` local.

```
main()
{
        parts_record resistors;     /* Structure variable.        */
```

Note that the structure variable `resistors` is used as its data type `parts_record`. Again, you still access each member of the structure by the member of operator; for example, `resistors.quantity` identifies the structure member `quantity`.

The various ways of declaring structures are illustrated in Figure 8.3.

Structure Pointers

You can declare a structure variable to be of type `pointer`. This is illustrated in Program 8.10. Notice that the C `typedef` is used to define the structure type. The program does exactly the same thing as before; however, it is now doing it by using a structure pointer.

Program 8.10

```
#include <stdio.h>

typedef struct
{
        char  manufacturer[20];         /* Resistor manufacturer.      */
        int   quantity;                 /* Number of resistors left.   */
```

```
        float price_each;               /* Cost of each resistor.       */
} parts_record;

main()
{
        parts_record *rcd_ptr;          /* Structure pointer.           */
        float total_value;              /* Total value of parts.        */

        /* Get name of manufacturer:         */

        printf("Name of manufacturer => ");
        gets(rcd_ptr -> manufacturer);

        /* Get number of parts left:          */

        printf("Number of parts left => ");
        scanf("%d",rcd_ptr -> quantity);

        /* Get cost of each part:             */

        printf("Cost of each part => ");
        scanf("%f",rcd_ptr -> price_each);

        /* Calculate total value:             */

        total_value = rcd_ptr -> quantity * rcd_ptr -> price_each;

        /* Display variables:                 */

        printf("\n\n");
        printf("Item:           Resistors\n\n");
        printf("Manufacturer:   %s\n",rcd_ptr -> manufacturer);
        printf("Cost each:      $%f\n",rcd_ptr -> price_each);
        printf("Quantity:       %d\n",rcd_ptr -> quantity);
        printf("Total value:    $%f\n",total_value);

}
```

Note that the structure pointer is declared the same way a pointer is normally declared:

```
main()
{
        parts_record *rcd_ptr; /* Structure pointer. */
```

However, `rcd_ptr` is now a pointer to a structure and as such, when an individual member of the structure is to be accessed, a special C symbol `->` must be used:

```
/* Get name of manufacturer: */

printf("Name of manufacturer => ");
gets(rcd_ptr -> manufacturer);
```

The symbol -> refers to the member of a structure pointed to by a pointer. Effectively, this replaces the member of operator (.) used in previous examples.

Since a structure may be defined as a pointer, it may also use the address of & operator as with any other variable. This means that the address of a structure may be assigned to a structure pointer.

A Function as a Structure

You may declare a C function to be of a structure type. This is demonstrated in Program 8.11. Again, the program does exactly the same thing as previous ones; the difference is that now `main()` is serving as the function that simply calls other functions. It is the other functions that do all the work. Note in the program that the function `get_input()` is declared as a type defined by `parts_record`. Thus, this function itself now has a record structure! Note also that a structure variable is used as a function argument to pass the structure data to the function `display_output()`.

Program 8.11

```
#include <stdio.h>

typedef struct
{
        char  manufacturer[20];        /* Resistor manufacturer.        */
        int   quantity;                /* Number of resistors left.     */
        float price_each;              /* Cost of each resistor.        */
} parts_record;

        parts_record get_input(void);  /* Get user input.               */
        void display_output(parts_record resistor_record);
                                       /* Display output.               */
main()
{
        parts_record resistor_box;     /* Declares a structure variable. */

        resistor_box = get_input();    /* Get user input. */
        display_output(resistor_box);  /* Display output. */
}

parts_record get_input()               /* Get user input. */
{
        parts_record resistor_information;

        /* Get name of manufacturer:        */

        printf("Name of manufacturer => ");
        gets(resistor_information.manufacturer);

        /* Get number of parts left:        */

        printf("Number of parts left => ");
```

```
        scanf("%d",&resistor_information.quantity);

        /* Get cost of each part:              */

        printf("Cost of each part => ");
        scanf("%f",&resistor_information.price_each);

        return(resistor_information);
}

void display_output(parts_record resistor_information)
{
        /* Display variables:                  */

        printf("\n\n");
        printf("Item:          Resistors\n\n");
        printf("Manufacturer:  %s\n",resistor_information.manufacturer);
        printf("Cost each:     $%f\n",resistor_information.price_each);
        printf("Quantity:      %d\n",resistor_information.quantity);
}
```

Program Analysis

The first new material in Program 8.11 is in the function prototypes:

```
parts_record get_input();        /* Get user input. */
```

The C `typedef` has been used to give the name `parts_record` to the structure. Now, the function `get_input` is being defined as a type `parts_record`.

The next item is the use of a structure type as a formal parameter:

```
void display_output(parts_record resistor_record); /* Display output. */
```

Here, the formal parameter `resistor_record` is of type `parts_record` (which is the name for the previously defined structure type). This means that the argument passed to this function must be of the same type. This is what happens in the body of `main()`.

```
main()
{
        parts_record resistor_box;      /* Declares a structure variable. */

        resistor_box = get_input();     /* Get user input. */
        display_output(resistor_box);   /* Display output. */
}
```

The variable `resistor_box` is defined as the type `parts_record` (as were the other structure variables). Now, since `resistor_box` and `get_input()` are both of the same type of structure, one may be assigned to the other.

Next, in the definition of the `get_input` function, the whole structure is returned back using the defined variable `resistor_information`, which is again of type `parts_record`.

The last function, `display_output`, simply takes the actual parameter from `main()` (which now contains the structure data) and uses it to display the output data.

Conclusion

This section demonstrated various forms of the C structure. Here you saw different ways of letting your C program know that you are constructing a structure. You also saw how to name function types, declare functions as structures, and pass structures between functions.

8.4 Section Review

1. Describe what is meant by a structure tag.
2. Give an example of a structure tag.
3. Can a structure be a variable type? Explain.
4. How is a structure member pointed to when a structure pointer is used?
5. What are the three operations that are allowed with structures?

8.5 The union and Structure Arrays

Discussion

This section presents the C **`union`**, which you will see is similar to a C structure in many ways but with an important difference. You will also be introduced to a method of bringing together the power of a C structure with your old friend, the C array. This combination will give you tremendous programming potential for handling many different data types. This section brings together much of the material that has been presented up to this point.

The C union

The C `union` is similar to the C structure. The difference is that a C `union` is used to store different data types in the same memory location.

The C `union` is

```
union tag
{

        type   member_identifier1
        type   member_identifier2
                 .
                 .
                 .
        type   member_identifierN
}
```

As you can see, a union declaration has the same form as the structure declaration. The difference is the keyword `union`. A union declaration names a union variable and states the set of variable values (members) of the union. These members can have different types. What happens in a `union` is that a union type variable will store one of the values defined by that type. Program 8.12 demonstrates the action of a C union.

Program 8.12

```
#include <stdio.h>

main()
{
        union    /* Define union. */
        {
                int   integer_value;
                float value;
        } integer_or_float;

        printf("Size of the union => %d bytes.\n", sizeof(integer_or_float));

        /* Enter an integer and display it: */

        integer_or_float.integer_value = 123;
        printf("The integer value is %d\n", integer_or_float.integer_value);
        printf("Starting address is => %d\n", &integer_or_float.integer_value);

        /* Enter a float and display it: */

        integer_or_float.value = 123.45;
        printf("The float value is %f\n", integer_or_float.value);
        printf("Starting address is => %d\n", &integer_or_float.value);
}
```

Execution of Program 8.12 produces

```
Size of the union => 4 bytes.
The integer value is 123
Starting address is => 7042
The float value is 123.45
Starting address is => 7042
```

What you see from the output above is that the size of the union is 4 bytes. This is because it takes 4 bytes to store a type `float`. First, an integer value is stored and then retrieved from this memory location (shown with a starting address of 7042). Then a `float` value is stored and retrieved from the same memory space.

As shown in Program 8.12, the union declarations have the same form as structure declarations. The difference is that the keyword `union` is used in place of the keyword `struct`. The rules covered up to this point for structures also apply to unions. The amount of storage required for a union variable is the amount of storage required for the largest member of the union. All members are stored in the same memory space (but not at the same time) with the same starting address.

Initializing the C Structure

A C structure may be initialized if it is a global or static variable. This is illustrated in Program 8.13.

Program 8.13

```
#include <stdio.h>

typedef struct
{
        char part[20];          /* Type of part.          */
        int quantity;           /* Number of parts left. */
        float price;            /* Cost of each part.    */
} parts_record;

main()
{
        static parts_record bin_1_contents =
        {
                "Resistors",
                25,
                0.05
        };

        static parts_record bin_2_contents =
        {
                "Capacitors",
                37,
                0.16
        };

        printf("Contents of bin #1:\n");
        printf("Item => %s\n",bin_1_contents.part);
        printf("Quantity => %d\n",bin_1_contents.quantity);
        printf("Cost each => $%f\n",bin_1_contents.price);

        printf("\nContents of bin #2:\n");
        printf("Item => %s\n",bin_2_contents.part);
        printf("Quantity => %d\n",bin_2_contents.quantity);
        printf("Cost each => $%f\n",bin_2_contents.price);
}
```

Execution of Program 8.13 produces

```
Contents of bin #1:
Item => Resistors
Quantity => 25
Cost each => 0.05

Contents of bin #2
Item => Capacitors
Quantity => 37
Cost each => 0.16
```

Program Analysis

Program 8.13 starts by naming a structure type called parts_record.

```
typedef struct
{
        char part[20];          /* Type of part.          */
        int quantity;           /* Number of parts left. */
        float price;            /* Cost of each part.     */

} parts_record;
```

Now parts_record means a structure that contains three members. Each member will contain information about parts in a storage bin. In the function main(), the data for each of the three members is assigned. Note that the new variable bin_1_contents is defined as a static of type parts_record:

```
main()
{
        static parts_record  bin_1_contents =
        {
                "Resistors",
                25,
                0.05
        };
```

Another variable, bin_2_contents, is also defined as a static of type parts_record. Again each member is assigned data. Note that each assignment corresponds to the member type.

```
static parts_record  bin_2_contents =
{
        "Capacitors",
        37,
        0.16
};
```

There are now two structures of the same type, bin_1_contents and bin_2_contents. Their structure members are the same, but the data assigned to each member is different.

The remainder of the program uses the member of operator to display the data contained in each structure:

```
printf("Contents of bin #1:\n");
printf("Item => %s\n",bin_1_contents.part);
printf("Quantity => %d\n",bin_1_contents.quantity);
printf("Cost each => $%f\n",bin_1_contents.price);

printf("\nContents of bin #2:\n");
printf("Item => %s\n",bin_2_contents.part);
printf("Quantity => %d\n",bin_2_contents.quantity);
printf("Cost each => $%f\n",bin_2_contents.price);
```

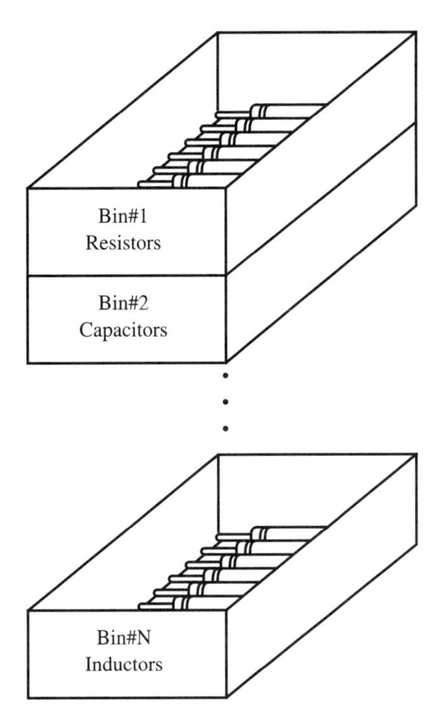

Figure 8.4 Storage Bins with Different Parts

Structure Arrays

The real power of using C structures comes when there are **structure arrays**. Consider an inventory program where there are many different storage bins for parts. For the contents of each storage bin you may want to know the name of the part, the quantity, and the price of each part. Thus you would want to know the same structure of information about many storage bins. This concept is illustrated in Figure 8.4.

Since the structure of information for each of these is the same, you can create a C program that uses an array of the same kind of structure. This concept is illustrated in Figure 8.5.

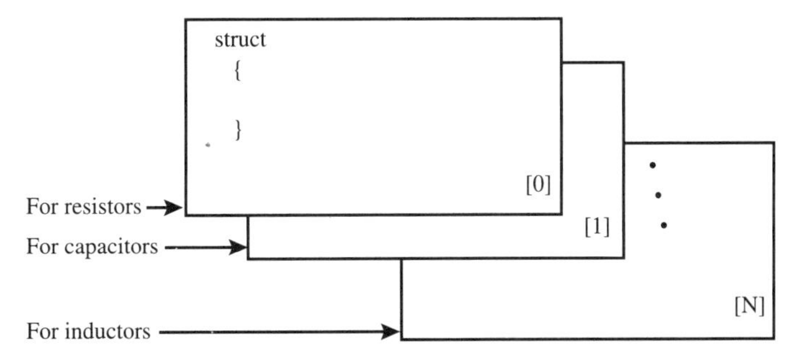

Figure 8.5 Creating an Array of Structures

Program 8.14 is a C program that contains an array of three structures. This program will allow the program user to input—for each of three part bins—the name of the part, the number of parts in the bin, and the cost of each part. The program will then display what the user has entered. This is achieved by defining one structure type and then making an array of this type.

Program 8.14

```
#include <stdio.h>
#include <string.h>

typedef struct
{
        char part[20];              /* Type of part.           */
        int quantity;               /* Number of parts left.   */
        float price;                /* Cost of each part.      */
        char initialized;           /* Test if record exists.  */
} parts_record;

main()
{
        static parts_record bin_contents[3];
        int record;

        do
        {
                /* Get bin number. */

                printf("Enter bin number from 1 to 3 (0 to quit) => ");
                scanf("%d",&record);

                --record;        /* Start at bin 0. */
                if(record < 0) continue;

                /* Get name of part. */

                printf("Part name => ");
                scanf("%s",bin_contents[record].part);

                /* Get number of parts left. */

                printf("Number of parts left => ");
                scanf("%d",&bin_contents[record].quantity);

                /* Get cost of each part. */

                printf("Cost of each part => ");
                scanf("%f",&bin_contents[record].price);

                /* Set record initialization TRUE. */
                bin_contents[record].initialized = 'T';
```

```
        } while (record >= 0);

        /* Output information. */

        for(record = 0; record <= 2; record++)
        {
                if(bin_contents[record].initialized == 'T')
                {
                        printf("Bin number %d contains:\n",record + 1);
                        printf("Part => %s\n",bin_contents[record].part);
                        printf("Quantity => %d\n",bin_contents[record].quantity);
                        printf("Cost each => $%f\n",bin_contents[record].price);
                }
        } /* End for. */
}
```

Execution of Program 8.14 produces

```
Enter bin number from 1 to 3 (0 to quit) => 1
Part name => Inductor
Number of parts left => 12
Cost of each part => 0.18

Enter bin number from 1 to 3 (0 to quit) => 3
Part name => Resistor
Number of parts left => 25
Cost of each part => 0.14

Enter bin number from 1 to 3 (0 to quit) => 0

Bin number 1 contains:
Part => Inductor
Quantity => 12
Cost each => $0.18

Bin number 3 contains:
Part => Resistor
Quantity => 25
Cost each => $0.14
```

Note that no data was entered for bin #2, and no data was displayed for it.

Program Analysis

Program 8.14 defines a type called `parts_record` that is a structure of four members.

```
typedef struct
{
        char part[20];          /* Type of part.          */
        int quantity;           /* Number of parts left.  */
```

```
        float price;            /* Cost of each part.     */
        char initialized;       /* Test if record exists. */
} parts_record;
```

The important point is what follows next. A new variable is defined as an array (consisting of three elements). The new variable is called `bin_contents[3]` and is defined as type `parts_record`. This means that you now have a three-element array that is a structure consisting of the four members defined in the `typedef`.

```
main()
{
        static parts_record  bin_contents[3];
```

This sets aside an array of three structures, all of which have the same number and type of members. The next part of the program asks the program user to input the bin number. This value will set the array number. However, since the C language starts its arrays at 0, the actual array number entered by the user has 1 subtracted from it. This is done using the C `--record` operation. User input is part of a C `do` loop that continues while the user input is not less than 0. (If the user inputs a 0 to stop input, 1 is subtracted from it, making the final input less than zero.) The `continue` statement acts like a jump to the end of the loop when activated.

```
do
{
        /* Get bin number. */

        printf("Enter bin number from 1 to 3 (0 to quit) => ");
        scanf("%d",&record);

        --record;        /* Start at bin 0. */
        if(record < 0) continue;
```

When input is taken, the format for selecting a specific structure from the array and a specific member of that structure is `structure_variable[N].structure_member`, where *N* is the index of the array.

```
/* Get name of part.  */

printf("Part name => ");
scanf("%s",bin_contents[record].part);

/* Get number of parts left. */

printf("Number of parts left => ");
scanf("%d",&bin_contents[record].quantity);

/* Get cost of each part. */

printf("Cost of each part => ");
scanf("%f",&bin_contents[record].price);
```

In C, an initialized record will contain some random data. (For example, in this program, if the random contents are *T* for the structure member `initialized`, then the program will fail. `static` prevents this from happening.) In order to avoid displaying this information when an initialized variable record is accessed, the last structure member (`char initialized`) is used to indicate if data has been placed in the array by the user. This is done by setting this member to the letter *T*:

```
/*  Set record initialization TRUE.  */
bin_contents[record].initialized = 'T';
```

The program uses a C `for` loop to output the data. It's important to note that no data will be displayed for a structure that has not been initialized (when `bin_contents[record].initialized` does not equal *T*):

```
for(record = 0; record <= 2; record++)
{
       if(bin_contents[record].initialized == 'T')
```

The output now simply uses C `printf` functions and identifies each member of a specific structure from the structure array:

```
{
       printf("Bin number %d contains:\n",record + 1);
       printf("Part => %s\n",bin_contents[record].part);
       printf("Quantity => %d\n",bin_contents[record].quantity);
       printf("Cost each => $%f\n",bin_contents[record].price);
}
```

Conclusion

This section presented the C `union`. You also saw the C structure and the C array brought together. This combination is a powerful new tool for handling complex data structures. In the next section, you will see how structures and arrays may be mixed to handle even more complex data structures. Check your understanding of this section by trying the following section review.

8.5 Section Review

1. Describe the purpose of a C `union`.
2. How is a C `union` declared?
3. Can a C structure be initialized by the program? Explain.
4. Explain how an array of structures may be declared.
5. Show how an individual member of an array of structures is accessed.

8.6 Ways of Representing Structures

Discussion

C structures are rich in the variety of ways they may be used. This section will demonstrate some of them. You will find many potential technical applications for this powerful feature of C.

Structures within Structures

You can have a C structure within another structure. Consider Program 8.15.

Program 8.15

```
#include <stdio.h>

typedef struct
{
        int member_1;
        char member_2;
        float member_3;
} first_structure;

struct second_structure
{
        first_structure second_member_1;
        int             second_member_2;
        char            second_member_3;
}

main()
{
}
```

As you can see from Program 8.15, the C structure `second_structure` contains the structure `first_structure` in the member called `second_member_1`. This is a case of a C structure containing another structure. The parts inventory program could use a structure within a structure such as this. For example, the manufacturer of the part could in itself be a structure that would contain the address and telephone number as members.

Arrays within Structures

You can also have C arrays within C structures. For example, consider Program 8.16. Here, one member of the structure is a two-dimensional array. Recall that you used character arrays to represent strings in previous C structures. This is no different.

Program 8.16

```
#include <stdio.h>

struct structure_tag
{
        float array_variable[5][6];
        int   second_member_2;
        char  second_member_3;
}

main()
{
}
```

As you can see from Program 8.16, the member `array_variable[5][6]` is a two-dimensional array contained within the C structure.

Multidimensional Structure Arrays

You may also have multidimensional arrays that are C structures. Consider Program 8.17. Here you have a two-dimensional array that is a structure that contains three members, one of which is itself a two-dimensional array.

Program 8.17

```
#include <stdio.h>

typedef struct
{
        float array_variable[5][6];
        int   second_member_2;
        char  second_member_3;
} record_1;

typedef record_1 complex_array[3][2];

main()
{
}
```

As shown in Program 8.17, `complex_array[3][2]` is a two-dimensional array of type `record_1`. Thus you have a data structure where there are actually $3 \times 2 = 6$ structures, each of which contains three members, one of which is an array itself.

Conclusion

A rich variety of structures and arrays are available with C. In this section you saw a few examples of the possibilities. Check your knowledge of this section by trying the following section review.

8.6 Section Review

1. Can a structure be an array? Explain.
2. Can a structure contain another structure? Explain.
3. Is it possible for an array to be a structure type that contains an array as one of its members? Explain.

8.7 Advanced Data Structures

Discussion

In this section we will examine a number of advanced data structures. These structures are all based on the utilization of *nodes* of data. A **node**, in its most basic form, consists of two parts: a data area and a pointer. Anything may be stored in the data area (integers, floats,

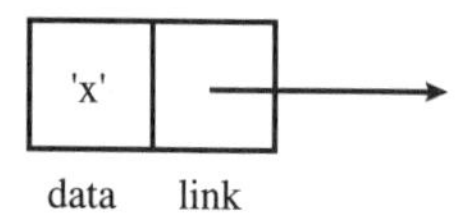

Figure 8.6 A Node Containing the Letter `'x'`

characters, strings, arrays, etc.). The pointer area contains at least one pointer (address) and *points* to another node of information. Figure 8.6 gives a structural representation of a node. The two portions of the node have been given the names *data* and *link* and are commonly referred to as the *data field* and the *link field*. The data field of the node contains the code for the letter `'x'`. Usually, a group of nodes is associated with a data structure. The data structures we will examine here are *linked-lists, stacks, queues,* and *binary trees*.

Linked-Lists

As previously mentioned, a data structure usually consists of a group of nodes. In an array, each data element occupies a consecutive memory location. Nodes, on the other hand, may be spread out over many different locations within memory. This is where the link field of the node becomes necessary. The link field contains a pointer (indicated by the arrow coming out of the link field) that shows to which other node (if any) the current node is connected. Figure 8.7 shows how three nodes are connected through their various link fields.

This collection of nodes is called a **linked-list**. Notice that the link field of the first node points to the second node, and that the link field of the second node points to the third node. The link field of the third node contains `NULL`. This indicates that the third node is the *end* of the linked-list. The entire linked-list is pointed to by the variable `list`.

To define a node for the linked-list, we use the following C structure:

```
typedef struct node
{
        char data;            /* data field */
        struct node *link;    /* link field */
} LNODE;
```

The data field in the `LNODE` structure reserves room for a single character. This part of the structure must be changed if the linked-list will be used to store integers, floats, arrays, or some other kind of data. The link field is written in such a way that the structure contains a pointer to a similar `LNODE` structure.

How are nodes created? To answer this question, let us review how a string of characters is created. The C statement

```
char string[20] = "Press Return!";
```

reserves 20 consecutive memory locations (plus a 21st for the `'\0'` terminator) for the string characters. This type of memory allocation is performed when the program is compiled.

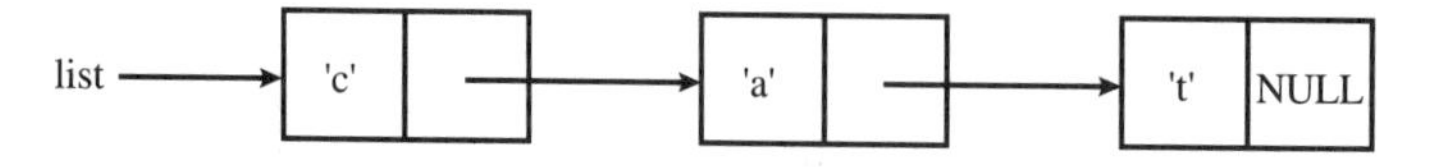

Figure 8.7 A Three-Node Linked-List

During execution, only the 20 reserved memory locations can be used for the string. If the program attempts to use more than 20 locations (during a `strcat()` operation) it is very possible that other reserved memory locations (or even the actual code of the program) will be overwritten. For this reason, nodes are created *dynamically* (whenever they are needed) and assigned unused memory locations during run time. This way, new nodes may be created and utilized as long as there is available free memory. Do you see the difference between compile time memory allocation and run time memory allocation?

The `<stdlib.h>` library contains a function called **`malloc`** which is used to perform dynamic storage allocation during run time. The number of new bytes required is passed to `malloc`, which returns a pointer (an address) to the block of new bytes if they are available. If the requested memory is not available, `malloc` returns a `NULL` pointer. So, to request memory space for a new node, we use the following statement:

```
ptr = malloc(sizeof(LNODE));
```

The `sizeof` function determines the size, in bytes, of the `LNODE` structure. The address of the memory allocated by `malloc` is saved in the pointer variable `ptr`, which must be defined as follows:

```
LNODE *ptr;
```

Allocating memory in this fashion allows the size of the data structure to be determined while the program is *executing*, instead of when it is compiled. The advantage is clear: we do not have to reserve memory in advance for our data structure. For example, a sorting program may not know in advance how many data items there will be. So, how many should be reserved? Ten? Twenty? One thousand? Even so, what if there is one more number inputted to the program than the programmer has reserved space for? Who knows what will happen then?

Program 8.18 shows how a three-element linked-list is created. Note how the nodes are allocated with `malloc`, and then assigned their respective data field and link field values.

Program 8.18

```
#include <stdio.h>
#include <stdlib.h>

        typedef struct node
        {
                char data;
                struct node *link;
        } LNODE;

void show_list(LNODE *ptr);

main()
{
        LNODE *n1, *n2, *n3;

        n1 = malloc(sizeof(LNODE));
        n2 = malloc(sizeof(LNODE));
        n3 = malloc(sizeof(LNODE));
```

```
        n1->data = 'c';
        n1->link = n2;
        n2->data = 'a';
        n2->link = n3;
        n3->data = 't';
        n3->link = NULL;

        printf("The linked-list is as follows: ");
        show_list(n1);
}

void show_list(LNODE *ptr)
{
        while(ptr != NULL)
        {
                printf("%c",ptr->data);
                ptr = ptr->link;
        }
        printf("\n");
}
```

For example, the first node is loaded with information as follows:

```
n1->data = 'c';
n1->link = n2;
```

Thus, the link field of the first node points to the second node in the list.

The function `show_list()` traverses the linked-list, beginning with the first node, printing the character stored in each data field, until the `NULL` pointer is encountered. Execution of Program 8.18 is as follows:

```
The linked-list is as follows: cat
```

Remember that *any* type of data can be stored in a node. The programs that follow will all use single-character data fields to demonstrate various operations on linked-lists.

Program 8.19 shows how a linked-list can be created by the user. The user simply enters characters one by one, which are placed into newly allocated nodes. The list is complete when the user enters return.

Program 8.19

```
#include <stdio.h>
#include <stdlib.h>

        typedef struct node
        {
                char data;
                struct node *link;
        } LNODE;

void show_list(LNODE *ptr);
void add_node(LNODE **ptr, char item);

main()
```

```
{
        LNODE *n1 = NULL;
        char item;

        do
        {
                printf("\nEnter item: ");
                item = getche();
                if(item != '\r')
                        add_node(&n1,item);
        } while(item != '\r');
        printf("\nThe new linked-list is: ");
        show_list(n1);
}

void show_list(LNODE *ptr)
{
        while(ptr != NULL)
        {
                printf("%c",ptr->data);
                ptr = ptr->link;
        }
        printf("\n");
}

void add_node(LNODE **ptr, char item)
{
        LNODE *p1, *p2;

        p1 = *ptr;
        if(p1 == NULL)
        {
                p1 = malloc(sizeof(LNODE));
                if(p1 != NULL)
                {
                        p1->data = item;
                        p1->link = NULL;
                        *ptr = p1;
                }
        }
        else
        {
                while(p1->link != NULL)
                        p1 = p1->link;
                p2 = malloc(sizeof(LNODE));
                if(p2 != NULL)
                {
                        p2->data = item;
                        p2->link = NULL;
                        p1->link = p2;
                }
        }
}
```

1. Get new node:

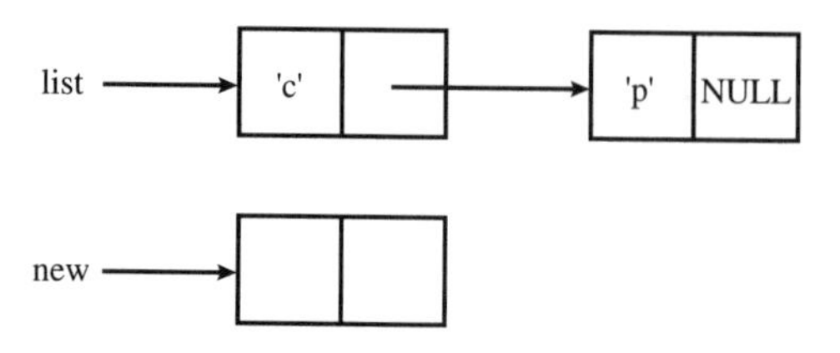

2. Assign data and link fields:

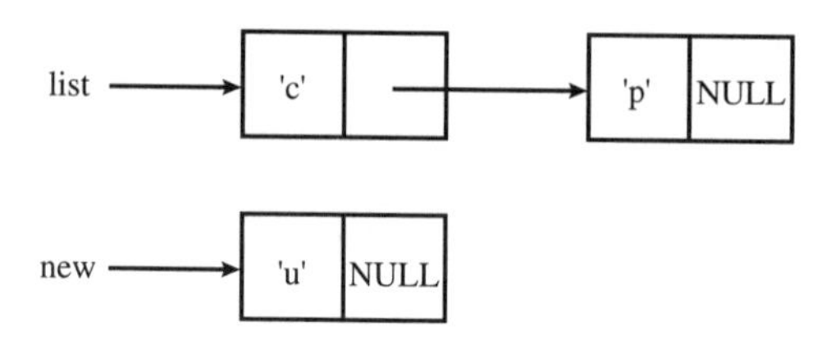

3. Establish link to existing list:

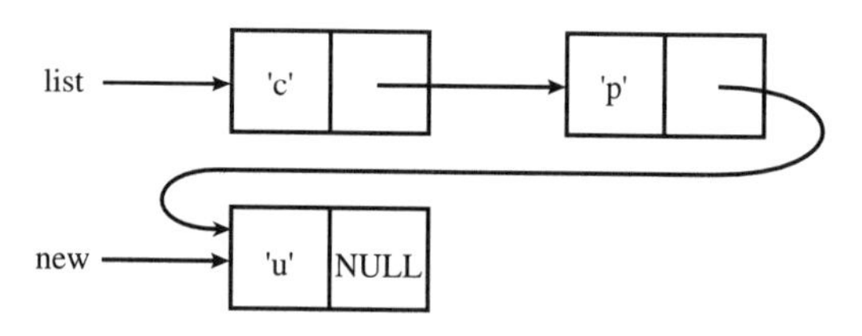

Figure 8.8 Adding a Node to a Linked-List

What is involved in adding a new node to a linked-list? Figure 8.8 shows the three-step process required to add a new node to an existing linked-list. Notice that the new node is added to the end of the list. If the list was originally empty, the process requires a different step 3, in which the `list` variable is assigned the address of the new node.

A sample execution of Program 8.19 is as follows:

```
Enter item: c
Enter item: p
Enter item: u
Enter item:
The new linked-list is: cpu
```

The next five programs contain predefined linked-lists for the purpose of illustrating the basic operations performed on linked-lists, which are `insert`, `delete`, and `search`.

Program 8.20 shows how the letter `'s'` is inserted into the beginning of a three-element linked-list `'cat'` to make the new linked-list `'scat'`. Care must be taken to correctly update the pointer to the list.

Program 8.20

```
#include <stdio.h>
#include <stdlib.h>

        typedef struct node
        {
                char data;
                struct node *link;
        } LNODE;

void show_list(LNODE *ptr);
void insert_at_head(LNODE **ptr, char item);

main()
{
        LNODE *n1, *n2, *n3;

        n1 = malloc(sizeof(LNODE));
        n2 = malloc(sizeof(LNODE));
        n3 = malloc(sizeof(LNODE));

        n1->data = 'c';
        n1->link = n2;
        n2->data = 'a';
        n2->link = n3;
        n3->data = 't';
        n3->link = NULL;

        printf("The linked-list is as follows: ");
        show_list(n1);
        insert_at_head(&n1,'s');
        printf("The new linked-list is: ");
        show_list(n1);
}

void show_list(LNODE *ptr)
{
        while(ptr != NULL)
        {
                printf("%c",ptr->data);
                ptr = ptr->link;
        }
        printf("\n");
}

void insert_at_head(LNODE **ptr, char item)
{
        LNODE *new;

        new = malloc(sizeof(LNODE));
```

```
        if(new != NULL)
        {
                new->data = item;
                new->link = *ptr;
                *ptr = new;
        }
}
```

Program 8.20 produces the following result during execution:

```
The linked-list is as follows: cat
The new linked-list is: scat
```

Program 8.21 also inserts a new node containing `'s'` into an existing list, but inserts it at the end of the list. This requires a search through the list to find the last node.

Program 8.21

```
#include <stdio.h>
#include <stdlib.h>

        typedef struct node
        {
                char data;
                struct node *link;
        } LNODE;

void show_list(LNODE *ptr);
void insert_at_tail(LNODE *ptr, char item);

main()
{
        LNODE *n1, *n2, *n3;

        n1 = malloc(sizeof(LNODE));
        n2 = malloc(sizeof(LNODE));
        n3 = malloc(sizeof(LNODE));

        n1->data = 'c';
        n1->link = n2;
        n2->data = 'a';
        n2->link = n3;
        n3->data = 't';
        n3->link = NULL;

        printf("The linked-list is as follows: ");
        show_list(n1);
        insert_at_tail(n1,'s');
        printf("The new linked-list is: ");
        show_list(n1);
}
```

```
void show_list(LNODE *ptr)
{
        while(ptr != NULL)
        {
                printf("%c",ptr->data);
                ptr = ptr->link;
        }
        printf("\n");
}

void insert_at_tail(LNODE *ptr, char item)
{
        LNODE *new;

        while(ptr->link != NULL)
                ptr = ptr->link;
        new = malloc(sizeof(LNODE));
        if(new != NULL)
        {
                ptr->link = new;
                new->data = item;
                new->link = NULL;
        }
}
```

This causes the original list `'cat'` to become `'cats'`, as shown in the following execution:

```
The linked-list is as follows: cat
The new linked-list is: cats
```

You may have noticed that the first node in the linked-list is referred to as the *head* and the last node as the *tail*. This is a common practice.

Figure 8.9 shows how a node is removed from the head of a linked-list. The address in the link field of the first node replaces the address in the `list` variable (which causes

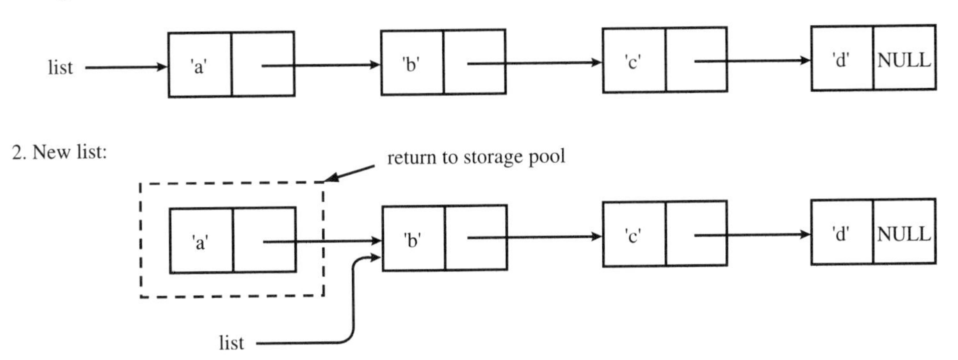

Figure 8.9 Removing a Node from the Head of a Linked-List

the list variable to now point to the second node). The first node is then returned to the storage pool. This is accomplished through the use of the `<stdlib.h>` function `free()`. If the first node is not returned, it stays allocated. But since the list variable has been changed, there is no way to access the first node anymore (without guessing its address). In this case the first node has become *garbage*. When many nodes are discarded in this fashion it is possible to run out of memory for new nodes. A cleverly written *garbage collection* routine must then be used to retrieve all lost nodes.

Program 8.22 shows how a node is deleted from the beginning of a linked-list.

Program 8.22

```
#include <stdio.h>
#include <stdlib.h>

        typedef struct node
        {
                char data;
                struct node *link;
        } LNODE;

void show_list(LNODE *ptr);
void delete_head(LNODE **ptr);

main()
{
        LNODE *n1, *n2, *n3, *n4;

        n1 = malloc(sizeof(LNODE));
        n2 = malloc(sizeof(LNODE));
        n3 = malloc(sizeof(LNODE));
        n4 = malloc(sizeof(LNODE));

        n1->data = 'a';
        n1->link = n2;
        n2->data = 'b';
        n2->link = n3;
        n3->data = 'c';
        n3->link = n4;
        n4->data = 'd';
        n4->link = NULL;

        printf("The linked-list is as follows: ");
        show_list(n1);
        delete_head(&n1);
        printf("The new linked-list is: ");
        show_list(n1);
}

void show_list(LNODE *ptr)
{
        while(ptr != NULL)
        {
                printf("%c",ptr->data);
```

```
                ptr = ptr->link;
        }
        printf("\n");
}

void delete_head(LNODE **ptr)
{
        LNODE *p;

        p = *ptr;
        if(p != NULL)
        {
                p = p->link;
                free(*ptr);
        }
        *ptr = p;
}
```

The original linked-list is `'abcd'`. Deleting the first node produces the new list, `'bcd'`. This is supported by the result of executing Program 8.22:

```
The linked-list is as follows: abcd
The new linked-list is: bcd
```

Deleting a node from the end of a linked-list requires a search to find the last node in the list. Figure 8.10 shows the necessary steps. It is important to maintain two pointers dur-

1. Original list:

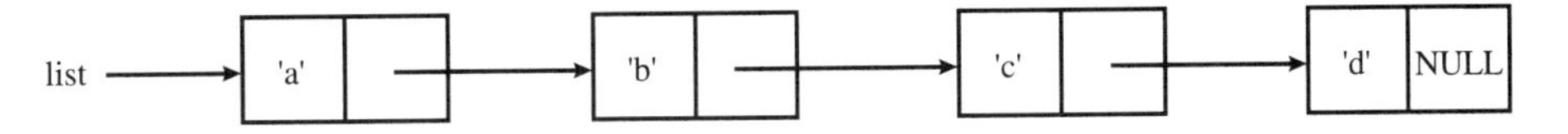

2. Find last node:

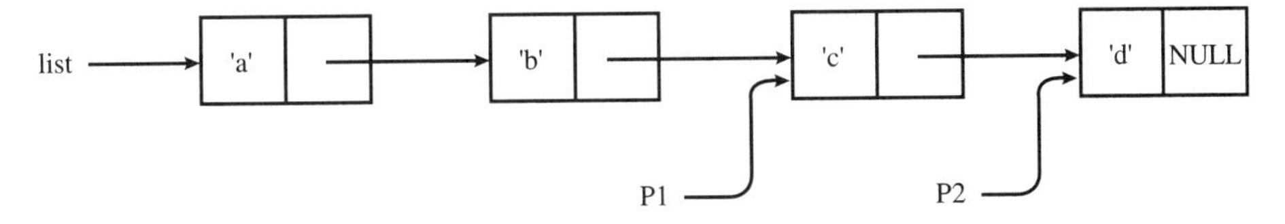

3. Reassign link field:

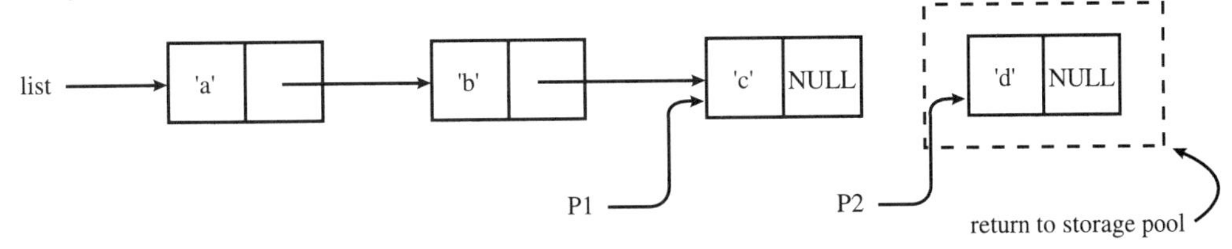

Figure 8.10 Removing a Node from the End of a Linked-List

ing the deletion process. One pointer to the last node is insufficient, since the link field of the second-to-last node must be changed to NULL. Program 8.23 shows how this process is accomplished.

Program 8.23

```
#include <stdio.h>
#include <stdlib.h>

        typedef struct node
        {
                char data;
                struct node *link;
        } LNODE;

void show_list(LNODE *ptr);
void delete_tail(LNODE **ptr);

main()
{
        LNODE *n1, *n2, *n3, *n4;

        n1 = malloc(sizeof(LNODE));
        n2 = malloc(sizeof(LNODE));
        n3 = malloc(sizeof(LNODE));
        n4 = malloc(sizeof(LNODE));

        n1->data = 'a';
        n1->link = n2;
        n2->data = 'b';
        n2->link = n3;
        n3->data = 'c';
        n3->link = n4;
        n4->data = 'd';
        n4->link = NULL;

        printf("The linked-list is as follows: ");
        show_list(n1);
        delete_tail(&n1);
        printf("The new linked-list is: ");
        show_list(n1);
}

void show_list(LNODE *ptr)
{
        while(ptr != NULL)
        {
                printf("%c",ptr->data);
                ptr = ptr->link;
        }
        printf("\n");
}
```

```
void delete_tail(LNODE **ptr)
{
        LNODE *p1, *p2;

        p1 = *ptr;
        if(p1 != NULL)
        {
                if(p1->link == NULL)
                {
                        free(*ptr);
                        *ptr = NULL;
                }
                else
                {
                        while(p1->link != NULL)
                        {
                                p2 = p1;
                                p1 = p1->link;
                        }
                        p2->link = NULL;
                        free(p1);
                }
        }
}
```

A sample execution of Program 8.23 is as follows:

```
The linked-list is as follows: abcd
The new linked-list is: abc
```

Note that in each of the previous four programs the complexity of the insert and delete functions was a result of the fact that the linked-list may contain 0, 1, or more elements. Different steps are required in each case.

The last operation on linked-lists we will examine is that of searching a list for a particular item. Program 8.24 allows the user to enter the character to search for in a given linked-list.

Program 8.24

```
#include <stdio.h>
#include <stdlib.h>

        typedef struct node
        {
                char data;
                struct node *link;
        } LNODE;

void show_list(LNODE *ptr);
int search_list(LNODE *ptr, char item);

main()
```

```
{
        LNODE *n1, *n2, *n3, *n4;
        char item;
        int found;

        n1 = malloc(sizeof(LNODE));
        n2 = malloc(sizeof(LNODE));
        n3 = malloc(sizeof(LNODE));
        n4 = malloc(sizeof(LNODE));

        n1->data = 'a';
        n1->link = n2;
        n2->data = 'b';
        n2->link = n3;
        n3->data = 'c';
        n3->link = n4;
        n4->data = 'd';
        n4->link = NULL;

        printf("The linked-list is as follows: ");
        show_list(n1);
        printf("Enter character to search for: ");
        item = getche();
        found = search_list(n1,item);
        printf("\nThe character %c was",item);
        found ? printf(" ") : printf(" not ");
        printf("found in the list: ");
        show_list(n1);
}

void show_list(LNODE *ptr)
{
        while(ptr != NULL)
        {
                printf("%c",ptr->data);
                ptr = ptr->link;
        }
        printf("\n");
}

int search_list(LNODE *ptr, char item)
{
        if(ptr == NULL)
                return(0);
        else
        {
                do
                {
                        if(ptr->data == item)
                                return(1);
```

```
                    ptr = ptr->link;
                } while(ptr != NULL);
                return(0);
        }
}
```

Two executions of Program 8.24 follow:

```
The linked-list is as follows: abcd
Enter character to search for: e
The character e was not found in the list: abcd

The linked-list is as follows: abcd
Enter character to search for: c
The character c was found in the list: abcd
```

Note that the search time for a character that is not in the list increases when the number of nodes in the list increases. This may be a consideration in time-critical applications. For this reason, efficient searches are accomplished through the use of a hash table (covered in Chapter 7). You may recall that the use of a hash table can result in a *collision*, where two or more variables hash to the same address. This problem can be solved through the use of linked-lists. If two or more variables hash to the same location, they can all be stored in a linked-list at that location.

Stacks and Queues

Many computations are greatly simplified by the use of a software-controlled stack or queue. Expression evaluation and round-robin algorithms are just two examples of applications of stacks and queues. The method of implementation is not critical, and in this section both are implemented as linked-lists.

One characteristic of a **stack** is that the last item pushed is always the first item popped. For this reason, stacks are commonly referred to as **LIFO** (Last In First Out) data structures. Programs 8.20 and 8.22 each performed operations on the first node in a linked-list. These operations are similar to the push and pop operations required by a stack. Figure 8.11 shows the contents of a stack during a number of push and pop operations. The `top` variable refers to the top of the stack, and changes with each push or pop. Notice that the pop operation

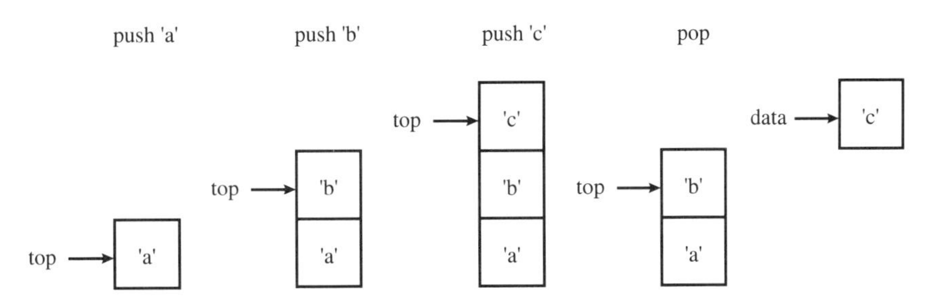

Figure 8.11 Stack Operation During Push and Pop

retrieves the last item pushed onto the stack (the 'c'). Program 8.25 shows how a character stack can be implemented. Since the stack is implemented as a linked-list, the push() and pop() routines manipulate the first node in the stack's linked-list.

Program 8.25

```
#include <stdio.h>
#include <stdlib.h>

        typedef struct node
        {
                char data;
                struct node *link;
        } LNODE;

void show_stack(LNODE *ptr);
int empty(LNODE *ptr);
void push(LNODE **ptr, char item);
void pop(LNODE **ptr, char *item);

main()
{
        LNODE *stacka = NULL;
        char item;

        push(&stacka,'a');
        push(&stacka,'b');
        push(&stacka,'c');
        printf("The stack is as follows: ");
        show_stack(stacka);
        pop(&stacka,&item);
        printf("\nThe first item popped is %c\n",item);
        pop(&stacka,&item);
        printf("The second item popped is %c\n",item);
        pop(&stacka,&item);
        printf("The third item popped is %c\n",item);
        pop(&stacka,&item);
}

void show_stack(LNODE *ptr)
{
        while(ptr != NULL)
        {
                printf("%c",ptr->data);
                ptr = ptr->link;
        }
        printf("\n");
}

int empty(LNODE *ptr)
{
```

```
        if(ptr == NULL)
                return(1);
        else
                return(0);
}

void push(LNODE **ptr, char item)
{
        LNODE *p;

        if(empty(*ptr))
        {
                p = malloc(sizeof(LNODE));
                if(p != NULL)
                {
                        p->data = item;
                        p->link = NULL;
                        *ptr = p;
                }
        }
        else
        {
                p = malloc(sizeof(LNODE));
                if(p != NULL)
                {
                        p->data = item;
                        p->link = *ptr;
                        *ptr = p;
                }
        }
}

void pop(LNODE **ptr, char *item)
{
        LNODE *p1;

        p1 = *ptr;
        if(empty(p1))
        {
                printf("Error! The stack is empty.\n");
                *item = '\0';
        }
        else
        {
                *item = p1->data;
                *ptr = p1->link;
                free(p1);
        }
}
```

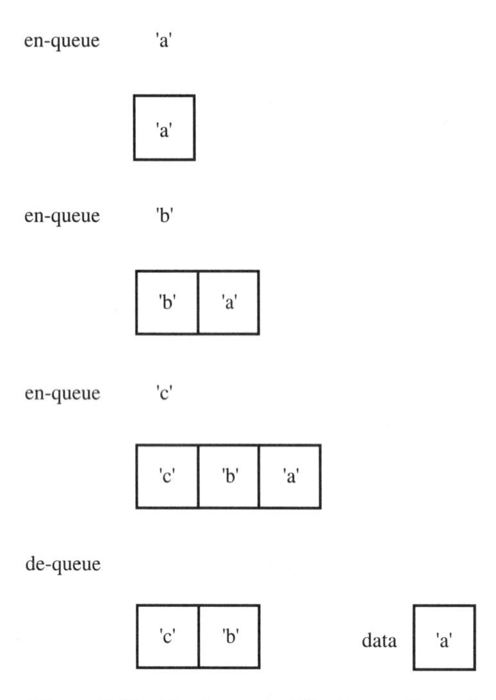

Figure 8.12 En-Queue and De-Queue Operations

Execution of Program 8.25 yields the following:

```
The stack is as follows: cba

The first item popped is c
The second item popped is b
The third item popped is a
Error! The stack is empty.
```

It is important to make use of the `empty()` function in `pop()`, since it makes no sense to pop from an empty stack.

A data structure similar to a stack is called a **queue**. The important difference between a queue and a stack is that the queue is a **FIFO** (First In First Out) data structure. Figure 8.12 shows a sample queue after a number of en-queue (put data into queue) and de-queue (take data from queue) operations.

Program 8.26 shows how data is placed onto a queue and how data is removed from a queue. Unlike the stack operation of Program 8.25, data comes off the queue in the order it was placed on it.

Program 8.26

```
#include <stdio.h>
#include <stdlib.h>

        typedef struct node
```

```
        {
                char data;
                struct node *link;
        } LNODE;

void show_queue(LNODE *ptr);
int empty(LNODE *ptr);
void en_queue(LNODE **head, char item);
void de_queue(LNODE **head, char *item);

main()
{
        LNODE *qhead = NULL;
        char item;

        en_queue(&qhead,'a');
        en_queue(&qhead,'b');
        en_queue(&qhead,'c');
        printf("The queue is as follows: ");
        show_queue(qhead);
        de_queue(&qhead,&item);
        printf("\nThe first item de-queued is %c\n",item);
        de_queue(&qhead,&item);
        printf("The second item de-queued is %c\n",item);
        de_queue(&qhead,&item);
        printf("The third item de-queued is %c\n",item);
        de_queue(&qhead,&item);
}

void show_queue(LNODE *ptr)
{
        while(ptr != NULL)
        {
                printf("%c",ptr->data);
                ptr = ptr->link;
        }
        printf("\n");
}

int empty(LNODE *ptr)
{
        if(ptr == NULL)
                return(1);
        else
                return(0);
}

void en_queue(LNODE **head, char item)
{
        LNODE *p;
```

```
        p = malloc(sizeof(LNODE));
        if(p != NULL)
        {
                p->data = item;
                p->link = *head;
                *head = p;
        }
}

void de_queue(LNODE **head, char *item)
{
        LNODE *p1, *p2;

        p1 = *head;
        if(empty(p1))
        {
                printf("Error! The queue is empty.\n");
                *item = '\0';
        }
        else
        {
                p2 = *head;
                while(p2->link != NULL)
                {
                        p1 = p2;
                        p2 = p2->link;
                }
                *item = p2->data;
                p1->link = NULL;
                free(p2);
                if(p1 == p2)
                        *head = NULL;
        }
}
```

Execution of Program 8.26 results in the following:

```
The queue is as follows: cba

The first item de-queued is a
The second item de-queued is b
The third item de-queued is c
Error! The queue is empty.
```

A queue may also be implemented as a *doubly* linked-list, as shown in Figure 8.13. Two pointers are used to access the queue. These two pointers point to the *head* (first node) and *tail* (last node) of the queue. Note that the nodes making up the queue's linked-list now contain two link fields. One link field is used to point to the next node in the list. The other link field is used to point to the previous node in the list. This allows easy access

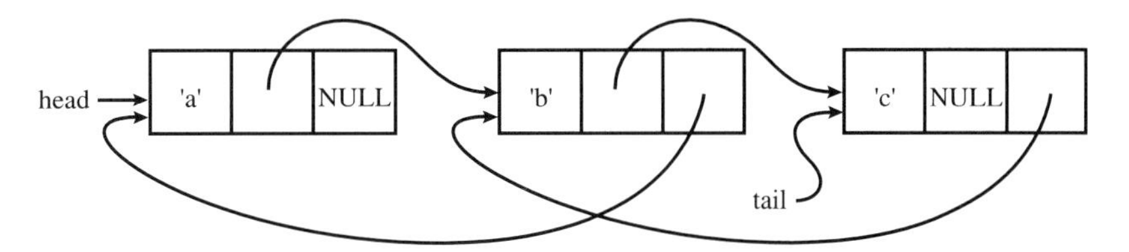

Figure 8.13 A Three-Element Queue

to data at both ends of the linked-list and eliminates the need to search for the end of the list when we are de-queuing. The required C structure might look like this:

```
typedef struct node
{
        char data;
        struct node *flink;
        struct node *blink;
} LNODE;
```

where `flink` and `blink` stand for forward-link and backward-link, respectively.

Binary Trees

A **binary tree** is a specific data structure composed of nodes containing a data field and two link fields, as shown in Figure 8.14. The link fields are called *left child* and *right child*. The term "binary" comes from the fact that each node has the capability of pointing to exactly two other nodes. The first node in the tree is commonly called the *root* node, and is located at the top of the tree's diagram. Figure 8.15 shows a sample binary tree. The root node of the tree is the node containing the `'*'` sign. The binary tree in Figure 8.15 was constructed in such a way that each link field points to one of three places: to a node containing a math operation, to a node containing a variable name, or to `NULL`. It is important to be able to traverse the binary tree and access the information contained within it. During a **traversal,** the data stored at each node is displayed or accessed.

There are three common forms of traversal: *pre-order, in-order,* and *post-order*. In all traversal methods, we attempt to go down the tree to the left as far as possible before going to the right. Our trip down the tree continues until we encounter `NULL` in a link field. The results of each traversal method are different, since each displays/accesses the node's data field at a different time.

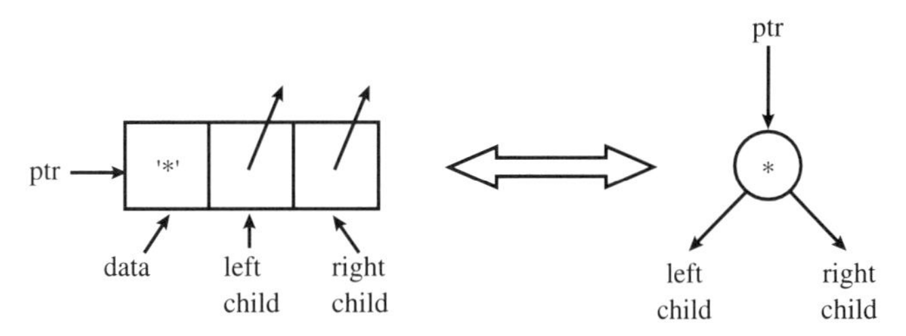

Figure 8.14 A Sample Node in a Binary Tree

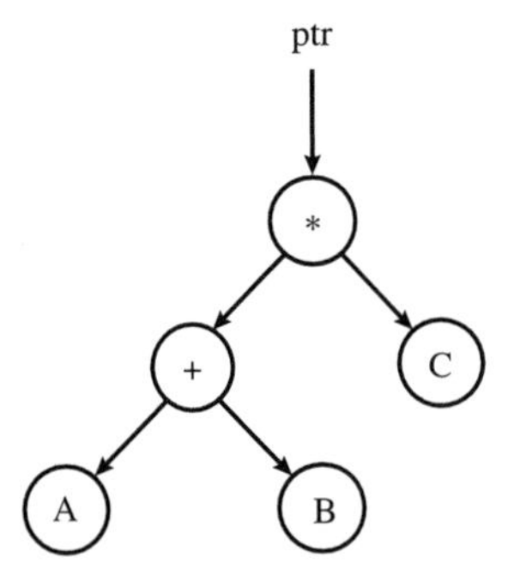

Figure 8.15 A Sample Binary Tree

For example, in a pre-order traversal, the following steps are performed at each node:

access data field
access left child
access right child

An in-order traversal accesses the data field like this:

access left child
access data field
access right child

A post-order traversal does the following:

access left child
access right child
access data field

Figure 8.16 shows the order in which the nodes in our sample binary tree are accessed by all three techniques. The results of each traversal are as follows:

```
Pre-order:     *+ABC
In-order:      A+B*C
Post-order:    AB+C*
```

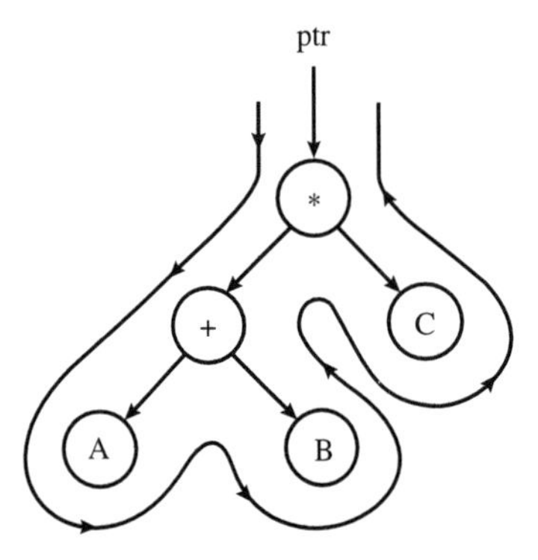

Figure 8.16 Binary Tree Traversal

These traversal techniques are easily implemented through the use of recursive function calls, as Program 8.27 illustrates.

Program 8.27

```
#include <stdio.h>
#include <stdlib.h>

        typedef struct node
        {
                char data;
                struct node *lchild;
                struct node *rchild;
        } LNODE;

void pre_order(LNODE *ptr);
void in_order(LNODE *ptr);
void post_order(LNODE *ptr);

main()
{
        LNODE *n1, *n2, *n3, *n4, *n5;

        n1 = malloc(sizeof(LNODE));
        n2 = malloc(sizeof(LNODE));
        n3 = malloc(sizeof(LNODE));
        n4 = malloc(sizeof(LNODE));
        n5 = malloc(sizeof(LNODE));
        n1->data = '*';
        n1->lchild = n2;
        n1->rchild = n3;
        n2->data = '+';
        n2->lchild = n4;
        n2->rchild = n5;
        n3->data = 'C';
        n3->lchild = NULL;
        n3->rchild = NULL;
        n4->data = 'A';
        n4->lchild = NULL;
        n4->rchild = NULL;
        n5->data = 'B';
        n5->lchild = NULL;
        n5->rchild = NULL;
        printf("Pre-order traversal  => ");
        pre_order(n1);
        printf("\nIn-order traversal   => ");
        in_order(n1);
        printf("\nPost-order traversal => ");
        post_order(n1);
}

void pre_order(LNODE *ptr)
{
```

```
        if(ptr != NULL)
        {
                printf("%c",ptr->data);
                pre_order(ptr->lchild);
                pre_order(ptr->rchild);
        }
}

void in_order(LNODE *ptr)
{
        if(ptr != NULL)
        {
                in_order(ptr->lchild);
                printf("%c",ptr->data);
                in_order(ptr->rchild);
        }
}

void post_order(LNODE *ptr)
{
        if(ptr != NULL)
        {
                post_order(ptr->lchild);
                post_order(ptr->rchild);
                printf("%c",ptr->data);
        }
}
```

Examine the method used to build the binary tree at the beginning of the program. The tree represents a mathematical expression whose original form was:

```
(A + B) * C
```

Usually, the binary tree for a mathematical expression is constructed by a function that follows some simple guidelines for operator precedence (which we saw in the Example Programs section of Chapter 6).

Execution of Program 8.27 produces

```
Pre-order traversal  => *+ABC
In-order traversal   => A+B*C
Post-order traversal => AB+C*
```

You are encouraged to devise your own trees using Program 8.27 and perform the traversals yourself.

The Example Programs section of this chapter will expand the concepts presented in this section. For now, see how well you do with the section review.

8.7 Section Review

1. Describe the basic structure of a node.
2. Explain the differences in memory usage in a linked-list of characters and a character string. Does either data structure have limitations?

3. What types of data can be stored in a node?
4. What is the importance of the NULL pointer?
5. How is run time memory allocation accomplished?
6. What are the basic operations performed on a linked-list, a stack, a queue, and a binary tree?
7. What is meant by FIFO? By LIFO?

8.8 Example Programs

In this section we will examine four more programs designed to show how linked-lists, stacks, queues, and binary trees are used.

Node Counting

Program 8.28 contains the statements needed to build the binary expression tree previously examined in Figure 8.15.

Program 8.28

```
#include <stdio.h>
#include <stdlib.h>

        typedef struct node
        {
                char data;
                struct node *lchild;
                struct node *rchild;
        } LNODE;

        int number_of_nodes = 0;
        void count_nodes(LNODE *ptr);

main()
{
        LNODE *n1, *n2, *n3, *n4, *n5;

        n1 = malloc(sizeof(LNODE));
        n2 = malloc(sizeof(LNODE));
        n3 = malloc(sizeof(LNODE));
        n4 = malloc(sizeof(LNODE));
        n5 = malloc(sizeof(LNODE));
        n1->data = '*';
        n1->lchild = n2;
        n1->rchild = n3;
        n2->data = '+';
        n2->lchild = n4;
        n2->rchild = n5;
        n3->data = 'C';
        n3->lchild = NULL;
        n3->rchild = NULL;
        n4->data = 'A';
        n4->lchild = NULL;
        n4->rchild = NULL;
```

```
        n5->data = 'B';
        n5->lchild = NULL;
        n5->rchild = NULL;
        printf("Number of nodes => ");
        count_nodes(n1);
        printf("%d", number_of_nodes);
}

void count_nodes(LNODE *ptr)
{
        if(ptr != NULL)
        {
                number_of_nodes++;
                count_nodes(ptr->lchild);
                count_nodes(ptr->rchild);
        }
}
```

A pointer to the root node of the tree is passed to the `count_nodes()` function, which counts the number of nodes contained in the tree. From Figure 8.15 it is clear that the tree contains five nodes. Program 8.28 agrees, as shown by its sample execution:

```
Number of nodes => 5
```

The technique used to count the nodes is similar to that used to perform a traversal. As each new node is reached, a counter is incremented. Then the left child and right child nodes (if any) are examined recursively.

Reversing a String with a Stack

Stacks have many uses when they are applied to string applications. For example, stacks can be used to verify that a given string is a palindrome, such as aabccbaa. In this example, Program 8.29 uses a stack to reverse the characters of a user-supplied string. Figure 8.17

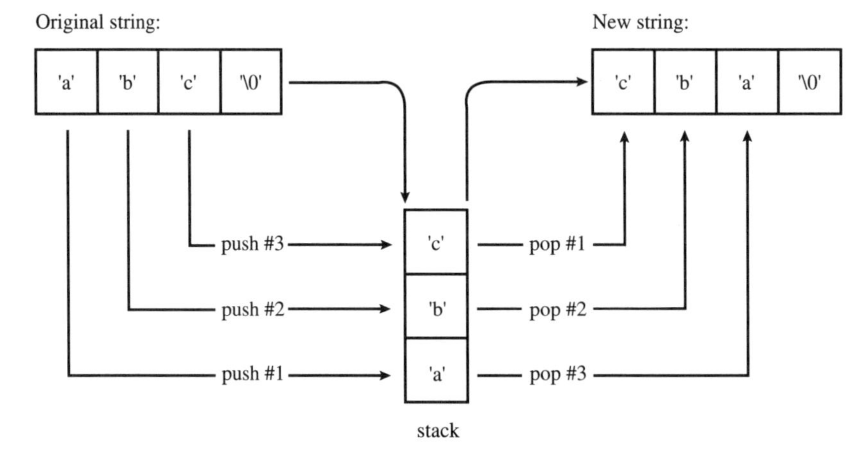

Figure 8.17 Reversing a String Using a Stack

shows the basic process. The string characters are pushed onto a stack one by one. When the entire string has been pushed onto the stack, the last character of the string is on top of the stack. As characters are now popped off the stack, they are written back into the string array beginning with the first position. Thus, the string is automatically reversed through the use of a single stack.

Program 8.29 makes use of the `strlen()` and `push()` functions to push all characters onto the stack, and then uses the `pop()` function to pop characters off the stack until it is empty. Note that the `NULL` character is never pushed, since it is not included in the length of the string determined by `strlen()`.

Program 8.29

```
#include <stdio.h>
#include <stdlib.h>

        typedef struct node
        {
                char data;
                struct node *link;
        } LNODE;

void show_stack(LNODE *ptr);
int empty(LNODE *ptr);
void push(LNODE **ptr, char item);
void pop(LNODE **ptr, char *item);

main()
{
        LNODE *stacka = NULL;
        char string[40], item;
        int i;

        printf("Enter a string: ");
        gets(string);
        for(i = 0; i < strlen(string); i++)
                push(&stacka,string[i]);
        printf("\nThe stack looks like this: ");
        show_stack(stacka);
        i = 0;
        while(!empty(stacka))
        {
                pop(&stacka,&string[i]);
                i++;
        }
        printf("\nThe new string is => %s",string);
}

void show_stack(LNODE *ptr)
{
        while(ptr != NULL)
        {
```

```
                printf("%c",ptr->data);
                ptr = ptr->link;
        }
        printf("\n");
}

int empty(LNODE *ptr)
{
        if(ptr == NULL)
                return(1);
        else
                return(0);
}

void push(LNODE **ptr, char item)
{
        LNODE *p;

        if(empty(*ptr))
        {
                p = malloc(sizeof(LNODE));
                if(p != NULL)
                {
                        p->data = item;
                        p->link = NULL;
                        *ptr = p;
                }
        }
        else
        {
                p = malloc(sizeof(LNODE));
                if(p != NULL)
                {
                        p->data = item;
                        p->link = *ptr;
                        *ptr = p;
                }
        }
}

void pop(LNODE **ptr, char *item)
{
        LNODE *p1;

        p1 = *ptr;
        if(empty(p1))
        {
                printf("Error! The stack is empty.\n");
                *item = '\0';
        }
        else
```

```
        {
                *item = p1->data;
                *ptr = p1->link;
                free(p1);
        }
}
```

Execution of Program 8.29 is as follows:

```
Enter a string: Microprocessors 123!

The stack looks like this: !321 srossecorporciM

The new string is => !321 srossecorporciM
```

Imagine what can be done if two or more stacks are used to process a string.

Binary Search

The binary search is a very efficient search technique commonly used with binary trees. Figure 8.18 shows a sample binary tree containing integer-based nodes. The tree is structured in such a way that the integers it contains have already been placed into their correct positions (for searching purposes). For example, notice that all nodes reachable from the left child of the root node have values (2, 3, 5, and 6) smaller than the root node value (7). All nodes reachable from the right child of the root node have values (12, 13, and 17) that are greater than the root node value. Searching the tree for a particular value—6 for example—involves the following tests:

1. Does the current node value equal the search value?

In this case the answer is no, since the current node value is 7.

2. Is the search value smaller than the current node value? If YES, make the current node the left child node.

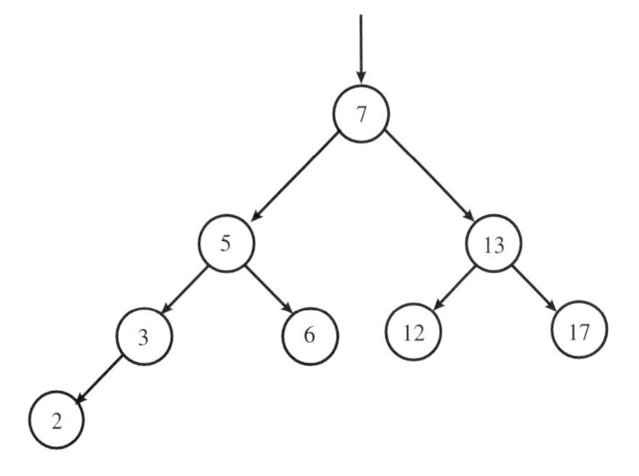

Figure 8.18 Binary Tree Containing Sorted Integers

In this case the answer is yes, since 6 is smaller than 7. So, the current node value becomes 5. Now step 1 is repeated again. Since 5 does not equal 6, we go on to step 2. The search value is not smaller than the current node value, which takes us to step 3.

3. Make the current node the right child node.

This causes the current node value to become 6. When step 1 is repeated again, we find our match. Thus, the procedure for a binary search involves taking left and right turns as we go down the tree looking for our search value.

In cases where the search value is not in the tree, steps 2 and 3 will eventually encounter `NULL` as the left or right pointer. If this happens, the search is terminated and we conclude that the search value does not exist within the tree.

Program 8.30 contains the function `binary_search()`, which implements the three-step search technique we just saw. Once again we use a recursive function call (`binary_search()` calls itself) to traverse the binary tree.

Program 8.30

```
#include <stdio.h>
#include <stdlib.h>

        typedef struct node
        {
                int data;
                struct node *lchild;
                struct node *rchild;
        } LNODE;

int binary_search(LNODE *ptr, int item);

main()
{
        LNODE *n1, *n2, *n3, *n4, *n5, *n6, *n7, *n8;
        int sval, found;

        n1 = malloc(sizeof(LNODE));
        n2 = malloc(sizeof(LNODE));
        n3 = malloc(sizeof(LNODE));
        n4 = malloc(sizeof(LNODE));
        n5 = malloc(sizeof(LNODE));
        n6 = malloc(sizeof(LNODE));
        n7 = malloc(sizeof(LNODE));
        n8 = malloc(sizeof(LNODE));
        n1->data = 7;
        n1->lchild = n2;
        n1->rchild = n3;
        n2->data = 5;
        n2->lchild = n4;
        n2->rchild = n5;
        n3->data = 13;
```

```
        n3->lchild = n6;
        n3->rchild = n7;
        n4->data = 3;
        n4->lchild = n8;
        n4->rchild = NULL;
        n5->data = 6;
        n5->lchild = NULL;
        n5->rchild = NULL;
        n6->data = 12;
        n6->lchild = NULL;
        n6->rchild = NULL;
        n7->data = 17;
        n7->lchild = NULL;
        n7->rchild = NULL;
        n8->data = 2;
        n8->lchild = NULL;
        n8->rchild = NULL;
        printf("Enter value to search for => ");
        scanf("%d",&sval);
        found = binary_search(n1,sval);
        printf("The value %d has",sval);
        found ? printf(" ") : printf(" not ");
        printf("been found in the binary tree.");
}

int binary_search(LNODE *ptr, int item)
{
        if(ptr == NULL)
                return(0);
        else
        if(ptr->data == item)
                return(1);
        else
        if(item < ptr->data)
                binary_search(ptr->lchild, item);
        else
                binary_search(ptr->rchild, item);
}
```

Two executions of Program 8.30 follow, one successful and the other unsuccessful.

```
Enter value to search for => 6
The value 6 has been found in the binary tree.

Enter value to search for => 22
The value 22 has not been found in the binary tree.
```

It is a rewarding programming challenge writing the function required to create the binary tree used in the search. You will really know how to use binary trees after doing so.

Depth-First and Breadth-First Searching

The depth-first and breadth-first search techniques are fundamentally different. The **depth-first search** is already familiar to us, since it is the search technique used to traverse the binary tree in Program 8.28 (and previously to accomplish pre-, in-, and post-order traversals). A depth-first search goes as far down the left side of the binary tree as it can. When `NULL` is encountered, the search switches to the bottom-most right child and resumes.

A **breadth-first search** accesses the nodes of the tree in a different order. As Figure 8.19 shows, nodes are accessed one level at a time, beginning with the first level (the root node). The child nodes of the root node (which occupy level 2) are accessed next. Then their child nodes are all accessed (the four nodes at level 3), and so on. This search technique is easily accomplished with the use of a queue. Only two steps are needed to control the queue:

1. Remove a node from the queue. This becomes the current node.
2. Place all child nodes of the current node onto the queue.

The queue is initially loaded with the root node of the tree. Steps 1 and 2 are repeated until the queue is empty. For the tree shown in Figure 8.19, the queue takes on the following values during a breadth-first search:

```
7                   -- initial node
13   5              -- 7's children added
6    3    13        -- 5's children added
17   12   6    3    -- 13's children added
2    17   12   6    -- 3's children added
2    17   12
2    17
2
```

Program 8.31 uses a queue of tree-node pointers to implement a breadth-first search. Note that two different structures are used in Program 8.31.

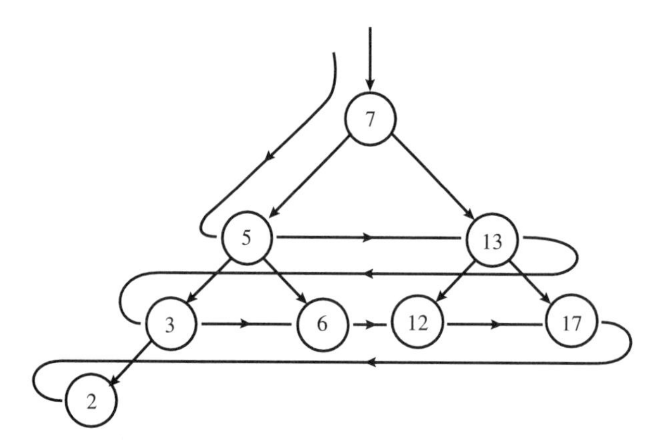

Figure 8.19 Breadth-First Search of a Binary Tree

Program 8.31

```
#include <stdio.h>
#include <stdlib.h>

        typedef struct tnode
        {
                int data;
                struct tnode *lchild;
                struct tnode *rchild;
        } TNODE;

        typedef struct qnode
        {
                struct tnode *tptr;
                struct qnode *link;
        } QNODE;

void bfs(TNODE *ptr);
void dfs(TNODE *ptr);
void show_queue(QNODE *ptr);
int empty(QNODE *ptr);
void en_queue(QNODE **head, TNODE *ptr);
void de_queue(QNODE **head, TNODE **ptr);


main()
{
        TNODE *n1, *n2, *n3, *n4, *n5, *n6, *n7, *n8;

        n1 = malloc(sizeof(TNODE));
        n2 = malloc(sizeof(TNODE));
        n3 = malloc(sizeof(TNODE));
        n4 = malloc(sizeof(TNODE));
        n5 = malloc(sizeof(TNODE));
        n6 = malloc(sizeof(TNODE));
        n7 = malloc(sizeof(TNODE));
        n8 = malloc(sizeof(TNODE));
        n1->data = 7;
        n1->lchild = n2;
        n1->rchild = n3;
        n2->data = 5;
        n2->lchild = n4;
        n2->rchild = n5;
        n3->data = 13;
        n3->lchild = n6;
        n3->rchild = n7;
        n4->data = 3;
        n4->lchild = n8;
        n4->rchild = NULL;
        n5->data = 6;
        n5->lchild = NULL;
        n5->rchild = NULL;
```

```
        n6->data = 12;
        n6->lchild = NULL;
        n6->rchild = NULL;
        n7->data = 17;
        n7->lchild = NULL;
        n7->rchild = NULL;
        n8->data = 2;
        n8->lchild = NULL;
        n8->rchild = NULL;
        printf("Depth-First Search   => ");
        dfs(n1);
        printf("\nBreadth-First Search => ");
        bfs(n1);
}

void dfs(TNODE *ptr)
{
        if(ptr != NULL)
        {
                printf("%4d",ptr->data);
                dfs(ptr->lchild);
                dfs(ptr->rchild);
        }
}

void bfs(TNODE *ptr)
{
        QNODE *qhead = NULL;
        TNODE *p;

        en_queue(&qhead, ptr);
        while(!empty(qhead))
        {
                de_queue(&qhead, &p);
                printf("%4d",p->data);
                if(p->lchild != NULL)
                        en_queue(&qhead, p->lchild);
                if(p->rchild != NULL)
                        en_queue(&qhead, p->rchild);
        }
}

int empty(QNODE *ptr)
{
        if(ptr == NULL)
                return(1);
        else
                return(0);
}

void en_queue(QNODE **head, TNODE *ptr)
```

```
{
        QNODE *p;

        p = malloc(sizeof(QNODE));
        if(p != NULL)
        {
                p->tptr = ptr;
                p->link = *head;
                *head = p;
        }
}

void de_queue(QNODE **head, TNODE **ptr)
{
        QNODE *p1, *p2;

        p1 = *head;
        if(empty(p1))
        {
                printf("Error! The queue is empty.\n");
                *ptr = NULL;
        }
        else
        {
                p2 = *head;
                while(p2->link != NULL)
                {
                        p1 = p2;
                        p2 = p2->link;
                }
                *ptr = p2->tptr;
                p1->link = NULL;
                free(p2);
                if(p1 == p2)
                        *head = NULL;
        }
}
```

These structures are the `TNODE` (tree node) and `QNODE` (queue node) structures. Both structures are utilized by the queuing functions. The depth-first and breadth-first searches are implemented via the `dfs()` and `bfs()` functions. A sample execution shows how the results of the two searches differ:

```
Depth-First Search   =>    7   5   3   2   6  13  12  17
Breadth-First Search =>    7   5  13   3   6  12  17   2
```

Each search technique has its own advantages and disadvantages. To illustrate, consider a chess-playing program. After only a few moves, the program will already be using a large decision tree to evaluate what move to make next. A depth-first search may spend a great deal of time looking at nodes on the left side of the search tree before getting to a more desirable choice node on the right side. A breadth-first search will usually get to the same node more quickly.

On the other hand, a program designed to pack items into a box might arrive at a solution right away via a depth-first search down the left side of a decision tree, whereas a breadth-first search might waste time considering many different partial solutions. Knowing when to use a particular search technique comes from practice.

Conclusion

In this section we examined many new applications that utilize advanced data structures. Test your understanding of this material with the following section review.

8.8 Section Review

1. When is `number_of_nodes` incremented in Program 8.28?
2. How is a stack used to reverse a string?
3. What are the recursive calls made when searching the binary tree of Figure 8.18 for the value 6? Use a depth-first search.
4. Repeat question 3 for a breadth-first search.

8.9 Troubleshooting Techniques

Discussion

In this section we will review a number of techniques that are designed to identify and eliminate program errors when you work with advanced data structures. They are meant to be used as a checklist when tracking down the type of errors unique to linked-lists and other dynamic structures.

Keep track of allocated memory. When requesting memory for a new node, it is good to make sure the allocation actually took place and was not denied by the operating system. For example, `malloc()` returns a NULL pointer if the requested memory could not be allocated. The pointer variable used with `malloc()` should always be checked before using it, to guarantee that a valid memory reference will take place. Unpredictable results are possible when the wrong locations are accessed using an invalid pointer.

Valid pointers can be tested like this:

```
MemPtr = malloc(sizeof(DataObj));
if (MemPtr == NULL)
{
      printf("Could not allocate!\n");
      exit(-1);
}
```

A NULL `MemPtr` value will case the condition to be TRUE in the `if()` statement.

Something potentially worse than not getting the requested memory is not giving it back when you are through with it. When a node is deleted from a linked-list, its memory space should be returned to the storage pool, via `free()`. If the memory is not returned, it is possible that the operating system will eventually run out of memory when a lot of node activity is encountered. For instance, the decision tree for the current move in a chess game may contain hundreds of nodes, spawning mini-decision trees as new moves are contemplated. It is easy to imagine a memory shortage as the game progresses, if the decision tree is never pruned.

Watch the pointers. When adding or deleting a node to a linked-list, care must be taken to guarantee that the pointer to the list is updated properly, as are any other pointers stored within the list. There are several cases to consider:

1 Adding a node to the head of a linked-list: The list pointer must point to the new node. The new node must point to the old head of the list.
2. Adding a node to the tail of a linked-list: The list pointer remains unchanged. The pointer stored in the last node of the list (tail) is changed to point to the new node. The new node's pointer must be set to NULL.
3. Deleting a node from the head of a linked-list: The list pointer must be changed to point to the second node in the list. The deleted node must be returned with `free()`.
4. Deleting a node from the tail of a linked-list: The list pointer remains unchanged. The pointer in the second to last node of the list must be set to NULL. The deleted node must be returned with `free()`.

It is not difficult to lose track of where a pointer is pointing when trying to analyze a program bug. It is good to draw a diagram of the list and the changes made to it as the program executes.

It is also necessary to properly initialize pointers (to NULL for an empty list), and test the value of the pointer accordingly. A function to add a node to a linked-list may require different steps for empty and nonempty lists. It is a good idea to consider all the possible actions that can be performed on the lists.

Use the right structure. Some data structures are better suited for one application than another. For example, as we saw in Chapter 6, an application that sorts strings must use a rectangular array for string storage, rather than a ragged array. A ragged array, with its efficient nonuniform storage requirements, is better suited for applications that do not require sorting.

In a similar fashion, linked-lists are better suited for some applications, and hash tables for others. Several factors must be considered when choosing one data structure over another. Storage space, access time to read an element, and the ability to add and delete information are three typical factors to think about when evaluating which structure to use. Sorting time, or the time needed to perform a comparison between two elements, may also be important. It pays to consider your choices before writing any code.

Conclusion

Working with complex data structures requires preplanning and attention to detail. Test your knowledge of these requirements with the following section review.

8.9 Section Review

1. Why should pointers initialized by `malloc()` be checked before being used?
2. Why should `free()` be used in a program that only uses a small number of nodes?

3. Why would a function used to delete a node from a linked-list need to know if the list pointer is NULL?
4. Explain how a binary tree might be more suitable for storing a 1000-element array than a linked-list would be.

8.10 Case Study: MiniMicro

Discussion

In this case study we will develop and examine the operation of a simple microprocessing system, called MiniMicro. The instruction set of the machine is shown in Table 8.1. The types of instructions included in MiniMicro are typical of those found on most microprocessors. The power of the `enum` data type allows us to define the LOAD, ADD, etc. instructions shown in Table 8.1 without having to assign any meaning or values to them. However, we can write a program for MiniMicro using combinations of instructions, and actually come up with a working machine.

The Problem

MiniMicro operates like any microprocessor, running the same sequence of operations over and over:

- Fetch instruction
- Decode instruction
- Execute instruction

Since the machine must read memory to fetch an instruction, we must provide memory for the program and data and also a pointer that allows MiniMicro to access memory. Some instructions use the pointer to read data during their execution phase (LOAD, ADD, SUB, MUL, and DIV), and other instructions load the pointer with a new value (JNZ, JMP, JSR, and RET).

Table 8.1 MiniMicro Instruction Set

Instruction Syntax	**Operation**
LOAD value	Load accumulator with value
ADD value	Add value to accumulator
SUB value	Subtract value from accumulator
MUL value	Multiply accumulator by value
DIV value	Divide accumulator by value
PRINT	Display accumulator contents
JNZ address	Jump to address if accumulator is not zero
JMP address	Jump to address
JSR address	Jump to subroutine beginning at address
RET	Return from subroutine
STOP	Stop execution

Developing the Algorithm

The steps required in the MiniMicro application are as follows:

1. Fetch instruction.
2. Read instruction operand from next location.
3. Decode instruction.
4. Execute instruction.

Each step requires knowledge about the machine's instruction pointer. The execute step utilizes the machine's accumulator. Recursion will be used to execute the JSR and RET instructions, since they both change and restore the instruction pointer during execution. In order to make the accumulator available to all levels of recursion (in case nested subroutines are used), the accumulator is defined as a *global* variable.

The Overall Process

Examine Program 8.32. The program contains a `do` loop that performs all actions necessary to execute a single instruction. A `switch()` statement is used to decode the fetched instruction. Each line of the `switch()` statement represents one of MiniMicro's enumerated instructions.

Program 8.32

```
#include <stdio.h>

/*****************************************************************/
/*                        MiniMicro                              */
/*****************************************************************/
/*                  Developed by A. MacHine                      */
/*                                                               */
/*****************************************************************/
/*     This program emulates the operation of a simple computer  */
/*     that contains a basic set of instructions and a single    */
/*     accumulator for results. Subroutines are provided through */
/*     the use of recursive function calls.                      */
/*****************************************************************/
/*                     Function Prototypes                       */
/*-------------------------------------------------------------*/
/*                                                               */
/* This function executes the instruction selected by pc.        */
/*-------------------------------------------------------------*/

void exec(int pc);

/*****************************************************************/
/*                     Global Variables                          */
/*-------------------------------------------------------------*/

        /* Define instruction set */
        enum iset {LOAD, ADD, SUB, MUL, DIV, PRINT,
```

```
                 JNZ, JMP, JSR, RET, STOP};

        /* Load program into memory */
        int mem[256] = {LOAD, 100, JSR, 6, PRINT, STOP,
                        MUL, 9, DIV, 5, ADD, 32, RET};

        int acc;        /* Accumulator */

/*------------------------------------------------------------------*/

main()
{
        exec(0);        /* Begin execution at address 0 */
}

/*------------------------------------------------------------------*/

void exec(int pc)
{
        int ip;         /* Instruction pointer */
        int inst;       /* Instruction */
        int iop;        /* Instruction operand */

        ip = pc;        /* Load instruction pointer */
        do
        {
                inst = mem[ip++];       /* Fetch instruction */
                iop = mem[ip++];        /* Read operand too */
                switch(inst)
                {
                        case LOAD  : acc = iop; break;
                        case ADD   : acc += iop; break;
                        case SUB   : acc -= iop; break;
                        case MUL   : acc *= iop; break;
                        case DIV   : acc /= iop; break;
                        case PRINT :
                        {
                                printf("ACC = %5d\n",acc);
                                ip--;  /* Fix ip */
                                break;
                        }
                        case JNZ   : ip = (acc != 0) ? iop : ip; break;
                        case JMP   : ip = iop; break;
                        case JSR   : exec(iop); break;
                        case RET   : return; break;
                        case STOP  : break;
                }
        } while (inst != STOP);
}
```

Notice how the JSR instruction is implemented:

```
case JSR : exec(iop); break;
```

A recursive call to `exec()` is used to handle subroutines. The address of the subroutine is passed to `exec()` via `iop`. This allows multiple instruction pointers to exist simultaneously.

When the RET instruction executes, its `switch()` statement:

```
case RET : return; break;
```

simply causes the current level of recursion to terminate and pass control back to the previous level, where the original value of the instruction pointer is restored.

The `do` loop is repeated until the STOP instruction is encountered.

Table 8.2 shows an example of a MiniMicro program that converts 100 degrees Celsius into Fahrenheit. Note that the actual conversion is performed in a subroutine.

A sample execution of Program 8.32 is as follows:

```
ACC =  212
```

which is the correct Fahrenheit temperature.

Conclusion

The MiniMicro machine is a good example of how enumeration is used in the real world. Even though our machine has a limited instruction set, adding more instructions is as simple as adding new enumerated types to our list, with the appropriate case statements within the `switch()`. Programs such as Program 8.32 are often called **emulators,** because they emulate (mimic) the operation of an actual machine, without the need for the actual machine to exist.

Test your understanding of the material in this section with the following section review.

8.10 Section Review

1. How are multiple program counters maintained when subroutines are used?
2. Explain how to add new instructions and data types to Mini-Micro.

Table 8.2 Example MiniMicro Program

Address	Instruction	Comment
0	LOAD 100	Load accumulator with 100
2	JSR 6	Jump to subroutine at address 6
4	PRINT	Display accumulator
5	STOP	Halt
6	MUL 9	Multiply accumulator by 9
8	DIV 5	Divide accumulator by 5
10	ADD 32	Add 32 to accumulator
12	RET	Return from subroutine

3. What does this program do?

```
LOAD   20
ADD    24
ADD    27
ADD    21
DIV     4
PRINT
STOP
```

Interactive Exercises

DIRECTIONS

These exercises require that you have access to a computer and software that supports C. They are provided here to give you valuable experience and immediate feedback on what the concepts and commands introduced in this chapter will do.

Exercises

1. Program 8.33 illustrates the C enum. Try it.

Program 8.33

```
#include <stdio.h>

enum numbers {One, Two, Three}

main()
{
        enum numbers value;

        value = One;
}
```

2. Program 8.34 illustrates the C enum with constants assigned to the identifiers. Do you think you know what the output will be? Try it.

Program 8.34

```
#include <stdio.h>

enum numbers {One = 1, Two = 2, Three = 3}

main()
{
        int result;

        result = One + Two;
        printf("Result = %d",result);
}
```

3. Program 8.35 illustrates a counting loop using the C `enum`. Predict what the output of the program will be, then try it.

Program 8.35

```
#include <stdio.h>

enum numbers {Zero, One, Two, Three}

main()
{
        enum numbers number;

        for(number = Zero; number <= Three; number++)
                printf("Digital value of enum type number = %d\n",number);
}
```

4. Program 8.36 attempts to illustrate the smallest possible structure. See if the output you get is what you would expect.

Program 8.36

```
#include <stdio.h>

main()
{
        struct {int number;} value;

        value.number = 5;
        printf("Value is %d",value.number);
}
```

5. Note how Program 8.37 declares the member declaration list. Will this work? Give it a try.

Program 8.37

```
#include <stdio.h>

main()
{
        struct {int number1, number2, number3;} value;

        value.number1 = 1;
        value.number2 = 2;
        value.number3 = 3;

        printf("Value of number1 is %d",value.number1);
}
```

6. Program 8.38 shows a use of the C `define` to replace a `long_identifier`. This is an interesting program to try.

Program 8.38

```
#include <stdio.h>

#define S long_identifier

main()
{
        struct {int number1, number2, number3;} long_identifier;

        S.number1 = 1;
        S.number2 = 2;
        S.number3 = 3;

        printf("Value of number1 is %d",S.number1);
}
```

7. Program 8.39 illustrates the use of a structure tag. See what the output is. Did you get what you expected?

Program 8.39

```
#include <stdio.h>

struct tag
{
        int first;
        int second;
}

main()
{
        struct tag structure;

        structure.first = 1;

        printf("Value of structure.first is %d",structure.first);
}
```

8. Program 8.40 shows a structure utilizing a C `typedef`. See if the output of this program is not the same as the last program.

Program 8.40

```
#include <stdio.h>

typedef struct
{
        int first;
        int second;
} name;

main()
```

```
{
        name structure;

        structure.first = 1;

        printf("Value of structure.first is %d",structure.first);
}
```

9. Do you see the differences between Program 8.40 and Program 8.41? Do you know why the outputs are different?

Program 8.41

```
#include <stdio.h>

typedef struct
{
        int first;
        int second;
} name;

main()
{
        name structure;

        structure.first = 11;

        printf("Value of structure.first is %d",structure.first);
}
```

10. Program 8.42 illustrates an example of a C `union`. Recall that a union allows the same memory location for variables of different types. Predict first what you think this program will do, then try it and check the results!

Program 8.42

```
#include <stdio.h>

typedef union
{
        int first;
        float second;
} value;

main()
{
        value number1;

        number1.first = 12;
        printf("Value of number1.first is %d\n",number1.first);

        number1.second = 34.5;
```

```
        printf("Value of number1.first is %d\n", number1.first);
}
```

11. Program 8.43 shows a structure within a structure. Analyze the program before trying it. Then check to see if it performs as expected.

Program 8.43

```
#include <stdio.h>

typedef struct
{
        int first;
} structure1;

typedef struct
{
        structure1 second;
} structure2;

main()
{
        structure2 new_structure;

        new_structure.second.first = 25;
        printf("The value is %d",new_structure.second.first);
}
```

12. Program 8.44 is an expansion of the previous program. It has increased the complexity of the data structure by making an array of a structure that contains another structure as its member. Study the program and see if it behaves as you expected.

Program 8.44

```
#include <stdio.h>

typedef struct
{
        int first;
} structure1;

typedef struct
{
        structure1 second;
} structure2;

typedef structure2 struct_array[5];

main()
{
        struct_array values;
```

```
        values[1].second.first = 54;
        printf("The value is %d",values[1].second.first);
}
```

13. Program 8.45 creates a three-node linked-list. Can you determine what will be displayed before you execute it?

Program 8.45

```
#include <stdio.h>
#include <stdlib.h>

        typedef struct node
        {
                char data;
                struct node *link;
        } LNODE;

void show_list(LNODE *ptr);

main()
{
        LNODE *n1, *n2, *n3;

        n1 = malloc(sizeof(LNODE));
        n2 = malloc(sizeof(LNODE));
        n3 = malloc(sizeof(LNODE));

        n1->data = 'c';
        n1->link = n2;
        n2->data = 'a';
        n2->link = n3;
        n3->data = 't';

        printf("The linked-list is as follows: ");
        show_list(n1);
}

void show_list(LNODE *ptr)
{
        while(ptr != NULL)
        {
                printf("%c",ptr->data);
                ptr = ptr->link;
        }
        printf("\n");
}
```

14. Program 8.46 creates a three-node linked-list. Predict what you will see on the screen before you execute the program.

Program 8.46

```
#include <stdio.h>
#include <stdlib.h>

        typedef struct node
        {
                char data;
                struct node *link;
        } LNODE;

void show_list(LNODE *ptr);

main()
{
        LNODE *n1, *n2, *n3;

        n1 = malloc(sizeof(LNODE));
        n2 = malloc(sizeof(LNODE));
        n3 = malloc(sizeof(LNODE));

        n1->data = 'c';
        n1->link = n2;
        n2->data = 'a';
        n2->link = n3;
        n3->data = 't';
        n3->link = n2;
        printf("The linked-list is as follows: ");
        show_list(n1);
}

void show_list(LNODE *ptr)
{
        while(ptr != NULL)
        {
                printf("%c",ptr->data);
                ptr = ptr->link;
        }
        printf("\n");
}
```

Self-Test

DIRECTIONS

Answer the following questions by referring to Program 8.32.

Questions

1. How is the instruction set defined?
2. Is there any significance to the order of the instructions defined by the `enum` statement?

3. Why does the accumulator need to be defined as a global?
4. What does `exec(50)` do?
5. What happens to the `iop` variable when PRINT, RET, or STOP execute?
6. How can the following instructions be added to the MiniMicro machine: AND, OR, XOR? Each instruction operates on the accumulator and a supplied operand.
7. How could a second accumulator be added to the emulator?
8. Explain how the JNZ instruction works.
9. What does this program do?

```
0: LOAD 10
2: PRINT
3: SUB 1
5: JNZ 2
7: STOP
```

End-of-Chapter Problems

General Concepts

Section 8.1

1. What data type in C is used to describe a discrete set of integer values?
2. Give the integer value of the first declared enumerated data type in C.
3. Does the C `enum` create a new data type? Explain.
4. Demonstrate how the C `enum` data type can be assigned an integer value.

Section 8.2

5. Does the C `typedef` create a new data type? Explain.
6. State the purpose of the C `typedef`.
7. Give the resulting new name of the data type for the following code:

```
typedef struct {
                    int value1;
               } structure;
```

Section 8.3

8. Define a C structure.
9. Give an example of a common everyday system that utilizes the concept of a C structure.
10. Illustrate the syntax of a C structure as presented in this section.
11. Explain what the C member of operator does. What symbol is used for this operator for a simple structure?

Section 8.4

12. What is the identifier that names the structure type defined in the member declaration list of the structure called?
13. Give an example of the use of a structure tag.
14. What is the purpose of the –> symbol in C as applied to C structures?
15. State the three operations that are allowed with structures.

Section 8.5

16. State how one or more different data types may be stored in the same memory location.
17. State the difference between declaring a C structure and a C union.
18. Can a C structure be initialized by the program in which it is written?

19. Explain what the following line of code represents:

```
variable[2].number
```

Section 8.6

20. Can a C structure have a member that is another structure? Explain.
21. Can a C array of structures contain an array as one of its members? Explain.
22. Can a C array of structures be a member of another array of structures? Explain.

Section 8.7

23. What determines the size of a node?
24. What are the links in a linked-list?
25. How are nodes inserted and deleted from each end of a linked-list?
26. Explain why stacks and queues are simply linked-lists accessed in special ways.
27. How is recursion used to access nodes in a binary tree?

Section 8.8

28. How can the depth of a binary tree be determined?
29. How can a stack be used to check whether an input string is a palindrome?
30. Why is the binary search so efficient for large groups of numbers?

Section 8.9

31. What happens if memory allocated with `malloc()` is never returned with `free()`?
32. Why does it matter what type of data structure is used in a program?

Section 8.10

33. What changes are required to add a STORE instruction to MiniMicro (Program 8.32)? The execution of STORE 100 causes the accumulator to be written to location 100.
34. What happens if STOP is used inside a subroutine instead of RET?
35. How could a bank of eight general purpose registers be added to MiniMicro? How would the instruction formats change?

Program Design

When writing the following C programs, utilize the new data structures discussed in this chapter. The use of pointers and recursion is also suggested, for those applications that make best use of them.

36. Write a C program that uses a structure to store four different resistor values for a series circuit. Compute the total resistance of the circuit by adding the individual resistor values together.
37. Repeat problem 36 for a parallel circuit structure.
38. Write a C program that allows the user to enter the following commands: HEAT, COOL, and SET. The program must maintain a temperature variable whose initial value is 60 degrees. HEAT adds 5 degrees and COOL subtracts 5 degrees from the temperature. SET allows the user to instantly change the temperature.
39. Write a C program that displays all combinations of the `enum` types RED, GREEN, and BLUE.
40. Develop a C program that uses a structure that will allow the program user to input the following data concerning a business client: name, address, phone number, and credit rating (good or bad).
41. Create a C program that demonstrates an arrayed structure that contains an arrayed structure element.
42. Write a C program that accepts ten floats from the user, stores them in a linked-list, and then searches the list for the largest number and displays it.

43. Write a C program that inserts a new integer into its correct position in an ordered linked-list. Use the following list in your program:

```
6  13  29  32  45
```

Test your program by inserting the following numbers: 3, 17, 40, and 50, one at a time.

44. Write a C program that deletes a node from the middle of a linked-list.
45. Write a C program that uses a stack to determine if a user string is a palindrome.
46. Write a C program that uses an integer stack to evaluate the following post-order expression:

```
5 4 3 + 7 * -
```

47. Write a C program that creates a binary expression tree when given an in-order expression as input. For example, (A + B) * C should result in the creation of the tree shown in Figure 8.15 from Section 8.7.
48. Design a C program that uses a structure to keep the following information on the status of a design project: ID Number, project name, client name, due date, and project completed (yes or no).
49. Develop a C program that uses a structure for the following information about different patients for a private practice: name, address, date of birth, sex, dates of visitation, amount owed, and medical problem.
50. Create a C program using a structure that will keep track of three different production schedules each of which contains the following information: production ID number, manufactured item, and customer name. The program is also to contain a structure on up to five employees within the production schedule structure. This should contain the following information: employee ID number, name, title, and hourly wages.

9 Disk I/O

Objectives

This chapter provides you the opportunity to learn:

1. How to create a disk file.
2. How to write data to a disk file.
3. How to read data from a disk file.
4. How to perform disk I/O with records.
5. How to supply information via DOS's command line.
6. The difference between sequential and random access files.

Key Terms

DOS
File Pointer
Character Stream
Text File
Binary Mode
Text Mode
Binary File
Record
Command Line
Stream
Buffer
Buffered Stream
Flush
Standard File
File Redirection
Piping
Filtering
Standard I/O
System I/O
Sequential Access File
Random Access File
File Pointer
Command Line Argument

Outline

Introduction

In this chapter we will add a whole new dimension to our programming applications by utilizing the personal computer's disk. The mechanics of file creation and access will be examined, along with methods for structuring disk data. We will also see how the parameters entered on DOS's command line can be used as input to our C programs.

9.1 Disk Input and Output

Discussion

The C programs you have developed up to this point did not allow the program user to save any entered data before the computer is turned off. This is a severe restriction for many different kinds of technology programs. In this section, you will discover how to save data to the disk and how to get it back. Doing this will allow you to create C programs that will let the program user store data entered into the program as well as get the data back when the program is used again.

Creating a Disk File with C

In order to store user input data on the computer disk, a file must be created on the disk for storing the data. This can be done in many different ways. One way is to have the C program do this automatically for the program user. In order to do this, you must observe the rules for the Disk Operating System (**DOS**) used by your system. The DOS used by the IBM PC and compatibles contains specific rules concerning the naming of a disk file. What follows is a summary of legal DOS file names.

A DOS Review

A legal DOS file name has the form

```
[Drive:]FileName[.EXT]
```

Where

Drive = The optional name of the drive that contains the disk which you want to access.
FileName = The name of the file (up to eight characters).
.EXT = An optional extension to the file name (up to three characters).

Both the `Drive:` and `.EXT` are optional. If `Drive:` is not specified, then the active drive will be used. The colon (:) following the drive letter is necessary. For example,

```
B:MYFILE.DAT
```

is the name of a disk file on drive `B:` with the file name `MYFILE` and the extension `.DAT`. It makes no difference if you enter a file name with uppercase or lowercase letters; DOS will always show all disk files in uppercase.

A File Creation Program

Program 9.1 shows a C program that creates a disk file called `MYFILE.DAT` on the active drive. The program does not store any information in the file. However, when the program is executed, a new file `MYFILE.DAT` will now exist on the disk in the active drive.

Program 9.1

```
#include <stdio.h>

main()
{
        FILE *file_pointer;      /* This is the file pointer. */

        /* Create a file called MYFILE.DAT and assign its address
           to the file_pointer: */

        file_pointer = fopen("MYFILE.DAT","w");

        /* Close the created file. */

        fclose(file_pointer);
}
```

Program Analysis

Program 9.1 starts by creating a **file pointer** that is of a data type called `FILE`:

```
FILE *file_pointer;    /* This is the file pointer.  */
```

This is a data type defined in most versions of C.

Next, the file pointer is set equal to the built-in C `fopen()` function. In order to use this function, the legal DOS name of the file and a file mode string are required in the argument (`w` means the file will be opened for writing into—it is a string because of the double quotes).

```
file_pointer = fopen("MYFILE.DAT", "w");
```

This will actually create a file named `MYFILE.DAT` on the active drive and open it for writing data into it.

The next statement is necessary when you are finished with the open files. If you don't use it, you may damage any data in your open files.

```
fclose(file_pointer);
```

This is the built-in C `fclose()`. Its argument is the file pointer.

Putting Data into the Created File

Program 9.1 created a disk file using a C program. However, nothing was put into the file. The program illustrates the minimum requirements for creating a disk file, but the file is empty. Program 9.2 illustrates the creation and opening of the same file, but this time the character C is put into the file.

Program 9.2

```
#include <stdio.h>

main()
{
        FILE *file_pointer;      /* This is the file pointer. */

        /* Create a file called MYFILE.DAT and assign its address
           to the file_pointer: */

        file_pointer = fopen("MYFILE.DAT","w");

        /* Put a letter into the opened file: */

        putc('C', file_pointer);

        /* Close the created file. */

        fclose(file_pointer);
}
```

Program Analysis

Program 9.2 does the same thing as Program 9.1 with one exception: it puts the letter C into the opened file. This is done by the built-in function `putc()`. This function puts a single character into an opened file. Its argument requires the character and the name of the file pointer.

You now have a file called `MYFILE.DAT` that contains the character C.

Reading Data from a Disk File

Program 9.2 showed you how to put a single character into a file created by a C program. What you now need is a method of allowing the C program to get data from an existing disk file. Program 9.3 shows the basic requirements for doing just this.

Program 9.3

```
#include <stdio.h>

main()
{
        FILE *file_pointer;      /* This is the file pointer.            */
```

```
    char file_character;    /* Character to be read from the file. */

    /* Open the existing file called MYFILE.DAT and assign its address
       to the file_pointer: */

    file_pointer = fopen("MYFILE.DAT","r");

    /* Get the first letter from the opened file: */

    file_character = getc(file_pointer);

    /* Display the character on the monitor. */

    printf("The character is %c\n",file_character);

    /* Close the opened file. */

    fclose(file_pointer);
}
```

For the file created in Program 9.2, Program 9.3, when executed, will display

```
The character is C
```

Program Analysis

Program 9.3, for reading data from an existing file, is very similar to Program 9.2, for creating and writing data into a file. As before, a pointer of type `FILE` is declared:

```
FILE *file_pointer;     /* This is the file pointer.          */
```

This is the same requirement as for creating and inputting data to a disk file. For this program, a variable to hold the character to be received from the file is declared:

```
char file_character;    /* Character to be read from the file. */
```

The file is opened using the same built-in C function `fopen()`. Again, its argument contains the file name and a different file operation command. This time `r` is used to mean that the file is being opened for reading.

```
file_pointer = fopen("MYFILE.DAT","r");
```

The file pointer is set from this function as before.

Next, the file character variable is loaded from the built-in C function `getc()` where the argument is the name of the file pointer.

```
file_character = getc(file_pointer);
```

Now the character read from the file is displayed on the monitor:

```
printf("The character is %c\n",file_character);
```

Last is the important step of closing the opened file when you are finished with it:

```
fclose(file_pointer);
```

Saving a Character String

The previous programs save and retrieve only a single character from a file on disk. However, they illustrate the fundamental requirements for a disk file with C.

Program 9.4 illustrates a method of saving a character string to a disk file. In this case, the program user may input any string of characters from the keyboard and have them automatically saved to a file called `MYFILE.DAT`. The program will continue getting characters from the program user until a carriage return is entered.

Program 9.4

```
#include <stdio.h>

main()
{
        FILE *file_pointer;      /* This is the file pointer.            */
        char file_character;     /* Character to be written to the file. */

        /* Create a file called MYFILE.DAT and assign its address
           to the file_pointer: */

        file_pointer = fopen("MYFILE.DAT","w");

        /* Put a stream of data into the opened file: */

        while((file_character = getche()) != '\r')
               file_character = putc(file_character, file_pointer);

        /* Close the created file. */

        fclose(file_pointer);
}
```

Note that like the other file programs, Program 9.4 uses the built-in C function `fopen()`, which includes the `w` character meaning that the file is being opened for writing into. Again, when all file activity is completed, the file is closed with the built-in C function `fclose()`.

The main difference between this program and the others is the line

```
file_character = putc(file_character, file_pointer);
```

The built-in C function `putc` inputs one character at a time into the file pointed to by the file pointer:

```
putc('C', file_ptr);
```

where C is the character and `file_ptr` is the file pointer.

Note that this is included in a C `while` loop:

```
while((file_character = getche()) != '\r')
       file_character = putc(file_character, file_pointer);
```

Here the loop will continue as long as the program user does not press the `Return` key (`\r`).

Reading a Character Stream

The concept of a **character stream** is a sequence of bytes of data being sent from one place to another (such as from computer memory to the disk).

To read a character stream from a disk file, you use the same tactics. The difference is that your program will continue to read a stream of characters until it encounters an end-of-file marker (EOF). This is automatically placed at the end of a disk file when the file is created. When this marker is read by the program, it indicates that the end of the file has been reached and there is nothing further to read. Program 9.5 illustrates.

Program 9.5

```
#include <stdio.h>

main()
{
        FILE *file_pointer;      /* This is the file pointer.            */
        char file_character;     /* Character to be read from the file. */

        /* Open an existing file called MYFILE.DAT and assign its address
           to the file_pointer: */

        file_pointer = fopen("MYFILE.DAT","r");

        /* Get a stream of characters from the opened file
           and display them on the screen.                  */

        while((file_character = getc(file_pointer)) != EOF)
               printf("%c",file_character);

        /* Close the opened file. */

        fclose(file_pointer);
}
```

Program 9.5 is very similar to the other file programs. The `fopen()` function is used with the `r` character directive, meaning that the file is being opened for reading. The main difference is in the C `while` loop:

```
while((file_character = getc(file_pointer)) != EOF)
       printf("%c",file_character);
```

Here, the C `while` loop continues until an EOF marker is reached, indicating the end of the file. While the loop is active, the characters received from the disk file are directed to the monitor screen for display.

Conclusion

This section introduced you to the concept of creating a disk file from a C program. Here you also saw how to save and retrieve characters from the created disk file. These programs did some very simple tasks, but they introduced a very powerful feature of current C systems. In the next section you will be presented a more detailed analysis of disk file creation using C. Check your understanding of this section by trying the following section review.

9.1 Section Review

1. Describe a legal DOS file name.
2. Explain what is meant by a file pointer in C. Give an example.
3. Name the built-in C function for opening a disk file.
4. Demonstrate how the file pointer is assigned to the DOS file name.
5. State the difference between the C file opening commands for writing to and reading from a file.
6. What must you always do when you are finished with an open file? Give an example.

9.2 More Disk I/O

Discussion

In the previous section you were introduced to file input/output (I/O) within a C program. This section presents a more detailed description of file I/O. Here you will discover more methods of developing disk I/O. These methods will allow you to store and retrieve complex data types such as arrays and structures.

The Possibilities

Table 9.1 shows the possible conditions you can encounter when working with disk data.

As you can see from Table 9.1, there are four possibilities when working with disk information. All of these possibilities will be presented in this section.

Besides the considerations presented in Table 9.1, in the C language there are four different ways of reading and writing data. These are listed in Table 9.2.

You have already used the first two methods in Table 9.2. These allow you to perform simple I/O where you can store and retrieve characters or strings. The programs in

Table 9.1 Disk File Conditions

Condition	Meaning
1	The disk file does not exist, and you want to create it on the disk and add some information.
2	The disk file already exists, and you want to get information from it.
3	The disk file already exists, and you want to add more information into it while preserving the old information that was already there.
4	The disk file already exists, and you want to get rid of all of the old information and add new information.

Table 9.2 Different Methods of Reading and Writing Data

Method	Comments
One character at a time	Inputs and outputs one character to the disk at a time.
Read and write data and strings	Inputs and outputs a string of characters to the disk.
Mixed mode	Used for I/O of characters, strings, floating points, and integers.
Structure or block method	Used for I/O of array elements and structures.

this section will concentrate on the last two methods: mixed data types and structure or blocks of data.

File Format

You may have noted in Section 9.1 that a C program which performs disk I/O has a specific structure. This is illustrated in Figure 9.1.

Referring to Figure 9.1, you can see that a C program designed for disk I/O uses the type `FILE` with a corresponding file pointer. The type `FILE` is a predefined structure declared in the header file `<stdio.h>`. This file must be included in all of your C programs designed for disk I/O. This predefined `FILE` structure helps establish the necessary link between your program and the disk operating system.

The next step in a disk I/O program is to open the file. As shown in Figure 9.1, the built-in C function `fopen("DOSNAME","command");` is used. As you saw before, `"DOSNAME"` is any legal DOS file name which may include an extension. (It should be noted that this may have also included an MS-DOS pathname such as `\MYFILES\MYFILE.DAT`. The purpose of the one-letter string is to identify the type of file that will be opened.)

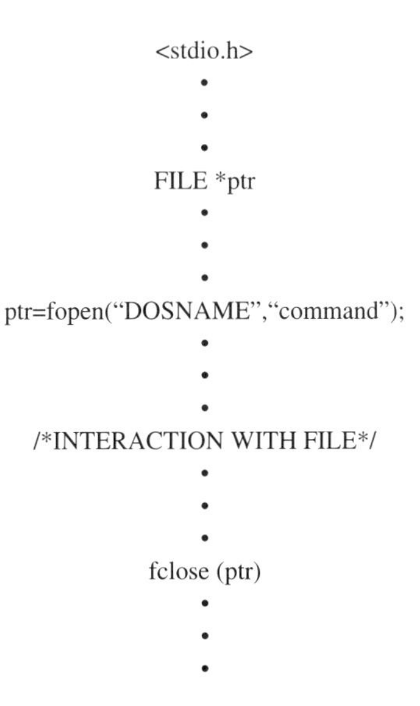

Figure 9.1 Standard C Structure for Disk I/O

Table 9.3 C File Type Commands

String Command	Meaning
"a"	Open for appending. New data is added to existing file data or a new file is created.
"r"	Open for reading. File must already exist.
"w"	Open for writing. Contents written over or new file created.
"a+"	Open for reading and appending. If file does not exist, it will be created.
"r+"	Open for both reading and writing—file must exist.
"w+"	Open for both reading and writing. Contents written over.

You have already used the `"r"` and `"w"` directives in the previous section. Note from Table 9.3 how these file directives relate to the different ways you can treat a disk file.

Example File Program

Program 9.6 demonstrates several important points about C disk files. This program allows the user one of the following options for creating and reading from a simple message file:

1. Create a new file by a user-supplied file name.
2. Create a new file by a user-supplied file name and place a message in it.
3. Read a message from an existing file.
4. Append a message to an existing file.
5. The user is told if a given file for reading or appending does not exist.

Program 9.6

```
#include <stdio.h>

main()
{
        char selection[2];       /* User file selection.          */
        char file_name[13];      /* Name of the disk file.        */
        char user_choice[2];     /* User choice for file activity. */
        int selection_value;     /* User selection number.        */
        int file_character;      /* File character to be saved.   */
        FILE *file_pointer;      /* File pointer.                 */

        /* Display user options. */

        printf("Select one of the following by number:\n");
        printf("1] Create a new file.    2] Write over an existing file.\n");
        printf("3] Add new data to an existing file.\n");
        printf("4] Get data from an existing file.\n");

        /* Get and act on user input. */

        do
```

```
{
        printf("Your selection => ");
        gets(user_choice);
        selection_value = atoi(user_choice);

        switch(selection_value)
        {

                case 1  :
                case 2  : strcpy(selection, "w");
                          break;
                case 3  : strcpy(selection, "a");
                          break;
                case 4  : strcpy(selection, "r");
                          break;
                default : {
                                printf("That was not one of the "
                                       "choices./n");
                                selection_value = 0;
                          }
        }
} while(selection_value == 0);

/* Get the file from the user. */

printf("Enter the name of the file => ");
gets(file_name);

/* Open the file for action. */

if((file_pointer = fopen(file_name, selection)) == NULL)
{
        printf("Cannot open file %s!", file_name);
        exit(-1);
}

/* Write to or read from the file. */

switch(selection_value)
{
        case 1 : break;
        case 2 :
        case 3 : {
                        printf("Enter string to be saved:\n");
                        while((file_character = getche()) != '\r')
                                file_character = putc(file_character,
                                                      file_pointer);
                 }
                 break;
        case 4 : {
                        while((file_character = getc(file_pointer)) !=
                                EOF)
```

```
                                        printf("%c",file_character);
                        }
                        break;
        }

        /* Close the opened file. */

        fclose(file_pointer);
}
```

Note: The function getch gets a character from the keyboard and does not echo it to the screen. The function getche does the same thing except it does echo the character to the screen.

Execution of Program 9.6 produces

```
Select one of the following by number:
1] Create a new file.   2] Write over an existing file.
3] Add new data to an existing file.
4] Get data from an existing file.
Your selection => 2
Enter the name of the file => MYFILE.01
Enter string to be saved: Saved by a C program.
```

The message above can be retrieved at a later date, added to, or written over. Note that the program will tell the user if the desired file does not exist. With this program, any new file must first be created by the user.

Program Analysis

The file pointer is declared in the declaration part of the program:

```
FILE *file_pointer;        /* File pointer. */
```

User options are then displayed using the printf() function:

```
printf("Select one of the following by number:\n");
printf("1] Create a new file.   2] Write over an existing file.\n");
printf("3] Add new data to an existing file.\n");
printf("4] Get data from an existing file.\n");
```

The user prompt for input becomes part of a C do loop. The reason for this is to repeat the prompt if the user input is not valid:

```
do
{
        printf("Your selection => ");
```

Note that next a C gets() function is used for receiving the user input. This is done to avoid the difficulties of using the scanf() function. The user input is then converted to a type int using the C atoi() function that converts a string to an integer.

```
gets(user_choice);
selection_value = atoi(user_choice);
```

A C `switch()` function is then used to take the converted integer value and select one of five possible conditions. Note that the first two conditions will both select the `"w"` file option for writing a file:

```
switch (selection_value)
{
        case 1   :
        case 2   : strcpy(selection, "w");
                   break;
        case 3   : strcpy(selection, "a");
                   break;
        case 4   : strcpy(selection, "r");
                   break;
        default  : {
                   printf("That was not one of the choices.\n");
                   selection_value = 0;
                   }
```

The C `default` condition will activate if one of the first four cases is not selected. Here the user is told that none of the possible selections was selected. Also note that in the `default` case, the variable `selection_value` is set to 0. This is done to cause a repeat of the C `do` loop and allow the user to try the selection again.

```
while(selection_value == 0);
```

Assuming the user makes a correct selection, the program will now ask the user for the name of the file. Note that again the C `gets()` function is used:

```
printf("Enter the name of the file => ");
gets(file_name);
```

The next part of the program makes use of the fact that if, for any reason, a file cannot be opened, the NULL value (`0`) will be returned to the file pointer.

```
if((file_pointer = fopen(file_name, selection)) == NULL)
```

In the statement above, the C `fopen()` function is being used to open the file. The variable `file_name` contains the name of the file entered by the user, and the variable `selection` will contain one of the file directives (`r`, `w`, or `a`). The file pointer `file_pointer` will be used to replace all references to the opened file. If the value of the pointer is `0`, the following compound statement will be executed:

```
{
        printf("Cannot open file %s!",file_name);
        exit(-1);
}
```

Otherwise, this statement is skipped by the program.

Next, a C `switch` is used to determine what action is to be taken with the now opened file. Note that in `case 1` the user opts to simply create a new file, so there is nothing to input or output:

```
switch(selection_value)
{
        case 1 : break;
```

In case 2 or 3, the program user has opted to save a string to the file (by writing over an existing string or appending a new string). Here, the C while loop is used to input the string using the C putc function, which writes a character file_character to the file pointed to by file_pointer. The while loop stays active until the program user presses the -RETURN/ENTER- key producing a carriage return (\r).

```
case 2 :
case 3 : {
        printf("Enter string to be saved:\n");
        while((file_character = getche()) != '\r')
                file_character = putc(file_character, file_pointer);
    }
    break;
```

In the fourth case, the user chooses to read a string from a selected file. Here, the getc() function is used to get a character from the file pointed to by file_pointer. This process is in a while loop that will continue until the EOF marker is reached. The printf() function is used to direct the resulting characters to the monitor screen.

```
case 4 : {
        while((file_character = getc(file_pointer)) != EOF)
                printf("%c",file_character);
    }
    break;
```

The program then terminates with the necessary file closing:

```
fclose(file_pointer);
```

Mixed File Data

The previous file programs were limited to working with strings. Program 9.7 illustrates a method of working with files that allows you to input numerical as well as string data. The program illustrates a process of entering the mixed data from the parts structure program of the previous section. This is accomplished with a new function called fprintf.

Program 9.7

```
#include <stdio.h>

main()
{
        char part_name[15];     /* Type of part.          */
        int quantity;           /* Number of parts left. */
        float cost_each;        /* Cost of each part.     */
        FILE *file_pointer;     /* File pointer.          */

        /* Open a file for writing. */

        file_pointer = fopen("B:PARTS.DAT","w");

        /* Get data from program user. */
```

```
        printf("Enter part type, quantity, cost each, separated by spaces:\n");
        printf("Press -RETURN- to terminate input.\n");
        scanf("%s %d %f", part_name, &quantity, &cost_each);
        fprintf(file_pointer, "%s %d %f", part_name, quantity, cost_each);

        /* Close the opened file. */

        fclose(file_pointer);
}
```

Execution of Program 9.7 produces:

```
Enter part type, quantity, cost each, separated by spaces:
Press -RETURN- to terminate input.
Resistor, 12, 0.05
```

Assuming that the program user carefully followed instructions, the `scanf()` function would receive the required data and the `fprintf()` function would store it in the opened file. The `fprintf()` function has the form:

```
fprintf(FILE *stream, const char *format, argument).
```

This function accepts a series of arguments, and outputs the formatted stream as specified by the format string. This is similar to `printf()`.

Program Analysis

Program 9.7 has several weaknesses. It doesn't protect user input (the `scanf()` function is used). It also doesn't let the user know if there was a problem opening the file. However, it does illustrate a simple C program that enters mixed data types into a file. The key to the program is the `fprintf()` function that allows the user to format the input data to the disk in many different ways—in this case as a string, then as an integer, and finally as a float.

Retrieving Mixed Data

Program 9.8 demonstrates how the mixed data stored on the disk may be retrieved and sent to the monitor screen.

The key to this program is the C `fscanf()` function. This function is similar to the `scanf()` function except that a pointer to `FILE` is used as its first argument.

Program 9.8

```
#include <stdio.h>

main()
{
        char part_name[15];      /* Type of part.          */
        int quantity;            /* Number of parts left. */
        float cost_each;         /* Cost of each part.    */
        FILE *file_pointer;      /* File pointer.         */
```

```
    /* Open a file for reading. */

    file_pointer = fopen("B:PARTS.DAT","r");

    /* Get data from disk file. */

    while(fscanf(file_pointer, "%s %d %f", part_name, &quantity,
                &cost_each) != EOF)
        printf("%s %d %f\n", part_name, quantity, cost_each);

    /* Close the opened file. */

    fclose(file_pointer);
}
```

Assuming that the file contained the data entered by the previous program, execution of Program 9.8 would yield:

```
Resistor 12 0.050000
```

Text vs. Binary Files

All of the file I/O you have been using up to this point involved **text files**. Figure 9.2 illustrates how a text file is stored on a disk.

As shown in Figure 9.2, numbers stored on the disk in string format are stored as strings instead of as numerical values. Because of this, the storage of numerical data as strings does not use disk space efficiently. One way to increase the disk storage efficiency is to use what is called the **binary mode** of disk I/O rather than the **text mode**. The **binary**

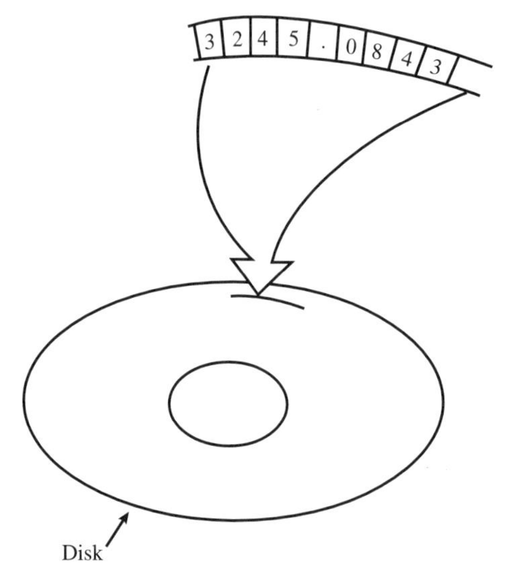

Figure 9.2 Storing Numbers in Text Format

file does not store numbers as a string of characters (as is done with text files). Instead, they are stored as they are in memory—two bytes for an integer, four for floating point, and so on for the rest. The only restriction is that a file stored in binary mode must also be retrieved in binary mode, or what you get will not make sense. All that you need to do is add the letter b after the file directive. Thus `fopen("MYFILE.01","wb")` means to open (or create) a file called `MYFILE.01` for writing in binary format. Likewise, `fopen("MYFILE.01","rb")` means to open the file `MYFILE.01` for reading in binary format. As you may have guessed, `fopen("MYFILE.01","ab")` means to open the file `MYFILE.01` for appending in binary format.

The `"ws"` directive could have been used to indicate that a file is to be opened for writing in string format. For the previous programs, the `s` could have been added to all of the file directives, resulting in a string format file. However, this is redundant, since, by default, files are opened for string format unless the directive is modified with the `b` for binary format.

Record Input

Program 9.9 shows an example of saving C data structures to a file. This is a powerful and important method of saving complex data to disk files. Observe that this program is similar to that of the mixed data type. However, the important difference is that it stores a data structure (called a **record**) to the file. This is done using a new function called `fwrite` which writes a complete record of data to the file. Data files are typically composed of equally-sized records of data.

Program 9.9

```
#include <stdio.h>

typedef struct
{
        char part_name[15];     /* Type of part.          */
        int quantity;           /* Number of parts left. */
        float cost_each;        /* Cost of each part.    */
} parts_structure;

main()
{
        parts_structure parts_data;     /* Parts structure variable. */
        FILE *file_pointer;             /* File pointer.             */

        /* Open a file for writing. */

        file_pointer = fopen("B:PARTS.DAT","wb");

        /* Get data from program user. */

        do
        {
                printf("\nName of part => ");
```

```
            gets(parts_data.part_name);
            printf("Number of parts => ");
            scanf("%d",&parts_data.quantity);
            printf("Cost per part => ");
            scanf("%f",&parts_data.cost_each);

            /* Write structure to opened file. */

            fwrite(&parts_data, sizeof(parts_data), 1, file_pointer);

            /* Prompt user for more input. */

            printf("Add more parts (y/n)? => ");
        } while (getche() == 'y');

        /* Close the opened file. */

        fclose(file_pointer);
}
```

Assuming that the user enters the given values, execution of Program 9.9 will yield

```
Name of part => Resistor
Number of parts => 12
Cost per part => 0.5
```

This data will now be saved to the disk as a block of information in a binary format under the file name of `PARTS.DAT` on drive B:.

Program Analysis

Program 9.9 first defines a type called `parts_structure` as a structure consisting of three members. This is the structure for the parts inventory presented in earlier sections of this chapter.

```
typedef struct
{
        char part_name[15];          /* Type of part.           */
        int quantity;                /* Number of parts left.   */
        float cost_each;             /* Cost of each part.      */
} parts_structure;
```

The program then declares the required file pointer as type `FILE` and the variable `parts_data` as the type `parts_structure`. For user input, a character string is also declared.

```
main()
{
        parts_structure parts_data;    /* Parts structure variable.   */
        FILE *file_pointer;            /* File pointer.               */
```

A file on drive B: called `PARTS.DAT` is opened for writing in the binary mode (if the file does not exist, it will be created).

```
/* Open a file for writing. */

file_pointer = fopen("B:PARTS.DAT","wb");
```

The program user is now prompted to enter data. Note the use of the member of (.) operator. This is done inside a C `do` loop so that the user may enter more than one structure of data:

```
/* Get data from program user. */

do
{
        printf("\nName of part => ");
        gets(parts_data.part_name);
        printf("Number of parts => ");
        scanf("%d",&parts_data.quantity);
        printf("Cost per part => ");
        scanf("%f",&parts_data.cost_each);
```

Data is then entered into the opened file using the C `fwrite()` function. This function has four arguments:

```
fwrite(buffer, size, n, pointer)
```

The `buffer` contains the address of the data that is to be written into the file. The `size` is the size of the buffer in bytes. The `n` is the number of items of this size, and `pointer` points to the file to be written to.

```
fwrite(&parts_data, sizeof(parts_data), 1, file_pointer);
```

Note that the C `sizeof()` function is used to automatically determine the size, in bytes, of the data to go to the file.

The user is prompted for more data. The input loop will end if the character `y` is not entered:

```
        printf("Add more parts (y/n)? => ");
} while(getche() == 'y');
```

The file is then closed.

```
/* Close the opened file. */

fclose(file_pointer);
```

Record Output

Program 9.10 illustrates the retrieval of the block of data entered by Program 9.9. Note that this program uses the same type definition of the parts structure as did Program 9.9.

Program 9.10

```
#include <stdio.h>

typedef struct
{
        char part_name[15];      /* Type of part.         */
        int quantity;            /* Number of parts left. */
        float cost_each;         /* Cost of each part.    */
} parts_structure;

main()
{
        parts_structure parts_data;     /* Parts structure variable. */
        FILE *file_pointer;             /* File pointer.             */

        /* Open a file for reading. */

        file_pointer = fopen("B:PARTS.DAT","rb");

        /* Get data from file and display. */

        while(fread(&parts_data, sizeof(parts_data), 1, file_pointer) == 1)
        {
                printf("\nPart name => %s\n",parts_data.part_name);
                printf("Number of parts => %d\n",parts_data.quantity);
                printf("Cost of part => %f\n",parts_data.cost_each);
        }

        /* Close the opened file. */

        fclose(file_pointer);
}
```

When executed, Program 9.10 will look for the file `PARTS.DAT` on drive B: and open it in the binary mode for reading (the file must already exist). The data is read using the C `fread()` function, which has the form

```
fread(buffer, size, n, pointer)
```

This is very similar to the `fwrite()` function. Again, `buffer` is a pointer to the data, `size` is the size of the data in bytes, `n` is the number of items of this size, and `pointer` points to the file to be written to.

The `fread()` function will return the number of items actually written to the disk. Since in the input program (Program 9.9) the value of `n` is 1, if this value differs in the output program, the file will no longer be read. This is accomplished by having the file read in a C `while` loop as long as this function is equal to 1.

Conclusion

This section demonstrated some key points concerning disk file I/O. Here you saw a new method of storing and retrieving different kinds of data to disk files. You also saw how to do this with blocks of data such as the structure. In the next section you will see some of the technical aspects of I/O. The case study section will present the development of a program that uses more sophisticated file techniques, such as record arrays with random access to each record.

Check your understanding of this section by trying the following section review.

9.2 Section Review

1. List the four disk file I/O conditions that are possible.
2. State the different methods of reading and writing data with the C language.
3. Give the three basic C file type commands. State their meanings.
4. What is the purpose of the C `fscanf()` function?
5. State the meaning of a block read or write function.

9.3 Streams, File Pointers, and Command-Line Arguments

Discussion

This section presents some of the technical details concerning I/O using C. Here you will also be introduced to some details concerning random access files. This section will help prepare you for the case study in Section 9.6.

Inside I/O

You can think of any sequence of bytes of data as a stream of data. This stream of data can be thought of as information being sent or received in a serial form. The stream concept is illustrated in Figure 9.3.

You have actually worked with two kinds of **streams**—a text stream and a binary stream (text files and binary files). A text stream consists of lines of characters. Each line of characters ends with a terminating newline character (`\n`). The important point about text streams (such as those stored to a disk file) is that they may not all be stored in the same way your C program stores them. This is the case with your MS-DOS, where, unlike standard C, the end of a line is terminated with both a newline character and a carriage return character (`\n\r`). This is not the case with a binary stream. The only requirement here is that you know the size of what was stored in the file.

In standard C, a file represents a source of data that is stored in some external media. These different types of media are illustrated in Figure 9.4.

The Buffered Stream

When a file I/O is performed, an association needs to be made between the stream and the file. This is done by using a section of memory referred to as a **buffer**. Think of a buffer

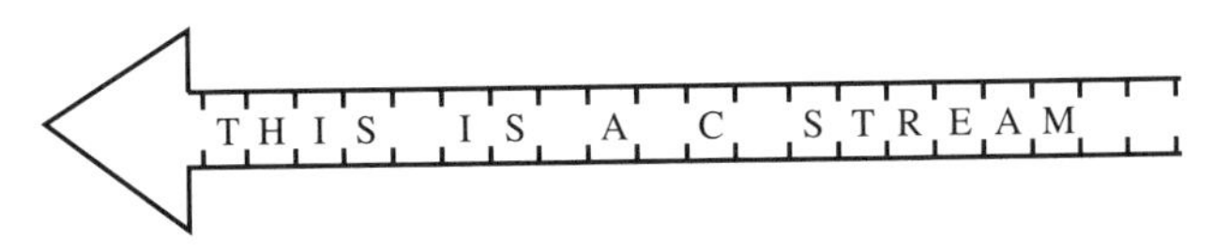

Figure 9.3 Concept of a C Stream

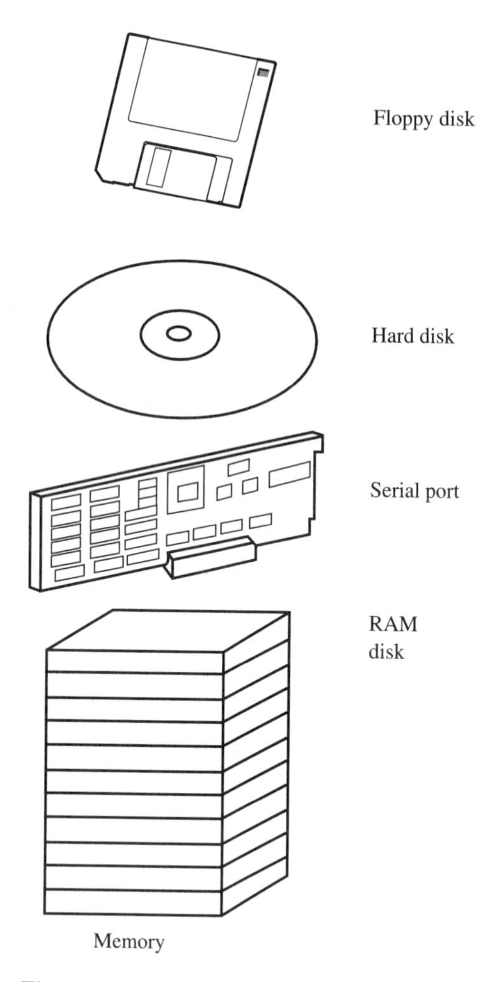

Figure 9.4 Different Types of C File Media

as a reserved section of memory that will hold data. This acts as a temporary storage place in memory. The bytes being read from or written to the file are stored here. In this way, when a file is read from the disk, it is first stored in the buffer as a fixed "chunk" of data. Each of these "chunks" is the same size. What happens is that a process that reads or writes data from or to the disk is actually first communicating with the buffer. Hence the name **buffered stream** for a stream of data that is stored in this buffer. Only when the buffer is **flushed** (data transferred from the buffer to the disk) is data actually stored to the disk from the buffer. Doing this reduces the amount of access time between the disk and the program.

A buffer may not be flushed if the program is abnormally terminated. It's of interest to note that the file pointer you declare and use in your disk file programs is actually a pointer to this buffer. Thus, for stream files, the file pointer replaces all references to the file after it has been opened.

In the standard header file <stdio.h>, the following constants are defined:

```
EOF = 1

NULL = 0

BUFSIZE = 512
```

As you may recall, EOF is the end-of-file marker placed at the end of a disk file. NULL is assigned to a file pointer if an error has occurred, and BUFSIZE is the size assigned to the I/O stream. You can redefine BUFSIZE from its assigned optimum value.

Standard Files

When you run your C program, it automatically opens three **standard files**. These are the standard input, standard output, and standard error files. On your PC, these files represent the keyboard (for standard input) and the monitor (for standard output and error). This idea is illustrated in Figure 9.5.

The reason for having a standard error output is to ensure that the program always has access to your monitor. This is necessary in case you have directed your standard output somewhere else (such as to the disk). Thus, if you have an error during this process, C still has access to your monitor to let you know that an error has taken place.

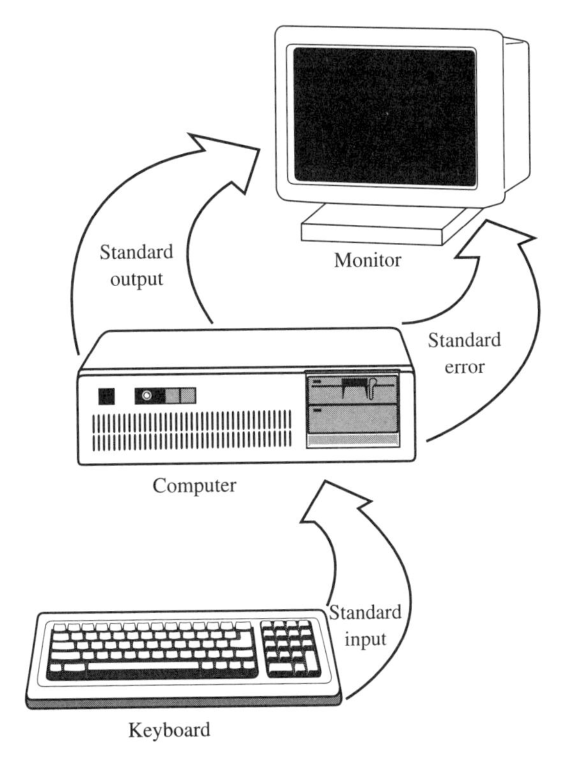

Figure 9.5 Illustration of the Three Standard Files

Redirecting Files

You can use DOS commands to carry out **file redirection** on your disk. As an example, suppose you have the following two disk files:

READFILE.EXE <= A C program that normally gets data from the keyboard (standard input).

DATAFILE <= A character file that contains data from the keyboard—this file could have been produced when a word processor was being used.

If you then enter the following from the DOS prompt (A>):

```
A> READFILE < DATAFILE
```

(the EXE extension is omitted), this will redirect standard input for the C READFILE to get its data from DATAFILE rather than from the keyboard.

The converse can also be done. Suppose that you have a C program called DATA-MAKE.EXE that produces data (say, the results of some calculations) and normally outputs the results to the monitor. Then

```
A> DATAMAKE > NEWDATA
```

will redirect the standard output so that instead of the data from the C program DATA-MAKE.EXE going to the monitor, it goes to the file called NEWDATA.

Piping of Files

Piping of a file simply means to cause two or more separate disk files to act as if they were joined together. The DOS piping command is the |. As an example, MS-DOS contains a utility program that will do an alphabetical sort (called SORT.COM). If you want to see your disk directory displayed alphabetically sorted you would enter

```
A> DIR | SORT
```

(The extension COM is not entered.)

Filtering

Filtering is the process of modifying data in some fashion while it is in the process of being transferred from one file to another. As an example, if you have a file that contains a list of words called WORDFILE.OLD, you can copy it to another file called WORD-FILE.NEW and alphabetize the words in the process. To do this, you need to have the MS-DOS SORT.COM file on your disk. Then do the following:

```
A> SORT < WORDFILE.OLD > WORDFILE.NEW
```

This will cause the information in WORDFILE.OLD to be stored alphabetically in the file called WORDFILE.NEW. What you have done here is filter the data through the SORT program in the process of copying it from one file to another. A helpful aid employed by many programmers is the use of filtering to display an alphabetically sorted disk directory one screen at a time. For this to be done, both of the MS-DOS programs SORT.COM and MORE.COM must be accessible. From the DOS prompt, you would enter

```
A> DIR | SORT | MORE
```

I/O Levels

There are actually two levels of I/O. The one you have been working with is called the **standard I/O**. The other is called the low-level I/O. The advantage of the standard I/O is that it requires less programming detail on your part. Its disadvantage is that you have less control over the details of the I/O process, and it is slower than low-level I/O. For most programming tasks, standard I/O will meet the I/O requirements. The standard I/O C functions actually use the low-level I/O. Low-level I/O is usually referred to as **system I/O** and is the topic of the next discussion.

Random Access Files

Up to this point, the types of files you have worked with have been **sequential access files**. This means that when you work with file data, you get all of the data in one chunk from the disk. In a **random access file**, a particular data item may be accessed while the rest of the file is ignored. Program 9.11 illustrates a random access file that uses the parts inventory program. To understand this program, you must first know what is meant by a file pointer.

Program 9.11

```
#include <stdio.h>

typedef struct
{
        char part_name[15];      /* Type of part.         */
        int quantity;            /* Number of parts left. */
        float cost_each;         /* Cost of each part.    */
} parts_structure;

main()
{
        parts_structure parts_data;      /* Parts structure variable. */
        FILE *file_pointer;              /* File pointer.             */
        int record_number;               /* Number of the record.     */
        long int offset;                 /* Offset of the record.     */

        /* Open the file for reading. */

        if((file_pointer = fopen("B:PARTS.DAT","r")) == NULL) /* Error check */
        {
                printf("Unable to open file B:PARTS.DAT");
                exit(-1);
        }

        /* Get record number from the user. */

        printf("Enter the record number => ");
        scanf("%d",&record_number);

        /* Compute the offset value of the selected record. */
```

```
        offset = record_number * sizeof(parts_data);

        /* Go to the required file. */

        if(fseek(file_pointer, offset, 0) != 0)    /* Error check */
        {
                printf("Pointer moved beyond file boundary.");
                exit();
        }

        /* Read the selected file data. */

        fread(&parts_data, sizeof(parts_data), 1, file_pointer);

        /* Display the file data. */

        printf("\nName of part => %s\n",parts_data.part_name);
        printf("Number of parts => %d\n",parts_data.quantity);
        printf("Cost of each part => %f\n",parts_data.cost_each);

        /* Close the opened file. */

        fclose(file_pointer);
}
```

File Pointers

A **file pointer** is simply a pointer to a particular place in the file. What a file pointer does is to point to the byte in the file where the next access will take place with the file. As an example, every time you access a disk file, the file pointer starts at position 0, the beginning of the file. Every time you write data to the file, the file pointer ends up at the end of the file. When you do an append operation to a file, the file pointer is first set to the end of the file before new data is written to the file. The concept of a file pointer is illustrated in Figure 9.6.

There is a C function called `fseek()` that moves the file pointer. This function will move the file pointer to the position in the file where the next file access will occur. It is used in Program 9.11 in order to access a single record from any place within the file.

Program Analysis

The first part of Program 9.11 is the same as the previous sequential access file I/O. The differences start with the variable declarations:

```
parts_structure parts_data; /* Parts structure variable. */
FILE *file_pointer;         /* File pointer.             */
int record_number;          /* Number of the record.     */
long int offset;            /* Offset of the record.     */
```

Here, two new variables, `record_number` and `offset`, are defined. The variable `record_number` will be the value of the record which the program user wishes to access. The

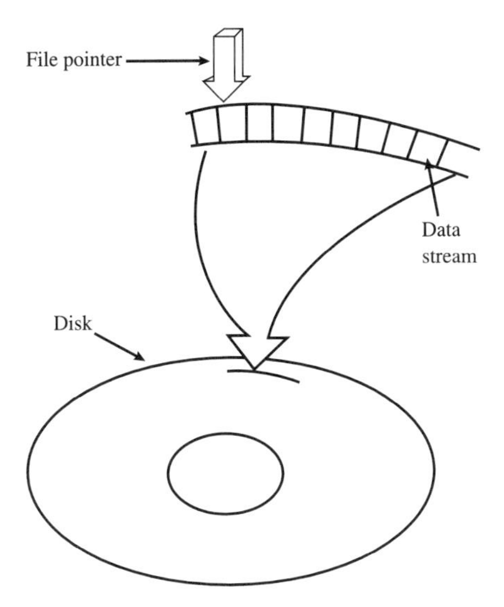

Figure 9.6 Concept of a File Pointer

variable `offset` will be the actual offset for the file pointer. This variable must be of type `long`. The offset is the number of bytes from a particular place in the file to start reading.

The file is first opened for reading with error checking:

```
if((file_pointer = fopen("B:PARTS.DAT","r")) == NULL) /* Error check.*/
{
        printf("Unable to open file B:PARTS.DAT");
        exit(-1);
}
```

This is no different from a sequentially accessed file. The difference starts when the program user is asked for the record number. What this means is that the user must already know how many records are in the disk file and the significance of each record number. For example, the record number could correspond to a parts bin number.

```
printf("Enter the record number => ");
scanf("%d",&record_number);
```

Next, the value of `offset` must be computed. Note that this is done by multiplying the record number by the C `sizeof()` function. The result will give the number of bytes the `offset` must have to pick up the record requested by the user.

```
offset = record_number * sizeof(parts_data);
```

Now the program uses the C `fseek()` function to actually move the file pointer to the position of the record and starts to read the data. An error trapping statement is attached.

```
if(fseek(file_pointer, offset, 0) != 0); /* Error check */
{
        printf("Pointer moved beyond file boundary.");
        exit();
}
```

The C `fseek()` contains three arguments. The first is the file pointer, next is the value of the offset, and last is the mode. There are three values for the C `fseek()` function mode. These are:

0 => Count from the beginning of the file.
1 => Start from the current pointer position.
2 => Start from the end of the file.

These values are defined as

0 => `SEEK_SET`,
1 => `SEEK_CUR`,
2 => `SEEK_END`.

The `fseek()` function returns a nonzero value only if it fails to perform its required operation.

In the case of our program, the pointer is starting its count from the beginning of the file. It should be noted that a positive offset value moves the pointer toward the end of the file whereas a negative offset value moves the pointer toward the beginning of the file.

Next, the statements

```
/* Read the selected file data. */
fread(&parts_data, sizeof(parts_data), 1, file_pointer);
```

read an entire `parts_data` structure from the file.

In the remaining part of the program, the data is displayed and the file is then closed:

```
printf("\nName of part => %s\n",parts_data.part_name);
printf("Number of parts => %d\n",parts_data.quantity);
printf("Cost of each part => %f\n",parts_data.cost_each);

/* Close the opened file. */

fclose(file_pointer);
```

Command Line Arguments—The Basic Idea

One of the powerful features of C is its ability to allow you to use **command line arguments**. What this means is that when you execute a C program from the DOS prompt (by entering the name of the C `.EXE` file), you can enter other information at the same time that will influence the operation of the program. For example, if you have information that must be kept confidential in a program that displays a parts inventory (called `PARTS.EXE`), from the DOS prompt you could execute the program by entering

```
A> PARTS
```

and the program would execute, allowing the user to access information about any part—except its wholesale price. In order to get the information about the wholesale cost of the part, the program user would have to know a code string (say it was `cost_code_1`). This would constitute a command line, and the same identical program would then be executed by

```
A> PARTS cost_code_1
```

Now when the program is executed, the program user can also get the information about the wholesale price of each part.

Developing Command Line Arguments

A command line argument is developed within function `main()`. This is what those parentheses following `main` are for, to place the command line arguments inside them. This is illustrated in Program 9.12.

Program 9.12

```
#include <stdio.h>

main(int argc, char *argv[])    /* Command line arguments.           */
{
        int counter;            /* For counting number of arguments. */

        /* Display the number of arguments entered by the user.      */

        printf("The number of arguments you entered is %d\n",argc);

        /* Show each of the entered arguments. */

        for(counter = 0; counter < argc; counter++)
                printf("Argument %d is %s\n", counter, argv[counter]);
}
```

If Program 9.12 is compiled and saved under the name of `MYSTUFF.EXE`, execution of the program from the DOS prompt with the following commands will produce the following output:

```
A> MYSTUFF ONE TWO THREE
The number of arguments you entered is 4
Argument number 0 is MYSTUFF.EXE
Argument number 1 is ONE
Argument number 2 is TWO
Argument number 3 is THREE
```

The format for a command line argument is shown in Figure 9.7.

Note that the two arguments `argc` and `argv` represent the number of command line arguments and an array of pointers. Observe that the first argument (`argv[0]`) is always the pathname of the program. The identifiers `argc` (for **arg**ument **c**ount) and `argv` (for **arg**ument **v**alue) are traditional. Any legal C identifier may be used instead.

Conclusion

This section presented some of the technical details concerning C I/O. Here you saw some more of the important disk file details. Material presented here will be used in the devel-

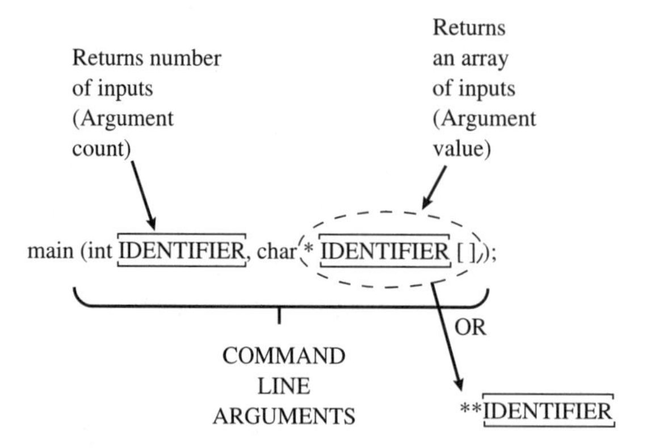

Figure 9.7 Format for Command Line Arguments

opment of the case study in Section 9.6. Check your understanding of this section by trying the following section review.

9.3 Section Review

1. Name the two kinds of data streams used in I/O operations.
2. Describe what is meant by a buffer in I/O.
3. Name the three standard files used by a C program.
4. What are the redirection operators used by DOS? Explain what they mean.
5. State what is meant by a command line argument.

9.4 Example Programs

In this section we will examine three more file applications. Program 9.13 estimates the number of 55-line single-space pages a text file will use if printed. It does so by counting the number of newline `\n` characters found in the file.

Program 9.13

```
#include <stdio.h>

main(int argc, char *argv[])
{
     FILE *fptr;
     char fchar;
     long int lines = 0;

     if (argc > 1)
     {
          fptr = fopen(argv[1],"r");
          if (fptr == NULL)
          {
               printf("Cannot open %s\n",argv[1]);
               exit();
          }
```

```
        }
        else
        {
            printf("No file specified.\n");
            exit();
        }
        fchar = getc(fptr);
        while(fchar != EOF)
        {
            if (fchar == '\n')
                lines++;
            fchar = getc(fptr);
        }
        printf("%s will require aproximately %3d printed pages.\n",
            argv[1], 1 + lines/55);
        fclose(fptr);
}
```

It is usually difficult to estimate the number of printed pages simply by looking at the size of a text file, since we do not know how many characters (on average) occupy a single line. For example, the following DOS command performs a wide directory listing of all files on hard drive D (including all subdirectories):

```
DIR D:\ /W/S > D-FILES
```

The directory listing is written to the file D-FILES. Suppose that drive D contains 10,381 files, resulting in a size of 183,725 bytes for the D-FILES text file. Assuming 80 characters per line and 55 lines per page, we might estimate 42 pages if D-FILES were printed. Running Program 9.13 with D-FILES as its input gives the following result:

```
d-files will require approximately  77 printed pages.
```

This is almost twice our estimate, and shows the usefulness of Program 9.13.

Since the input file is specified in the command line, the user may run Program 9.13 on any text file, as in:

```
P9-13 <filename>
```

A better approach would be to rename P9-13 as something like PAGEST, and copy PAGEST into your \DOS directory. Then, PAGEST can be executed from any directory in the system, via

```
PAGEST <filename>
```

The next program is a companion to Program 9.13, and is used to display a text file one screen at a time. Program 9.14 pauses at the bottom of each screen until the user hits a key, and then moves on to the next page. This is useful when long text files must be viewed, and is similar to the operation of this DOS command:

```
TYPE <FILENAME> | MORE
```

which uses a pipe to feed the output of the TYPE command into the MORE utility.

Program 9.14

```
#include <stdio.h>

main(int argc, char *argv[])
{
     FILE *fptr;
     char fchar, temp;
     int lines = 0;

     if (argc > 1)
     {
          fptr = fopen(argv[1],"r");
          if (fptr == NULL)
          {
               printf("Cannot open %s\n",argv[1]);
               exit();
          }
     }
     else
     {
          printf("No file specified.\n");
          exit();
     }
     fchar = getc(fptr);
     while(fchar != EOF)
     {
          printf("%c",fchar);
          if (fchar == '\n')
          {
               lines++;
               if (lines == 23)
               {
                    lines = 0;
                    printf("\nMore...");
                    temp = getch();

                    /* Erase "More..." from screen */
                    printf("\b\b\b\b\b\b\b");
                    printf("       ");
                    printf("\b\b\b\b\b\b\b");
               }
          }
          fchar = getc(fptr);
     }
     fclose(fptr);
}
```

Like Program 9.13, Program 9.14 also counts newline characters found in the input file. Each time the count reaches 23, the message:

```
More...
```

is displayed on the bottom of the screen. When the user presses a key, the `More...` message is erased and the next 23 lines of the text file are displayed. This continues until the entire file has been shown. Program 9.14 requires the input file to be specified on the command line. If Program 9.14 is useful to you, rename it as something like `PAGER` and place it in the `\DOS` directory.

The last program in this section is used to provide a measure of security to a text file by *scrambling* it. Program 9.15 changes all upper- and lowercase characters in the input file into different upper- and lowercase characters. The input file must have a `.TXT` extension. The output file created by Program 9.15 has the same name as the input file, but uses an `.SCR` extension.

Program 9.15

```
#include <stdio.h>
#include <ctype.h>
#include <string.h>

main(int argc, char *argv[])
{
     FILE *fin, *fout;
     char fsrc[12] = "", fdst[12] = "", fchar;

     if (argc > 1)
     {
          strcat(fsrc,argv[1]);
          strcat(fsrc,".txt");    /* Use ".scr" for unscrambler */
          fin = fopen(fsrc,"r");
          if (fin == NULL)
          {
               printf("Cannot open %s\n",fsrc);
               exit();
          }
     }
     else
     {
          printf("No file specified.\n");
          exit();
     }
     strcat(fdst,argv[1]);
     strcat(fdst,".scr");    /* Use ".txt" for unscrambler */
     fout = fopen(fdst,"w");
     if (fout == NULL)
     {
          printf("Cannot open %s\n",fdst);
          exit();
     }
     fchar = getc(fin);
     while(fchar != EOF)
     {
          if (0 != isupper(fchar))
          {
```

```
                fchar += 13;
                fchar = (0 == isupper(fchar)) ? fchar - 26 : fchar;
            }
            if (0 != islower(fchar))
            {
                fchar += 13;
                fchar = (0 == islower(fchar)) ? fchar - 26 : fchar;
            }
            putc(fchar,fout);
            fchar = getc(fin);
        }
        fclose(fin);
        fclose(fout);
}
```

The technique used to scramble the file characters is called a *Caesar-shift*, and simply shifts the letters of the alphabet 13 positions forward (with wraparound to `A/a` if we go past `Z/z`).

Consider the following input file:

```
KU,
I've been working on the 68000 emulator.
The TRAP instructions work now.
JU
```

The name of this file is `TEST.TXT`. When it is used as the input to Program 9.15, as in:

```
P9-15 TEST
```

the program creates `TEST.SCR`, which looks like this:

```
XH,
V'ir orra jbexvat ba gur 68000 rzhyngbe.
Gur GENC vafgehpgvbaf jbex abj.
WH
```

Clearly, the meaning of the second file is not obvious, and would require a good deal of time to decode.

To unscramble a file, just run Program 9.15 again, with the scrambled file as the input. However, since the input file must always have a `.TXT` extension, you will have to rename the `.SCR` file into a `.TXT` file. A nice addition to this program would be a few statements that examine the input file extension and create the appropriate output file accordingly. For example, if the input file has an `.SCR` extension, a `.TXT` output file is created.

As in the previous two programs, you may wish to rename `P9-15` as something like `SCRAMBLE` and place it in your `\DOS` directory.

9.4 Section Review

1. How does each program check for proper opening of the input file?
2. How is the `More...` message erased from the screen?
3. Why is it necessary to include the test `(argc > 1)` at the beginning of each program?
4. How does Program 9.15 perform wraparound when scrambling a letter?

9.5 Troubleshooting Techniques

Always make sure the operation was successful. This statement really sums up the best course of action to take when using file operations. When you attempt to open a file, make sure it is opened. When you write data to a file, make sure the data was written completely. The values returned by the file functions must be examined to determine if the operation was successful. In general, it is a good programming habit to include code to handle every case that might be encountered during a file operation.

Know your data. It is also important to keep in mind that data is interpreted differently depending on the mode of a file. For example, the byte values `0x0d` (ASCII carriage return) and `0x0a` (ASCII line feed) from a file opened in text mode are interpreted as the single `\r` character. The same two values are treated as individual bytes if the file is opened as a binary file. Some word processors use `Control-Z` (`0x1a`) to denote the end of a file. If this value happens to occur in a data file, it might incorrectly be interpreted as the end-of-file character. To be safe, always check the description of the file function being used, and open the file in the proper mode. This will help keep the data flowing smoothly between the disk and your program.

9.5 Section Review

1. List the return values of `fopen()` and `fseek()`. Why are these values important?
2. Name one difference between a text file and a binary file.

9.6 Case Study: Parts Database

Discussion

The case study for this chapter shows the development of a simple database system. The system illustrates some of the major points of database management, such as saving data to a disk file, retrieving it from the file, and an example security system.

The Problem

Create a C database program that will store data on an assortment of up to 25 parts. Each assortment is to contain the name of the part, quantity of parts, and price of each part. The program will save this data to a disk file on drive B:. The program user may randomly access this file to get information on any one of the part assortments. Only an authorized program user may input data to be stored in the disk file. This prevents the files from being purged by unauthorized users.

First Step—Stating the Problem

Stating the case study in writing yields

- Purpose of the Program: Provide for the creation of an array of 25 structures. Each structure shall contain record number, name of item, quantity of items, and cost of each item. Structure array is to be stored on disk file in drive B: to be modified by authorized users only. Any user may have random access to any single record in the disk file.

- Required Input: User identification (to select personnel authorized to store data in file). User selection to read file, and (if authorized) to enter new data to file. User selection for getting any stored file from the disk. Authorized user input of: Item name, Quantity, and Cost.
- Process on Input: Store arrayed data to a disk file on drive B: for authorized personnel. Random access records from same disk file for any program user.
- Required Output: Two menus, one for users authorized to store new files and one for any other program user. Prompts for program repeats. Display of file contents randomly accessed by user.

Developing the Algorithm

The steps in this database management program are

1. Check user authorization.
2. Explain program to user.
3. Get user options.

   If authorized => Enter new data.
   Any user => View any record in the file.

4. Display output record.

This program will use a file writing function that writes arrays to the disk file and a file reading function that will allow random access to any single array stored in the file.

Program Development

Good programming practice discourages the use of global variables. Good programming practice encourages the use of block structure and top-down design. The programmer's block is developed first.

The complete program is shown in Program 9.16.

Program 9.16

```
#include <stdio.h>
#include <conio.h>
#include <string.h>
#include <math.h>

#define TRUE 1
#define FALSE 0

/**********************************************************************/
/*                        Parts Inventory Program                    */
/*-------------------------------------------------------------------*/
/*                     Developed by A. C. Programmer                 */
/*-------------------------------------------------------------------*/
/*                                                                   */
/**********************************************************************/
/*      This program will inventory up to 25 different assortments    */
```

```
/*  of parts.  Each part assortment contains data about the name of      */
/*  the part, the number of parts, and the price of each part.           */
/*      The program will store and retrieve this information to a        */
/*  disk file called PARTS.DAT.  The program will allow only             */
/*  authorized users to enter data to the disk file.  No special         */
/*  authorization is required to read from the file.                     */
/*      Authorization for writing to the file is obtained by the         */
/*  command line argument "Parts Access 123"                             */
/************************************************************************/
/*                        Type Definitions                               */
/*----------------------------------------------------------------------*/

typedef struct
{
        char part[20];          /* Type of part.                         */
        int quantity;           /* Number of parts left.                 */
        double price;           /* Cost of each part.                    */
        int record_number;      /* Number of the parts record.           */
} parts_record;

/* This type definition defines each part assortment.                    */
/************************************************************************/
/*                        Function Prototypes                            */
/*----------------------------------------------------------------------*/
        void explain_program(void);
/*      This function explains the operation of the program to the       */
/*      program user.                                                    */
/*----------------------------------------------------------------------*/
        char read_menu(int authorization);
/*      This function displays the selection menu for the program        */
/*      user and returns the selected value.                             */
/*----------------------------------------------------------------------*/
        void enter_data(void);
/*      This function allows the program user to enter new data into    */
/*      the file.                                                        */
/*----------------------------------------------------------------------*/
        void read_data(int record_number);
/*      This function retrieves a given record from the disk file.       */
/*----------------------------------------------------------------------*/

main(int argc, char *argv[])
{
        char in_string[10];             /* User input string.  */
        char ch;                        /* Utility character.  */
        int repeat = TRUE;              /* Repeat flag.        */
        int index;                      /* Record number.      */
        int authorization;              /* User authorization. */

        /* Test for user file writing authorization. */

        if(strcmp(argv[1], "Parts Access 123") == 0)
```

```
		authorization = TRUE;
	else
		authorization = FALSE;

	explain_program();			/* Explain program to user. */

	while(repeat)
	{
		ch = read_menu(authorization);	/*  Get user selection. */
		printf("\n\n");

		/* Act on input. */

		switch(ch)
		{
			case 'R' : {
					printf("\n\nEnter record number => ");
					gets(in_string);
					index = atoi(in_string);
					read_data(index);
				   }
				   break;
			case 'X' : repeat = FALSE;	/* Set flag false. */
				   break;
			case 'E' : if (authorization)
				   {
					enter_data();	/* Enter new data. */
					break;
				   }
			default  : printf("That was not a correct selection:\n");
		}
	}
}

/*----------------------------------------------------------------------*/

void explain_program()
{
	printf("\n\n\n");
	printf("   This program will allow you to store and retrieve\n");
	printf("data  on  up  to 25 assortments of parts.  Each part\n");
	printf("assortment  lets  you enter and retrieve the name of\n");
	printf("the part, number of parts left, and price of each.\n\n");
	printf("Entering  new data is  only available  to authorized\n");
	printf("personnel.\n");
}

/*----------------------------------------------------------------------*/

char read_menu(int authorization)
```

```
{
        char ch;         /* User selection. */

        /* Display the menu. */

        printf("\n\nSelect one of the following by letter:\n\n");
        if(authorization)
                printf(" E] Enter new data\n");

        printf(" R] Read data files. \n");
        printf(" X] Exit this program.\n");

        /* Get user selection. */

        printf("Choice: ");
        ch = toupper(getche());
        return(ch);
}

/*------------------------------------------------------------------------*/

void enter_data()
{
        FILE *file_ptr;                 /* File pointer. */
        parts_record parts_array[25];   /* Up to 25 assortments.     */
        char in_string[10];             /* User input string.        */
        int index;                      /* Record number.            */
        int repeat;                     /* Continue data input flag. */
        char ch;                        /* User input response.      */

        repeat = TRUE;                  /* Set continue data flag.   */
        index = 1;                      /* Initialize the index.     */

        while(repeat)
        {
                /* Get name of part. */

                printf("\nPart name => ");
                gets(parts_array[index].part);

                /* Get number of parts left. */

                printf("Number of parts left => ");
                gets(in_string);
                parts_array[index].quantity = atoi(in_string);

                /* Get cost of each part. */

                printf("Cost of each part => ");
                gets(in_string);
```

```
                parts_array[index].price = atof(in_string);

                /* Set record number. */

                parts_array[index].record_number = index;

                /* Ask for more data. */

                printf("Enter more data (Y/N) => ");
                ch = toupper(getche());

                index++;

                if(ch != 'Y')
                        repeat = FALSE;         /* Break out of loop. */
        }

        /* Enter data into disk file. */

        if((file_ptr = fopen("B:PARTS.DAT","wb")) == NULL)
        {
                printf("Can't open parts file.");
                exit();
        }

        /* Enter the file data. */

        fwrite(parts_array, sizeof(parts_array), index, file_ptr);

        /* Close the opened file. */

        fclose(file_ptr);
}

/*---------------------------------------------------------------------------*/

void read_data(int record_number)
{
        FILE *file_ptr;                 /* The file pointer.            */
        parts_record record_file;       /* Structure for reading file. */
        long int offset;                /* Offset for random access.   */

        /* Open the file for reading. */

        if((file_ptr = fopen("B:PARTS.DAT", "rb")) == NULL)
        {
                printf("Can't open parts file.");
                exit();
        }

        /* Compute the offset. */
```

```
        offset = record_number * sizeof(record_file);

        /* Move the file pointer. */

        if(fseek(file_ptr, offset, SEEK_SET) != 0)
        {
                printf("Can't find the file data.");
                exit();
        }

        /* Retrieve the file data. */

        fread(&record_file, sizeof(record_file), 1, file_ptr);

        /* Close the opened file. */

        fclose(file_ptr);

        printf("\nRecord %d\n", record_file.record_number);
        printf("Item => %s\n", record_file.part);
        printf("Quantity => %d\n", record_file.quantity);
        printf("Cost each => %f\n", record_file.price);
}
```

Program Analysis

The arguments in function `main()` get command line input from the user. This will allow authorization for putting data into the disk files. Notice that the command line arguments use double quotes. This is done so that spaces can be placed on the input string.

In the C `switch` function of `main()`, the `case 'E'` placed before the `default` is there so that if `authorization` is FALSE, then the message in the `default` section will be displayed.

Observe the use of the function `toupper()`. This allows the program user to enter either upper- or lowercase in response to menu prompts. The member `price` must be of type `long` because the `atof()` function converts the user input to this type. Information in the disk file is in binary mode. This is because the `fopen()` function for entering and receiving data uses the `"b"` extension in its file directive.

Conclusion

This section presented the development of a simple database system that uses disk file access and file security. The Self-Test at the end of this chapter includes specific questions about this program.

For now, test your understanding of this material with the following section review.

9.6 Section Review

1. How is the user's authorization checked?
2. What happens when the `PARTS.DAT` file cannot be opened?
3. When is `PARTS.DAT` opened and closed?

Interactive Exercises

DIRECTIONS

These exercises are provided to help you review the operation of the various file functions discussed in this chapter.

Exercises

1. Program 9.17 will place a new file on your active disk. After you try the program, look at your disk files. What new file do you see?

Program 9.17

```
#include <stdio.h>

main()
{
        FILE *file_ptr;

        file_ptr = fopen("NEWFILE.DAT","w");

        fclose(file_ptr);
}
```

2. Program 9.18 puts something in the new file created on your disk by Program 9.17. Make sure to execute Program 9.17 before trying Program 9.18.

Program 9.18

```
#include <stdio.h>

main()
{
        FILE *file_ptr;

        file_ptr = fopen("NEWFILE.DAT","w");

        putc('C', file_ptr);

        fclose(file_ptr);
}
```

3. Program 9.19 gets from the file what Program 9.18 put into it. Try it and see if you get what you expected.

Program 9.19

```
#include <stdio.h>

main()
{
        FILE *file_ptr;
        char ch;
```

```
        file_ptr = fopen("NEWFILE.DAT","r");

        ch = getc(file_ptr);
        printf("The character is %c",ch);

        fclose(file_ptr);
}
```

4. What changes are made to `NEWFILE.DAT` when Program 9.20 executes?

Program 9.20

```
#include <stdio.h>

main()
{
     FILE *file_ptr;

     file_ptr = fopen("NEWFILE.DAT","a");
     putc('!',file_ptr);
     fclose(file_ptr);
}
```

5. What does Program 9.21 do? Are any errors generated? Is it necessary for `MORENEW.DAT` to exist before running Program 9.21?

Program 9.21

```
#include <stdio.h>

main()
{
     FILE *file_ptr;

     file_ptr = fopen("MORENEW.DAT","a");
     putc('!',file_ptr);
     fclose(file_ptr);
}
```

Self-Test

DIRECTIONS

Answer the following questions by referring to Program 9.16 in the case study section of this chapter.

Questions

1. How many different structure members are contained in the program? Name them.
2. What type of a function is `read_menu()`? Explain.
3. Are any global variables used in the program? If so, name them.

4. State the purpose of the arguments in function `main()`. What are they called?
5. In the C `switch` in function `main()`, why was the `case 'E'` placed just before the `default`?
6. Are any `scanf()` functions used for user input in the program? Explain.
7. Will the program respond to the user's selection if uppercase or lowercase letters are entered during the menu prompt? Explain.
8. Why is it necessary to declare the `price` member as a type `double` in the structure?
9. Is the information in the disk file in binary or text mode? How did you determine this?

End-of-Chapter Problems

General Concepts

Section 9.1

1. Name one advantage of using a disk file.
2. Is the following a legal DOS reference to a file name?

```
B:FILER.02
```

3. Explain what the DOS file name reference of problem 2 above means.
4. How is a file pointer declared in C? Show by example.
5. When disk file I/O is being performed with C, what must always be done before the program is terminated?

Section 9.2

6. State the different methods of reading and writing data with the C language.
7. Give the three basic C file command strings.
8. If the command for writing to a disk file in C is given, and the file does not exist, what will happen?

Section 9.3

9. State the two kinds of data streams used by C.
10. Name the three standard files used by C.
11. What is the physical representation of the three standard files used by C?

Section 9.4

12. How are lines counted in Program 9.13?
13. How are the file extensions generated in Program 9.15?

Section 9.5

14. Why is it important to make sure a file operation was successful?
15. Why do the values `0x0d` and `0x0a` have different interpretations when encountered in text and binary files?

Section 9.6

16. How many bytes are used to store a single `parts_record` variable? How is this determined?
17. What happens to the old contents of `PARTS.DAT` each time Program 9.16 is executed?

Program Design

In developing the following C programs, use the methods developed in the case study section of this chapter. This means to use top-down design and block structure with no global variables. The function `main()` should do little more than call other functions. Use pointers when necessary to pass variables between functions. Be sure to include all of the documentation in your final program. This should consist of, but not be limited to, the programmer's block, function prototypes, and a description of each function as well as any formal parameters you may use.

18. Modify the case study program (Program 9.16) so that the total cost of any bin is computed and displayed. This should be an added option on the menu.
19. Modify the case study program (Program 9.16) so that a second command line authorization code is necessary in order to get any pricing data about the parts (this is in addition to the command line argument for the existing authorization code for entering new data to the disk files).
20. Expand the case study program (Program 9.16) so that new data may be added without purging the existing data to the disk file.
21. Change the case study program (Program 9.16) so that the program user may have the records displayed in alphabetical order according to the name of the part.
22. Modify the case study program (Program 9.16) so that there is another member of the structure. This member will contain the minimum number of parts allowed for each type before reordering is necessary. If the quantity of parts falls below this number, then every time the program is activated, the program user is automatically alerted to which parts need to be reordered.
23. Create another structure for the case study program (Program 9.16) that will contain the name of the parts manufacturer, street address, city, state, and phone number. This new structure will now become a member of the existing structure.
24. Develop a C program that will allow the program user to use the computer as a simple word processor. The program must allow the user to save and retrieve text files, name each file, and delete old files.
25. Write a C program that searches the input file for every occurrence of a string. The input file and search string must be specified on the command line, as in:

    ```
    SEARCH <filename> <string>
    ```

26. Write a C program that makes a list of all words found in an input file. Write the sorted list to an output file, along with the number of times each word appeared in the input file.

10 Color and Technical Graphics

Objectives

This chapter provides you the opportunity to learn:

1. How to produce text color.
2. How computer screens are used to represent graphical displays.
3. What is needed in your computer system to produce graphics and how to find out what graphic capabilities it has.
4. The concept of a pixel and how to use it in graphics.
5. How color is generated on the IBM and most compatibles.
6. The method of drawing lines on the graphic screen.
7. Some of the special built-in graphics functions.
8. Ways of creating bar graphs for technical analysis.
9. Methods of generating different graphic fonts.
10. Methods of graphing mathematical functions.

Key Terms

Monochrome
Default
Color Constants
Text Screen
Text Mode
Graphics Screen
Graphics Mode
Graphics Adapter
Graphics Driver
Autodetection
Pixel
Mode
Color Palette
Area Fill
Aspect Ratio
Fill-Style
Font
Bit Mapped

Stroked Font
Scaling
Coordinate Transformation
Ray Casting
Texture Mapping

Outline

Introduction

This chapter introduces the exciting world of color. Because of the diversity of commands, Turbo C is explained in this section.

As you will learn in this chapter, this C software possess the power to give text screens a new dimension by using color.

This chapter also introduces the power of the computer graph. One of the fastest growing areas in computers is the world of computer graphics! Understanding computer graphics will help you present information as drawings and graphs.

You will also learn the secrets of adding color to graphics. The use of different text fonts is discussed. This chapter will require patience and practice—but the personal and professional rewards of this new skill will more than compensate for the time you will invest. Programmers who understand graphic programming skills are on the cutting edge of a new technology. Starting this chapter is your exciting first step.

10.1 C Text and Color

Discussion

This section demonstrates how to employ color in your C programs. Technical programs that use color can take advantage of a new dimension of useful information. For example, imagine a program for checking the progress of an automated assembly line or for monitoring a power plant. Green text could be used to indicate that all systems are normal, yellow text could indicate that something requires attention, red text could mean that immediate attention is needed, and blinking red text could mean possible danger.

What Your System Needs

In order to produce color using Turbo C, your computer system must meet the following three requirements:

1. Have a color monitor.
2. Have a color graphics card installed.
3. Be an IBM PC or true compatible.

The color commands for text are defined in `conio.h`. This means that you must use `#include <conio.h>` in programs that use these features.

The Color Monitor

There are two different basic types of monitors used with your personal computer. One is called a **monochrome** monitor (meaning one color); the other is called a color monitor and is capable of producing a rich variety of different colors all at the same time.

System monitors have a coating of phosphor on the inside of the display screen. Electrical currents controlled by the computer strike this phosphor and cause it to glow. This glow emits light, the color of which depends on the type of phosphor used to coat the inside of the display screen. It is this action that allows you to see images displayed on your monitor screen.

In a monochrome monitor, only one type of phosphor is used. Thus, depending on the type of phosphor, the screen of a monochrome monitor may be green or amber or any other single color. With this type of monitor, it is not possible to change its single color.

A color monitor has three different colors of phosphors placed on the inside of the glass face in the form of thousands of tiny triads. These triads consist of the red, green, and blue light-emitting phosphors. These are the primary colors used for color mixing. With a color monitor of this type, 16 different colors may be displayed using text at the same time.

Display Capabilities

If your system can produce color, then it is also capable of producing two different sizes of text. One size is the 80-column-by-25-row standard that you are used to. This is the **default** text size, meaning the size automatically used by your system when you turn it on. The second size produces larger text. It is 40 columns by 25 rows. All of these variations, including the color capabilities for text, are illustrated in Figure 10.1.

The mode Function

Turbo C has a built-in function for selecting either 40- or 80-column text. Your system normally comes up in 80 columns; Program 10.1 will cause it to switch to 40 columns (and will work only if you have a color graphics card installed in your computer system).

Program 10.1

```
#include <stdio.h>
#include <conio.h>

main()
{
        textmode(0);    /* Changes to 40 column wide text. */

        printf("This is 40 columns wide.\n");
        printf("It works only on systems with\n");
        printf("a graphics card installed.\n");
}
```

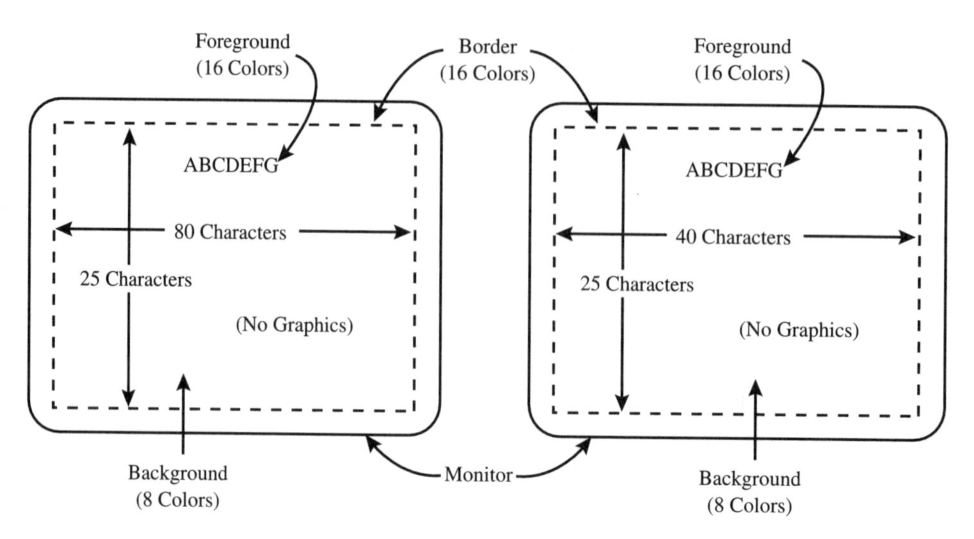

Figure 10.1 Text Display Capabilities

Table 10.1 Results of Integer Values for `textmode`

Mode Value	Results
0	Produces 40 × 25 monochrome text. An installed graphics card is required.
1	Produces 40 × 25 color text. An installed color graphics card is required.
2	Produces 80 × 25 monochrome text. An installed color graphics card is required.
3	Produces 80 × 25 color text. An installed color graphics card is required.
7	Produces 80 × 25 monochrome text. No color graphics card is installed in the system.

The built-in C function is

```
textmode(int mode);
```

Where

Mode = An integer as described in Table 10.1.

Note from Table 10.1 that you cannot get 40-column text without a graphics card installed in your system. Program 10.2 gives an example of the use of all variations for the built-in `textmode()`.

Program 10.2

```
#include <stdio.h>
#include <conio.h>

main()
{
        printf("This text is in the default 80 columns wide\n");
        printf("by 25 rows down.\n");
```

```
    getch();

    textmode(0);     /* Changes to 40 X 25 Monochrome. */
    printf("This text is now 40 columns wide\n");
    printf("but still 25 rows down.\n");
    getch();

    textmode(1);     /* Changes to 40 X 25 Color. */
    printf("Still a 40 column text, but on a\n");
    printf("color monitor display, it may reduce\n");
    printf("color streaking.\n");
    getch();

    textmode(2);     /* Changes to 80 X 25 Monochrome. */
    printf("This is the 80 column standard you are used to.\n");
    printf("It is 25 rows down.\n");
    getch();

    textmode(3);     /* Changes to 80 X 25 Color. */
    printf("Still 80 columns but on a color monitor display, it\n");
    printf("may reduce color streaking.\n");
    getch();

    textmode(7);     /* 80 X 25 Monochrome - no graphics card required. */
    printf("Still 80 columns, but for those systems that do not have\n");
    printf("a graphics card installed.\n");
    getch();
}
```

Whenever the `textmode()` is used, the entire screen is cleared. Thus, for Program 10.2, the screen is cleared every time the mode is changed by `textmode()`.

For convenience, Turbo C has defined constants that represent the numerical values of the `textmode` procedure. These are defined as follows:

Mode Number	*Mode Identifier*
0	BW40
1	CO40
2	BW80
3	CO80
7	MONO

Thus, Program 10.2 could have been written with `textmode(BW40)` instead of `textmode(0)`. The same is true for the other modes. Using the mode identifier rather than the mode number makes the code easier to read.

Text Color

The built-in Turbo function called `textcolor()` will change the color of your text, as shown in Program 10.3.

Program 10.3

```
#include <stdio.h>
#include <conio.h>

main()
{
        int color_number;
        char color_name[20];

        for(color_number = 0; color_number <= 15; color_number++)
        {
                switch(color_number)
                {
                        case 0  : strcpy(color_name, "black");
                                  break;
                        case 1  : strcpy(color_name, "blue");
                                  break;
                        case 2  : strcpy(color_name, "green");
                                  break;
                        case 3  : strcpy(color_name, "cyan");
                                  break;
                        case 4  : strcpy(color_name, "red");
                                  break;
                        case 5  : strcpy(color_name, "magenta");
                                  break;
                        case 6  : strcpy(color_name, "brown");
                                  break;
                        case 7  : strcpy(color_name, "light gray");
                                  break;
                        case 8  : strcpy(color_name, "dark gray");
                                  break;
                        case 9  : strcpy(color_name, "light blue");
                                  break;
                        case 10 : strcpy(color_name, "light green");
                                  break;
                        case 11 : strcpy(color_name, "light cyan");
                                  break;
                        case 12 : strcpy(color_name, "light red");
                                  break;
                        case 13 : strcpy(color_name, "light magenta");
                                  break;
                        case 14 : strcpy(color_name, "yellow");
                                  break;
                        case 15 : strcpy(color_name, "white");
                                  break;
                } /* End of switch. */

                textcolor(color_number);        /* This changes the color
                                                   of the text. */
                cprintf("This text is in %s\n\r",color_name);
        }
}
```

Table 10.2 Turbo C Built-in Color Constants

Color Number	Color Constant
0	BLACK
1	BLUE
2	GREEN
3	CYAN
4	RED
5	MAGENTA
6	BROWN
7	LIGHTGRAY
8	DARKGRAY
9	LIGHTBLUE
10	LIGHTGREEN
11	LIGHTCYAN
12	LIGHTRED
13	LIGHTMAGENTA
14	YELLOW
15	WHITE
+128	BLINK

In Program 10.3, you will not see the `cprintf()` function `This text is in black` because the background of your monitor is black. However, all the other colors will be displayed against this background.

Note that the `printf()` function is not used when text is to be presented in color. Instead, the `cprintf()` is required. Unlike the `printf()` function, `cprintf()` does not translate the `(\n)` linefeed into a carriage return/linefeed command. Thus, the `\n\r` is required in the `cprintf()` function.

Again, for convenience, Turbo C has predefined **color constants** for the `textcolor` procedure, as shown in Table 10.2.

Note the addition of BLINK in Table 10.2. This command causes the text to flash on the screen. Thus `textcolor(RED + BLINK);` will cause flashing red text for the next `cprintf()` function. Program 10.4 uses the color constants in place of the color number.

Program 10.4

```
#include <stdio.h>
#include <conio.h>

main()
{
        int color_number;
        char color_name[20];

        for(color_number = BLACK; color_number <= WHITE; color_number++)
        {
              switch(color_number)
```

```
    {
            case BLACK        : strcpy(color_name, "black");
                                break;
            case BLUE         : strcpy(color_name, "blue");
                                break;
            case GREEN        : strcpy(color_name, "green");
                                break;
            case CYAN         : strcpy(color_name, "cyan");
                                break;
            case RED          : strcpy(color_name, "red");
                                break;
            case MAGENTA      : strcpy(color_name, "magenta");
                                break;
            case BROWN        : strcpy(color_name, "brown");
                                break;
            case LIGHTGRAY    : strcpy(color_name, "light gray");
                                break;
            case DARKGRAY     : strcpy(color_name, "dark gray");
                                break;
            case LIGHTBLUE    : strcpy(color_name, "light blue");
                                break;
            case LIGHTGREEN   : strcpy(color_name, "light green");
                                break;
            case LIGHTCYAN    : strcpy(color_name, "light cyan");
                                break;
            case LIGHTRED     : strcpy(color_name, "light red");
                                break;
            case LIGHTMAGENTA : strcpy(color_name, "light magenta");
                                break;
            case YELLOW       : strcpy(color_name, "yellow");
                                break;
            case WHITE        : strcpy(color_name, "white");
                                break;
    } /* End of switch. */

  textcolor(color_number);        /* This changes the color
                                     of the text. */
  cprintf("This text is in %s\n\r",color_name);
 }
}
```

You can cause any text to blink just by adding the value of 128 to the color value. Program 10.5 illustrates.

Program 10.5

```
#include <stdio.h>
#include <conio.h>

main()
{
        textcolor(RED);
        cprintf("This text is now in red.\n\r");
```

```
        textcolor(4);
        cprintf("This text is also in red.\n\r");

        textcolor(RED + BLINK);
        cprintf("This is blinking red text.\n\r");

        textcolor(4 + 128);
        cprintf("This is also blinking red text.\n\r");

        textcolor(RED);
        cprintf("This stopped the red text from blinking.\n\r");

        textcolor(LIGHTGREEN + 128);
        cprintf("This text is in blinking light green.\n\r");
}
```

Changing the Background

To change the background color in Turbo C, use the built-in Turbo function `textbackground(int color)`. This is illustrated in Program 10.6.

Program 10.6

```
#include <stdio.h>
#include <conio.h>

main()
{
        int color_number;
        int color_name[10];

        for(color_number = 0; color_number <= 7; color_number++)
        {
                switch(color_number)
                {
                        case 0 : strcpy(color_name, "black");
                                 break;
                        case 1 : strcpy(color_name, "blue");
                                 break;
                        case 2 : strcpy(color_name, "green");
                                 break;
                        case 3 : strcpy(color_name, "cyan");
                                 break;
                        case 4 : strcpy(color_name, "red");
                                 break;
                        case 5 : strcpy(color_name, "magenta");
                                 break;
                        case 6 : strcpy(color_name, "brown");
                                 break;
                        case 7 : strcpy(color_name, "lightgray");
                                 break;
```

```
            } /* End of switch. */

            textbackground(color_number);

            if(color_number == 7)
                    textcolor(BLACK);

            gotoxy(20,12);
            cprintf("This background is in %s. \n\r",color_name);
            getch();
      }
}
```

Note from Program 10.6 that there are only eight background colors possible. The `textbackground()` function is:

`textbackground(int color);`

and is defined in the `conio.h` file. Also notice that the `textcolor()` was changed when the background was light gray.

As with the `textcolor()` function, Turbo C has identifier constants that represent the names of the colors. Thus `textbackground(BLACK)` has the same effect as `textbackground(0)`.

Conclusion

This section presented the important aspects of using color with text. Here you saw how to change the size of the text as well as how to invoke 16 different colors of text and eight different background colors. Check your understanding of this section by trying the following section review.

10.1 Section Review

1. State what is needed in order to obtain text color.
2. What is needed to change the text size from 80 columns to 40 columns?
3. How many different colors can text have? Can these colors all appear at the same time on the screen?
4. How many different background colors are available? What command is used to bring the screen to the desired color?
5. What choices are available to indicate the desired color? Give an example.

10.2 Graphics Mode

Discussion

This section introduces you to the fascinating and challenging world of Turbo C graphics. In this section you will see what your system, software, and programming needs are in order to get started. You will also be introduced to the most fundamental graphics commands.

Basic Idea

If you have an installed color graphics adapter in your computer, then your monitor is capable of displaying two different kinds of screens. One screen is the **text screen** (sometimes referred to as **text mode**). The text screen is the one that you have been using up to this point. It does nothing more than display text—essentially characters of a preassigned shape, such as the letters of the alphabet, numbers, punctuation marks, and the like. The other screen is called the **graphics screen** (sometimes referred to as the **graphics mode**). The graphics screen allows you to define your own shapes. Thus, when in graphics mode, you can create lines, graphs, charts, diagrams, pictures, and animation—essentially almost anything you can put on paper. Figure 10.2 illustrates the difference between the two modes of operation.

You can be in only one mode at a time, either in text mode or graphics mode. Your system comes up in text mode. In order to get into graphics mode, you must use special built-in Turbo C functions.

What You Need to Know

Before you can get into graphics mode with Turbo C, you must know what kind of a color **graphics adapter** your system is using. You must know this because the C program needs this information in order to use the correct built-in code. You will see how this can also be done automatically.

Turbo C has six different programs, made especially for six different graphics adapters. This is why you need to know what kind of graphics adapter your system has—so Turbo C will know which of the six different programs (called **graphics drivers**) to use. These graphics drivers all have the extension of `.BGI` for Borland Graphic Interface.

Fortunately, Turbo C has a built-in procedure that will allow you to find out which kind of installed graphics adapter is in your system. This is demonstrated by Program 10.7.

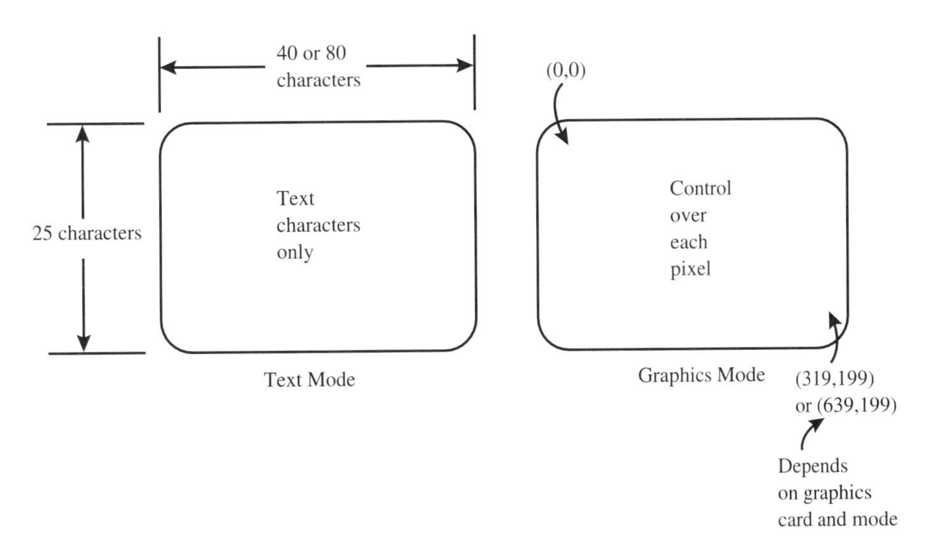

Figure 10.2 Difference between Text Mode and Graphics Mode

Program 10.7

```
#include <stdio.h>
#include <graphics.h>           /* Necessary for graphics. */

main()
{
        int graph_driver;       /* Specifies the graphics driver.      */
        int graph_mode;         /* Specifies the graphics mode.        */
        char string[10];        /* Gives the installed graphics driver. */

        /* Automatically detect the installed graphics driver. */

        detectgraph(&graph_driver, &graph_mode);

        switch(graph_driver)
        {
                case 1 : strcpy(string, "CGA");
                         break;
                case 2 : strcpy(string, "MCGA");
                         break;
                case 3 : strcpy(string, "EGA");
                         break;
                case 4 : strcpy(string, "EGA64");
                         break;
                case 5 : strcpy(string, "EGAMONO");
                         break;
                case 6 : strcpy(string, "IBM8514");
                         break;
                case 7 : strcpy(string, "HERCMONO");
                         break;
                case 8 : strcpy(string, "ATT400");
                         break;
                case 9 : strcpy(string, "VGA");
                         break;
               case 10 : strcpy(string, "PC3270");
                         break;
        } /* End of switch. */

        if(graph_driver == -2)
                printf("\n\nNo hardware detected.\n");
        else
                printf("\n\nYour installed graphics system is %s\n",string);
}
```

Program 10.7 automatically detects the kind of graphics hardware in your system. Using the program to do this is called **autodetection.** The Turbo built-in function used in Program 10.7 is:

```
void far detectgraph(int far *driver_address, int far *mode_address);
```

The C `detectgraph()` function is defined in the Turbo C `graphics.h` file. It takes two arguments that are addresses into which the system can place the type of adapter and the mode numbers (you'll see what these mean shortly).

Selecting Your Graphics Driver

Once you know which installed graphics adapter you have in your system, you must copy the appropriate graphics driver program from the original Turbo C disks to your working disk. The Turbo C system disks have the following graphics drivers:

ATT.BGI = Graphics driver for AT&T 6300 graphics.
CGA.BGI = Graphics driver for CGA and MCGA graphics.
EGAVGA.BGI = Graphics driver for EGA and VGA
HERC.BGI = Graphics driver for Hercules monographics.
IBM8514.BGI = Graphics driver for IBM 8514 graphics.
PC3270.BGI = Graphics driver for 3270PC graphics.

The Graphics Screen

Your graphics screen is divided into **pixels**. You can think of a pixel as the smallest point possible that can be displayed on your graphics screen. Figure 10.3 presents the concept of pixels.

How many pixels your graphics screen has depends on the kind of graphics adapter installed in your system. For example, CGA, MCGA, and ATT400 graphics adapters can have 320 pixels horizontally and 200 vertically. Each pixel is identified by a coordinate system of (X,Y) where X is the horizontal value of the pixel and Y is the vertical value of the pixel. The graphics screen for this hardware is shown in Figure 10.4.

Note that the top left pixel is identified as (0,0), the top right pixel as (639,0), the bottom left pixel as (0,199), and the bottom right pixel as (639,199). Thus, each pixel on the graphics screen has a unique location that can be identified by the (X,Y) system.

Lighting up a Pixel

The simplest graphics command is to light up a pixel on the graphics screen. To do this, your C program must do three things:

1. Get your system into the graphics mode with the correct graphics driver.
2. Activate the selected pixel.
3. Leave the graphics mode.

As you may suspect, Turbo C has three built-in functions for doing these three things. They:

1. Get you into the graphics mode:

```
void far initgraph(int far *driveradd, int far *modeadd, int far *pathadd);
```

This is a `far` function defined in `graphics.h`. The three arguments hold the addresses of the driver type, the mode address, and the address of the directory path where the graphics drivers are stored. If they are stored on the default drive, then this need not be specified.

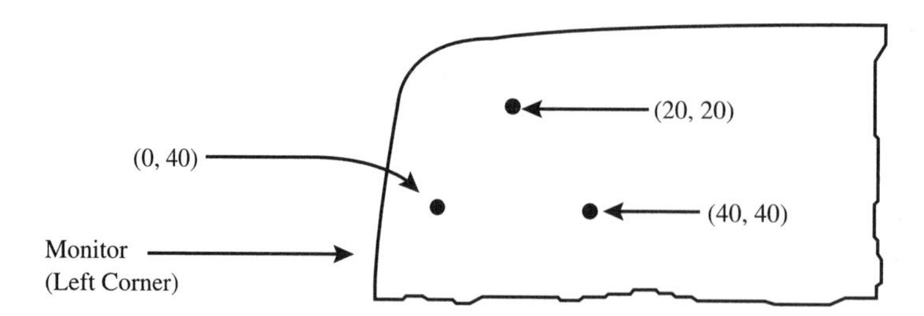

Figure 10.3 Concept of Pixels

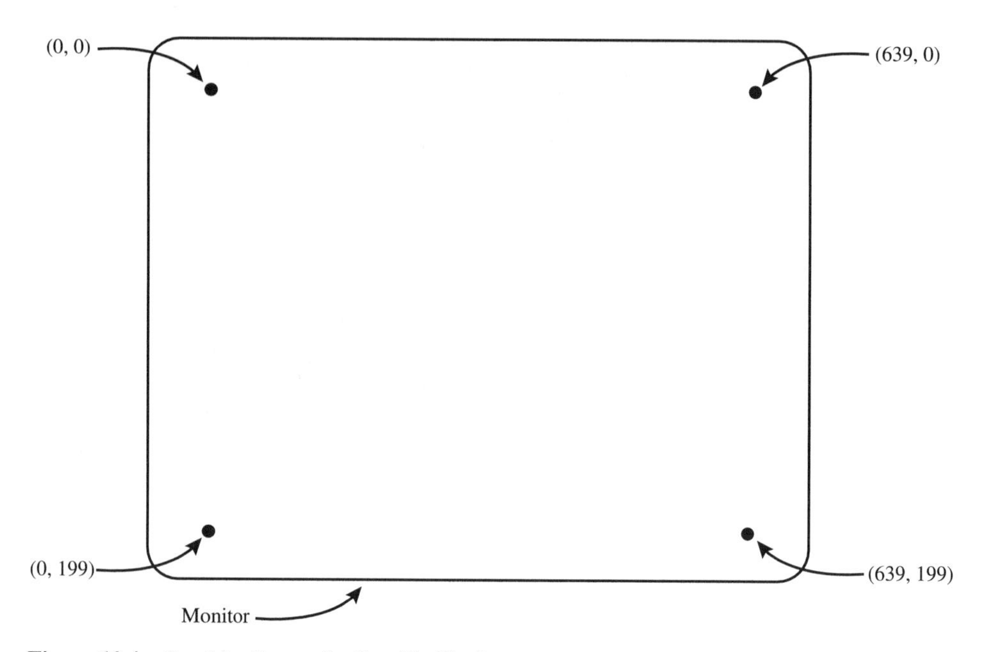

Figure 10.4 Graphics Screen for Specific Hardware

2. Light up a pixel on the graphics screen:

```
int putpixel(int x, int y, int color);
```

Where

x = An `int` representing the horizontal value of the pixel.
y = An `int` representing the vertical value of the pixel.
color = An `int` for number (or predefined color constant) indicating the color of the pixel.

3. Leave the graphics mode:

```
closegraph();
```

This function shuts down the graphics system and restores the original screen mode before graphics was initialized.

Program 10.8 is for a system that has a CGA color graphics adapter installed.

Program 10.8

```
#include <stdio.h>
#include <graphics.h>          /* Necessary for graphics. */

main()
{
        int graph_driver;       /* Specifies the graphics driver.  */
        int graph_mode;         /* Specifies the graphics mode.    */
        graph_driver = CGA;     /* For systems with IBM color graphics
                                   adapter. */
        graph_mode = 2;         /* Select a 320 X 200 screen with green,
                                   red, and brown. */

        /* Get into graphics with selected driver and mode. */
        initgraph(&graph_driver, &graph_mode, "");

        /* Plot points on the monitor screen. */
        putpixel(160,100,0);    /* Black dot near center of screen. */
        putpixel(162,100,1);    /* Green dot near center of screen. */
        putpixel(164,100,2);    /* Red dot near center of screen.   */
        putpixel(166,100,3);    /* Brown dot near center of screen. */

        getchar();              /* Hold graphics for observation.   */

        /* Shut down the graphics system. */

        closegraph();
}
```

Program 10.8 places four color dots on the graphics screen. The function `put-pixel-(X,Y,COLOR)` puts a dot at the *X* and *Y* coordinates in the value indicated by COLOR.

Note that the `getchar()` function is used so that you may view the graphics. After you press the Return/Enter key, the `closegraph()` function is initialized.

Program 10.8 uses `graph_mode = 2`. The reason for this is explained below.

Using Color

The kind and amount of color available in graphics mode is much less than the rich color combinations you could get while in text mode. As a matter of fact, the kind and amount of color (if any) depends upon the installed graphics adapter in your system. Some graphics adapters have different **modes** of operation. As an example, the CGA graphics adapter for Program 10.8 has five different graphics modes numbered 0 through 4. Table 10.3 lists the modes available for different graphics hardware.

As you can see from Table 10.3, the CGA driver that will be used in the following example programs can display a maximum of only three colors at a time, from four different modes.

Table 10.3 Graphics Modes for Given Installed Hardware

Hardware	Mode # Const	Screen Size (Pixels)	Colors
CGA	0 CGAC0	320 × 200	LightGreen, LightRed, Yellow
	1 CGAC1	320 × 200	LightCyan, LightMagenta, White
	2 CGAC2	320 × 200	Green, Red, Brown
	3 CGAC3	320 × 200	Cyan, Magenta, LightGray
	4 CGAHI	640 × 200	(Color on black)
MCGA	0 MCGAC0	320 × 200	LightGreen, LightRed, Yellow
	1 MCGAC1	320 × 200	LightCyan, Light Magenta, White
	2 MCGAC2	320 × 200	Green, Red, Brown
	3 MCGAC3	320 × 200	Cyan, Magenta, LightGray
	4 MCGAMED	640 × 200	(Color on black)
	5 MCGAHI	640 × 480	(Color on black)
EGA	0 EGALO	640 × 200	16 colors
	1 EGAHI	640 × 350	16 colors
EGA64	0 EGA64LO	640 × 200	16 colors
	1 EGA64HI	640 × 350	4 colors
	3 EGAMONO	640 × 350	(White on black)
HERCMONO	0 HERCMONOHI	720 × 348	(White on black)
ATT400	0 ATT400C0	320 × 200	LightGreen, LightRed, Yellow
	1 ATT400C1	320 × 200	LightCyan, LightMagenta, White
	2 ATT400C2	320 × 200	Green, Red, Brown
	3 ATT400C3	320 × 200	Cyan, Magenta, LightGray
	4 ATT400MED	640 × 200	(Color on black)
	5 ATT400HI	640 × 400	(Color on black)
VGA	0 VGALO	640 × 200	16 colors
	1 VGAMED	640 × 300	16 colors
	2 VGAHI	640 × 480	16 colors
PC3270	0 PC3270HI	720 × 350	(White on black)

It should be pointed out that in using the built-in Turbo C function `initgraph()`, Turbo has defined graphics driver constants corresponding to the number of the driver. These are listed as follows:

DETECT = 0
CGA = 1
MCGA = 2
EGA = 3
EGA64 = 4
EGAMONO = 5
IBM8514 = 6
HERCMONO = 7

ATT400 = 8
VGA = 9
PC3270 = 10

Thus `initgraph(CGA,0, "");` means initialize graphics mode using the CGA.BGI graphics driver in mode 0 (320 X 200 screen, colors light green, light red, and yellow). The CGA.BGI driver program will be on the active drive.

```
initgraph(1, 0, "");
```

means exactly the same thing, and so does

```
initgraph(CGA, CGAC0, "");
```

Making an Autodetect Function

Program 10.9 illustrates a user defined function that will automatically detect the type of graphics hardware in your system and then automatically select the proper graphics driver. In order for this to work on any system, you must have all of the graphics drivers on the active disk or indicate where they are located with the `initgraph()` function.

Program 10.9

```
#include <stdio.h>
#include <graphics.h>           /* Necessary for graphics. */

void auto_initialization(void); /* Function to initialize to the
                                   graphics mode. */

main()
{
        auto_initialization();  /* Call to initialize graphics.   */

        /* Graphic functions... */

        closegraph();           /* Shut down the graphics system. */
}

void auto_initialization()      /* Initialize to graphics mode.   */
{
        int graph_driver;       /* Specifies the graphics driver. */
        int graph_mode;         /* Specifies the graphics mode.   */

        graph_driver = DETECT;  /* Initializes autodetection.     */

        initgraph(&graph_driver, &graph_mode, "");
}
```

The function `auto_initialization()` will be used with all of the graphics programs in the remainder of this chapter.

Conclusion

This section presented the necessary information you need in order to get into graphics modes using Turbo C and IBM systems or true compatibles. Once you can get past this point, you will be able to enjoy the world of technical graphics available to you. Test your understanding of this section by trying the following section review.

10.2 Section Review

1. Name the two different kinds of screens available on your monitor.
2. State the difference between the graphics mode and text mode.
3. Before you can get your system into graphics operation with Turbo C, what one piece of hardware and two kinds of software will you need?
4. State how many different colors are available with color graphics.
5. Define the term *pixel*.

10.3 Knowing Your Graphics System

Discussion

This section will help you become familiar with your graphics system. Here you will learn about the different modes of operation for your graphics system and how to use them to your advantage in the creation of technical graphics. This section will also introduce some of the most fundamental graphics concepts and commands.

Your System's Screen Size and Colors

It's important that you know about your system's screen size and colors. If you don't, you might mistakenly try to draw lines on your graphics screen in black and never see anything against a black background. Or you might try drawing images that do not fit on your graphics screen. The size of the graphics screen and colors available (if any) depend largely on the kind of graphics hardware installed in your system.

Recall from the last section that the amount of color you can get in the graphics mode depends on the type of graphics hardware installed in your system. Table 10.3 lists the kinds of hardware and their various modes of operation. For example, from Table 10.3, you can see that a CGA (Color Graphics Adapter) has five modes of operation, as repeated in Table 10.4.

Table 10.4 shows that for the first four modes of operation (modes 0 through 3) the screen size is 320 pixels horizontally and 200 pixels vertically. The first four modes have only three colors available at any one time. These different color modes are referred to as the **color palette** or simply the palette. Thus the palette for mode 0 of the CGA hardware is light green, light red, and yellow, whereas the palette for mode 1 is light cyan, light magenta, and white. The important point is that you cannot have all palettes at the same time (meaning you must be in one of the four color modes). Black is also available in all modes.

Now note that the fifth mode (mode 4) has more pixels horizontally (640) and the same number vertically (200). The background will be black and visible graphics will be of a single color. Because of the makeup of the CGA adapter, this foreground color is actually set by the `setbkcolor()` function. In this mode you can get more detail (for a CGA adapter), but you lose the ability to produce more than one color at a time.

Table 10.4 Excerpt from Table 10.3 (Graphics Modes for Given Installed Hardware)

Hardware	Mode # Const	Screen Size (Pixels)	Colors
CGA	0 CGAC0	320 × 200	LightGreen, LightRed, Yellow
	1 CGAC1	320 × 200	LightCyan, LightMagenta, White
	2 CGAC2	320 × 200	Green, Red, Brown
	3 CGAC3	320 × 200	Cyan, Magenta, LightGray
	4 CGAHI	640 × 200	(Color on black)

From the information given above, you can see how important it is to be able to

1. Determine the modes of your system.
2. Change the modes of your system.

First, look at how to determine the modes of your system.

Your System's Modes

Program 10.10 will determine what mode your graphics system is in as well as other useful graphics information.

Program 10.10

```
#include <stdio.h>
#include <graphics.h>          /* Necessary for graphics. */

void auto_initialization(void); /* Function to initialize to the
                                   graphics mode. */

main()
{
     int high_mode;          /* Highest mode range.           */
     int low_mode;           /* Lowest mode range.            */
     int current_mode;       /* The current graphics mode.    */
     int max_color;          /* Highest value color for current
                                driver and mode. */
     int max_horiz;          /* Maximum horizontal value.     */
     int max_vert;           /* Maximum vertical value.       */

     auto_initialization();  /* Call to initialize graphics.  */

     /* Determine mode range: */

     getmoderange(CGA, &low_mode, &high_mode);
     printf("\n\nCGA mode range is %d high to %d low.\n",high_mode, low_mode);

     /* Determine current mode: */
```

```
    current_mode = getgraphmode();
    printf("The current graphics mode is %d\n",current_mode);

    /* Determine maximum color value: */

    max_color = getmaxcolor();
    printf("The maximum color value for this mode is %d\n",max_color);

    /* Determine maximum coordinates: */

    max_horiz = getmaxx();
    max_vert  = getmaxy();
    printf("The maximum horizontal value is %d\n",max_horiz);
    printf("and the maximum vertical value is %d\n",max_vert);

    getchar();              /* Hold output for observation. */

    /* Shut down the graphics system. */
    closegraph();
}


void auto_initialization()     /* Initialize to graphics mode.   */
{
    int graph_driver;       /* Specifies the graphics driver. */
    int graph_mode;         /* Specifies the graphics mode.   */

    graph_driver = DETECT;  /* Initializes autodetection.     */

    initgraph(&graph_driver, &graph_mode, "");
}
```

The results you get with Program 10.10 will depend on the graphics hardware installed in your system. If your system has an IBM CGA, then the output will be

```
CGA mode range is 4 high to 0 low.
The current graphics mode is 4.
The maximum value color for this mode is 1.
The maximum horizontal value is 639
and the maximum vertical value is 199.
```

The important point about Program 10.10 is that it lets you know exactly where you are. What follows is an explanation of each of the functions used in Program 10.10.

- Auto Initialization: This function automatically detects the type of graphics hardware and gets you into graphics mode. It will be used with every graphics program in this text. The function was explained in the last section.
- Mode Range: This function gives you the range of the modes available with your installed graphics hardware. The built-in Turbo C function used here is

```
getmoderange(int graph_driver, int far *low_mode, int far *high_mode);
```

Where

`graph_driver`	=	The graph driver for your system (such as 1 or CGA).
`low_mode`		Returns the value of the lowest permissible mode.
`hi_mode`		Returns the value of the highest permissible mode.

- Current Mode: This function identifies the active mode which your system is currently in. The built-in Turbo C function that returns the numerical value of the current mode is

```
int far getgraphmode(void);
```

Where

The value returned is the graphics mode set by `initgraph()` or `setgraph()`. (Refer to Table 10.3 for the significance of each graphics mode.)

- Maximum Colors: This function identifies the maximum number of colors available in the current mode. The built-in Turbo C function that returns this value is

```
int far getmaxcolor(void);
```

Where

The value returned is the largest-value valid color for the current graphics driver and mode that can be used. (In those modes where only white graphs will appear on a black background, 0 represents black and 1 represents white. In any mode, black is always 0, whereas the other numbers indicate colors that depend on the installed graphics hardware and active palette.)

- Graphics Screen Size: This function gives the maximum size of the current graphics screen. There are two built-in Turbo C functions that allow you to get these values:

```
getmaxx();
getmaxy();
```

Where

The value returned is the maximum screen coordinate for the driver and mode.

Creating Lines

In Turbo C there is a built-in function that allows you to create a line on the graphics screen. In order to use this function, your system must be in graphics mode:

```
line(int X1, int Y1, int X2, int Y2);
```

Where

$X1$, $Y1$ = Values of the starting point of the line.
$X2$, $Y2$ = Values of the ending point of the line.

Program 10.11

```
#include <stdio.h>
#include <graphics.h>          /* Necessary for graphics. */

void auto_initialization(void); /* Function to initialize to the
                                   graphics mode. */

main()
```

```
{
        auto_initialization();  /* Call to initialize graphics.   */

        line(0, 50, 299, 50);   /* Draw a line.                   */

        getchar();              /* Hold output for observation.   */

        closegraph();           /* Shut down the graphics system. */
}


void auto_initialization()      /* Initialize to graphics mode.    */
{
        int graph_driver;       /* Specifies the graphics driver.  */
        int graph_mode;         /* Specifies the graphics mode.    */

        graph_driver = DETECT;  /* Initializes autodetection.      */

        initgraph(&graph_driver, &graph_mode, "");

} /* End of auto_initialization. */
```

Program 10.11 causes a horizontal line to be created on the graphics screen. How far the line extends across the screen will depend on your installed graphics hardware. If you want to ensure that the line goes all the way across the screen, do the following modification:

```
line(0,50,getmaxx(),50);
```

Now the line will always extend across the whole graphics screen regardless of the current mode.

Program 10.12 makes a box around the graphics screen, using the built-in Turbo C functions `getmaxx()` and `getmaxy()`.

Program 10.12

```
#include <stdio.h>
#include <graphics.h>            /* Necessary for graphics. */

void auto_initialization(void); /* Function to initialize to the
                                    graphics mode. */

main()
{
    auto_initialization();  /* Call to initialize graphics.    */

    line(0,0,getmaxx(),0);                      /* Line across top of screen.      */
    line(0,getmaxy(),getmaxx(),getmaxy());      /* Line across bottom.             */
    line(0,0,0,getmaxy());                      /* Line along left side of screen. */
    line(getmaxx(),0,getmaxx(),getmaxy());      /* Line along right side.          */

    getchar();              /* Hold output for observation.    */
```

```
    closegraph();           /* Shut down the graphics system.  */
}


void auto_initialization()      /* Initialize to graphics mode.    */
{
    int graph_driver;       /* Specifies the graphics driver.  */
    int graph_mode;         /* Specifies the graphics mode.    */

    graph_driver = DETECT;  /* Initializes autodetection.      */

    initgraph(&graph_driver, &graph_mode, "");

} /* End of auto_initialization. */
```

Conclusion

This section presented some very important built-in Turbo C functions for understanding your graphics system. You were also introduced to some fundamental graphics commands. In the next section, you will learn some of the powerful graphics commands of Turbo C that will allow you to create different shapes in a variety of colors and line styles. For now, check your understanding of this section by trying the following section review.

10.3 Section Review

1. State why it's important to know about your graphics system's modes and colors.
2. Explain what is meant by a color palette.
3. How many different color palettes are available for the IBM CGA color graphics card? How many colors are in each palette?
4. Are there any modes in which color is not available? Explain how you would find this information.
5. How could you find out how many modes your system has?

10.4 Built-in Shapes

Discussion

In this section you will learn about Turbo C built-in shapes. You will discover how to easily create different line styles, rectangles, and circles. You will also see how to fill areas of the graphics screen with different patterns and colors. This is an exciting section. With the information you learn here, you can begin to construct programs that will produce powerful technical graphics.

Changing Graphics Modes

Recall from Table 10.3 that the graphics hardware in your system may offer several different modes of operation. In the last section, you were introduced to built-in Turbo C functions that allow you to find how many different modes your particular installed graphics system has.

For the example programs in this section, it will be assumed that you have IBM CGA graphics hardware in your system. The commands are the same no matter what acceptable system you have. For the purposes here, if your system is different, only the modes and pixel sizes may be different.

Program 10.13 demonstrates how to change graphics modes.

Program 10.13

```
#include <stdio.h>
#include <graphics.h>            /* Necessary for graphics. */

void auto_initialization(void); /* Function to initialize to the
                                   graphics mode. */

main()
{
        auto_initialization();  /* Call to initialize graphics.             */

        /* IMPORTANT!  The following comments apply only to an IBM
           CGA color graphics adapter.  The results with your system's
           installed graphics hardware may be different.  Refer to
           Table 10.3.    */

        setgraphmode(CGAC0);    /* 320 X 200 pixel screen, palette 0.       */
        printf("Current color is %d\n",getcolor());
        line(0,0,319,199);      /* Produces a yellow line.                  */
        getchar();

        setgraphmode(CGAC1);    /* 320 X 200 pixel screen, palette 1.       */
        printf("Current color is %d\n",getcolor());
        line(0,0,319,199);      /* Produces a white line.                   */
        getchar();

        setgraphmode(CGAC2);    /* 320 X 200 pixel screen, palette 2.       */
        printf("Current color is %d\n",getcolor());
        line(0,0,319,199);      /* Produces a brown line.                   */
        getchar();

        setgraphmode(CGAC3);    /* 320 X 200 pixel screen, palette 3.       */
        printf("Current color is %d\n",getcolor());
        line(0,0,319,199);      /* Produces a light gray line.              */
        getchar();

        setgraphmode(CGAHI);    /* 640 X 200 pixel screen, white on black. */
        printf("Current color is %d\n",getcolor());
        line(0,0,319,199);      /* Produces a white line.                   */
        getchar();

        closegraph();           /* Shut down the graphics system.  */
}

void auto_initialization()      /* Initialize to graphics mode.    */
```

```
{
        int graph_driver;       /* Specifies the graphics driver.  */
        int graph_mode;         /* Specifies the graphics mode.    */

        graph_driver = DETECT;  /* Initializes autodetection.      */

        initgraph(&graph_driver, &graph_mode, "");

} /* End of auto_initialization. */
```

Program 10.13 is commented for a system with an IBM CGA color graphics adapter. As seen in Table 10.3, this adapter has five modes of operation. Each of these modes may be referred to by a number (0 through 4) or by a built-in predefined Turbo graphics constant (CGAC0 through CGAHI). The built-in Turbo C function that sets the graphics mode is

`setgraphmode(int mode)`

Where

Mode = An `int` that must be a valid mode for the installed graphics hardware. You may use a number or one of the predefined graphics constants.

Each time `setgraphmode()` is activated, the graphics screen is cleared and the graph is redrawn in the default graphics color. In Program 10.13, `setgraphmode(CGAC0)` produces the same result as `setgraphmode(0)` and `setgraphmode(CGAHI)` produces the same results as `setgraphmode(4)`.

Line Styles

Program 10.13 produces a diagonal line across the screen. Turbo C allows you to select from several different line styles. You can choose to have your lines drawn as solid lines, dotted lines, dashed lines, or center lines. You can also select one of two line thicknesses. This is demonstrated by Program 10.14.

Program 10.14

```
#include <stdio.h>
#include <graphics.h>           /* Necessary for graphics. */

void auto_initialization(void); /* Function to initialize to the
                                   graphics mode. */

main()
{
        auto_initialization();  /* Call to initialize graphics.           */

        setgraphmode(CGAC0);    /* 320 X 200 pixel screen, palette 0.     */

        setlinestyle(SOLID_LINE, 0, NORM_WIDTH);
        line(0,10,150,10);      /* Draw a solid line of normal width.     */

        setlinestyle(DOTTED_LINE, 0, NORM_WIDTH);
```

```
        line(0,20,150,20);      /* Draw a dotted line of normal width.      */

        setlinestyle(CENTER_LINE, 0, NORM_WIDTH);
        line(0,30,150,30);      /* Draw a center line of normal width.      */

        setlinestyle(DASHED_LINE, 0, NORM_WIDTH);
        line(0,40,150,40);      /* Draw a dashed line of normal width.      */

        setlinestyle(SOLID_LINE, 0, THICK_WIDTH);
        line(0,50,150,50);      /* Draw a solid line of thick width.        */

        setlinestyle(DOTTED_LINE, 0, THICK_WIDTH);
        line(0,60,150,60);      /* Draw a dotted line of thick width.       */

        setlinestyle(CENTER_LINE, 0, THICK_WIDTH);
        line(0,70,150,70);      /* Draw a center line of thick width.       */

        setlinestyle(DASHED_LINE, 0, THICK_WIDTH);
        line(0,80,150,80);      /* Draw a dashed line of thick width.       */

        getchar();

        closegraph();           /* Shut down the graphics system.   */
}

void auto_initialization()      /* Initialize to graphics mode.     */
{
        int graph_driver;       /* Specifies the graphics driver.   */
        int graph_mode;         /* Specifies the graphics mode.     */

        graph_driver = DETECT;  /* Initializes autodetection.       */

        initgraph(&graph_driver, &graph_mode, "");

} /* End of auto_initialization. */
```

The results of Program 10.14 are illustrated in Figure 10.5.

The built-in Turbo C function for setting the line style is

```
setlinestyle(int linestyle, unsigned pattern, int thickness);
```

Where

linestyle = A type `int` that is one of the following:

Number	Constant	Result
0	SOLID_LINE	Graphic lines will be solid.
1	DOTTED_LINE	Graphic lines will be dotted.
2	CENTER_LINE	Graphic lines will be center lines (long line followed by a short one).
3	DASHED_LINE	Graphic lines will be dashes.
4	USERBIT_LINE	A user defined line style.

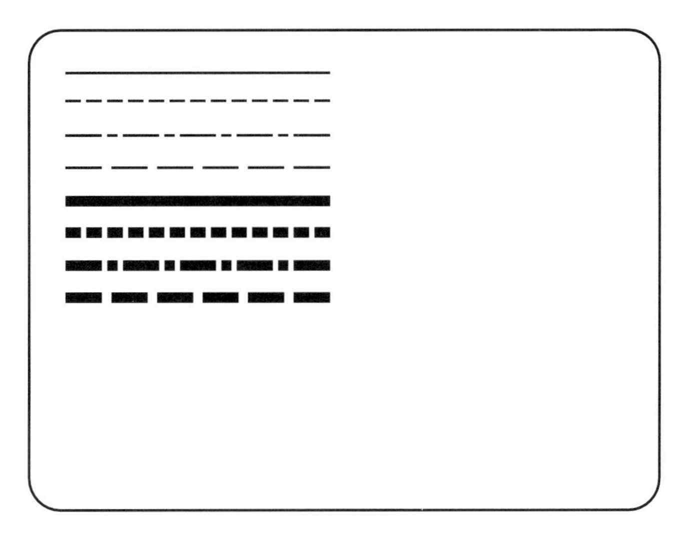

Figure 10.5 Different Graphic Line Styles

pattern = The bit pattern for the line. For the purpose of this text, set this to 0.
thickness = A type `int` that is used to determine one of two available line thicknesses: 1 = `NORM_WIDTH` (a line of "normal" width); 2 = `THICK_WIDTH` (a line thicker than the "normal" width).

For example, `setlinestyle(SOLID_LINE,0,NORM_WIDTH)` is identical to the command `setlinestyle(0,0,1)`.

Making Rectangles

Turbo C has built-in graphics procedures for making rectangles. This is a great time-saving feature. As you may suspect, Turbo C allows you to select the line styles for each of these rectangles. This is illustrated in Program 10.15.

Program 10.15

```
#include <stdio.h>
#include <graphics.h>           /* Necessary for graphics. */

void auto_initialization(void); /* Function to initialize to the
                                   graphics mode. */

main()
{
    auto_initialization();      /* Call to initialize graphics.             */

    setgraphmode(CGAC0);        /* 320 X 200 pixel screen, palette 0.       */

    setlinestyle(SOLID_LINE, 0, NORM_WIDTH);
    rectangle(0,10,100,30);     /* Draw a solid rectangle of normal width.  */

    setlinestyle(DOTTED_LINE, 0, NORM_WIDTH);
```

```
    rectangle(0,40,100,60);      /* Draw a dotted rectangle of normal width. */

    setlinestyle(CENTER_LINE, 0, NORM_WIDTH);
    rectangle(0,70,100,90);      /* Draw a center rectangle of normal width. */

    setlinestyle(DASHED_LINE, 0, NORM_WIDTH);
    rectangle(0,100,100,120);    /* Draw a dashed rectangle of normal width. */

    setlinestyle(SOLID_LINE, 0, THICK_WIDTH);
    rectangle(110,10,210,30);    /* Draw a solid rectangle of thick width.   */

    setlinestyle(DOTTED_LINE, 0, THICK_WIDTH);
    rectangle(110,40,210,60);    /* Draw a dotted rectangle of thick width.  */

    setlinestyle(CENTER_LINE, 0, THICK_WIDTH);
    rectangle(110,70,210,90);    /* Draw a center rectangle of thick width.  */

    setlinestyle(DASHED_LINE, 0, THICK_WIDTH);
    rectangle(110,100,210,120); /* Draw a dashed rectangle of thick width.  */

    getchar();

    closegraph();            /* Shut down the graphics system.  */
}

void auto_initialization()  /* Initialize to graphics mode.    */
{
    int graph_driver;        /* Specifies the graphics driver.  */
    int graph_mode;          /* Specifies the graphics mode.    */

    graph_driver = DETECT;   /* Initializes autodetection.      */

    initgraph(&graph_driver, &graph_mode, "");

} /* End of auto_initialization. */
```

Figure 10.6 illustrates the results of Program 10.15.

The built-in Turbo C function for drawing a rectangle on the graphics screen is

```
rectangle(int X1, int Y1, int X2, int Y2);
```

Where

$X1, Y1$ = `int` type coordinates of the top left corner of the rectangle.
$X2, Y2$ = `int` type coordinates of the bottom right corner of the rectangle.

As you can see from Program 10.15, the line style of the rectangle is determined by the last `setlinestyle()` function used in the program.

Creating Circles

As you may have suspected, Turbo C also has a built-in function for creating circles. However, unlike the `rectangle()` function, the `circle()` function is affected by the

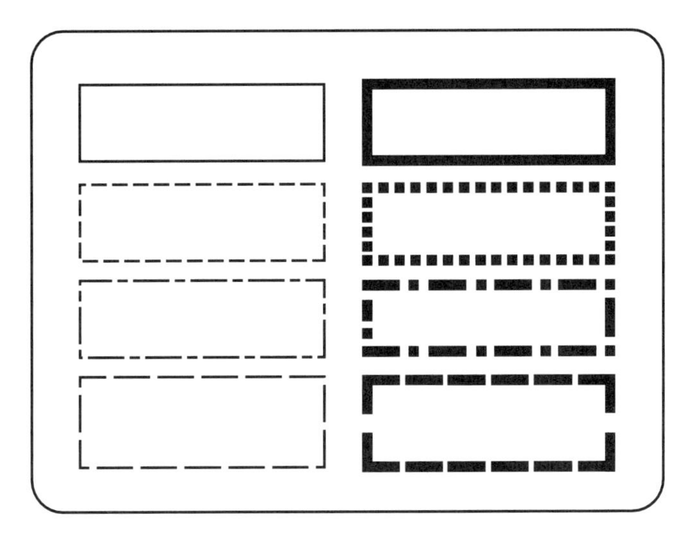

Figure 10.6 Built-in Rectangle Shapes with Different Line Styles

`setlinestyle()` function in only one of two ways. The circle will always be drawn with a solid line that is one of two thicknesses. This is illustrated in Program 10.16, where the resulting circles ignore the types of lines issued by the `setlinestyle()` functions.

Program 10.16

```
#include <stdio.h>
#include <graphics.h>            /* Necessary for graphics. */

void auto_initialization(void); /* Function to initialize to the
                                   graphics mode. */

main()
{
        auto_initialization();      /* Call to initialize graphics.       */

        setgraphmode(CGAC0);        /* 320 X 200 pixel screen, palette 0. */

        setlinestyle(SOLID_LINE, 0, NORM_WIDTH);
        circle(50,30,40);    /* Draw a solid circle of normal width.  */

        setlinestyle(DOTTED_LINE, 0, NORM_WIDTH);
        circle(50,60,40);    /* Draw a dotted circle of normal width. */

        setlinestyle(CENTER_LINE, 0, NORM_WIDTH);
        circle(50,90,40);    /* Draw a center circle of normal width. */

        setlinestyle(DASHED_LINE, 0, NORM_WIDTH);
        circle(50,120,40);   /* Draw a dashed circle of normal width. */

        setlinestyle(SOLID_LINE, 0, THICK_WIDTH);
```

```
        circle(140,30,40);    /* Draw a solid circle of thick width.   */

        setlinestyle(DOTTED_LINE, 0, THICK_WIDTH);
        circle(140,60,40);    /* Draw a dotted circle of thick width.  */

        setlinestyle(CENTER_LINE, 0, THICK_WIDTH);
        circle(140,90,40);    /* Draw a center circle of thick width.  */

        setlinestyle(DASHED_LINE, 0, THICK_WIDTH);
        circle(140,120,40);   /* Draw a dashed circle of thick width.  */

        getchar();

        closegraph();            /* Shut down the graphics system.   */
}

void auto_initialization()       /* Initialize to graphics mode.     */
{
        int graph_driver;        /* Specifies the graphics driver.   */
        int graph_mode;          /* Specifies the graphics mode.     */

        graph_driver = DETECT;   /* Initializes autodetection.       */

        initgraph(&graph_driver, &graph_mode, "");

} /* End of auto_initialization. */
```

Figure 10.7 shows the result of Program 10.16. Again, note that there are only two line styles for the circle.

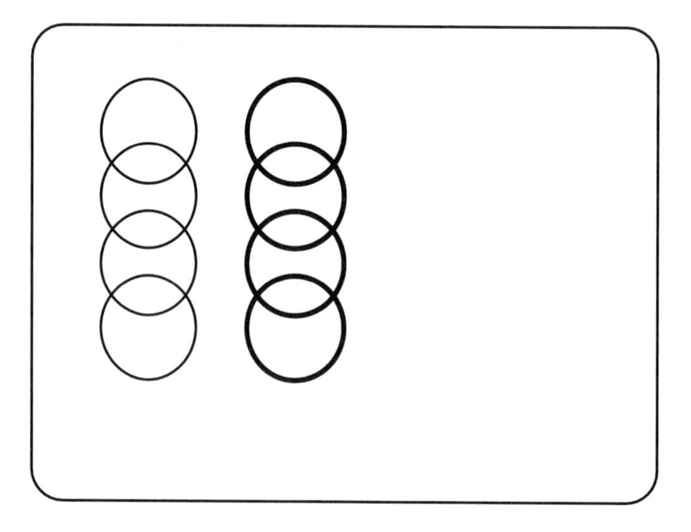

Figure 10.7 Circles Created by the `circle()` Function

The built-in Turbo C function for creating circles on the graphics screen is

```
circle(int X, int Y, int radius);
```

Where

X, *Y* = The `int` coordinates of the circle.
radius = The circle radius of type `int`.

Area Fills

There may be times when you want the area within an enclosed shape such as a circle, rectangle, or any shape you may create to be of a different color from the background. Turbo C has a built-in graphics function for filling in enclosed areas thus making the area inside a shape different from its background. This is demonstrated by Program 10.17.

Program 10.17

```
#include <stdio.h>
#include <graphics.h>          /* Necessary for graphics. */

void auto_initialization(void); /* Function to initialize to the
                                   graphics mode. */

main()
{
        auto_initialization();

        setgraphmode(CGAC0);

        setlinestyle(SOLID_LINE, 0, NORM_WIDTH);
        circle(30,30,30);
        floodfill(30,30,3);
        getchar();
        closegraph();
}

void auto_initialization()
{
        int graph_driver;        /* Specifies the graphics driver.  */
        int graph_mode;          /* Specifies the graphics mode.    */

        graph_driver = DETECT;   /* Initializes autodetection.      */

        initgraph(&graph_driver, &graph_mode, "");

} /* End of auto_initialization. */
```

The result of Program 10.17 is shown in Figure 10.8.

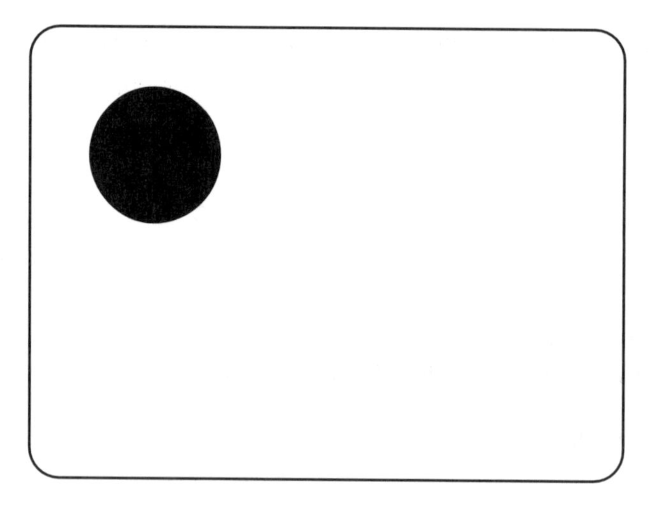

Figure 10.8 Results of Filling an Area of an Object

The built-in Turbo function for an **area fill** (filling in an enclosed area) is

```
floodfill(int X, int Y, int border);
```

Where

X, *Y* = Any coordinates inside the area of the object to be filled.
border = Color of the fill (must be the color of the object border).

Aspect Ratios

The ratio of a screen's height to its width is called its **aspect ratio**. A square screen has an aspect ratio of 1.0. The aspect ratios of different color adapters are not the same. This means that a circle may look like an ellipse if the aspect ratio is something different from 1.0. The CGA adapter has an aspect ratio of 1.6, whereas the aspect ratio of the EGA is 1.8286.

Turbo C tries to automatically compensate for these differences to produce true circles and squares when you want them. You can get the aspect ratio of your system from the built-in Turbo C function,

```
getaspectratio(int far *X_asp, int far *Y_asp);
```

Where

`X_asp` = The horizontal aspect factor.
`Y_asp` = The vertical aspect factor.

The built-in Turbo C function

```
setaspectratio(int X_asp, int Y_asp);
```

will change the default aspect ratio of the graphics system.

Conclusion

This section presented the basic graphics building blocks offered by Turbo C. Here you saw how to create different line styles, rectangles that use the line styles, and circles that use only some of the line styles. You also saw how to fill areas while in the graphics mode.

In the next section, you will see how easy it is to create technical bar graphs using the built-in features of Turbo C. For now, test your understanding of this section by trying the following section review.

10.4 Section Review

1. State what effect changing a graphics mode could have on the graphics screen.
2. How many different line styles are available in Turbo C? State what they are.
3. Give the built-in Turbo C shapes introduced in this section. Which one(s), if any, have limited line styles?
4. What does a flood fill do? How do you determine what area will be filled?

10.5 Bars and Text in Graphics

Discussion

This section shows you how to create bar charts and how to put text into your graphics. Being able to manipulate text by changing its size and style (font) as well as being able to write vertically will greatly enhance your technical graphics. In addition, you will see how to create the bars used in bar charts and how to change color and built-in bar-fill patterns.

Creating Bars

Bar charts are a common method of presenting information. An example bar chart is shown in Figure 10.9.

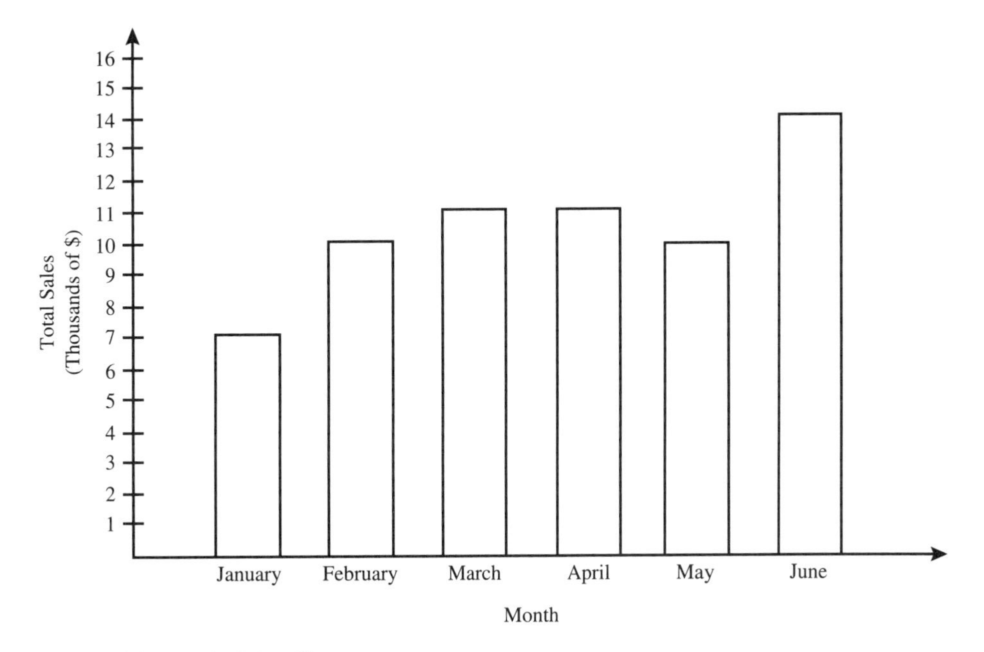

Figure 10.9 Typical Bar Chart

As you can see from the figure, the basic graphic for the bar chart is a rectangle. You could construct a rectangle in Turbo C graphics by simply using the `rectangle()` function. However, there is an easier way, as shown in Program 10.18.

Program 10.18

```
#include <stdio.h>
#include <graphics.h>           /* Necessary for graphics. */

#define TRUE 1
#define FALSE 0
void auto_initialization(void); /* Function to initialize to the
                                   graphics mode. */

main()
{
        auto_initialization();
        setgraphmode(0);

        bar(20,20,70,100);
        bar3d(90,20,140,100,10,FALSE);
        bar3d(160,20,220,100,10,TRUE);

        getchar();

        closegraph();
}

void auto_initialization()
{
        int graph_driver;       /* Specifies the graphics driver.   */
        int graph_mode;         /* Specifies the graphics mode.     */

        graph_driver = DETECT;  /* Initializes autodetection.       */

        initgraph(&graph_driver, &graph_mode, "");

} /* End of auto_initialization. */
```

The graph generated by Program 10.18 is shown in Figure 10.10.

Note from the figure that three bars are generated. One is "flat," and the other two have some depth, one without a top and the last one with a top. The built-in Turbo function that produces these bars is

```
bar(int X1, int Y1, int X2, int Y2);
```

Where

$X1, Y1$ = The top left point of the bar.
$X2, Y2$ = The bottom right point of the bar.

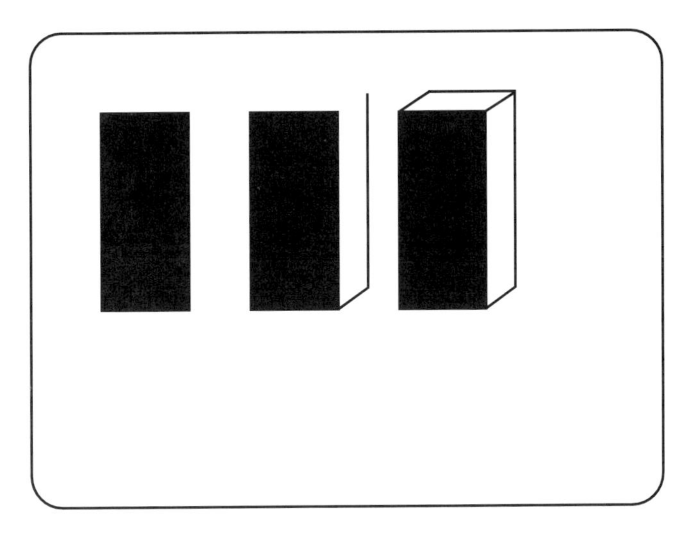

Figure 10.10 Resulting Graph for Program 10.18

```
bar3d(int X1, int Y1, int X2, int Y2, int depth, int top);
```

Where

X1, Y1 = The top left point of the bar.
X2, Y2 = The bottom right point of the bar.
depth = The number of pixels deep for the three-dimensional outline.
top = If nonzero, a top is put on the bar.

The bars are drawn in the current color and fill-style. A discussion of fill-styles follows.

Filling in the Bars

There may be times when many bars in a bar graph display must be used to represent many different types of information on the same graph. Since the use of color is limited on some graphics adapters, Turbo has a built-in function for automatically filling the area within a bar with one of 12 different **fill-styles**. One of the fill-styles is demonstrated by Program 10.19.

Program 10.19

```
#include <stdio.h>
#include <graphics.h>          /* Necessary for graphics. */

#define TRUE 1
#define FALSE 0

void auto_initialization(void); /* Function to initialize to the
                                   graphics mode. */

main()
```

```
{
        auto_initialization();

        setgraphmode(0);
        setfillstyle(LINE_FILL,2);
        bar(20,20,70,100);
        bar3d(90,20,140,100,10,FALSE);
        bar3d(160,20,220,100,10,TRUE);

        getchar();

        closegraph();
}

void auto_initialization()
{
        int graph_driver;          /* Specifies the graphics driver.  */
        int graph_mode;            /* Specifies the graphics mode.    */

        graph_driver = DETECT;     /* Initializes autodetection.      */

        initgraph(&graph_driver, &graph_mode, "");

} /* End of auto_initialization. */
```

The graph generated by Program 10.19 is shown in Figure 10.11.

As you can see from the figure, each of the bars has a distinctive pattern. The commands for each of the bars are exactly the same as before, but now the pattern inside the bar is different. This is done by the built-in Turbo C function `setfillstyle()`.

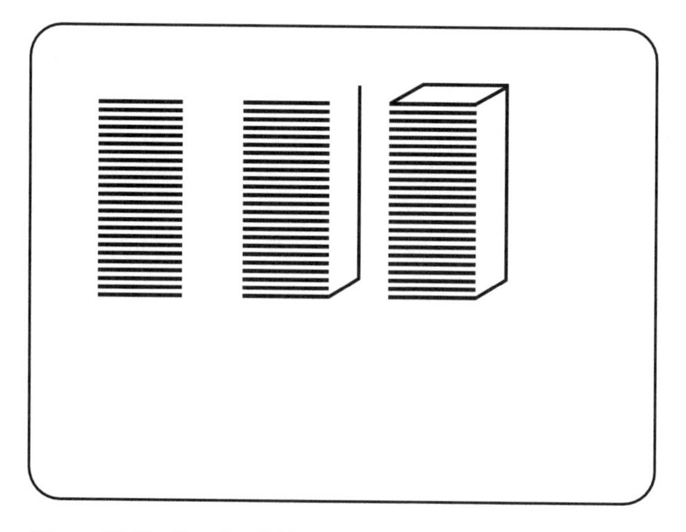

Figure 10.11 Bar Graph Generated by Program 10.19

Table 10.5 Fill Pattern Constants

Number	Constant	Bar is filled:
0	Empty_Fill	in the background color.
1	Solid_Fill	with a solid color.
2	Line_Fill	with horizontal lines.
3	LtSlash_Fill	with light slashes.
4	Slash_Fill	with thick slashes.
5	BkSlash_Fill	with thick backslashes.
6	LtBkSlash_Fill	with light backslashes.
7	Hatch_Fill	with light hatch marks.
8	XHatch_Fill	with heavy cross hatches.
9	InterLeave_Fill	with an interleaving line.
10	Wide_Dot_Fill	with widely spaced dots.
11	Close_Dot_Fill	with closely spaced dots.

```
setfillstyle(int pattern, int color);
```

Where

pattern = An `int` from 0 to 11 that sets the style of the fill pattern. (A value of 12 allows a user-defined fill pattern. Refer to the Turbo C manuals.)

color = An `int` that sets the color of the fill pattern.

Turbo C has defined a set of constants in the graphics library, as shown in Table 10.5.

Program 10.20 illustrates the use of several of the constants listed in Table 10.5 and the reason for having tops on or off of three-dimensional bars.

Program 10.20

```
#include <stdio.h>
#include <graphics.h>          /* Necessary for graphics. */

#define TRUE 1
#define FALSE 0

void auto_initialization(void); /* Function to initialize to the
                                    graphics mode. */

main()
{
        auto_initialization();

        setgraphmode(0);

        setfillstyle(SLASH_FILL, 3);
        bar3d(10,100,50,150,10,FALSE);
```

```
        setfillstyle(LINE_FILL, 1);
        bar3d(10,50,50,100,10,FALSE);

        setfillstyle(HATCH_FILL, 2);
        bar3d(10,10,50,50,10,TRUE);

        setfillstyle(WIDE_DOT_FILL, 2);
        bar3d(50,85,90,150,10,TRUE);

        getchar();

        closegraph();
}

void auto_initialization()
{
        int graph_driver;       /* Specifies the graphics driver.  */
        int graph_mode;         /* Specifies the graphics mode.    */

        graph_driver = DETECT;  /* Initializes autodetection.      */

        initgraph(&graph_driver, &graph_mode, "");

} /* End of auto_initialization. */
```

The graph resulting from Program10.20 is shown in Figure 10.12. Note that the two bottom stacked bars used the top as FALSE (equal to 0) and the top one had this item TRUE (equal to 1).

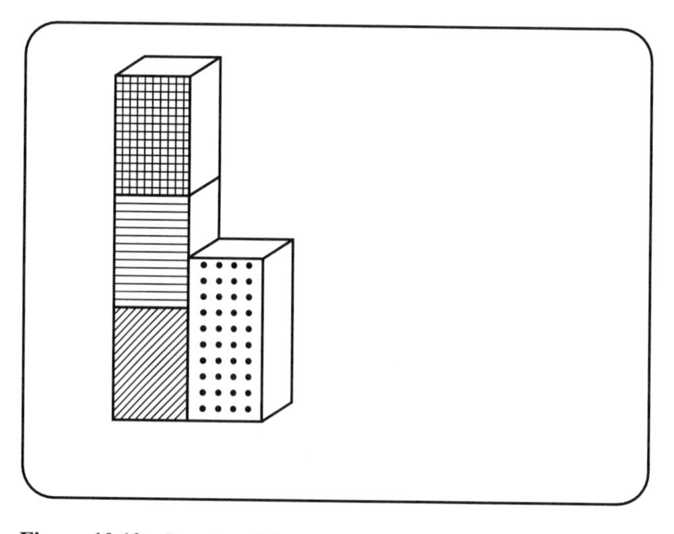

Figure 10.12 Results of Program 10.20

Text with Graphics

The use of text with graphics is an important feature of technical graphics. Turbo C provides many advanced methods for presenting text on your graphics screens. The basic text graphic command is illustrated in Program 10.21.

Program 10.21

```
#include <stdio.h>
#include <graphics.h>           /* Necessary for graphics. */

void auto_initialization(void); /* Function to initialize to the
                                   graphics mode. */

main()
{
        auto_initialization();

        outtext("This is the default font...\n");
        outtext("See where this is...");

        getchar();

        closegraph();
}

void auto_initialization()
{
        int graph_driver;       /* Specifies the graphics driver.  */
        int graph_mode;         /* Specifies the graphics mode.    */

        graph_driver = DETECT;  /* Initializes autodetection.      */

        initgraph(&graph_driver, &graph_mode, "");

} /* End of auto_initialization. */
```

Execution of Program 10.21 results in display of the following text, starting at the top left of the monitor:

```
This is the default font...See where this is...
```

Note that there is no carriage return (as you would find in a `printf()` function). Instead, the built-in Turbo function `outtext()` sends a string of characters to the current position of the graphics pointer. The built-in function is

```
outtext(char far *textstring);
```

The graphics pointer is the point on the graphics screen from which some action will take place. It is similar to the text cursor, only it isn't visible. The `outtext()` function will produce a string of text from the position of the graphics pointer and leave the pointer at the end of the text string. One method of moving the text string is to use the built-in Turbo C function `outtextxy()`.

```
outtextxy(int X1, int Y1 char far *textstring);
```

Where

$X1, Y1$ = The position on the graphics screen from which the string is to start.

Changing Text Fonts

Turbo C allows you to select different text styles (called **fonts**). Turbo C actually has two major types of graphic text styles. One is **bit mapped**; this is the default font. The other type is a **stroked font**. The difference is that characters in bit mapped fonts are composed of rectangular pixel arrays. Characters in stroked fonts are composed of line segments whose sizes and directions are defined relative to some starting point.

Turbo C allows you to change the size of graphic text. When you enlarge bit mapped fonts, the pixels are simply enlarged. Doing this makes the text look "blocky" with large stairstep effects. However, with a stroked font, text enlargement is done by making each component line segment longer; thus, these fonts look more natural. The difference is illustrated in Figure 10.13.

The main advantage of bit mapped fonts is speed. Stroked fonts take longer because of their many line segments.

The built-in Turbo C function that selects the font is

```
settextstyle(int font, int direction, int size);
```

Where

font = The style of the font.
direction = The direction of the text to be displayed (0 left to right, 1 bottom to top).
size = Size of the text (1 to 10). A size of 1 produces a character of 8×8 pixels; a size of 2 produces a character of 16×16 pixels, and so on.

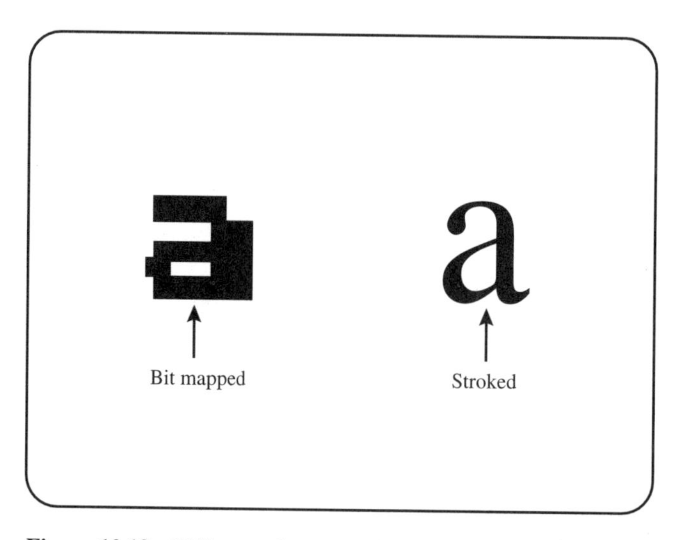

Figure 10.13 Difference between Bit Mapped and Stroked Fonts

The different text styles supported by Turbo C are

0 = DEFAULT_FONT (Bit mapped)
1 = TRIPLEX_FONT (All the rest are stroked)
2 = SMALL_FONT
3 = SANSSERIF_FONT
4 = GOTHIC_FONT

The active disk must contain the following Turbo C fonts:

```
TRIP.CHR   LITT.CHR   SANS.CHR   GOTH.CHR
```

These come on your Turbo C distribution disks. Other built-in constants for the `settextstyle()` function are

```
0 = HORIZ_DIR
1 = VERT_DIR
```

Thus, `settextstyle(TRIPLEX_FONT, HORIZ_DIR, 1)` means exactly the same thing as `settextstyle(1, 0, 1)`.

Program 10.22 illustrates the use of graphic text with a bar graph.

Program 10.22

```
#include <stdio.h>
#include <graphics.h>           /* Necessary for graphics. */

#define TRUE 1
#define FALSE 0

void auto_initialization(void); /* Function to initialize to the
                                   graphics mode. */

main()
{
        auto_initialization();
        setgraphmode(0);

        setfillstyle(SLASH_FILL, 3);
        bar3d(30,100,70,150,60,FALSE);

        setfillstyle(LINE_FILL, 1);
        bar3d(30,50,70,100,10,FALSE);

        setfillstyle(HATCH_FILL, 2);
        bar3d(30,10,70,50,10,TRUE);

        setfillstyle(WIDE_DOT_FILL, 2);
        bar3d(80,85,120,150,10,TRUE);

        /* Set the text. */
```

```
        settextstyle(DEFAULT_FONT, VERT_DIR, 1);
        setcolor(1);
        outtextxy(20,10,"TOM  DICK  HARRY");
        setcolor(2);
        outtextxy(145,85, "OTHERS");

        getchar();

        closegraph();
}

void auto_initialization()
{
        int graph_driver;        /* Specifies the graphics driver.  */
        int graph_mode;          /* Specifies the graphics mode.    */

        graph_driver = DETECT;   /* Initializes autodetection.      */

        initgraph(&graph_driver, &graph_mode, "");

} /* End of auto_initialization. */
```

Figure 10.14 shows the result of Program 10.22. Note that the vertical text displays actually *start* at the given coordinates and are drawn down.

The built-in Turbo C function `setcolor()` determines the graphic drawing color:

```
setcolor(int color);
```

Where

color = The selected color, defined within the limits of your system's color graphics adapter and the mode you select.

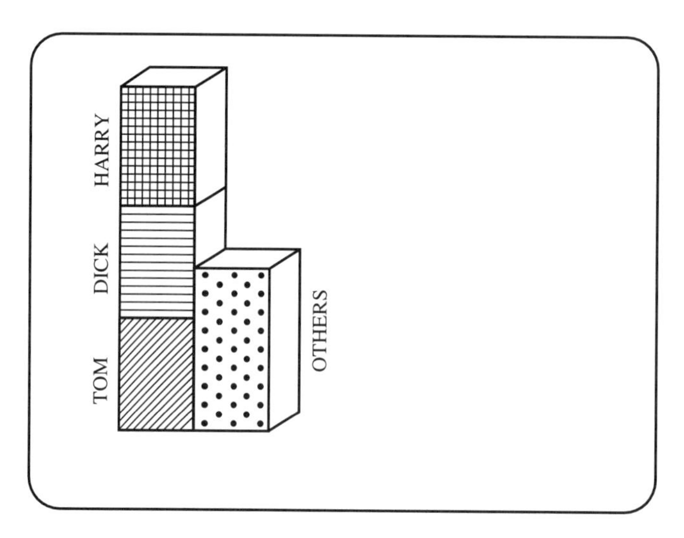

Figure 10.14 Use of Vertical Text in Graphics Mode

Conclusion

This section presented important information concerning the creation of bars used in bar graphs. Here you saw how to create three-dimensional bars and fill in their areas with different pre-made fill patterns. You also saw how to add text to graphics. The different fonts were presented as well as how to change the size of text and make it appear either vertically or horizontally.

Test your knowledge of this section by trying the following section review.

10.5 Section Review

1. Why does Turbo C have a built-in function for creating bars?
2. State what kinds of bars can be displayed using built-in Turbo C functions.
3. What options are available for presenting a three-dimensional bar using Turbo C built-in functions? Why is this available?
4. Describe how you can distinguish one bar from another using built-in Turbo C functions.
5. State some of the options available to you when using text in Turbo C graphics.

10.6 Graphing Functions

Discussion

Turbo C is a powerful language system for graphing mathematical functions. Recall that a mathematical function shows the relationship between two or more quantities. This section presents the fundamentals for programming this type of technical graph. Knowing how to do this greatly enhances your technical programming skills.

Fundamental Concepts

Scaling is the use of numerical methods to ensure that all of the required data appears on the graphics screen and utilizes the full pixel capability. Remember that for any computer system, the number of pixels is limited. For example, in the high-resolution mode of the CGA adapter there are 640 horizontal pixels and 200 vertical pixels.

To understand how to use scaling to produce practical graphics, consider the graph of the function: $Y = X + 2$. The plot of this function is shown in Figure 10.15.

The graph shown in Figure 10.15 uses three of the four quadrants of the Cartesian coordinate system. Suppose you need to develop a computer program that will display such a graph and you want the full graphics screen to be used in this display. This means that the actual values used to plot the graph would not be the values used in the equation. Figure 10.16 illustrates this important point.

Observe from Figure 10.16 that the extreme left of the graph (the point that represents $X = -5$ and $Y = -3$) must actually have the plotted values of $X_P = 0$ and $Y_P = 199$ (the P subscript is used to denote the actual plotted values). The process used to achieve this transformation is called scaling. Also note from Figure 10.16 that the origin of the coordinate system is not in the exact center of the screen. This is done in order to make practical use of the full size of the monitor. So, not only has scaling been used, but so has **coordinate transformation**.

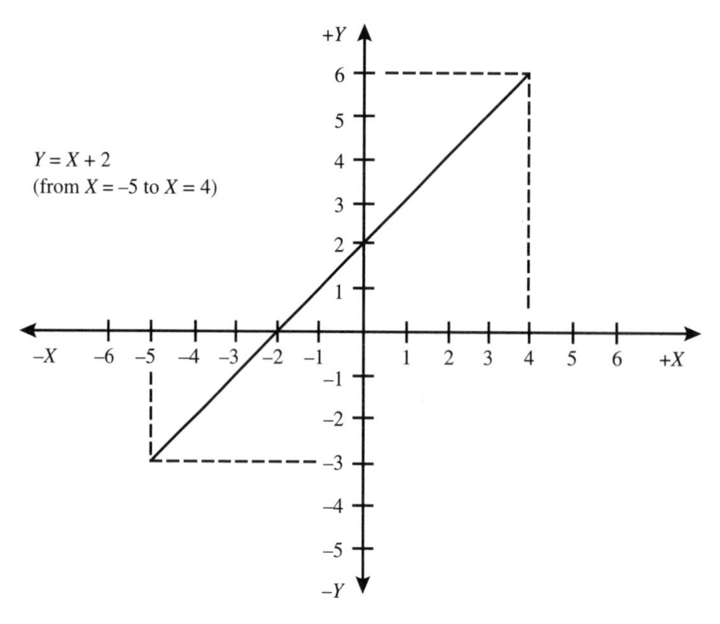

Figure 10.15 Graph for Scaling Example

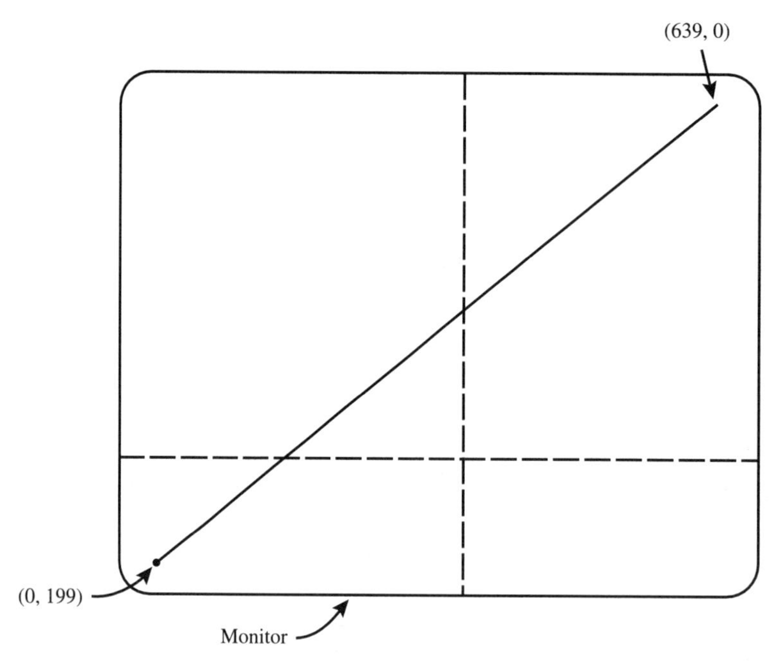

Figure 10.16 Actual Plotting Values for Graphical Display

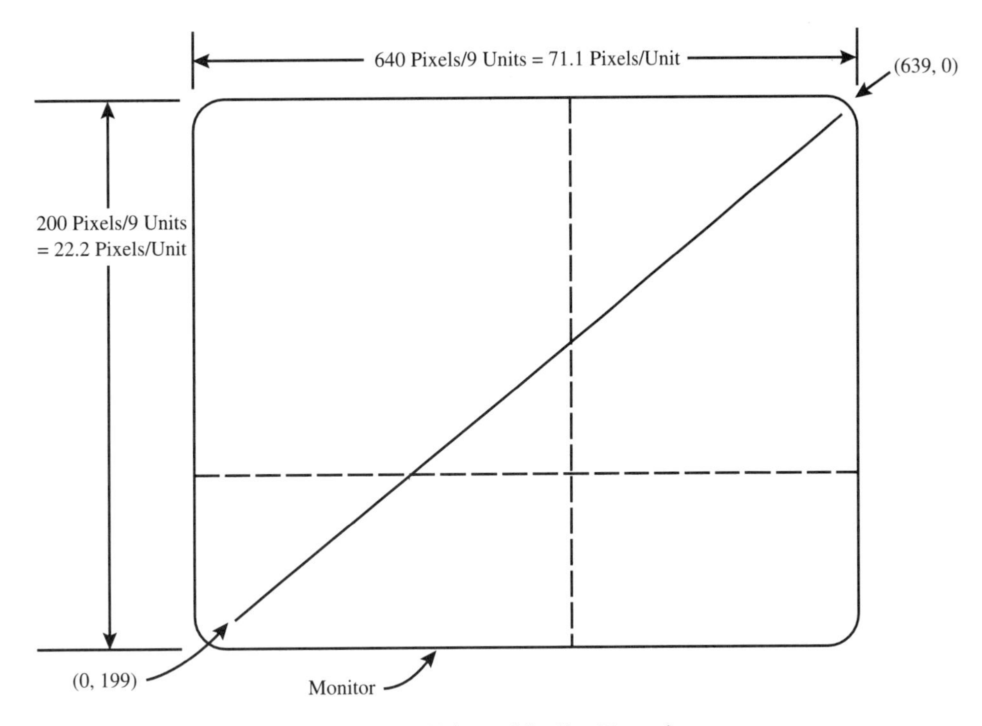

Figure 10.17 Minimum and Maximum Values of Scaling Example

Scaling

Observe Figure 10.17.

The horizontal scaling factor can be expressed mathematically as

$$HS = P_H/(X_2 - X_1)$$

Where

HS = The horizontal scaling factor.
P_H = Number of pixels in the horizontal direction.
X_1 = Minimum value of X.
X_2 = Maximum value of X.

The vertical scaling factor can be expressed mathematically as

$$VS = P_V/(Y_2 - Y_1)$$

Where

VS = The vertical scaling factor.
P_V = Number of pixels in the vertical direction.
Y_1 = Minimum value of Y.
Y_2 = Maximum value of Y.

To calculate the scaling factors for the graph in Figure 10.15:

For the horizontal:

$$HS = P_H / | X_2 - X_1 |$$
$$HS = 640/((4) - (-5))$$
$$HS = 640/(4 + 5) = 640/9 = 71.1$$

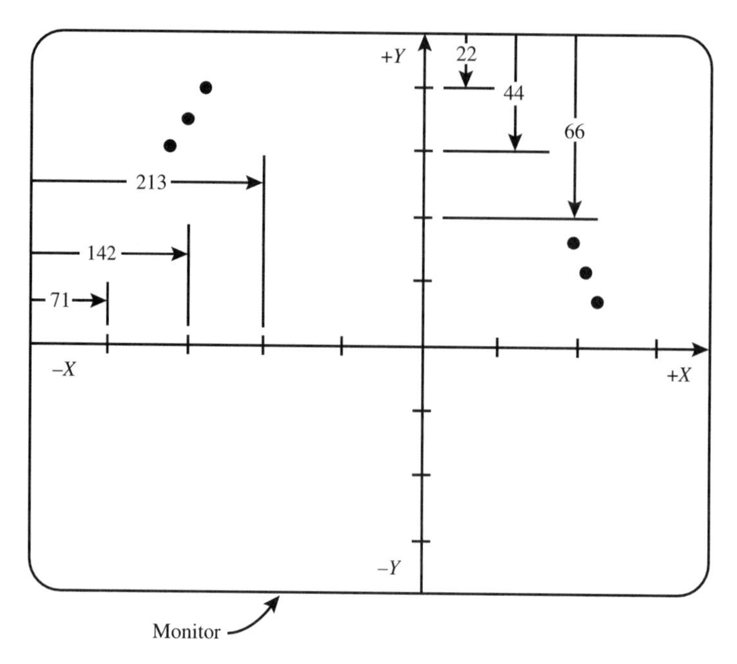

Figure 10.18 Meaning of Scaling for Example Graph

For the vertical:

$$VS = P_V / |Y_2 - Y_1|$$
$$VS = 200/((6) - (-3))$$
$$VS = 200/(6 + 3) = 200/9 = 22.2$$

What these calculations mean is that every major division along the X axis will be 71 pixels, and every major division along the Y axis will be 22 pixels. Doing this will use the full capabilities of the graphics screen. This is shown in Figure 10.18.

Coordinate Transformation

Note from Figure 10.18 that the origin of the coordinate system used to display the example graph is not in the exact center of the graphics screen. This was done intentionally in order to display the full range of required data utilizing the full graphics screen. In order to accomplish this, coordinate transformation was required. Coordinate transformation, mentioned in the last section, is the process of using numerical methods to cause the origin of the coordinate system to appear at any desired place on the graphics screen.

The process of transforming coordinates can be expressed mathematically as follows:

For the horizontal transformation:

$$XT = |(HS)(X_1)|$$

Where

XT = Horizontal transformation.
HS = Horizontal scaling factor.
X_1 = Minimum value of X.

For the vertical transformation:

$$YT = |(VS)(Y_2)|$$

Where

YT = Vertical transformation.
VS = Vertical scaling factor.
Y_2 = Maximum value of Y.

For our example this becomes

Horizontal transformation:

$$XT = |(HS)(X_1)|$$
$$XT = |(71.1)(-5)| = |-355|$$
$$XT = 355.$$

Vertical transformation:

$$YT = |(VS)(Y_2)|$$
$$YT = |(22.2)(6)| = |133|$$
$$YT = 133.$$

What these calculations mean is that the origin of the coordinate system will be at the locations of $X_G = 355$ and $Y_G = 133$. This is illustrated in Figure 10.19.

The only thing left to do now is to develop equations that can be used to display the data on the graphics screen.

Putting It Together

To display the graph of any continuous function of two variables when scaling and coordinate transformations are used, the following equations are necessary:

Horizontal value of graph:

$$XG = XT + (X)(HS)$$

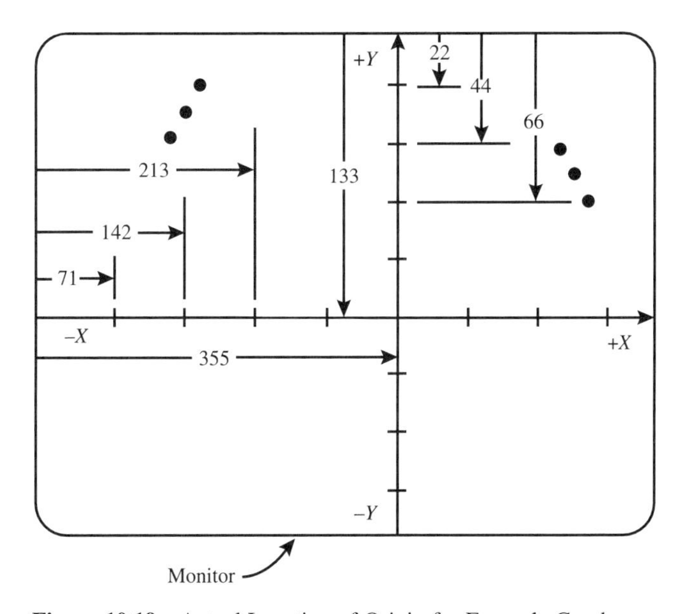

Figure 10.19 Actual Location of Origin for Example Graph

Where

XG = Graphical X value.
XT = Horizontal transformation.
HS = Horizontal scaling factor.

Vertical value of graph

$$YG = YT - (Y)(VS)$$

Where

YG = Graphical Y value.
YT = Vertical transformation.
VS = Vertical scaling factor.

Program 10.23 plots the graph of the function $X = Y + 2$. The program asks the user for the maximum and minimum values of X. It then draws the transformed coordinate system and scales the plot of the graph. Note that it makes no difference what pixel size the graphics screen has as long as Turbo C recognizes it.

Program 10.23

```
#include <stdio.h>
#include <graphics.h>          /* Necessary for graphics.        */

void auto_initialization(void); /* Function to initialize to the
                                   graphics mode.         */
        float Y_value(float X); /* Function for display.           */

main()
{
        float max_X;            /* Maximum horizontal value.       */
        float max_Y;            /* Maximum vertical value.         */
        float min_X;            /* Minimum horizontal value.       */
        float min_Y;            /* Minimum vertical value.         */
        float hor_scale;        /* Horizontal scaling factor.      */
        float vert_scale;       /* Vertical scaling factor.        */
        float hor_trans;        /* Horizontal transformation.      */
        float ver_trans;        /* Vertical transformation.        */
        int hor_pixel;          /* Horizontal pixel position.      */
        int ver_pixel;          /* Vertical pixel position.        */
        float X_cord;           /* X coordinate for graphing.      */

        auto_initialization();  /* Get into graphics mode.         */

        printf("\n\nMaximum X = ");
        scanf("%f",&max_X);

        max_Y = Y_value(max_X); /* Calculate maximum Y value.      */

        /* Calculate horizontal scale. */
        if(max_X == min_X)
                hor_scale = getmaxx();
        else
                hor_scale = getmaxx()/(max_X - min_X);
```

```
        /* Calculate vertical scale. */
        if(max_Y == min_Y)
                vert_scale = getmaxy();
        else
                vert_scale = getmaxy()/(max_Y - min_Y);

        /* Calculate horizontal and vertical transformations. */
        hor_trans = (hor_scale * min_X);
        if(hor_trans < 0)       /* Convert to a positive value. */
                hor_trans *= -1;
        ver_trans = (vert_scale * max_Y);
        if(ver_trans < 0)      /* Convert to a positive value. */
                ver_trans *= -1;

        /* Display calculated results. */
        printf("Horizontal scale = %f\n", hor_scale);
        printf("Vertical scale = %f\n", vert_scale);
        printf("Horizontal transformation = %f\n", hor_trans);
        printf("Vertical transformation   = %f\n", ver_trans);
        printf("Enter X to eXit => ");

        /* Draw the coordinate system. */
        setlinestyle(SOLID_LINE, 0, THICK_WIDTH);
        line(0, (int)ver_trans, getmaxx(), (int) ver_trans);
        line((int)hor_trans, 0, (int)hor_trans, getmaxy());

        /* Draw the graph. */

        X_cord = min_X;

        do
        {
                hor_pixel = (int)(hor_trans + X_cord * hor_scale);
                ver_pixel = (int)ver_trans - Y_value(X_cord) * vert_scale;
                putpixel(hor_pixel, ver_pixel, 1);
                X_cord = X_cord + 0.1;
        } while(hor_pixel < getmaxx());

        while('X' != toupper(getchar()));

        closegraph();
}

float Y_value(float X_value)
{
        return(X_value + 2.0);
}

void auto_initialization()
{
        int graph_driver;       /* Specifies the graphics driver.  */
```

```
    int graph_mode;          /* Specifies the graphics mode.     */

    graph_driver = DETECT;   /* Initializes autodetection.       */

    initgraph(&graph_driver, &graph_mode, "");

} /* End of auto_initialization. */
```

The graph resulting from Program 10.23 is shown in Figure 10.20.

Analyzing the Program

First, the function `auto_initialization` takes place. This is done so that the Turbo system knows what kind of graphics system the computer has (if any at all). The equation to be graphed is put in the form of a function. This is done so that it is easy to change the relationship between *X* and *Y* if you wish to plot a different formula.

The program computes the horizontal and vertical scales as well as the horizontal and vertical transformations necessary for the graphics screen. Note that if the maximum and minimum values of *X* or *Y* are equal (this could happen if plotting a quadratic), then the scales are changed to the maximum pixel values of the given graphics driver.

As an aid in seeing the chosen values, `printf()` statements are used to display the transformation and scaling information on the upper left portion of the screen. The graph is then constructed using the built-in Turbo C function `putpixel`. However, the `putpixel` function requires a type `int`, and all of the graph values up to this point have been of type `float`. In order to convert these numbers to type `int`, the process of type casting is used. For example, if `hor_trans` is of type `float`, it can be converted to a type `int` by

```
(int)hor_trans;
```

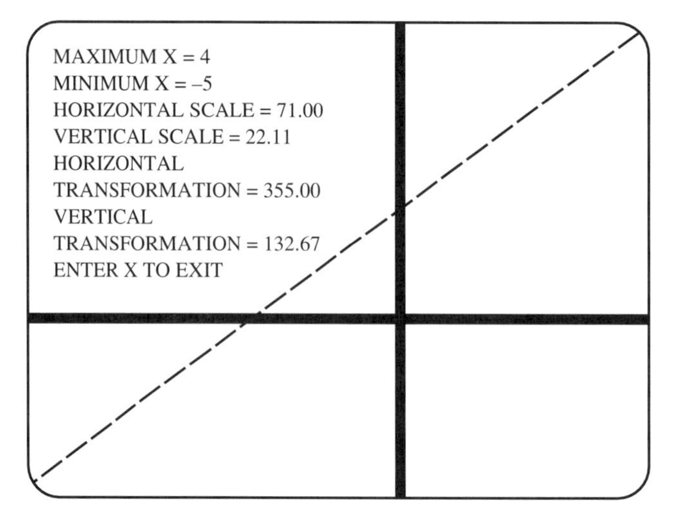

Figure 10.20 Result of Computer Generated Graphic

You will note that this was done several times in this program.

The value of X is incremented by 0.1 each time. This can be changed depending on the resolution you want (and the time it takes to generate the graph).

Conclusion

This section introduced you to the practical requirements for displaying the graph of a formula. Here you were introduced to the concepts of scaling and coordinate transformation. These techniques will prepare you for plotting technical data that relates one variable to another.

Test your understanding of this section by trying the following section review.

10.6 Section Review

1. Define scaling as it applies to computer graphics.
2. Define coordinate transformation as it applies to computer graphics.
3. State the factors that determine the values of the scaling factors.
4. State the factors that determine the values of the coordinate transformation values.

10.7 Example Program: Virtual Maze

The popularity of first-person virtual reality games (such as DOOM) is largely due to their realistic real-time color graphics. The player actually moves around in the game world, as if the game were looking through the player's eyes. Simulating a three-dimensional room or landscape can be a very time-consuming and computationally complex exercise. Doing the simulation in real-time is even harder. The programmers behind these types of games know how to squeeze the most performance out of the processor. They must pay attention to execution time at every step. In addition, they need to have a good understanding of algebra and trigonometry (as well as physics) to write the necessary graphic routines.

In this section we will examine the development, and theory behind, a simple 3-D, first-person game called *Virtual Maze*. The player "walks" through the maze by pressing the arrow keys or by moving the mouse. The goal is to reach an "Exit" door.

The Virtual Maze game changes the computer's video mode to VGA and uses a 320 by 200 pixel resolution, with 256 colors. Figure 10.21 shows a screen shot of the game in action.

The technique used to draw the game screen is called *ray casting*. The ray casting must be done as fast as possible so that the screen updates in real-time. Let us see how ray casting works.

Ray Casting

The basic method behind ray casting follows this sequence of steps:

1. Pick a direction for the ray.
2. Find the distance to the first horizontal intersection of the ray with an object.
3. Find the distance to the first vertical intersection of the ray with an object.
4. Pick the intersection with the smallest distance.
5. Use the distance to calculate the height of a vertical strip to draw on the screen.
6. Draw the strip.

Figure 10.21 Virtual Maze Screen Shot

The objects in Virtual Maze are walls. So, steps 2 and 3 are actually finding the distance to the closest wall in front of the player. The height of the wall is based on its distance from the player. The larger the distance, the smaller the height. Figure 10.22 shows a sample ray being cast.

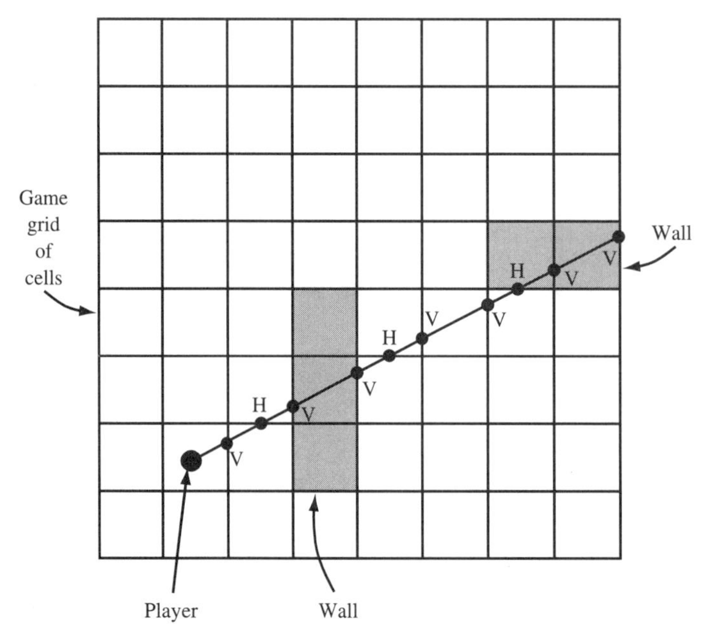

Figure 10.22 Casting a Ray

The ray begins at the player's position and extends outward until it eventually hits the edge of the game world, which is composed of a rectangular grid of *cells,* where each cell represents a 64 by 64 section of the overall game world. The 8 by 8 game grid in Figure 10.22 actually represents a 512 by 512 world. Each cell contains one of three things: the player, a floor section, or a wall section.

Finding the intersection of a ray involves checking horizontal intersections (H in Figure 10.22) and vertical intersections (V). Notice that there are only three horizontal intersections in Figure 10.22, and seven vertical ones. Some of these intersections hit cells that contain wall sections. The second vertical intersection (the first to hit a wall) is closer to the player than the third horizontal intersection (also the first to hit a wall). A strip from the wall associated with the closer vertical intersection is drawn.

To generate a screen, 320 rays are cast. The player's direction is used to cast out rays beginning at 30 degrees minus the player's direction, and going to 30 degrees plus the player's direction. This sweeps a 60 degree arc over the game field. Each ray cast results in one column of pixels drawn on the screen (the strip from steps 5 and 6). So, casting 320 rays results in an entire screen of the game world being drawn.

Representing the Game World

As previously mentioned, the game world is represented by a rectangular grid of cells, each of which contains information about a specific 64 by 64 section of the virtual game world. The game world rendered by Virtual Maze is stored in an ASCII file called `GAMEMAP.DAT`, which looks like this:

```
1111111111111111111111111111111111114444444444444444444444444444
1...................................4..........................4
1...................................4..........................4
1...................................4..........................4
1...222222...2222222222222222...22222222222222222222222....44444
1........2.......4..........2...2..............................1
1........2.......4..........2...2..........................11111
1....1...2...222222222222...........2222222222222222222....13E33
1....1...2............................................2....1...3
144441...2....................P.......................244441...3
4....1...2............................................2....1...3
4....1...2222222222222222...........22222222222.......2....1...3
4............4..............2...2..........................1...3
4............4..............2...2..........................1...3
4...44444...22222222222222222...2222222222222222222222222221...3
4......4.................1........1.......3.......3........3...3
4......4.................1111111111.......3.......3........3...3
4......4..............................3.......3.......3........3
4......4..............................3.......3.......3........3
4444444411111111111111111111111111111133333333333333333333333333
```

where the characters have the following meaning:

`.`	Floor
`1-4`	Wall patterns
`P`	Player
`E`	Exit

Virtual Maze reads the `GAMEMAP.DAT` file at the beginning of the game. The cell containing `P` is the initial player position. The player is only allowed to move into cells that are floor cells, or into the Exit cell (which ends the game).

The game grid is stored in ASCII so that you may create your own virtual maze with a simple text editor, such as the EDIT utility supplied with DOS.

The Walls

The patterns for the walls are stored in the `WALLPATT.DAT` file. Each wall (there are five) is represented by a block of 4096 bytes, organized into a two-dimensional array that contains data for a 64 by 64 pixel pattern. One column of the array specifies one column of pixels (a strip) on the screen. Each pixel can be one of the 256 standard VGA colors.

The wall patterns can be modified, or new patterns created, using the `WE.EXE` utility supplied on the companion disk. `WE.EXE` creates and stores a single wall pattern into a data file. Use this command line to edit or create a pattern:

```
C> WE IMAGE.DAT
```

When you have five pattern files, create a new `WALLPATT.DAT` file with this DOS command:

```
C> COPY P1.DAT + P2.DAT + P3.DAT + P4.DAT + P5.DAT   WALLPATT.DAT
```

The five wall patterns are loaded into a global three-dimensional array at the beginning of the game. The fifth pattern is always the Exit wall.

Drawing a strip of a wall is done after the nearest intersection is found and the distance to it has been calculated. The distance is used to scale the height of the strip so that its height changes in proportion to distance. The maximum height is limited to 200 pixels, since the resolution of the game screen is 320 by 200.

When the height of the strip is known, a technique called *texture mapping* is used to transfer the graphical strip data from the global pattern array to the screen. The character stored in the game map cell that is associated with the intersection specifies the wall pattern to use.

Three cases are possible when drawing a strip:

1. The height of the strip is less than 64.
2. The height of the strip equals 64.
3. The height of the strip is greater than 64.

Figure 10.23 shows how strips are mapped in each case. As Figure 10.23(a) shows, when the strip is scaled down, pixels in the original strip (stored in the global walls array) are skipped at regular intervals. Instead of skipping pixel values, Figure 10.23(c) shows pixel values being duplicated in the scaled strip, the essential technique behind a zoom operation. To draw the scaled strip as fast as possible, a custom function is used, rather than one of the built-in graphic functions (which do not perform texture mapping either).

The Floor and Ceiling

To add to the overall sense that we are looking at a real environment, it is necessary to simulate the floor and ceiling as well. Texture mapping can be used to map lights onto

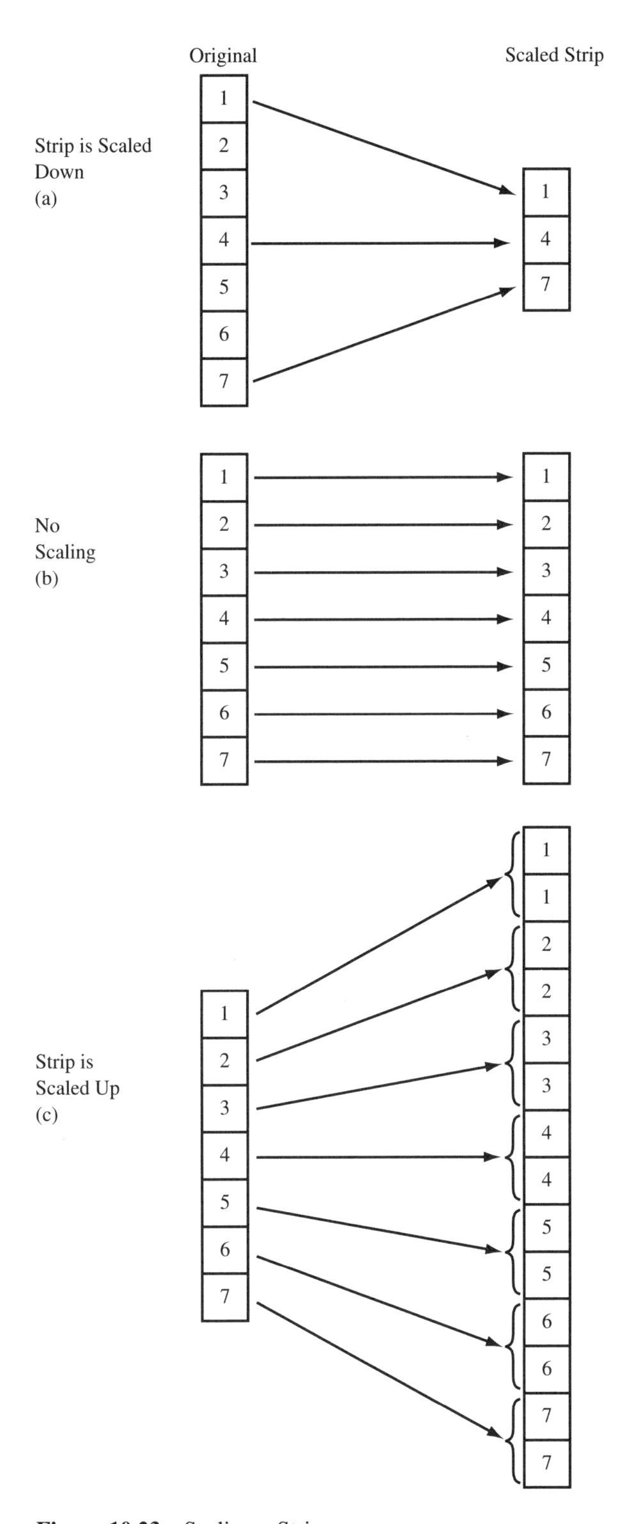

Figure 10.23 Scaling a Strip

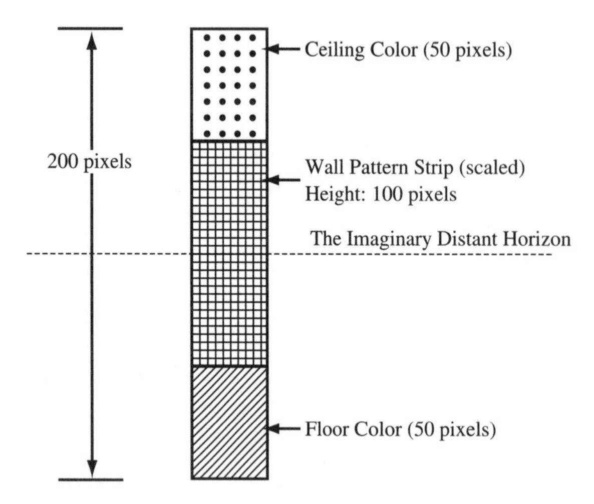

Figure 10.24 One Column of Pixels in the Game Screen

the ceiling and panels onto the floor (or carpeting, or pools of radioactive material). In Virtual Maze, two different colors are used to create the effect of a floor and ceiling. No texture mapping is performed.

Figure 10.24 illustrates the components of an entire column of pixels in the game screen. In this example, the 100 pixel wall strip is centered around line 100. The remaining 100 pixels of the column are divided equally, giving 50 pixels each for the ceiling and floor colors. Every strip has its height centered around line 100 to create the illusion of a horizon far off in the distance.

Keeping Things Fast

A number of techniques are used to help minimize the execution time of the code. Let us examine them.

The functions that find the horizontal and vertical intersections use a rearranged line equation to solve for the next horizontal or vertical intercept. The equations use the tangent (and inverse tangent) of the ray's angle in their solution. Calculating a trigonometric function during ray casting will slow things down significantly. To eliminate this problem, an array of precomputed tan and inv-tan values are used as a lookup table during ray casting. The tables are computed for a specific set of angles at the beginning of the game, before ray casting begins. The set of angles begins at 0 degrees and goes up in fixed increments all the way to 360 degrees. The size (in angles) of the player's field of view and the number of columns in the game screen determine the size of the increments between angles. Rays are only cast out at angles that are represented in the tables.

For every pixel written to the screen, a calculation must be performed to find its address within the video display memory used by the VGA hardware. The 320 by 200 pixel resolution requires a display memory of 64,000 bytes. So, pixels have addresses between 0 and 63,999. One way to generate the address looks like this:

```
pixel_address = (row * 320) + column;
```

This method requires a multiply operation, which really means 64,000 multiply operations when drawing an entire screen. This could be very time consuming. Instead, pixel addresses are generated like this:

```
pixel_address = (row << 8) + (row << 6) + column;
```

This equation takes advantage of the fact that the sum of 64 and 256 is 320. In addition, shifting a binary number to the left 6 bits (or 8) is equivalent to multiplying the number by 64 (or 256). The shift left operation is much faster than the multiply, and helps keep the address generation time to a minimum.

The third technique involves the use of a `#define`, rather than a function, to actually plot a pixel on the screen. A pixel-plotting function would require three parameters (row, column, and color) to be pushed onto the run time stack, along with other necessary information. Once again, this is really 64,000 sets of pushes (and pops) for an entire game screen. To speed things up, a `#define` is used instead. The `#define` will be expanded into code *in place*, and will not require use of the run time stack for parameter passing. The `#define` code is essentially the same code that would be in the function; we have just eliminated the need to call the function.

The `#define` that plots pixels in the Virtual Maze program is as follows:

```
#define pp(r,c,color) vbuff[(((int)r<<8)+((int)r<<6))+(int)c]=(unsigned char)color;
```

Note the use of type casting. This is necessary with the `r` and `c` variables, which are declared as `floats` to more accurately keep track of the pixels position.

`vbuff` is a pointer initialized to the base address of the VGA display memory. This address is typically `0xA0000000` on a DOS machine.

The Overall Game Loop

Virtual Maze uses a very simple loop:

1. Get player movements.
2. Ray cast the new scene.

Player movements are controlled by the arrow keys on the keyboard, or by moving the mouse. The up arrow moves the player forward, the down arrow backward. The left and right arrows rotate the player counterclockwise and clockwise, respectively. Similar movements on the mouse have the same effect.

Once the player's new position and direction have been calculated (assuming the player has not tried to walk through a wall), a new scene is rendered. If the player stands still, the ray caster does not render a new scene. This helps to reduce flicker and is acceptable since Virtual Maze is a static environment. A dynamic environment contains things that move, lights that flash, and so forth and must be continuously rendered.

The Program

Program 10.24 is the Virtual Maze game. It is quite lengthy and will require a good deal of time to examine. A number of its parts will require you to skip ahead to Chapter 11 for some information on interrupts and 80x86 microprocessor architecture. These sections of code are kept very small and should be straightforward if you read the comments and think about what the game is doing while the code is executing.

Program 10.24

```
/* Virtual Maze */

#include <bios.h>
#include <conio.h>
#include <dos.h>
#include <fcntl.h>
#include <io.h>
#include <math.h>
#include <memory.h>
#include <stdio.h>
#include <stdlib.h>
#include <string.h>

/* ------------------------------------------------------------------------ */
/* Various constants */

/* Boolean */

#define FALSE      0
#define TRUE       1

/* Video */

#define VGA        0x13
#define TEXT       0x03
#define FLOOR      66
#define CEILING    34
#define STRIP_SIZE 200

/* Math */

#define PI         3.1415926536
#define TSIZE      1920
#define ANGLEINC   ((double)360/TSIZE)
#define D90        TSIZE/120*90/3
#define D180       2*D90
#define D270       3*D90
#define INFINITY   1e10
#define NOTZERO    1e-10
#define SCALE      14000

/* Keyboard */

#define UP         72
#define DN         80
#define LT         75
```

```
#define RT          77

/* Mouse */

#define RESET_MOUSE 0
#define GET_STATUS  3
#define READ_MOUSE  0x0B
#define MOUSE_INT   0x33

/* Time */

#define READ_TIME   0x2c
#define DOS_INT     0x21

/* Game */

#define MAPCOLS     64
#define MAPROWS     20
#define CELLSIZE    64
#define RMAX        MAPROWS*CELLSIZE
#define CMAX        MAPCOLS*CELLSIZE
#define NUMWALLS    5

/* Stuff to do on-the-fly */

#define shl(r,n) ((int)r << n)
#define pp(r,c,color) vbuff[shl(r,8)+shl(r,6)+(int)c]=(unsigned char)color
#define cellval(x,y) game_map[(int)x/CELLSIZE][(int)y/CELLSIZE]
#define sqr(z) (z)*(z)
#define is_wall(cell) ((('1' <= cell) && (cell <= '4')) || (cell == 'E'))
#define dist(x1,y1,x2,y2) sqrt(sqr(x1-x2) + sqr(y1-y2))

/* ------------------------------------------------------------------ */
/* Define player info. */

typedef struct
{
        int cellr, cellc;
        float r,c;
        float dir;
} player_type;

        player_type player;

/* Precomputed tables */

        float tantab[TSIZE];
        float invtan[TSIZE];
        float corr[320];

/* Wall variables */
```

```
        unsigned char wall_maps[NUMWALLS][CELLSIZE][CELLSIZE];
        int tptr,slice,hslice,vslice,wall;
        unsigned char hcolor,vcolor;

/* Game control */

        unsigned char game_map[MAPROWS][MAPCOLS];
        int stopped;
        long starttime;
        int gotdir;
        int mousedr,mousedc;
        unsigned char oldfloor;
        float angle,ta;

/* Pointer to base of video RAM */

unsigned char far *vbuff = (char far *)0xA00000001;

/* ---------------------------------------------------------------------- */
/* Function Prototypes */

float hit_hwall(void);
float hit_vwall(void);
void ray_cast(void);
void draw_strip(int col, int size, int pos);
int move_player(void);
void build_tables(void);
void load_map(void);
void load_walls(void);
int use_mouse(int op);
void video_mode(int mode);

/* ---------------------------------------------------------------------- */
/* Here we go... */

void main()
{
        int k,still_playing,mins,secs;
        long stoptime,gametime;
        union REGS regs;

        regs.h.ah = READ_TIME;
        int86(DOS_INT, &regs, &regs);
        starttime = regs.h.ch * 3600 + regs.h.cl * 60 + regs.h.dh;
        video_mode(VGA);
        load_map();
        load_walls();
        build_tables();
        use_mouse(RESET_MOUSE);
        ray_cast();
        stopped = FALSE;
```

```
        still_playing = TRUE;
        while (!stopped && still_playing)
        {
                still_playing = move_player();
                ray_cast();
        }
        video_mode(TEXT);
        regs.h.ah = READ_TIME;
        int86(DOS_INT, &regs, &regs);
        stoptime = regs.h.ch * 3600 + regs.h.cl * 60 + regs.h.dh;
        gametime = stoptime - starttime;
        if (gametime < 0)
                gametime += 24*3600;
        mins = gametime / 60;
        secs = gametime % 60;
        printf("You played for %d minutes and %d seconds.\n",mins,secs);
}

/* --------------------------------------------------------------- */
/* Set the video mode (using inline assembly language) */
/* See Chapter 11 */

void video_mode(int mode)
{
        asm mov ah,0
        asm mov al, byte ptr mode
        asm int 0x10
}

/* --------------------------------------------------------------- */
/* Read the game map into memory */

void load_map()
{
        FILE *fp, *fopen();
        int r,c;
        unsigned char mdat;

        fp = fopen("gamemap.dat","rb");
        for(r = 0; r < MAPROWS; r++)
        {
                for(c = 0; c < MAPCOLS; c++)
                {
                        mdat = getc(fp);
                        game_map[r][c] = mdat;
                        if (mdat == 'P')
                        {
                                player.cellr = r;
                                player.cellc = c;
                                player.r = r*CELLSIZE + CELLSIZE/2;
                                player.c = c*CELLSIZE + CELLSIZE/2;
```

```
                                player.dir = (starttime % 8) * 45;
                                oldfloor = game_map[r][c-1];
                        }
                }
                mdat = getc(fp); /* eat cr */
                mdat = getc(fp); /* eat lf */
        }
        fclose(fp);
}

/* ------------------------------------------------------------------ */
/* Precompute tangent and distance-correction tables */

void build_tables()
{
        int k;
        double angle;

        for(k = 0; k < TSIZE; k++)
        {
                angle = k*ANGLEINC*PI/180.0;
                switch(k)
                {
                case    0: tantab[k] = 0;
                           invtan[k] = INFINITY;
                           break;
                case  D90: tantab[k] = INFINITY;
                           invtan[k] = 0;
                           break;
                case D180: tantab[k] = 0;
                           invtan[k] = INFINITY;
                           break;
                case D270: tantab[k] = INFINITY;
                           invtan[k] = 0;
                           break;
                default:   tantab[k] = (float)tan(angle);
                           invtan[k] = (float)(1.0/tantab[k]);
                }
        }
        angle = -30.0*PI/180.0;
        for(k = 0; k < 320; k++)
        {
                corr[k] = (float)1.0/cos(angle);
                angle += 2*30/320.0*PI/180.0;
        }
}

/* ------------------------------------------------------------------ */
/* Calculate new player position based on keyboard/mouse inputs */

int move_player()
```

```
{
        int mr,mc,left_click,button;
        player_type temp;
        char kyb;
        unsigned char spot;

        left_click = use_mouse(READ_MOUSE);
        if (left_click)
                stopped = TRUE;
        temp.r = player.r;
        temp.c = player.c;
        temp.dir = player.dir;
        gotdir = FALSE;
        if (kbhit())
        {
                kyb = getch();
                if (!kyb)
                {
                        kyb = getch();
                        gotdir = TRUE;
                        switch(kyb)
                        {
                              case UP:    temp.r -= 24*sin(temp.dir*PI/180.0);
                                          temp.c += 32*cos(temp.dir*PI/180.0);
                                          break;
                              case DN:    temp.r += 24*sin(temp.dir*PI/180.0);
                                          temp.c -= 32*cos(temp.dir*PI/180.0);
                                          break;
                              case LT:    temp.dir += 3; break;
                              case RT:    temp.dir -= 3; break;
                              default:    gotdir = FALSE;
                        }
                }
        }
        if (mousedr || mousedc)
        {
                temp.r += mousedr*sin(temp.dir*PI/180.0);
                temp.c -= mousedr*cos(temp.dir*PI/180.0);
                temp.dir -= mousedc/12.0;
                button = use_mouse(RESET_MOUSE);
                gotdir = TRUE;
        }
       if(temp.r < 0) temp.r = 0;
       if(temp.r > RMAX) temp.r = RMAX;
       if(temp.c < 0) temp.c = 0;
       if(temp.c > CMAX) temp.c = CMAX;
       if(temp.dir < 0)
                temp.dir += 360;
       if(temp.dir > 360)
                temp.dir -= 360;
       temp.cellr = temp.r / CELLSIZE;
```

```
        temp.cellc = temp.c / CELLSIZE;
        spot = game_map[temp.cellr][temp.cellc];
        if((spot == '.') || (spot == 'P') && gotdir)
        {
                game_map[player.cellr][player.cellc] = oldfloor;
                player.cellr = temp.cellr;
                player.cellc = temp.cellc;
                player.r = temp.r;
                player.c = temp.c;
                player.dir = temp.dir;
                oldfloor = game_map[player.cellr][player.cellc];
                game_map[player.cellr][player.cellc] = 'P';
        }
        else
        if ((spot == 'E') && gotdir)
                stopped = TRUE;
        return(kyb != 'q');
}

/* ------------------------------------------------------------------- */
/* Reset/Read the mouse */

int use_mouse(int op)
{
        union REGS regs;

        switch(op)
        {
                case RESET_MOUSE: regs.h.ah = RESET_MOUSE;
                        int86(MOUSE_INT, &regs, &regs);
                        return(0);
                case READ_MOUSE: regs.x.ax = READ_MOUSE;
                        int86(MOUSE_INT, &regs, &regs);
                        mousedr = regs.x.dx;
                        if (mousedr > 50) mousedr = 50;
                        if (mousedr < -50) mousedr = -50;
                        mousedc = regs.x.cx;
                        if (mousedc > 50) mousedc = 50;
                        if (mousedc < -50) mousedc = -50;
                        regs.x.ax = GET_STATUS;
                        int86(MOUSE_INT, &regs, &regs);
                        return(regs.x.bx & 1);
                default:
                        return(0);
        }

}

/* ------------------------------------------------------------------- */
/* Look for first horizontal intersection with a wall */

float hit_hwall()
```

```
{
        float rpos,rstep,cpos,cstep;
        int found;

        rpos = (player.cellr * CELLSIZE) - 1;
        cstep = CELLSIZE * invtan[tptr];
        if (angle < 180)
                rstep = -CELLSIZE;
        else
        {
                rstep = CELLSIZE;
                rpos += (1+CELLSIZE);
                cstep *= -1;
        }
        cpos = player.c + (player.r - rpos) * invtan[tptr];
        hcolor = cellval(rpos,cpos);
        if (is_wall(hcolor))
        {
                if (hcolor == 'E')
                        hcolor = NUMWALLS - 1;
                else
                        hcolor -= '1';
                hslice = ((int)cpos) % CELLSIZE;
                return dist(player.r,player.c,rpos,cpos);
        }
        found = 0;
        while ((cpos >= 0) && (cpos < CMAX) && !found)
        {
                rpos += rstep;
                cpos += cstep;
                hcolor = cellval(rpos,cpos);
                if (is_wall(hcolor))
                {
                        found = 1;
                        if (hcolor == 'E')
                                hcolor = NUMWALLS - 1;
                        else
                                hcolor -= '1';
                        hslice = ((int)cpos) % CELLSIZE;
                }
        }
        if (found)
                return dist(player.r,player.c,rpos,cpos);
        else
        {
                hcolor = 0;
                hslice = -1;
                return INFINITY;
        }
}

/* ------------------------------------------------------------------ */
```

```
/* Look for first vertical intersection with a wall */

float hit_vwall()
{
        float rpos,rstep,cpos,cstep;
        int found;

        cpos = (player.cellc * CELLSIZE) - 1;
        rstep = CELLSIZE * tantab[tptr];
        if ((angle >= 90) && (angle < 270))
                cstep = -CELLSIZE;
        else
        {
                cstep = CELLSIZE;
                cpos += (1+CELLSIZE);
        }
        if ((angle < 90) || (angle >= 270))
                rstep *= -1;
        rpos = player.r + (player.c - cpos) * tantab[tptr];
        vcolor = cellval(rpos,cpos);
        if (is_wall(vcolor))
        {
                if (vcolor == 'E')
                        vcolor = NUMWALLS - 1;
                else
                        vcolor -= '1';
                vslice = ((int)rpos) % CELLSIZE;
                return dist(player.r,player.c,rpos,cpos);
        }
        found = 0;
        while ((rpos >= 0) && (rpos < RMAX) && !found)
        {
                rpos += rstep;
                cpos += cstep;
                vcolor = cellval(rpos,cpos);
                if (is_wall(vcolor))
                {
                        found = 1;
                        if (vcolor == 'E')
                                vcolor = NUMWALLS-1;
                        else
                                vcolor -= '1';
                        vslice = ((int)rpos) % CELLSIZE;
                }
        }
        if (found)
                return dist(player.r,player.c,rpos,cpos);
        else
        {
                vcolor = 0;
                vslice = -1;
                return INFINITY;
```

```
        }
}

/* ---------------------------------------------------------------- */
/* Ray cast an entire scene (320 columns) */

void ray_cast()
{
        float xdist,ydist,dist,temp;
        int i,r,c,ray,size;
        unsigned char color;

        color = 0;
        for(ray = 0; ray < 320; ray++)
        {
                angle = player.dir - ray*ANGLEINC + 30;
                if (angle < 0)
                        angle += 360;
                if (angle > 360)
                        angle -= 360;
                ta = (angle/ANGLEINC);
                tptr = ta;
                xdist = hit_hwall();
                ydist = hit_vwall();
                if (xdist < ydist)
                {
                        dist = xdist;
                        color = hcolor;
                        wall = color;
                        slice = hslice;
                }
                else
                {
                        dist = ydist;
                        color = vcolor;
                        wall = color;
                        slice = vslice;
                }
                temp = (SCALE/(dist + NOTZERO)*corr[ray]);
                size = (int)(temp + 0.5);
                if (slice == -1) slice = 0;
                draw_strip(ray,size,slice);
        }
}

/* ---------------------------------------------------------------- */
/* Draw a scaled, texture-mapped vertical strip */

void draw_strip(int col, int size, int pos)
{
        int i,j,k;
        float pixel_inc,pixel_row;
```

```
        unsigned char color;

        pixel_inc = (float)CELLSIZE / size;
        if (size > STRIP_SIZE)
        {
                pixel_row = ((size - STRIP_SIZE) / 2) * pixel_inc;
                size = STRIP_SIZE;
        }
        else
                pixel_row = 0;
        j = (STRIP_SIZE - size) / 2;
        for(i = 0; i < j; i++)
                pp(i,col,CEILING);
        for(k = 0; k < size; k++)
        {
                color = wall_maps[wall][(int)pixel_row][pos];
                pixel_row += pixel_inc;
                pp(j,col,color);
                j++;
        }
        for(i = j; i < STRIP_SIZE; i++)
                pp(i,col,FLOOR);
}

/* ------------------------------------------------------------------ */
/* Read the wall patterns into memory */

void load_walls()
{
        FILE *fp, *fopen();
        int wall,r,c;

        fp = fopen("wallpatt.dat","rb");
        for(wall = 0; wall < NUMWALLS; wall++)
                for(r = 0; r < CELLSIZE; r++)
                        for(c = 0; c < CELLSIZE; c++)
                                wall_maps[wall][r][c] = getc(fp);
        fclose(fp);
}
```

A large number of global variables are used to pass information back and forth between various functions. While it is good to keep global variable usage to a minimum, in this case we take advantage of the global accessibility to avoid time-consuming parameter passing on the run time stack.

What's Next?

Once you have a complete understanding of how the program works, you can make many improvements to challenge your programming skills. Here is a short list of features you might try to add:

- Textured ceilings and floors
- Flashing lights
- Other objects (people, robots, monsters) to interact with the player
- Transporters
- Doors that open and close
- Stairs and elevators to different levels (almost a true 3-D world)
- Furniture
- A view of the sky

If you are interested in further details of game design, a large number of books are available to satisfy your needs. The inspiration for this material came from an excellent chapter in *Tricks of the GAME Programming Gurus.* Chapter 6, "The Third Dimension," written by Andre LaMothe, is a detailed introduction to ray casting, with many illustrations to help you understand the material.

Conclusion

This section was a brief introduction to a simple first-person virtual reality game. Many of the topics presented in this book are employed in the Virtual Maze game. Test your understanding of this material with the following section review.

10.7 Section Review

1. What is a first-person game?
2. Describe the ray casting technique.
3. Why is it important to cast every ray as fast as possible?
4. How is the game world represented?
5. How are walls represented?
6. What is texture mapping?
7. Why are the trig tables built at the beginning of the game?

10.8 Troubleshooting Techniques

Discussion

Troubleshooting a problem in a graphical application can be very tricky. Here is a short list of things that often go wrong:

1. The screen goes blank and the computer locks up.
2. The screen is a meaningless jumble of graphics.
3. The colors are wrong.
4. An image slants to the left or the right.
5. The proportions of the display are incorrect.

Finding the cause behind any of these problems requires patience and careful examination of every statement. In graphical applications the calculations that generate coordinates (like those for the endpoints of a line) must be checked with all possible combinations of input values. If an input requires an angle, you must be sure the trigonometry operations on the angle give the proper results for all quadrants.

It is often useful to print out the results of individual calculations as the program executes. Unfortunately, the information printed shows up on the graphics screen and interferes with the real graphical work being done.

One way to fix this problem is to leave the print statements in the program but *redirect* the output to a data file. This generates a permanent record of what went on during the program's execution. Redirection is easily specified in the DOS command line that launches the graphical application. For example, Program 10.24 (Virtual Maze) from Section 10.7 could be executed like this:

```
C> P10-24 > GAME.OUT
```

Any data printed during execution will be placed into the `GAME.OUT` file, which can then be viewed with a text editor or printed.

Conclusion

Having a permanent record of the results is much more useful than trying to read the results from the graphical screen before they are overwritten by other graphics. Test your knowledge of this material by trying the following section review.

10.8 Section Review

1. Why are `printf()` statements useful when troubleshooting a graphical application?
2. What is output redirection?

10.9 Case Study: Sinewave Generator

Discussion

The case study for this chapter demonstrates the use of text color, graphics, and different text fonts. Here you will see the development of a technical program used to generate a sine wave. The sine wave has a wide range of applications to many areas of technology, especially in the area of electronics technology.

The Problem

Create a demonstration program in C that will generate a sine wave. The program user may input the values of the amplitude, number of cycles, and the phase. The program is not to use any global variables, and it must demonstrate the use of the C `typedef struct` where appropriate. The user has the option of repeating the program.

First Step—Stating the Problem

Stating the case study in writing yields

- Purpose of the program: Provide a program that will generate a sine wave. User input is the amplitude, number of cycles, and phase of the sine wave. User has program repeat option.

- Required Input: Amplitude, number of cycles, and phase of the sine wave. Is the program to be repeated?
- Process on Input: Perform required calculations.
- Required Output: Sine wave which has the amplitude, number of cycles, and phase of the user input.

Developing the Algorithm

Developing the algorithm requires the following steps:

1. Explain program to user.
2. Prompt user for input:
   Amplitude
   Number of cycles
   Phase angle
3. Display the sine wave.
4. Ask for program repeat (do not repeat program explanation).

Arithmetic Functions

This program requires the use of one of the C built-in math functions. (The ANSI C standard math functions are listed in Appendix B.) The one that will be used for this program is the built-in `sin()` function, defined as

```
double sin(double X);
```

where *X* is the angle expressed in radians.

Program Development

In keeping with the idea that this program is to be a model program with C `typedef struct`, Program 10.25 has been developed. Note that there are actually two C `typedef` structures used in the program. The program also makes extensive use of text color and displays the instructions on the 40-character-wide screen.

Observe the block structure and the fact that a whole structure of information is passed from the calling function to the called function. This is another advantage of using structures in C: many different data types may be easily passed between functions by using structures in their arguments.

Program 10.25

```
#include <stdio.h>
#include <graphics.h>          /* Necessary for graphics.           */
#include <conio.h>
#include <math.h>

#define PI  3.14159
#define TRUE 1
#define FALSE 0
```

```
/**********************************************************************/
/*                   Sine Wave Generation Program                     */
/**********************************************************************/
/*                   Developed by: A. C. Student                      */
/**********************************************************************/
/*      This program will generate a sine wave.  The amplitude,       */
/*  number of cycles, and the phase may be determined by the          */
/*  program user.                                                     */
/**********************************************************************/
/*                      Type Definitions                              */
/*--------------------------------------------------------------------*/

typedef struct
{
        float degrees;          /* Number of degrees.                 */
        float radians;          /* Number of radians.                 */
        int X_value;            /* Number of X coordinate.            */
        int Y_value;            /* Number of Y coordinate.            */
} wave_values;
/*  This type definition defines the values of the sine wave.         */
/*--------------------------------------------------------------------*/

typedef struct
{
        char user_input[10];    /* User input string.                 */
        int amplitude;          /* Height of the sine wave.           */
        int cycles;             /* Number of cycles to be displayed.  */
        int phase;              /* Phase of the sine wave.            */
} user_input;
/* This type definition defines the values inputted by the user.      */
/**********************************************************************/
/*                      Function Prototypes                           */
/*--------------------------------------------------------------------*/

void explain_program(void);
/*      This function explains the purpose of the program to the      */
/*  user.                                                             */
/*--------------------------------------------------------------------*/
void press_return(void);
/*      This function holds the current screen for the program user   */
/*  until the -RETURN- key is pressed.                                */
/*--------------------------------------------------------------------*/
void auto_initialization(void);
/*      This function initializes the graphics screen.                */
/*--------------------------------------------------------------------*/
void sine_wave_display(user_input values);
/*      values = structure of user input values.                      */
/*  This function displays the actual sine wave.                      */
/*--------------------------------------------------------------------*/
int program_repeat(void);
```

```
/*      Gives program user the option of repeating the program.              */
/**************************************************************************/

main()
{
        user_input in_values;   /* Values inputted by user.                    */

        explain_program();      /* Explain program to user.                    */

        do
        {
                /* Get user input. */
                printf("\nAmplitude => ");
                gets(in_values.user_input);
                in_values.amplitude = atoi(in_values.user_input);
                printf("\nCycles => ");
                gets(in_values.user_input);
                in_values.cycles = atoi(in_values.user_input);
                printf("\nPhase => ");
                gets(in_values.user_input);
                in_values.phase = atoi(in_values.user_input);

                sine_wave_display(in_values);   /* Display sine wave.    */
        } while(program_repeat());

        closegraph();           /* Shut down the graphics system.              */
        textmode(C80);          /* Return to normal text mode.                 */

}  /* End of main. */

/*----------------------------------------------------------------------*/

void explain_program()          /* Explain program to user.                    */
{
        char *string;           /* String to hold program execution.     */

        textmode(C40);          /* Make sure you're in text mode.              */
        textbackground(BLUE);   /* Set the color of the text background. */
        clrscr();               /* Clear the screen. */

        textcolor(RED);         /* Sets the color of the text.                 */

        string = "\n\n\n"
                "This program will display a sine wave\n\r"
                "in graphics mode.  You may input the\n\r"
                "values of the amplitude, phase and the\n\r"
                "number of cycles of the sine wave.\n\r"
                "\n"
                "In order to keep all of the wave\n\r"
                "on your monitor screen, keep the\n\r"
```

```
                "amplitude less than 75.";

        cprintf("%s", string);              /* Print the above string. */
        press_return();                     /* Holds screen for user.  */
} /* End of explain_program() */

/*------------------------------------------------------------------*/

void sine_wave_display(user_input values)
{
        wave_values display;    /* Structure for the sine wave display. */

        /* Get into graphics mode: */
        auto_initialization();

        setgraphmode(3);
        setbkcolor(1);

        /* Draw the graph: */

        for(display.X_value = 0; display.X_value <= 319; display.X_value++)
        {
                display.degrees = display.X_value;
                display.radians = display.X_value * PI / 180;
                display.Y_value = (int)(values.amplitude)*
                sin((double)(values.cycles*display.radians + values.phase));
                putpixel(display.X_value, 75 - display.Y_value, WHITE);
        }
}  /* End of sine_wave_display() */

/*------------------------------------------------------------------*/

void auto_initialization()
{
        int graph_driver;       /* Specifies the graphics driver.      */
        int graph_mode;         /* Specifies the graphics mode.        */

        graph_driver = DETECT;  /* Initializes autodetection.          */

        initgraph(&graph_driver, &graph_mode, "");

} /* End of auto_initialization. */

/*------------------------------------------------------------------*/

void press_return()             /* Hold screen for user.               */
{
        gotoxy(10,25);          /* Place the cursor at bottom of screen. */
        textcolor(YELLOW);      /* Sets color of text.                 */
        cprintf("Press -RETURN- to continue.");
```

```
        getchar();              /* Wait for user response.                */

        clrscr();               /* Clears the screen.                     */
        textcolor(WHITE);       /* Returns text color to white.           */

} /* End of press_return() */

/*----------------------------------------------------------------------*/

int program_repeat()            /* Repeat program option.                 */
{
        char response;          /* User input response.                   */

        settextstyle(SMALL_FONT, HORIZ_DIR, 3);
        setcolor(RED);
        outtextxy(120,140,"Do you want to repeat the program(Y-N)? => ");

        do
        {
                response = toupper(getche());
        } while((response != 'Y') && (response != 'N'));

        if(response == 'Y')
                return(TRUE);
        else
                return(FALSE);
} /* End of program_repeat() */

/* End of program. */
```

Program Analysis

The reason that the string functions in `explain_program()` are terminated with a `\n\r` is because the `cprintf()` function is used. Recall that unlike the `printf()` function, the `cprintf()` does not give an automatic carriage return with just the `\n`. The `cprintf()` function is used in order to display text color.

Observe how the C `while` is used in the function `program_repeat()` to ensure that the only acceptable input response by the user is a *Y* or *N* (upper or lowercase). Note that the program does not use any global variables. There are two structures in the program. Their identifiers are `wave_values` and `user_input`. Notice that the user input variables are passed to the function `sine_wave_display()` by passing the name of the structure as an argument.

A sample output from Program 10.25 is shown in Figure 10.25.

Conclusion

In this section you had the opportunity to see the development of a graphics technology program. There is no section review here since the questions for this section are covered in the Self-Test for this chapter.

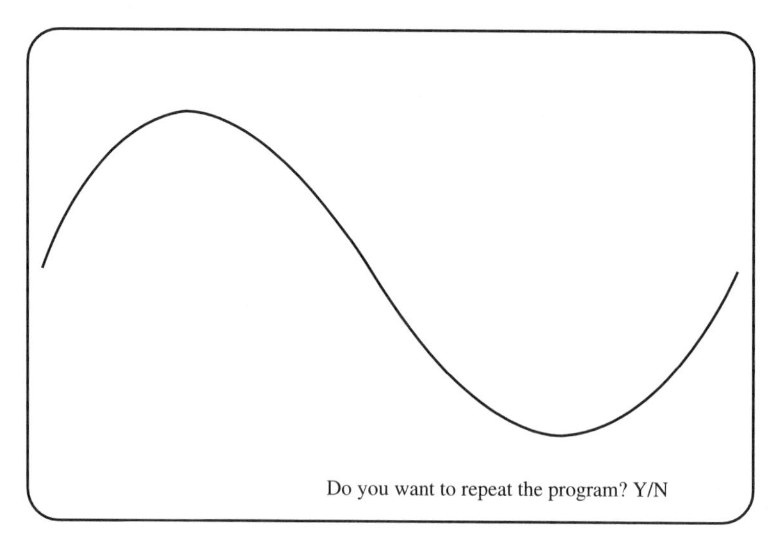

Figure 10.25 Typical Output from Case Study Program

Interactive Exercises

DIRECTIONS

Because of the nature of this chapter and the differences in computer systems that display color text and graphics, this section differs from the other interactive exercise sections.

This section asks questions about the computer system(s) to which you have access. The questions are designed as a guide to help you find out important information that will be useful in developing the programs in the end-of-chapter problems.

Exercises

1. Is your system capable of displaying text in color? How did you find this out?
2. How many different colors of text can your system display? Can all these different colors be displayed at the same time?
3. If your system can display text in color, can it also display different background colors? How many?
4. What kind of graphics adapter does your system have? How did you find this out?
5. For the graphics adapter in your system, how many different modes does it have? How did you find this out?
6. What `.BGI` file do you need on your disk in order to operate your graphics system?
7. For your graphics system, what is the maximum number of horizontal and vertical pixels available? How did you determine this?
8. What is the maximum number of drawing colors available with your system? How was this determined?
9. How many palettes does your system have? How did you determine this?
10. State the maximum number of colors that can be displayed on your graphics screen at the same time. How was this determined?

Self-Test

DIRECTIONS

Answer the following questions by referring to Program 10.25 in the case study section of this chapter.

Questions

1. How many structures are used in the program? What are their identifiers?
2. Are there any global variables in the program? If so, name them.
3. How many function prototypes are in the program? Name them.
4. What are the local variables declared in function `main()`?
5. State why the function `program_repeat()` is of type `int`.
6. In the function `explain_program()`, why is each of the strings terminated with `\n\r`?
7. What is the reason for using the `cprintf()` function in the function `explain_program()`?
8. How are the user input variables passed to the function `sine_wave_display()`?
9. State the purpose of the C `while` in the function `program_repeat()`.

End-of-Chapter Problems

General Concepts

Section 10.1

1. State what your computer system must contain in order to produce text in color using Turbo C.
2. How many columns of text can be displayed with an installed graphics adapter?
3. What is the maximum number of colors text may have? Can all of these text colors be displayed at the same time?
4. Explain how you would cause colored text to blink.

Section 10.2

5. Explain the difference between a text screen and a graphics screen.
6. State what your system needs, in terms of hardware and software, in order to get a Turbo C program into the graphics mode of operation.
7. Define the term *pixel*.
8. State which color graphics adapters supported by Turbo C can produce 16 colors in graphics.

Section 10.3

9. What built-in Turbo C function would you use to determine the modes available on your system?
10. State what Turbo C command you would use in order to determine the number of available colors for any given graphics mode of your system.
11. Explain the purpose of the built-in Turbo C functions `getmaxx()` and `getmaxy()`.
12. State the meaning of the term *palette* as used in Turbo C.

Section 10.4

13. Explain how you would change the palette on the graphics screen using Turbo C.
14. What built-in Turbo C function would you use to cause dotted lines to be created?
15. Name two ways of creating a rectangle using Turbo C.

16. When creating a rectangle in Turbo C, can you change the style of the line used to create the rectangle? Explain.
17. What are the line styles for creating a circle in Turbo C?

Section 10.5

18. Three varieties of bars are built into Turbo C. What are they?
19. Explain what a Turbo C fill-style is.
20. State when you would want the top of a three-dimensional bar not to appear.
21. How many different fill-styles for bars are available in Turbo C?
22. What built-in Turbo C function determines the fill-style of a bar?

Section 10.6

23. State what is meant by *scaling* in the graphing of a mathematical function.
24. Explain what is meant by *coordinate transformation* in the graphing of a mathematical function.
25. Why is coordinate transformation used?
26. How are the values of the scaling factors determined?
27. State how the coordinate transformation values are determined.

Section 10.7

28. What are the actual row/column dimensions of the Virtual Maze game world (Program 10.24)?
29. The Virtual Maze is a static game world. What does this mean?
30. A scaled strip is 160 pixels high. How many pixels are used for the floor and ceiling?
31. Repeat problem 30 for a scaled strip that is 30 pixels high.
32. What techniques are used to speed up the ray casting in the Virtual Maze?
33. What does each cell of the game map contain?

Section 10.8

34. What would you do if your new graphical application blanks the screen and locks up the computer?
35. Why is output redirection helpful in a graphics application?

Section 10.9

36. How is the user's input string converted into a value?
37. Why is it necessary to convert from degrees to radians in the `sine_wave_display()` function?

Program Design

For the following programs, use the structure that is assigned to you by your instructor. Otherwise, use a structure that you prefer. Remember, your program should be easy for anyone to read and understand if it ever needs modification. Note that for these programs, you will need a system with color graphics capabilities.

38. Develop a C program that makes use of text color to display the wattage value of a resistor. The user input is the resistance of the resistor and the voltage across the resistor. The color of the answer is to be influenced by the resulting wattage values as given below:

microwatts = blue.
milliwatts = purple.
1 to 10 watts = green.
10 to 100 watts = yellow.
100 to 1KW = red.
over 1 KW = flashing red.

The relationship for power is

$$P = I^2R$$

39. Create a C program that will display a graph of the relationship of the voltage across a resistor vs. the current in the resistor and the value of the resistor. User input is the value of the resistor and the range of currents to be graphed. The X axis represents the current, the Y axis the voltage. This relationship is known as Ohm's Law: $E = IR$.
40. Make a C program that will display the reactance of an inductor for a given range of frequencies. The user input is the value of the inductor and the range of frequencies to be graphed. The X axis represents the frequency, the Y axis the inductive reactance. The relationship is $X_L = 2\pi FL$.
41. Develop a C program that will display the impedance of a series RLC circuit. User input is the values of the resistor, inductor, and capacitor. The X axis represents frequency, the Y axis impedance. The relationship is

$$Z_t = \sqrt{(R^2 + (X_L - X_C)^2)}$$

42. Create a C program that will display the *resultant wave of two sine waves.* The user input is the amplitude, frequency, and relative phase of each sine wave.
43. Make a C program that will display a bar graph of the amount of sales, measured in dollars, for five salespersons in a given year. The X axis is to represent each salesperson and the Y axis the amount of money. The user input is the amount of sales for each salesperson.
44. Develop a C program that will place a coordinate axis system anywhere on the graphics screen. User input is the line style and location of the coordinates.
45. Create a C program that will display a rectangle with dimension lines. User input is the size of the rectangle in pixels. The dimension lines are to contain the size of the rectangle in pixels.
46. Develop a C program that displays the land defined by four posts. The program user may enter the coordinates of each post, and the program returns with a display of lines connecting each of the posts.
47. The department head needs a C program that will display the temperature readings of three different patients taken five times during the day. The X axis represents the time a temperature reading is made; the Y axis represents the temperature reading in degrees F. Use a different line style for each patient.
48. A machine shop requires a C program that will display a stacked bar graph of the number of different parts used in an assembly process each day (during the assembly process, some of the parts are lost or damaged). There are five different parts used in the process, and the graph is to display the results for five days. The X axis represents the day and the Y axis the number of parts. User input is the day and the number of parts used on that day.
49. Expand the program in problem 43 so that the name of each salesperson may be displayed vertically along the bar representing his/her sales for the time period.
50. The built-in Turbo C function for developing an arc is

```
arc(X,Y, StartAngle, EndAngle, Radius);
```

Where

X, Y	=	`int`
StartAngle, EndAngle	=	`int`
Radius	=	`int`

The function draws a circular arc from the coordinates X, Y with a radius of Radius. The arc travels from StartAngle to EndAngle and is drawn in the current drawing color.

Create a C program that will allow the program user to select the variables for such an arc. The program is to display the arc along with the values of the variables selected by the program user.

51. The built-in Turbo C function for creating an ellipse is

```
ellipse(X,Y, StartAngle, EndAngle, XRadius, YRadius);
```

Where

X,*Y*	=	`int`
StartAngle, EndAngle	=	`int`
*X*Radius, *Y*Radius	=	`int`

The function draws an elliptical arc with *X* and *Y* as the center point. *X*Radius and *Y*Radius are the horizontal and vertical axes. The ellipse travels from StartAngle to EndAngle and is drawn in the current color.

Create a C program that will allow the program user to select the variables for such an ellipse. The program is to display the ellipse along with the values of the variables selected by the program user.

52. Expand the program in problem 46 so that the program user may enter any number of posts. The program will then display the land enclosed by the posts.
53. Expand the program in problem 47 so that each patient's blood pressure (taken the same number of times each day) is displayed on the same graph with the temperature. Here, use a different line color to represent the patients' blood pressures.
54. Create a C program that will display a bar graph of the production output of ten factory workers for a given month. The *Y* axis is to represent each individual factory worker and the *X* axis the amount of production output (assume 100 units maximum). The program user is to enter the production output for each of the ten workers along with the name of each worker. Make sure that each bar in the graph is easily distinguished from the others and that the name of the worker appears horizontally along the corresponding bar.

11 Hardware and Language Interfacing

Objectives

This chapter gives you the opportunity to learn:

1. The basic concepts of assembly language programming.
2. Microprocessor architecture as it applies to assembly language programming.
3. Methods of interfacing between C and assembly language programs.
4. The basic concepts of the IBM BIOS system.
5. Using BIOS interrupts with the C language.
6. Modifying the C library using Turbo or Microsoft C environments.
7. Developing your own library using Turbo or Microsoft C environments.
8. Some of the common utility programs used by both the Turbo and Microsoft C environments, such as the MAKE utility.
9. An introduction to the concepts of hardware interfacing.
10. Interfacing between data and printer ports using C.

Key Terms

Architecture
Internal Registers
Write Operation
Read Operation
General Purpose Registers
Pointer and Index Registers
Special Purpose Registers
Stack
LIFO
Segmented Memory Architecture
Segment Register
Flag Register
Assembler
Calling Convention
Memory Models
Near Pointer

Far Pointer
Pseudo Variables
Inline Assembly
Inline Machine Code
Utility Program
MAKE Utility
Library Utilities
BIOS
DOS Call
Interrupt
Interrupt Handler
Interrupt Service Routine
Interrupt Vector
Vector Table
Interrupt Number

Outline

Introduction

This chapter of the book presents some important advanced topics. These topics include the introduction of assembly language and how to let your C programs interact with this language. You will discover how to accomplish this using either the Turbo C or Microsoft C environment.

This chapter also introduces you to the IBM BIOS. You can think of this as the software that interfaces between your program and the actual circuits that make up your computer. There are many useful tricks to be learned here. These can be used to increase your ability to control the operation of the computer using the C language.

Also presented is how you can create your own library of useful C functions that you have developed. Again, both the Turbo and Microsoft C environments are presented.

The case study demonstrates the development of a program that allows you to use your knowledge of the C language to create a program that will allow interfacing.

11.1 Inside Your Computer

Discussion

Before you can appreciate what assembly language does and how it can interface with your C programs, you need to know some things about the inside of your computer. That is what this section will discuss. Here you will get enough of an introduction to the **architecture** of your IBM computer so that the other sections that follow will "make sense."

Basic Architecture

You don't need to know every detail about the microprocessor (μP) in your computer—there are whole books devoted to just that subject—but you do need to know something about its major features. The parts of the μP (80x86) you need to know about to develop a simple assembly language program are outlined in Figure 11.1.

As shown in Figure 11.1, the major parts of the μP consist of **internal registers**. You can think of a register as a place for storing bit patterns just like they can be stored in memory. The difference is that a register inside the μP can change these bit patterns by shifting, incrementing, decrementing, as well as performing arithmetic and logical operations. Recall that you were introduced to the concept of bit manipulation using C in Chapter 3.

The microprocessor interacts with the computer's memory in two fundamental ways:

1. It selects a memory location and copies a bit pattern from one of its internal registers to that memory location. This is called a **write operation**.
2. It selects a memory location and copies a bit pattern from that memory location into one of its internal registers. This is called a **read operation**.

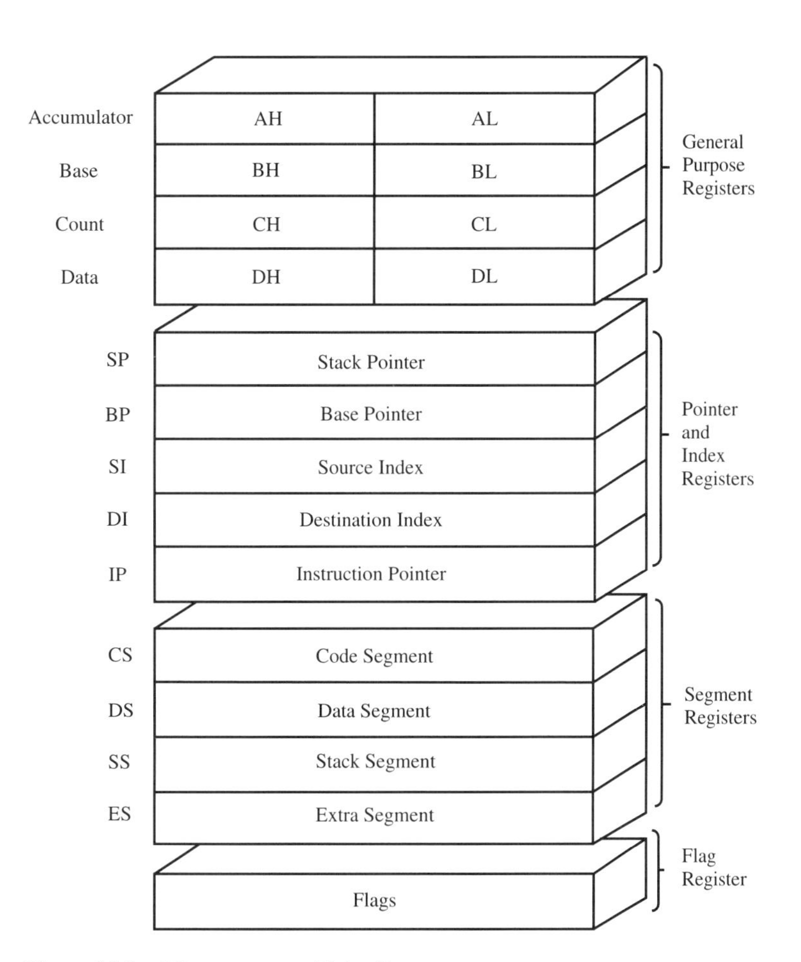

Figure 11.1 Microprocessor Major Parts

As you can see, the internal registers in the μP can be divided into four groups: the *general purpose registers,* the *pointer and index registers,* the *segment registers,* and the *flag register*. Each of these register groupings will be presented in this section.

General Purpose Registers

Figure 11.2 shows some of the details of the **general purpose registers**.

Each of these registers can be accessed either as four separate 2-byte (16-bit) registers or as eight separate 1-byte (8-bit) registers. This means they can store a 16-bit variable or an 8-bit variable.

When one of these registers is accessed as a 16-bit register, it is referred to by the letter representing the first letter of the register name followed by the letter X. Thus, if an accumulator is referenced as AX, this means to treat the accumulator as a single 16-bit register. When a register is accessed as an 8-bit register, it is referred to again by using the first letter of the register name, but this time the letter is followed by either an H or an L. As you can see in Figure 11.2, each of the registers is divided into a lower half (representing the eight least significant bits of a 16-bit number) and the upper half (representing the eight most significant bits of a 16-bit number).

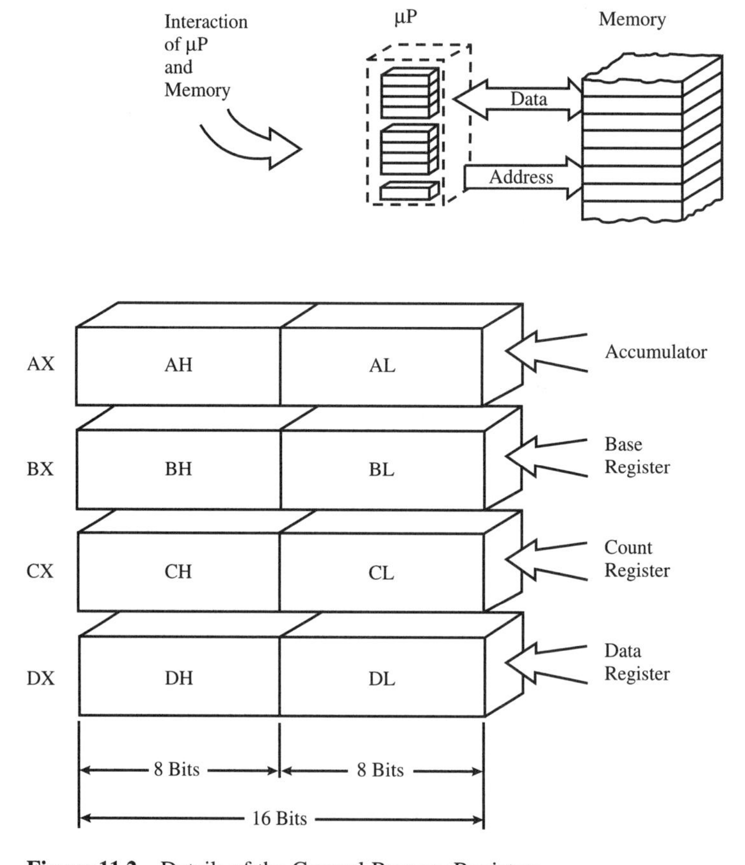

Figure 11.2 Details of the General Purpose Registers

Table 11.1 General Purpose Registers

Register	Function
Accumulator	Stores the temporary result of an arithmetic or logical operation. Also used to interact directly with other hardware called I/O ports.
Base Register	Used to indicate specific memory locations. It is often used to store the value of a pointer; thus it will point to a memory location.
Count Register	Used as a counter for specific instructions, including bit shift and rotate instructions, or as a counter in a loop operation.
Data Register	A general register. It is usually used to hold parts of answers for multiplication and division as well as I/O port numbers when involved in I/O operations.

Thus, when the accumulator is to be represented as two separate 8-bit registers, the designation is either AH or AL (for a reference to the first half [AH] or last half [AL] of this register). The purpose of each of these registers is given in Table 11.1.

Pointer and Index Registers

The registers just presented are primarily used to interact with data. The next set of registers, the **pointer and index registers**, are primarily used to interact with memory locations used by the data. The details of these registers are shown in Figure 11.3.

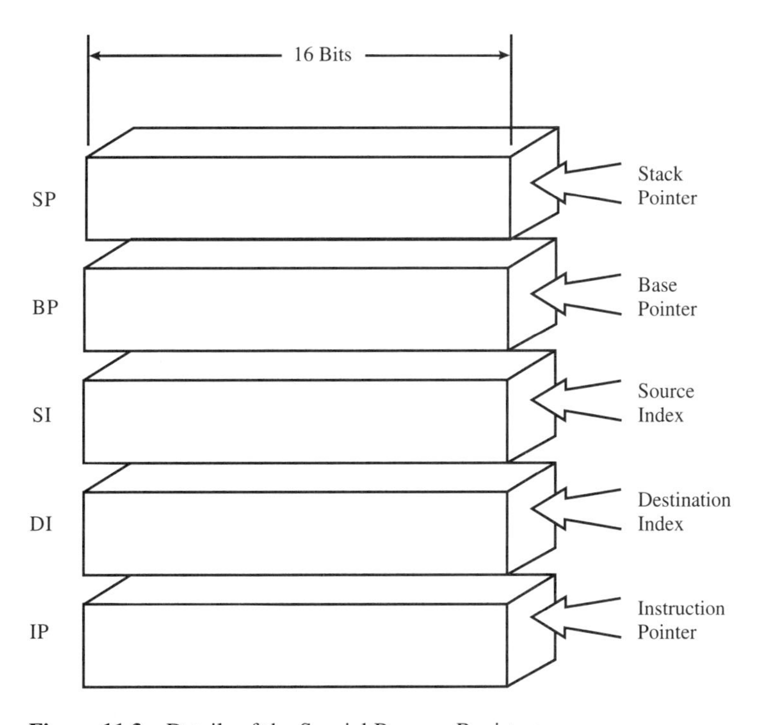

Figure 11.3 Details of the Special Purpose Registers

Table 11.2 Special Purpose Registers

Register	Function
Stack Pointer	Used to sequentially address data in a part of memory (called the stack) that is set aside especially for this purpose.
Base Pointer	A general purpose register sometimes used to hold an address that indicates the beginning (base) of the stack.
Source Index	Usually used to address source data with the string instructions.
Destination Index	Usually used to address the destination of data required by string instructions.
Instruction Pointer	Used to select the next instruction to be executed.

As shown in Figure 11.3, all of the pointer and index registers (also called **special purpose registers**) are 16-bit registers. The purpose of each of these registers is presented in Table 11.2.

Figure 11.4 gives you an idea of how the stack pointer (SP) and the base pointer (BP) are used to identify the area of memory used as the **stack**.

This arrangement allows the μP to access data in memory by the order in which the data is stored. This process is referred to as **L**ast **I**n **F**irst **O**ut or **LIFO**. What this means is that data is stored (pushed) into the stack in sequential order (one memory location after the other). This order goes toward lower memory locations, meaning the stack grows down. When data is to be copied from the stack (referred to as *popping* the stack), it is copied in sequential order—only this time from a lower memory location toward a higher memory location. In other words, the last piece of data to be pushed into the stack is the first piece of data to be popped from the stack. More will be said about the operation of the stack in the next section of this chapter.

Memory Segmentation

The 80x86 μP uses what is referred to as **segmented memory architecture**. What this means is that since the internal addressing registers of the μP are only 16 bits wide, they are capable of addressing only from $0000\ 0000\ 0000\ 0000_2$ to $1111\ 1111\ 1111\ 1111_2 =$ 65,535. This represents a total of 65,536 memory locations. (This is normally stated as 64K, where a K represents 1,024. Thus 65,536/1,024 = 64K.) If this were the maximum amount of memory that the μP could access, then the capabilities of your computer would be quite limited. However, those with an IBM or compatible can address up to 1Mb (1 megabyte = 1,048,576 memory locations) in real mode. This requires a binary number consisting of 20 bits ($1111\ 1111\ 1111\ 1111\ 1111_2 = 1{,}048{,}575$). To accomplish this task with only 16-bit registers, a process known as *segmentation* is used. This process requires **segment registers**. These registers are illustrated in Figure 11.5.

The purpose of each of the segment registers is outlined in Table 11.3.

Calculating the Address

The actual address is determined by using two values produced by the μP. One value is called the *segment address* and the other is called the *offset*. The segment address is

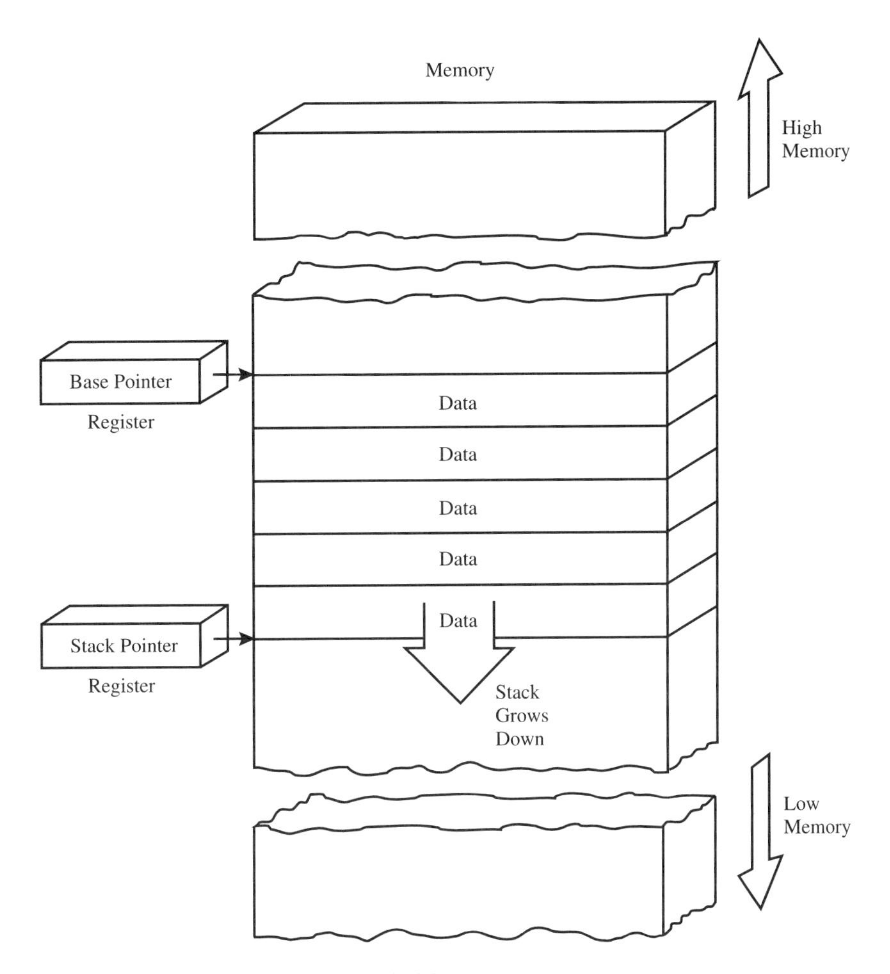

Figure 11.4 Use of the SP and BP in Memory

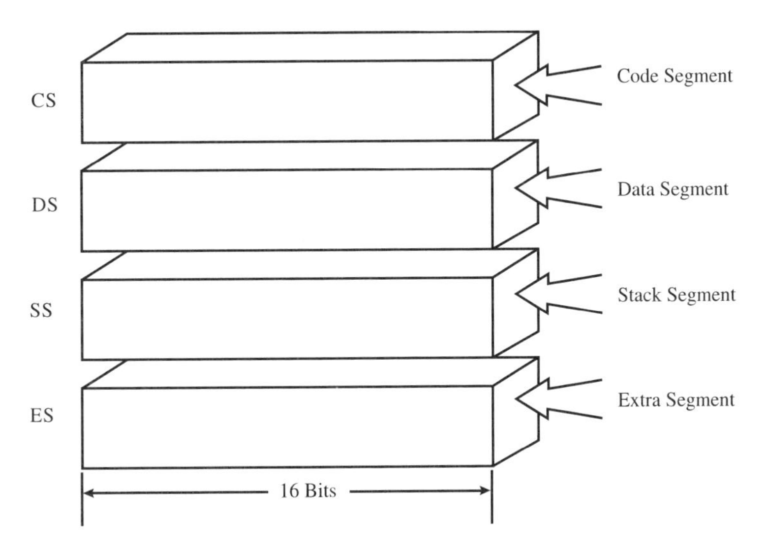

Figure 11.5 The Segment Registers

Table 11.3 Segment Registers

Register	Function
Code Segment	Used to interact with the section of memory that contains the program code. Works with the instruction pointer (IP) register to address memory.
Data Segment	Used to interact with the data register to access data in memory.
Stack Segment	Interacts with the stack pointer (SP) to help access data in the stack.
Extra Segment	Special segment register that is normally used to help with the addressing of string instructions.

contained in a segment register, and the offset is taken from one of the index or pointer registers. For example, the stack segment register would work with the stack pointer register.

To get the actual address in memory, the bit pattern in the segment register is shifted left four bits and then added to the contents of the register containing the offset. Thus if the stack segment register contains $3F86_{16}$ and the offset contains $05DA_{16}$, then the absolute address would be found as shown in Figure 11.6.

Because the segment register is always shifted left 4 bits, its representation as a 20-bit binary number makes its last 4 bits zero. Thus, memory segments can only start at every 16 bytes through memory.

Figure 11.7 shows how memory is partitioned by the use of the segment registers.

Flag Register

The **flag register** uses individual bits to keep track of the results of many of the operations performed by the microprocessor. This register is sometimes referred to as the *status register,* and its contents are called the *program status word.* For example, one of the bits indicates if the result of the last arithmetic or logic operation was a zero. Another bit indicates if the last arithmetic operation produced a carry. The details of this register are not necessary for an understanding of what is to follow in this chapter.

Conclusion

This section presented the foundation you will need in order to understand the basics of assembly language for your computer. This information will also be helpful in understanding

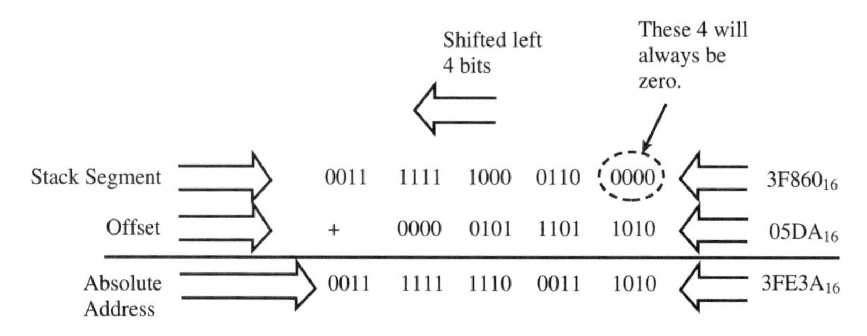

Figure 11.6 Calculation of Absolute Address

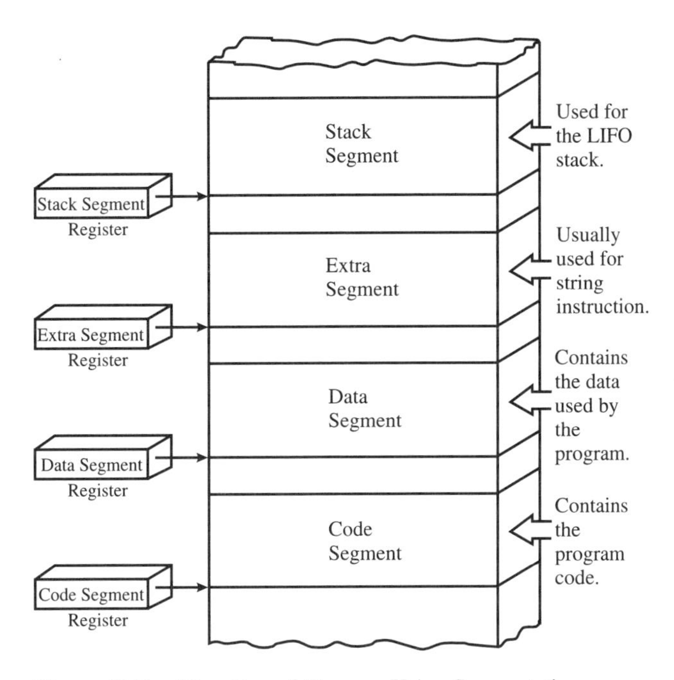

Figure 11.7 Allocation of Memory Using Segmentation

how the C language itself interacts with your computer's internal workings. Check your understanding of this section by trying the following section review.

11.1 Section Review

1. Name the four major groupings of the registers in the 80x86 μP.
2. Describe the difference between a register inside the μP and a memory location in the computer's memory.
3. State how the μP interacts with the computer's memory.
4. Which registers inside the μP can be treated as 8-bit or 16-bit registers?
5. Explain how a 16-bit μP can access up to 1Mb of memory.

11.2 Assembly Language Concepts

Discussion

Having knowledge of assembly language is about as close as you can get to having complete and absolute control over your computer without the use of a soldering iron. Assembly language programs run faster than equivalent programs written in other languages, and assembly language programs take up less memory than high-level languages trying to do the same thing. But there is a big price to pay for all of this. You must be intimately familiar with the inner workings of your computer, and you must be a very, very careful programmer. Another disadvantage of assembly language programs is that they work for only one specific kind of microprocessor family and are not very portable.

With all this, there may be times where you need the advantages of an assembly language program as a part of your C program. And, if for no other reason, as a technically oriented person you should have an idea of what assembly language looks like and how it interacts with your C program.

Basic Idea

Use of assembly language requires an intimate knowledge of the structure (architecture) of your computer and its microprocessor. When speed, control, and compactness are required, then assembly language is used.

What Assembly Language Looks Like

An assembly language is specific to a family of microprocessors. An assembly language uses mnemonics in place of the 1s and 0s of a machine language. A mnemonic is a short word used as a memory aid to indicate a process on a specific microprocessor. For example, the assembly language instruction

```
mov ah,04
```

means (for the 80x86 μP family) to put the number 4 into the AH register (the high byte of the accumulator).

As with any language above machine language, an assembly language program must be converted to machine language before it can be executed. This is done by a program called an **assembler**. This idea is illustrated in Figure 11.8.

The name *assembly language* comes from the assembler program that assembles the mnemonics into a machine code.

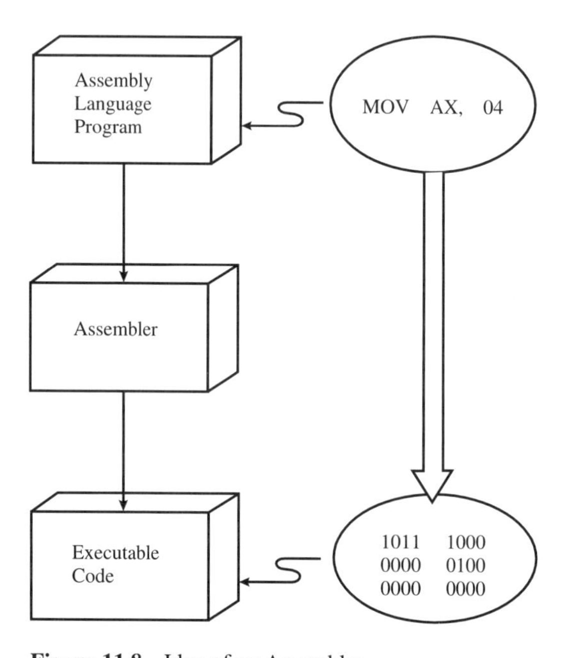

Figure 11.8 Idea of an Assembler

Types of Assembly Language Instructions

There are over 200 different instructions available for the microprocessor in your PC. Each of these instructions has a unique mnemonic to represent it. Generally speaking, these instructions can fall into one of the categories shown in Figure 11.9.

Converting C to Assembly Language Code

Both Microsoft and Turbo C can convert your C program source code into assembly language. As a matter of fact, this can be a real help in developing your own assembly language programs. First write what you want to have happen in C (or as close as you can get to what you want to have happen), then use your Microsoft or Turbo operating system to convert what you have written in C to assembly language. In order for this to be useful, you need to know some of the working details of how C uses your computer's memory. In this way, the assembly language programs generated by your C programs will make sense.

C at the Machine Level

When a C function uses memory, there are certain tasks that it must perform. Keep in mind that your C program, when executed, is actually causing the μP inside your system to perform specific tasks. It will cause bit patterns to be transferred between the μP's internal registers and memory as well as many other details.

A C function will use the stack to store key parts of the function. For this process, the stack pointer and base pointer are used. This process is illustrated in Figure 11.10.

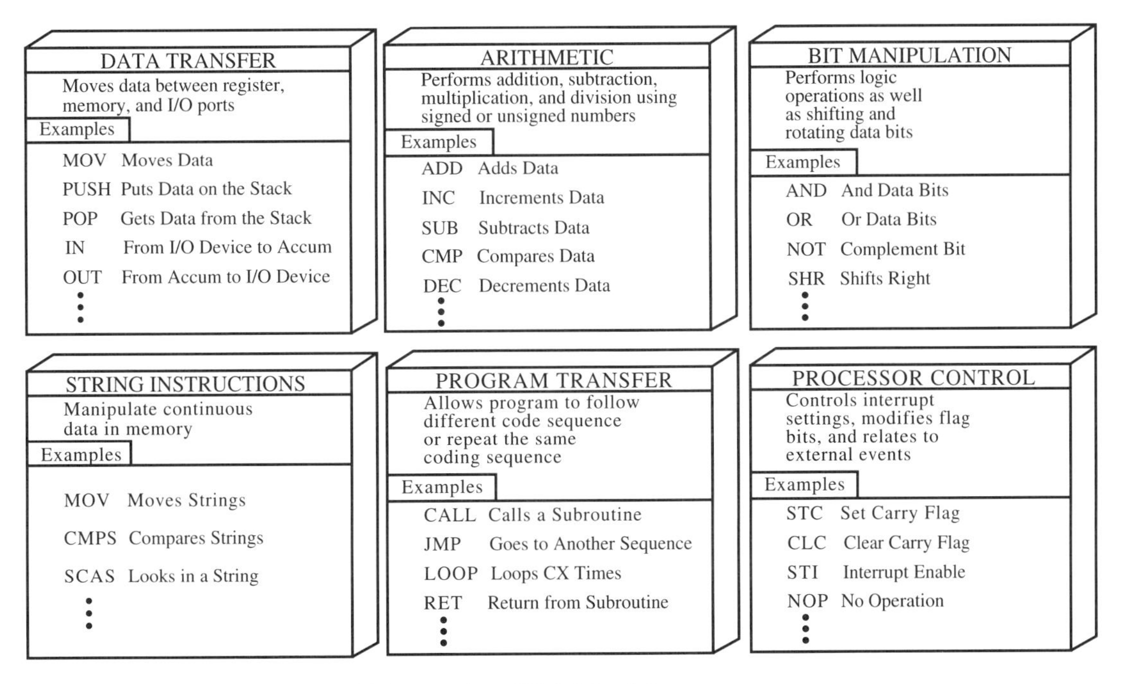

Figure 11.9 General Catalogs of μP Instructions (80x86 Family)

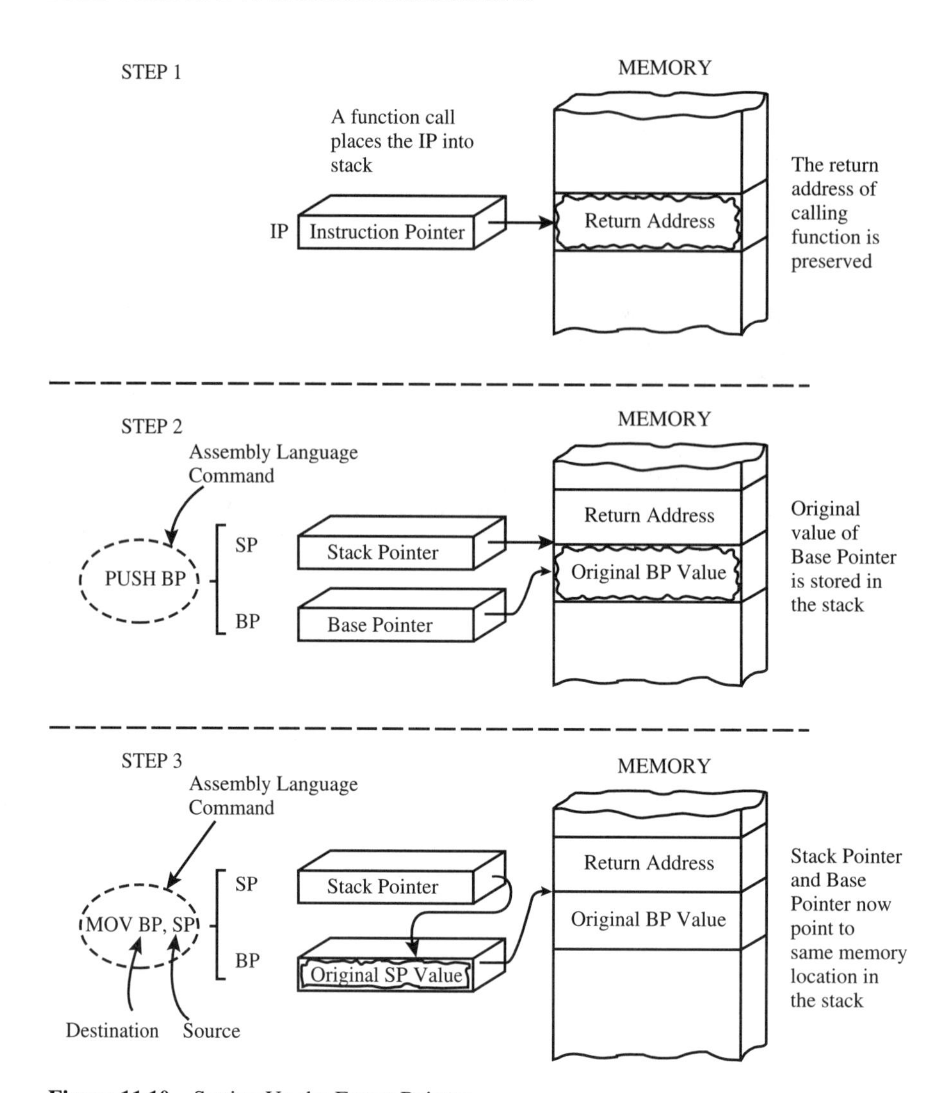

Figure 11.10 Setting Up the Frame Pointer

The sequence shown in Figure 11.10 establishes a reference that will be used to access data located on the stack. The space just above the base pointer value contains the contents of the instruction pointer so that the program knows where to return to when the called function is completed. Note in the figure that the equivalent assembly language commands are given for each of these processes.

Saving Space for Data

A C function will save memory space for any locally declared variables in the stack. This is done by decreasing the value of the stack pointer (called *going toward the top of the stack*). This is shown in Figure 11.11.

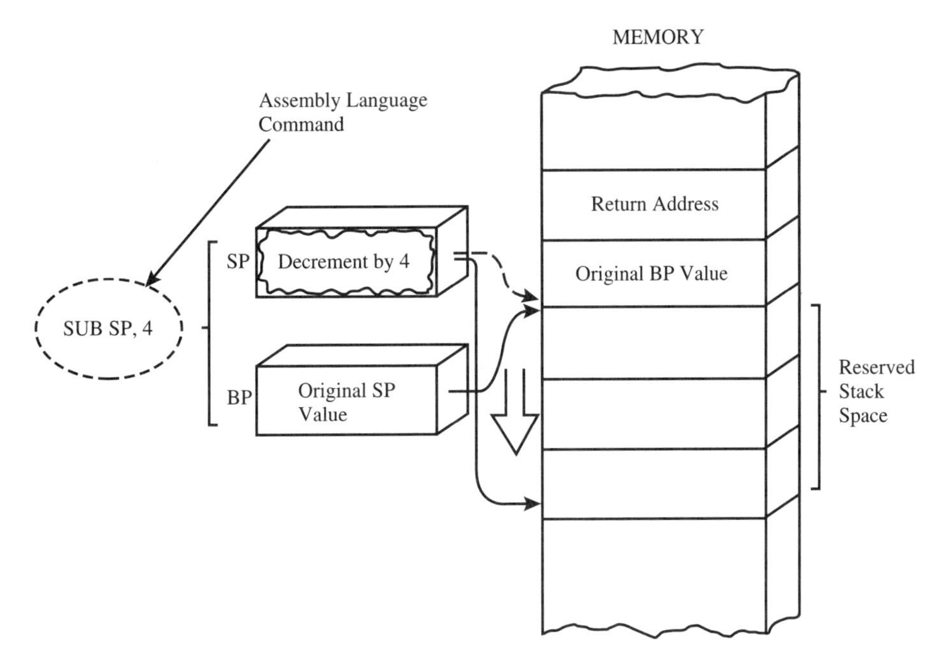

Figure 11.11 Saving Memory for Locally Declared Data

If, for example, the called C function has three locally declared `int` type variables, then four memory locations will be preserved for their values.

Saving µP Contents

There are times when C may need the contents of some of the µP registers to be preserved. This is usually because the values of these registers will be needed by the calling function, and the called function may modify them in some way in order to perform its required operation. Figure 11.12 shows how C would preserve the contents of the SI and DI registers onto the top of the stack.

Note that a `push` command is used to store these contents. When returning from the called function, a `pop` command will be used to set these internal registers to their original values.

C Arguments

So far, you have seen how a call to a C function sets up the top of the stack. The bottom of the stack is also used for function arguments. Suppose, for example, that the prototype for the called C function looks like this:

```
void function1(int val1, int val2);
```

What is known as the C **calling convention** will cause the function parameters to be pushed onto the stack in a right-to-left order. This will immediately be followed by the

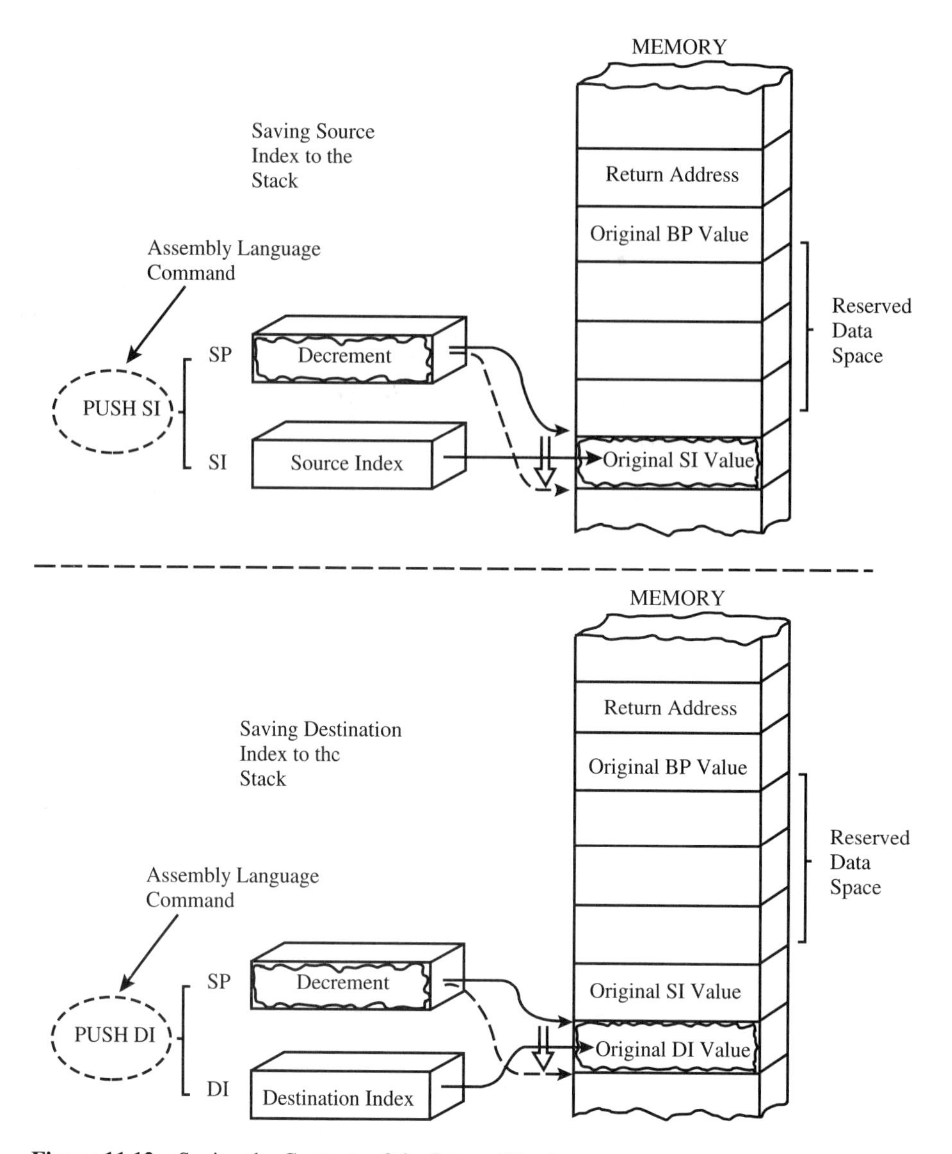

Figure 11.12 Saving the Contents of the Internal Registers

address to which the C program must return when the called function is completed. This process is illustrated in Figure 11.13.

The complete stack of a called C function is shown in Figure 11.14.

When the called function has completed its execution, the original value of the IP will be returned to that register, and the data in the stack will no longer be used. This data is never erased, but it will be written over by the next called function. This is why data that is local to a function is available only when the function is active. This feature protects local data from being modified by another part of the program.

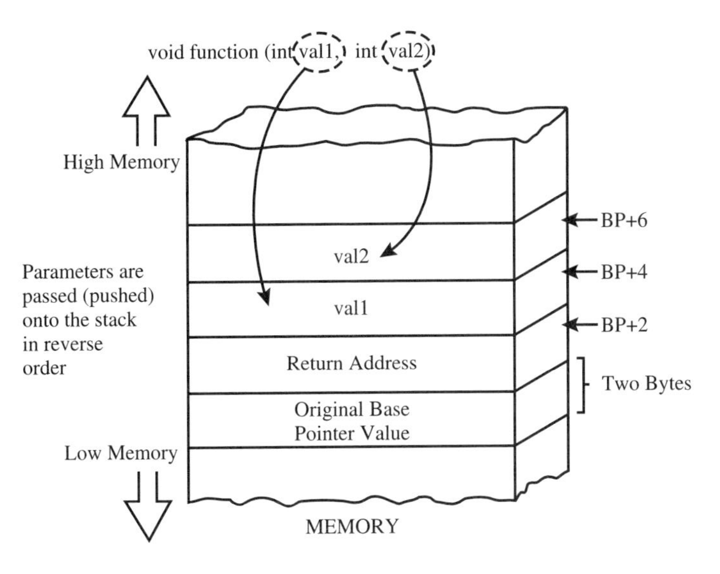

Figure 11.13 The C Calling Convention

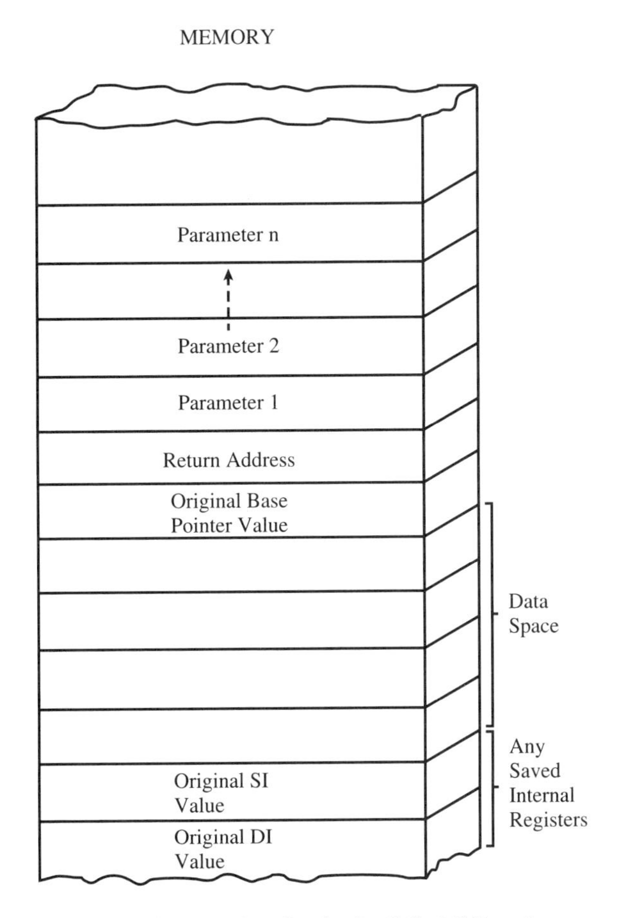

Figure 11.14 Complete Stack of a Called C Function

Storing Static Variables

Remember that a static variable has a lifetime that exceeds that of the function in which it is declared. Because of this, a variable of this type cannot be stored in the stack. Instead, variables that have a life longer than the function that called them are stored in the data segment of memory. In this manner, their values can be preserved.

Conclusion

In this section you got a basic idea of what assembly language looks like. Here you also saw how a called C function uses memory inside your computer. You will need this information in order to analyze assembly language programs produced from your C source code. Check your understanding of this section by trying the following section review.

11.2 Section Review

1. State three advantages of assembly language programs.
2. Give some disadvantages of assembly language programs.
3. What is a mnemonic?
4. When a C function is called, what is the first internal register placed on the stack?
5. Where on the stack does C store locally declared variables? Function parameters?

11.3 Memory Models

Discussion

This section shows how C handles segmented memory. The purpose of this section is to make you aware of options you have that allow you to use all of the memory available in your computer. Recall from the first section of this chapter that the way the microprocessor in your PC gets around the limitations of just 64K of memory is by using segment registers and offsets. This allows it to access up to 1Mb of memory. You can control how much memory your C program may access by using different kinds of **memory models**.

Basic Idea

If all of your executable code was confined to one 64K segment, then all of the segment registers could have a fixed value. This would mean that only the 16-byte offset would need to be changed as different locations (addresses) in memory were accessed. As an example, if all of the data in your program was limited to a single 64K segment, then the DS register could stay fixed and only the 16-bit (2-byte) offsets would have to change. This idea is illustrated in Figure 11.15.

Confining segments to only 64K of memory has the advantage of speed since only the offset needs to be changed. This also results in more compact code because the address of your variables needs to occupy only two bytes of memory.

The disadvantage of this system is that not all of your computer's memory is usable. Confining the four available segments to 64K allows you a total of 64K $\times$ 4 = 256K of

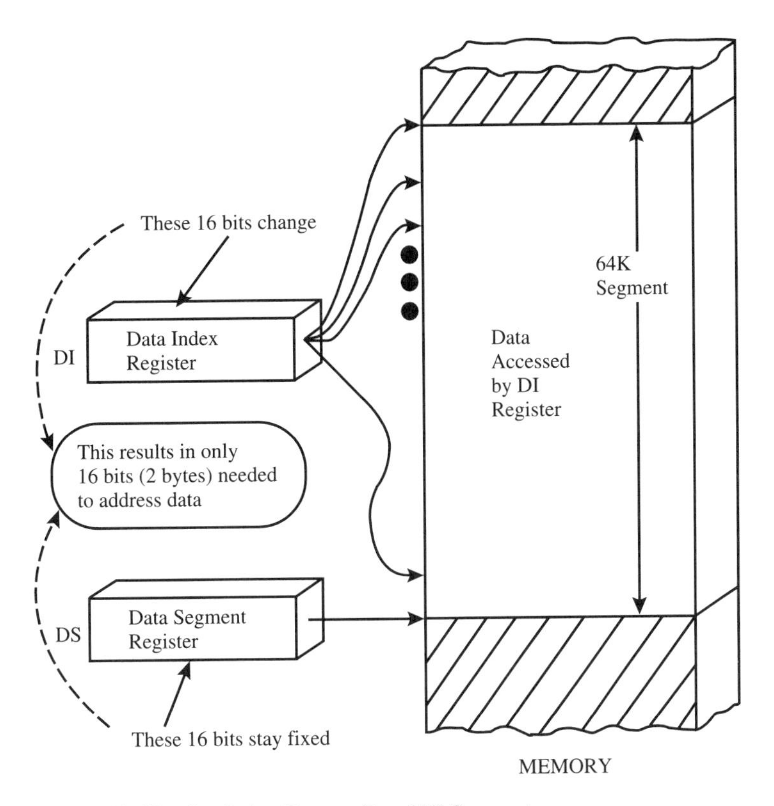

Figure 11.15 Confining Data to One 64K Segment

memory access—only 25% of its full 1Mb potential. For example, some C programs may require more than a 64K segment for data. This may come about with a large database management system where it is necessary to store large amounts of data at the same time. The concept of extending data access to more than one 64K segment is illustrated in Figure 11.16.

As you can see from the discussion above, confining your segments to a maximum size of 64K requires only a 16-bit (2-byte) size pointer. To expand beyond this restriction requires a 32-bit (4-byte) size pointer—because the 16-bit segment address must be given as well as the 16-bit offset. As you will see, the 16-bit pointer is called a **near pointer** and a 32-bit pointer is called a **far pointer**. Up to this point, almost all of your C programs have used near pointers.

It's important that your C program know what type of pointers to use. Recall from the previous section how a C function uses the stack. Storing the values of far pointers requires twice as much stack space as storing the values of near pointers (4 bytes vs. 2 bytes).

Memory Models

Both Turbo C and Microsoft C offer options as to how memory will be allocated to your C program. The memory models for Turbo C are listed in Table 11.4.

The memory models for Microsoft C are listed in Table 11.5.

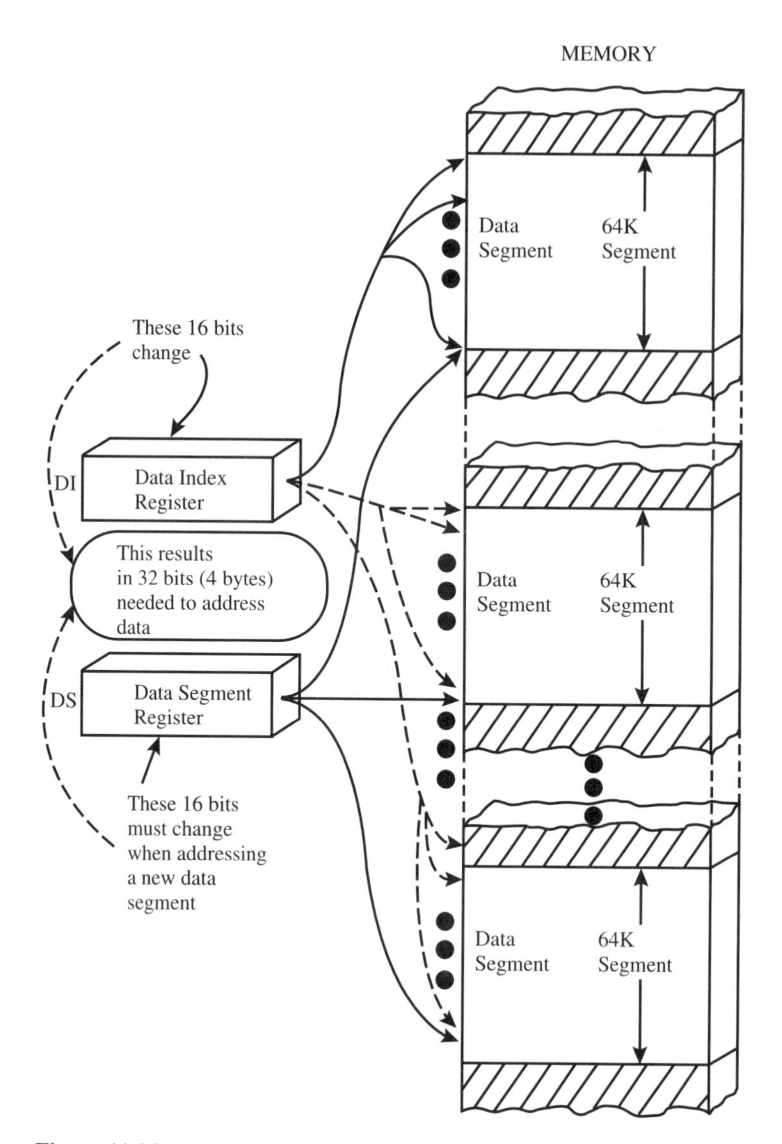

Figure 11.16 Expanding Data Access to More Than One 64K Segment

Near and Far Pointers

You can declare C types to be near or far by using the keyword `near` or `far`. As an example, if a pointer is to hold a 16-bit address it may be declared as `int near *ptr`. If, on the other hand, the pointer is to hold a 32-bit address, it may be declared as `int far *ptr`. However, it is usually best to use the default values of the program models. It's important to note that the keywords `near` and `far` are not part of the ANSI C standard. They are useful only in systems (such as the 80x86 family) that need to use memory segmentation. This is important if you want to create C code that can be used on other systems that may not have this requirement.

Table 11.4 Turbo C Memory Models

Model Name	What It Does
TINY	This produces a total of 64K for the whole program. The reason for this is that all four segment registers are placed at the same address. This means that code, data, arrays, and the stack all share the same 64K of memory. All pointers are always near pointers. This is a good model to use when you have a system of only 64K of memory.
SMALL	In this model, your program code has one 64K segment and your program's static data has one 64K segment. Static data does not create more memory as more data space is needed. Stack and extra segments start at the same address. Pointers are near pointers. Use this model for most of your general purpose programming applications.
MEDIUM	Here the code segment is allowed to use up to the full 1Mb of addressable memory. The data segment is still limited to a single 64K segment. In this mode, far pointers are used for code, and near pointers are used for data. Use this model for large programs that do not keep much data in memory.
COMPACT	This model is just the opposite of the MEDIUM mode. Here, the code is limited to a single 64K segment whereas data is allowed to use up to the full 1Mb of addressable memory. Use this model when you have a limited amount of code with large quantities of data. Far pointers are used for data, and near pointers are used for code.
LARGE	In this model, both the MEDIUM and COMPACT models are combined. Here you have the potential of a full 1Mb for both code and data. Far pointers are used for both code and data. This model should be used for large programs that must process large amounts of data.
HUGE	The difference between this model and the LARGE is that static data may exceed 64K. In the other models, static data is limited to 64K.

Table 11.5 Microsoft C Memory Models

Model Name	What It Does
SMALL	Code and data each have their own 64K segments. In this model, the total size of the program cannot exceed 128K. Near pointers are used for both code and data. This is a good model to use for most general applications.
MEDIUM	This model limits data to one 64K segment but allows multiple segments of code. Here, data pointers are near, and code pointers are far. This is a good model to use when you have large amounts of code and small amounts of data.
COMPACT	This model is just the opposite of the MEDIUM. Here code is limited to one 64K segment, but data can occupy multiple segments. Far pointers are used for data, and near pointers are used for code. Use this model when you have large amounts of data and limited amounts of code that can be confined to one 64K segment.
LARGE	In this model, both code and data segments may occupy as many 64K segments as needed up to the full 1Mb of addressable memory. Both data and code pointers are far. Use this model for when you have large amounts of code and data in your executed program.

Conclusion

This section presented the concept of a memory model and the options offered to you by both Turbo C and Microsoft C. Here you saw the significance of the different memory models and the requirements for addressing within these different models. The Turbo C and Microsoft C manuals contain much more detailed information concerning these memory models and should be consulted if you need to change pointer characteristics from the defaults offered by the different models. Check your understanding of this section by trying the following section review.

11.3 Section Review

1. State the advantage of confining all of your C code to a 64K segment.
2. Describe what is meant by a memory model.
3. State the difference between the MEDIUM and COMPACT memory models.
4. What is the difference between a far pointer and a near pointer?

11.4 C Source Code to Assembly Language

Discussion

In this section you will discover how to convert your C source code into assembly language. Here also you will learn something about reading the resulting assembly language output. This is an important section because you get a built-in tutorial on assembly language programming. How? By writing what you want to do in C and then converting it to assembly language. This will result in a template that you can then modify or expand for your own programming needs.

Doing the Conversion

If you have Turbo C, you can convert your C source code into assembly language. To do this you must have the Turbo line compiler on your disk (`TCC.EXE`). This is not the Turbo C Integrated Development Environment (IDE) that has all of the pull-down menus (`TC.EXE`). The Turbo C command line (`TCC.EXE`) offers several options that are not available with the integrated development environment. There isn't enough space in this chapter to cover all of the details concerning the use of the command line compiler. For the purposes of this section, you will be shown how to use the line compiler to convert a C function into assembly language source code.

Assume that you have written a short C function, as shown in Program 11.1.

Program 11.1

```
/* FUNCT1.C */

function_1(int value)
{
        return value = value + 5;
}
```

The function in Program 11.1 could be used as a part of a C program and called from `main()`. Assume that this function is stored on your disk as `FUNCT1.C`. Once this is done, from the DOS prompt enter the following:

```
TCC -S -EFUNCT1.ASM FUNCT1.C
```

Essentially what this does is to evoke `TCC.EXE`; the `-S` states that TCC is to convert a C source code to assembly source code. The `-EFUNCT1.ASM` directs TCC to create a new file for the assembly source code and store it on the disk with the name `FUNCT1.ASM`, and it will find the C source code in the file named `FUNCT1.C`. It's important to note that whereas C source code files are given an extension of `.C`, assembly language source code files are given an extension of `.ASM`.

Once you enter the command above, the `TCC` will take a few seconds while it does the job you asked of it. Then you will find a new file called `FUNCT1.ASM` on your disk. What it contains is shown in Program 11.2.

Program 11.2

```
        ifndef   ??version
?debug  macro
        endm
$comm   macro   name,dist,size,count
        comm    dist name:BYTE:count*size
        endm
        else
$comm   macro   name,dist,size,count
        comm    dist name[size]:BYTE:count
        endm
        endif
        ?debug  S " d:\cbook\ch9\p9-1.c"
        ?debug  C E977A9641C13643A5C63626F6F6B5C6368395C70392D312E63
_TEXT   segment byte public 'CODE'
_TEXT   ends
DGROUP  group   _DATA,_BSS
        assume  cs:_TEXT,ds:DGROUP
_DATA   segment word public 'DATA'
d@      label   byte
d@w     label   word
_DATA   ends
_BSS    segment word public 'BSS'
b@      label   byte
b@w     label   word
_BSS    ends
_TEXT   segment byte public 'CODE'
   ;
   ;    function_1(int value)
   ;
        assume  cs:_TEXT
_function_1     proc        near
        push    bp
        mov     bp,sp
```

```
        push    si
        mov     si,word ptr [bp+4]
   ;
   ;  {
   ;            return value = value + 5;
   ;
        mov     ax,si
        add     ax,5
        mov     si,ax
        jmp     short @1@58
@1@58:
   ;
   ;  }
   ;
        pop     si
        pop     bp
        ret
_function_1     endp
        ?debug  C E9
_TEXT   ends
_DATA   segment word public 'DATA'
s@      label       byte
_DATA   ends
_TEXT   segment byte public 'CODE'
_TEXT   ends
        public   _function_1
_s@     equ      s@
        end
```

Wow! That's a lot of code for such a small C function. For the purposes of this chapter, concentrate on the assembly code (`push`, `pop`, `add`, `mov`, `ret`). The rest of the information is essential overhead for the assembler and debugger. The assembly code that does the work required by the function is shown in Program 11.3.

Program 11.3

```
push        bp
mov         bp,sp
push        si
mov         si,word ptr [bp+4]
mov         ax,si
add         ax,5
mov         si,ax
pop         si
pop         bp
ret
```

Meaning of the Instructions

What happens in Program 11.3 is based on an understanding of the material presented in Section 11.2. Each of the assembly language instructions and the resulting process is

push bp	Saves BP register value on the stack.
mov bp,sp	Puts SP value into BP register.
mov ax,[bp+4]	Copies the contents of memory that is the BP register value plus 4 into the AX register.
add ax,5	Adds 5 to the contents of the AX register. This is the value returned.
pop bp	Gets original BP value back from the stack.
ret	Causes program to return back to main sequence.

Figure 11.17 Assembly Language Process

explained in Figure 11.17. As you can see from the figure, the processes involved represent the actions explained in Section 11.2.

Conclusion

This section gave you an opportunity to see how you can convert C source code into assembly language source code. You also saw how to interpret the resulting assembly language code and relate it to the C function. It was not our intention in this section to explain every detail about the resulting assembly code. That is a subject beyond the scope of this book.

Test your understanding of this section by trying the following section review.

11.4 Section Review

1. Briefly describe what is needed to convert C source code to assembly language source code.
2. What is the extension normally used for an assembly language source file?
3. Explain what is meant by the assembly language instruction `push bp`.
4. State the difference between the meanings of `bp+4` and `[bp+4]`.

11.5 Pseudo Variables and Inline Assembly

Discussion

There are other options available for getting some of the power offered by assembly language from your C programs. For example, Turbo C offers the choice of using **pseudo variables** or **inline assembly**. This section will introduce you to both of these methods.

Introduction to Pseudo Variables

In Turbo C, a pseudo variable is an identifier that corresponds to one of the 80x86 family internal registers. These are listed in Table 11.6.

What the pseudo variables allow you to do is to access the internal registers of your μP directly from your Turbo C program. This gives you the option of either loading any

Table 11.6 Pseudo Variables—Turbo C

Pseudo Variable	Type	μP Internal Register
_AX	unsigned int	Accumulator (16-bit) AX
_AL	unsigned char	Accumulator Lower Byte AL
_AH	unsigned char	Accumulator Upper Byte AH
_BX	unsigned int	Base Register (16-bit) BX
_BL	unsigned char	Base Lower Byte BL
_BH	unsigned char	Base Upper Byte BH
_CX	unsigned int	Count Register (16-bit) CX
_CL	unsigned char	Count Lower Byte CL
_CH	unsigned char	Count Upper Byte CH
_DX	unsigned int	Data Register (16-bit) DX
_DL	unsigned char	Data Lower Byte DL
_DH	unsigned char	Data Upper Byte DH
_SP	unsigned int	Stack Pointer SP
_BP	unsigned int	Base Pointer BP
_DI	unsigned int	Destination Index DI
_SI	unsigned int	Source Index SI
_CS	unsigned int	Code Segment CS
_DS	unsigned int	Data Segment DS
_SS	unsigned int	Stack Segment SS
_ES	unsigned int	Extra Segment ES

of these registers with a given value before calling a function or testing and seeing their current values.

Program 11.4 illustrates an application using this feature.

Program 11.4

```
#include <stdio.h>

main()
{
        unsigned char lower;
        unsigned char upper;
        unsigned int total;

        _AL = 6;
        _AH = 10;
        lower = _AL;
        upper = _AH;
        total = _AX;

        printf(" Value in AH register is %X\n" , upper);
```

```
    printf(" Value in AL register is %X\n" , lower);
    printf(" Value in AX register is %X\n" , total);
}
```

Execution of Program 11.4 produces

```
Value in AH register is A
Value in AL register is 6
Value in AX register is A06
```

Using Pseudo Variables

Even though the Turbo C pseudo variables can be used as if they were global variables, you must keep in mind that they are interacting with the internal registers of your computer's μP, not any locations in memory. Because of this, you cannot use the address (&) operator because the internal registers of the μP do not have an address. It's also necessary to keep in mind that your C program (or any other program) is constantly using these registers. Thus, whatever you put into one of them may not be there very long, and whatever value you read from them may not stay that way for long. The point here is that you should make use of the interaction with the register as soon as possible in your source code.

One note of caution. Storing values directly into the CS, BP, SP, or SS can cause your C program to crash, because the contents of these registers are used by the program. Note that these pseudo variables are not ANSI C standards.

Introduction to Inline Assembly

Turbo C allows you to write assembly language code directly into your C source code. This means that you can have the power of assembly language without the hassle of first assembling a separate .ASM file and then linking it to your C code.

To use this feature, you must have a copy of the Turbo Assembler (TASM). This comes in a package separate from Turbo C and includes the Turbo Debugger. To let your C program know that you are using inline assembly, you can use the -B compiler option or put the statement

```
#pragma inline
```

in your source code.

When using inline assembly code, the following format must be observed:

```
asm <opcode> <operand>, <operand>
```

Where

opcode = an allowable 80x86 instruction.
operand = data for the corresponding opcode.

As an example, Program 11.5 is an assembly language program that will activate the speaker by actually sending on/off pulses to it.

Program 11.5

```
main()
{
        asm     in      al,0x61
        asm     and     al,0xfc
        asm     mov     bl,255
        start:
        asm     xor     al,02
        asm     out     0x61,al
        asm     mov     cx,0x1000
        place:
        asm     loop    place
        asm     dec     bl
        asm     jnz     start
}
```

In order to jump to a particular place in the program, Turbo C inline assembly requires that you use a label (such as `start:`), as shown in Program 11.5.

If you are familiar with assembly language programming, then you are aware of the practice of using the semicolons for the beginning of a comment:

```
mov ah,3   ;This is a comment.
```

The semicolon is not allowed for this purpose when used as a part of inline assembly code in C, because the semicolon there has another meaning. To use comments in inline assembly code, you must use the standard C symbols:

```
asm  mov ah,3   /* This is a comment.  */
```

Note that no semicolon is needed in the inline assembly statement above. This makes statements in inline assembly language the only C statements that do not require semicolons. Keep in mind that inline assembly code is not an ANSI C standard.

Inline Machine Code

Turbo C offers the option of using **inline machine code**. The advantage of this is that you do not need a separate compiler, and the program may be compiled and executed from the IDE (the inline assembly code cannot do this—it must use the command line TCC). A machine code simply consists of the hex value of the assembly instruction. For example, the machine code for the assembly instruction `mov ah,08` is `B408`, where `B4` is the machine code for the move immediate instruction and the `08` is the value to be moved into the AH register.

The Turbo C function that does this is

```
_ _emit _ _(argument,...)
```

Where

`argument` = generally a single-byte machine instruction.

This requires `dos.h` and is not an ANSI C standard.

Program 11.6 shows a use of this function. The program uses the inline machine code to place a value of 8 into the AH register and then uses the pseudo variable _AH to verify that the machine code did indeed do what was intended.

Program 11.6

```
#include <stdio.h>
#include <dos.h>

main()
{
        unsigned char value;

        __emit__(0XB4,0X08);     /* mov ah,08 */

        value = _AH;

        printf(" The value in AH is %0X\n" , value);
}
```

Conclusion

This section introduced two powerful features available with Turbo C. These are the pseudo variables that allow you to directly access the internal registers of your μP, and inline assembly code by which you can include assembly language instructions as a part of your C program.

Check your understanding of this section by trying the following section review.

11.5 Section Review

1. Explain what is meant by a pseudo variable in Turbo C.
2. Give an example of the use of a pseudo variable.
3. Can the address operator (&) be used with a pseudo variable? Explain.
4. State what is meant by inline assembly.
5. Are pseudo variables and inline assembly accepted ANSI C standards? Explain.

11.6 Programming Utilities

Discussion

A typical C development system usually consists of many different programs. For example, one of the programs on your Learn C disk is LC.EXE, which is the integrated development environment (IDE) that contains the built-in editor, filer, linker, and so on. In a like manner, both Turbo C and Microsoft C have other programs on their disks that can assist your C program development in many different ways. These ways range from searching for a word on one or more disk files to helping you create your own C libraries.

In this section, you will get an overview of some of the useful **utility programs** that may be available with your C system. Having a general understanding of them will help pave the way for the more detailed discussions in the systems manuals that accompany the respective C development systems.

Utility Overview

Table 11.7 lists some of the more common utility programs that may be available with a C development system.

As you can see from Table 11.7, an extensive list of possible utilities is available for your use in the development of a C program. This section will not replace the extensive presentations available in the manuals that accompany professional C development systems. But it will make you aware of what is available, the reason for this availability, and an overview of what the utilities can do.

The MAKE Utility

The **MAKE utility** is a separate program contained on the systems disks for both Microsoft C and Turbo C. The reason for having such a utility available is discussed below.

During development of a large C program, it will usually be broken up into several different files of which only one file at a time is being worked on. Figure 11.18 illustrates this concept by showing a parts inventory program under development.

As shown in Figure 11.18, the executable program `INVENT.EXE` consists of the three object files, `INVENT.OBJ`, `EXPLAIN.OBJ`, and `GETIT.OBJ`. Further, `INVENT.OBJ` uses functions contained in `EXPLAIN.OBJ` and `GETIT.OBJ`.

The `.OBJ` files were compiled from their respective .C source code files. As an example, the `GETIT.OBJ` file was compiled from the `GETIT.C` file. In turn, the `.EXE` file was compiled from the respective `.OBJ` files. This means that `INVENT.EXE` consists of the code in `INVENT.OBJ`, `EXPLAIN.OBJ`, and `GETIT.OBJ`. All of this results in seven different disk files.

A typical file directory for such an arrangement is shown in Figure 11.19.

As you will see, it is very important to make sure that the date and time entered into your computer are correct. Note from the date/time stamps of the directory illustrated in Figure 11.19 that the `.EXE` file is the most recent and that the `.OBJ` files are more recent

Table 11.7 Common Utility Programs Used with C

Utility Type	Purpose
MAKE Utilities	Separate programs that help keep all of your source, object, and .EXE files current.
File Date/Time Change (TOUCH Utilities)	Separate programs that allow you to change the date and time stamp of one or more files.
Linker	A separate program used to do the linking work when source code is compiled outside of IDE.
Library Utilities	Separate programs that allow you to create your own libraries that can then be accessed by your C programs.
File Search Utilities	Separate programs that let you search for specified text in several files at once.
Conversion Utilities	Separate programs that allow you to link graphics driver files and fonts directly into your C program.
Cross Reference	A separate program that allows you to get specific information concerning the contents of your object and library files.

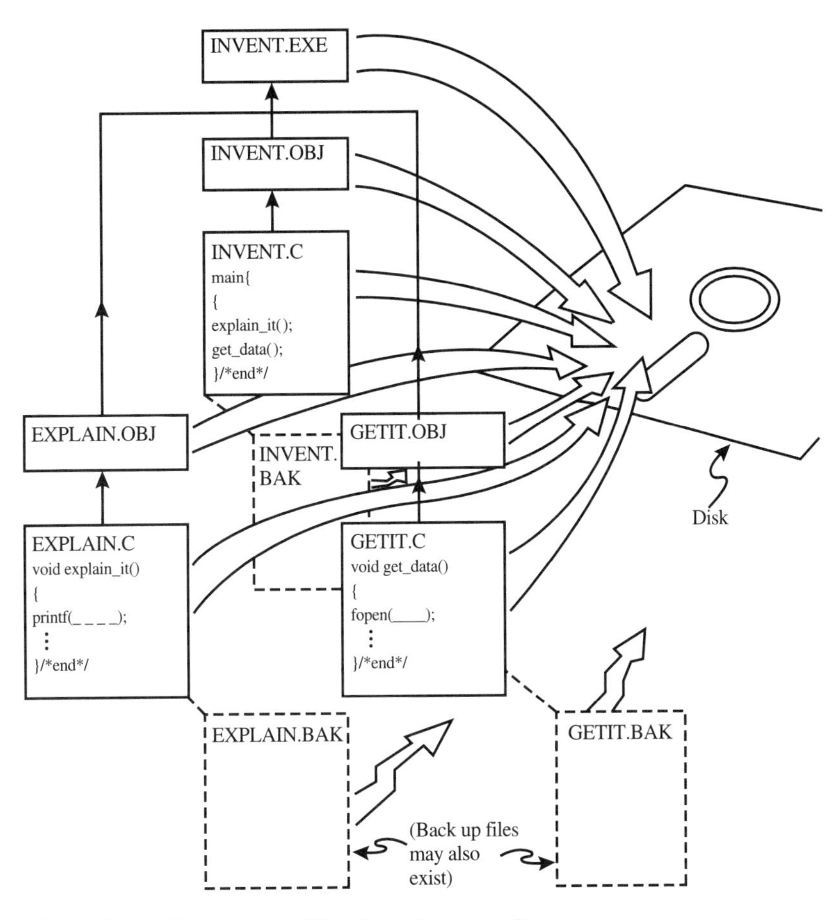

Figure 11.18 Development Files for an Inventory Program

than their related C source code files. Of course, you should know that the one `.EXE` file is the only file of the seven needed in this example to execute the program. However, while the program is under development, all the other files are needed. Figure 11.20 illustrates the dependency that one file has on another.

Filename	Extension	File Size	Date Stamp	Time Stamp
INVENT	EXE	10187	9-12-97	1:15p
INVENT	OBJ	430	9-12-97	1:08p
INVENT	C	642	9-12-97	1:05p
EXPLAIN	OBJ	983	9-12-97	12:45p
EXPLAIN	C	1246	9-12-97	12:36p
GETIT	OBJ	10083	9-12-97	10:23a
GETIT	C	12436	9-12-97	10:15a

Figure 11.19 File Directory for Inventory Program

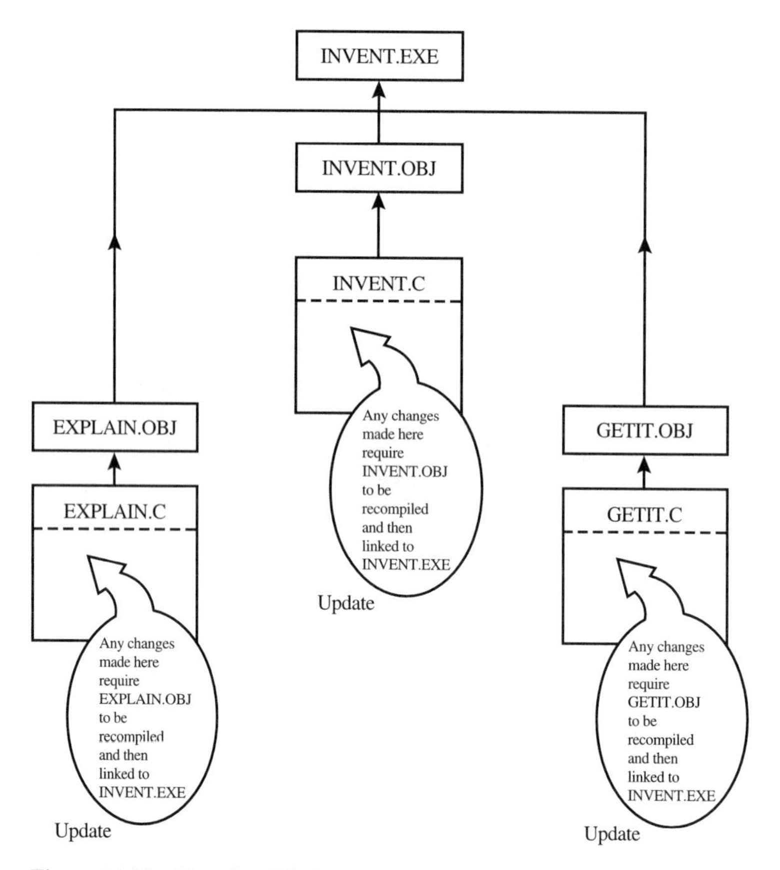

Figure 11.20 How One File Depends on Another

As you can see from Figure 11.20, changing any of the source files requires an update of other files that depend on it. Whenever you change any of your source code files and save them to the disk, their date/time stamps will now be later than these of their corresponding `.OBJ` files and later than those of the `.EXE` files. This means that the system of files must be updated. This situation is shown in Figure 11.21.

As you can see from Figure 11.21, the `.EXE` file does not include the latest changes that were made in the .C source code file. This means that if the program is now executed, it will not contain the latest changes from the source code file.

INVENT	EXE	10187	9-12-97	1:15p	← Does not contain most recent change
INVENT	OBJ	430	9-12-97	1:08p	
INVENT	C	642	9-12-97	1:05p	
EXPLAIN	OBJ	983	9-12-97	12:45p	← No longer current
EXPLAIN	C	1287	9-12-97	3:06p	← Recent update
GETIT	OBJ	10083	9-12-97	10:23a	
GETIT	C	12436	9-12-97	10:15a	

Figure 11.21 Differences in Date/Time Stamps for Recently Changed Files

INVENT	EXE	10195	9-12-97	3:15p	← Now updated to include changes
INVENT	OBJ	430	9-12-97	1:08p	
INVENT	C	642	9-12-97	1:05p	
EXPLAIN	OBJ	998	9-12-97	3:11p	← Now current
EXPLAIN	C	1287	9-12-97	3:06	← Changed file
GETIT	OBJ	10083	9-12-97	10:23a	
GETIT	C	12436	9-12-97	10:15a	

Figure 11.22 Resulting Directory of Updated Files

Of course you could, from the IDE, compile the changed source code, updating its `.OBJ` file, and then link all the `.OBJ` files to create an updated `.EXE` file. The resulting directory could then appear as shown in Figure 11.22.

Doing this in the IDE is fine for a small program such as the one illustrated here. However, for large programs consisting of many different source files with their resulting object files, this could become a major task. This is especially true if you are also developing your own custom .H (header) files as well as custom library files. Imagine making a change on a custom header file on which one or more C source files depend. Making all the required updates could be a major undertaking. This is where the MAKE utility can be very helpful. Table 11.8 lists some of the functions that can be performed by this powerful utility.

Other features may be available with the MAKE utility, depending on the development system you are using.

How MAKE Works

To use the MAKE utility, you first create a file, called a *description file* or MAKEFILE, that contains the commands you want the MAKE utility to perform. Figure 11.23 illustrates the relationship between the MAKE description file and your program files.

You create a description file (`MAKEFILE`) by using a word processor (such as the one in the IDE) and simply type in the instructions you want the MAKE utility to perform.

As an example, for the Turbo C MAKE utility, to give the instruction that the file `INVENT.EXE` depends on the files `INVENT.OBJ`, `EXPLAIN.OBJ`, and `GETIT.OBJ`, you would enter

```
invent.exe: invent.obj explain.obj getit.obj
```

and simply save it on your disk, giving the file the name `MAKEFILE`. Then, when you're ready to use the MAKE utility, simply enter MAKE, and the MAKE utility will automatically

Table 11.8 Major Functions of the MAKE Utility

Environment	Capabilities
Program Development	Automatically updates executable files whenever any dependent object or source files have been modified.
Library Management	Automatically rebuilds a library file any time one of the library modules is altered.
Networking	Automatically makes current a copy of a local program or file that is stored on the network whenever the master copy is updated.

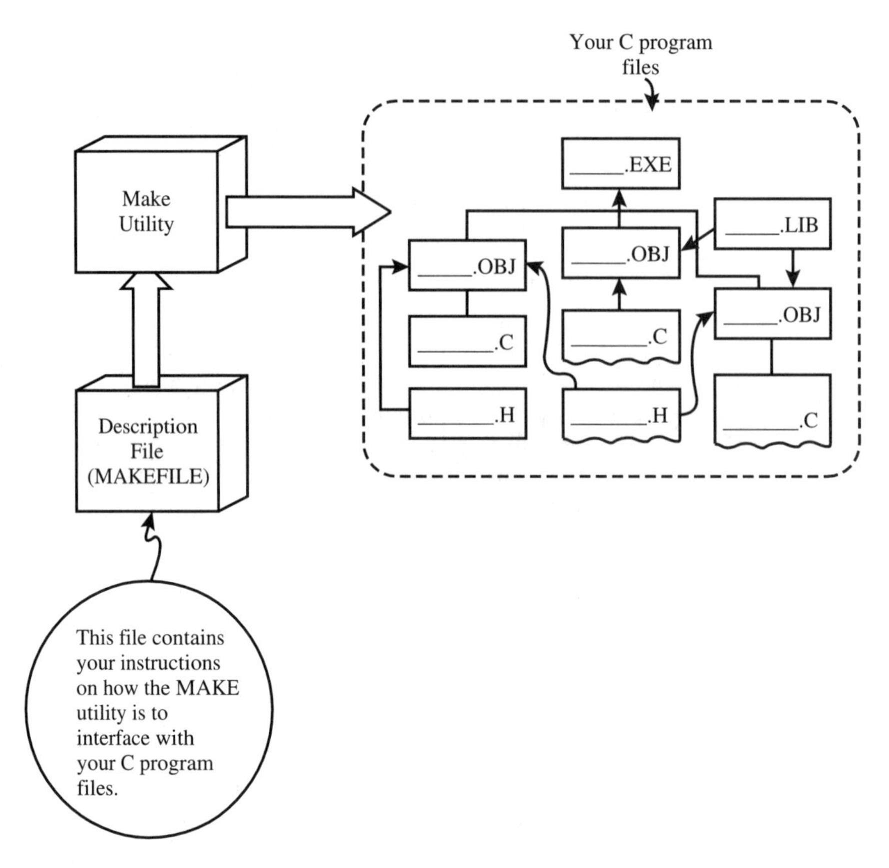

Figure 11.23 Relationship between MAKE Description File (MAKEFILE) and C Files under Development or Modification

search for a file called `MAKEFILE`, read it, and then act accordingly. What MAKE then does is to compare the time and date of each of the `.OBJ` files with the time and date of the resultant `.EXE` file (and any header files you may have indicated in your `MAKEFILE`). If any of these are later than the `.EXE` file, MAKE will know that the `.EXE` file is not current and will automatically call `TLINK`, the Turbo C linker, and relink the dependent `.OBJ` files.

The MAKE utility is a powerful one. The presentation here sets the foundation of information you need to know in order to understand the specific details and features of the MAKE utility that comes with your system. It is suggested at this time that you refer to your C system manuals for specific details and then actually try a simple application of this useful utility. Newer IDEs provide the ability to create project files that work the same as MAKE's MAKEFILE.

Date/Time Utilities

Turbo C contains a stand-alone utility that can force a particular file to be recompiled or rebuilt. This can be done even though you may not have made any changes to the file. The

Turbo C TOUCH utility will change the date and time of one or more files to the current date and time. Doing this will of course make these files newer than they were.

Library Utility

Recall that a program library is a collection of programs that have been compiled or assembled to produce a set of object modules that can be used by the C programs. Both Microsoft C and Turbo C provide the tools necessary to produce your own C library. In Turbo C this utility is a separate program on the system disk called `TLIB.EXE`, and in Microsoft C it is called `LIB.EXE`.

Basically what **library utilities** can do is to make a single library from one or more `.OBJ` files, add or delete `.OBJ` files from existing libraries, get an `.OBJ` file from an existing library, and, depending on the system, combine libraries and list the contents of a library.

There are many advantages of creating your own C libraries. First of all, as you develop C programs in your particular area of interest, you will have certain functions that you use over and over again in more than one program. These are prime candidates for storing in a library. When linking your program, the linker automatically looks for data needed from your library. Using a library can speed up the action of the linker and also results in the use of less disk space than a collection of many different `.OBJ` files.

Using the Turbo C Library Utility

To create a library using the Turbo C library utility, you must have the program `TLIB.EXE` on your active disk drive.

The syntax of the library utility commands is

```
TLIB library-name [/C] [/E] [operations] [, listfile]
```

Where

`TLIB` = the command that activates the library utility.
`library-name` = the name you wish to give to your library.
This file will automatically have the extension `.LIB` added to it (or you may add it yourself). In order for your library file to function properly, the file must have the `.LIB` extension.
`/C, /E` = options for case sensitivity and creation of an extended dictionary. Essentially the `/C` option is not normally used and you should refer to your Turbo C Reference Guide for details on the advanced `/E` feature.
`operations` = the list of operations you want the TLIB utility to perform.
`listfile` = an option that allows you to view the contents of the library.

To create a library named `TECH.LIB` that consists of modules `OHMS.OBJ`, `REACT.OBJ`, and `IMPED.OBJ`, enter

```
TLIB TECH +OHMS +REACT +IMPED
```

The new library will then be created and stored on the active drive with the name `TECH.LIB`.

Using the Quick C Library Utility

In the Microsoft C environment, you must have the file `LIB.EXE` on your active drive. When you enter `LIB`, the following prompts will appear:

```
Library name:
Operations:
List file:
Output library:
```

Where

`Library name` = name of the library you intend to modify or create.
`Operations` = command symbols and object files stating what changes to make in the library.
`List file` = the name of a cross reference listing file. No listing file is created if you do not enter a name for this prompt.
`Output library` = name of the changed library to be created as output. For your first-time use of this Quick C utility, it is suggested that you simply use the default values for each of the prompts.

More details for the creation of library files are contained in the respective system manuals for Microsoft C and Turbo C.

File Search Utilities

Turbo C offers a powerful search utility that can search for text in several files at once. This is especially useful for large source code programs or several source code files where you must change a word and are not sure exactly where in the source code (or in which file) the word is contained.

To operate this utility, you must have the file `GREP.EXE` on your active drive. The command is

```
GREP [options] searchstring [filespec ...]
```

Where

`options` = one or more single characters (as given in the Turbo C Reference Guide) that state how the files are to be searched.
`search-string` defines the pattern `GREP` will search for.
`filespec` = specification that tells `GREP` which file(s) to search.

As an example,

```
GREP impedance *.C
```

instructs the file search utility to search all files on the active drive with the .C extension for the string `impedance`.

Conclusion

It was our intent in this section to make you aware of the many other features that may be available with your C system. Here you were introduced to some of the most common ones in two of the most popular C development systems, Microsoft C and Turbo C.

The details of operation for these utilities are covered in the respective Microsoft C and Turbo C manuals along with many suggestions and examples. You are encouraged to refer to the manuals that come with your C system. Check your understanding of this section by trying the following section review.

11.6 Section Review

1. Briefly state the purpose of a MAKE utility.
2. What does a MAKE utility look for when working to keep your files updated?
3. State the use of the Turbo C TOUCH utility.
4. What is a library utility?
5. Give the name of the Turbo C utility that will search files for a given string.

11.7 BIOS and DOS Interfacing

Discussion

This section is specifically for the IBM PC and compatibles. As you will discover, there are many different powerful programs built into the permanent memory (usually referred to as ROM for **R**ead **O**nly **M**emory). These permanently built-in programs are referred to as the **B**asic **I**nput/**O**utput **S**ystem or simply **BIOS**, or the IBM ROM BIOS.

You will also discover in this section that the DOS (**D**isk **O**perating **S**ystem) also contains many different and powerful programs, that, once loaded into your computer's read and write memory (referred to as RAM for **R**andom **A**ccess **M**emory), are also available for your use.

This section will introduce you to the advantages and disadvantages of using the built-in routines in both of these systems. Here you will also see some of the built-in C functions used for these purposes as well as a short application program for the BIOS and DOS.

BIOS and DOS

Figure 11.24 illustrates the roles that the BIOS and DOS play in your computer system.

Figure 11.24 is an important figure. As shown, both BIOS and DOS contain programs that control the Input/Output (I/O) of your computer. Most application software programs (including your C operating system) will use the I/O routines already existing in DOS rather than having to write these routines themselves. (That's why you must load DOS into your computer before you can use commercial application programs). Also, as shown in the figure, you can have your C program use either the BIOS or the DOS for performing I/O operations. Table 11.9 lists the merits of each system.

Some BIOS Functions

Both Turbo C and Microsoft C support BIOS interfacing. Some of the more common I/O functions are listed in Table 11.10.

DOS Calls

Both Turbo C and Microsoft C provide functions that interface directly with the DOS routines. Some of these are listed in Table 11.11.

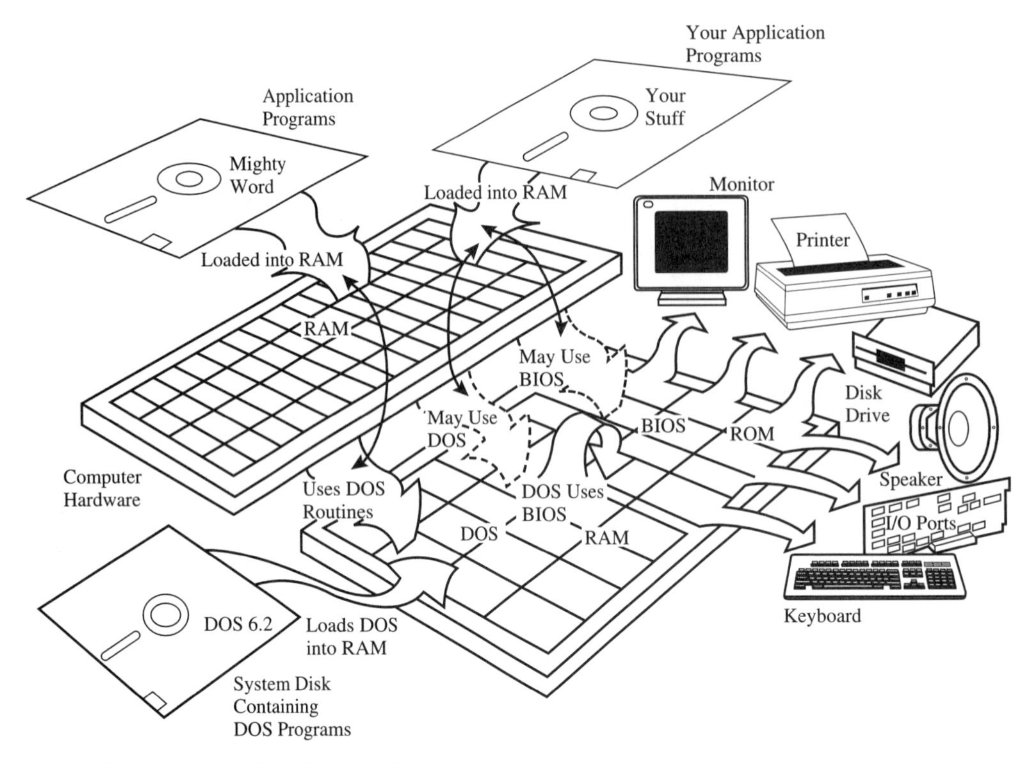

Figure 11.24 Roles of BIOS and DOS

Table 11.9 BIOS and DOS I/O

I/O Type	Comments
BIOS	Faster and more direct than DOS. However, because these routines are contained in a specific computer machine architecture, they may not be configured in exactly the same way on a system made by a different manufacturer. Thus, any C program that uses BIOS directly may not be portable to these other systems.
DOS	Not as fast as accessing BIOS directly and not as versatile. However, the advantage is that your program will be more portable, because DOS can adjust for some hardware differences.

Table 11.11 lists just some of the C DOS functions. There are about 40 different DOS function calls used by Turbo C and Microsoft C.

BIOS Application

Program 11.7 illustrates a BIOS application. This program uses the Turbo C `biosequip()` function that returns specific information about your IBM PC (or true compatible).

The `biosequip()` function returns an integer whose value depends on the equipment connected to your system. For the meaning of the returned number to be understandable,

Table 11.10 Common BIOS Functions in C—Uses <bios.h>

Function		Description
Turbo C	Microsoft C	
bioscom	_bios_serialcom	Performs serial I/O.
biosdsk	_bios_disk	Provides many different disk access functions. Generally, this function controls the movement of the disk read/write head on a specified drive.
biosequip	_bios_equipment	This function determines the hardware and peripherals currently used by your system.
bioskey	bios_keybrd	Accesses the keyboard directly through the BIOS keyboard routines. Generally, this function checks to see if the keyboard is ready to be read, reads it, then gets the character.
biosmemory	_bios_memory	Returns the total amount of memory available on your system.
biosprint	_bios_printer	Performs a variety of printer functions, such as checking if the printer is out of paper.

NOTE: These are not ANSI C compatible and will work only on IBM PCs and true compatibles.

Table 11.11 Common DOS Functions in C—Uses <dos.h>

Function		Description
Turbo C	Microsoft C	
inport		Reads a word from a specified port. (Microsoft C has a similar function called inp—it uses the <conio.h> header file.)
outport		Outputs a word to a specified hardware port. (Microsoft C has a similar function called outp—it uses the <conio.h> header file.)
peek		Returns a word at a memory location specified by an offset and segment value.
poke		Stores an integer value at a memory location specified by an offset and segment value.
keep	_dos_keep	Used to install a terminate-and-stay resident program in memory.
getdate	_dos_getdate	Gets the system date.

NOTE: These functions are not ANSI C standards and require the presence of DOS (3.0 or higher for some of the C DOS functions).

it must be converted to a 16-bit binary number. In this binary number, each bit or group of bits indicates a hardware detail about your system. Figure 11.25 shows the significance of each bit for this function.

Program 11.7 illustrates the biosequip() function in action.

Program 11.7

```
#include <stdio.h>
#include <bios.h>

main()
{
        printf(" The biosequip number for this system is %0X" , biosequip());
}
```

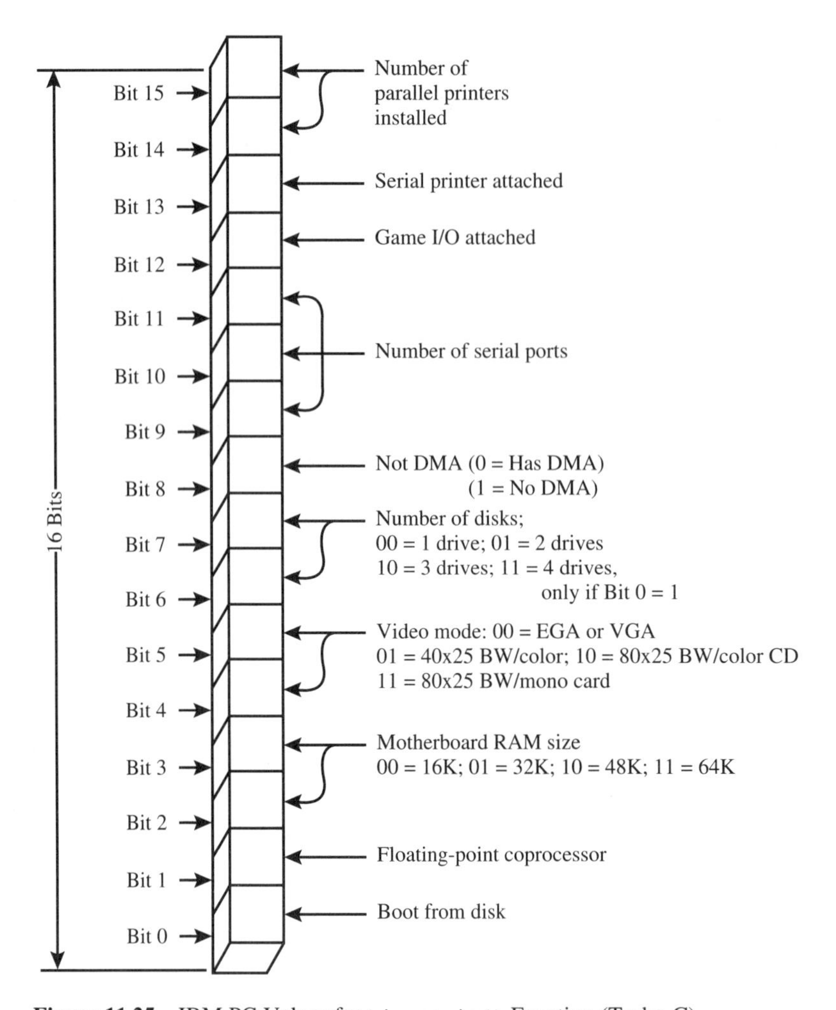

Figure 11.25 IBM PC Values for `biosequip()` Function (Turbo C)

Note that Program 11.7 causes the decimal output to be expressed in hexadecimal notation. This is done so that the resulting number can easily be converted to binary.

A typical output for Program 11.7 could be:

```
406D
```

When `406D` is converted to a binary number, you have:

```
0100 0000 0110 1101
```

As you can see, because bits 15 and 14 are 01, this system has one parallel printer installed. With bit 13 at 0, there is no serial printer attached, and bit 12 at 0 indicates that there is no game I/O attached. Bits 6 and 7 are 01, indicating that there are two drives attached (a 00 indicates one drive). Bits 2 and 3 are 11, indicating that the motherboard RAM size is 64K.

DOS Call Application

Program 11.8 illustrates an application using a Microsoft C **DOS call**. The function used here is `_dos_getdate(date)`, and it has the following structure defined in the `dos.h` header file:

```
struct dosdate_t
    {
        unsigned char day;          /* Value 1 - 31   */
        unsigned char month;        /* Value 1 - 12   */
        unsigned int year;          /* Value 1980 - 2099   */
        unsigned char dayofweek;    /* Sunday = 0, through 6 */
    }  *date;
```

Program 11.8

```
#include <stdio.h>
#include <dos.h>

main()
{
        struct dosdate_t date;

        /* Get the current system date. */

        _dos_getdate(&date);

        /* Print out the results. */

        printf(" \nThe month is => %d\n" , date.month);
        printf(" The day is   => %d\n" , date.day);
        printf(" The year is  => %d\n" , date.year);
        printf(" Day of the week is => %d\n" , date.dayofweek);
}
```

Conclusion

This section introduced you to direct hardware BIOS or DOS interfacing using Turbo C or Microsoft C. Here you saw the similarities and differences of both interfacing methods.

You also saw some of the most common functions used for this type of interfacing. These include a BIOS and a DOS interfacing application. Check your understanding of this section by trying the following section review.

11.7 Section Review

1. Briefly state the differences between BIOS and DOS routines.
2. What are the advantages of using BIOS routines? The disadvantage?
3. What are the advantages of using DOS routines? The disadvantage?
4. State a BIOS function and briefly describe its purpose.
5. State a DOS function and briefly describe its purpose.

11.8 Troubleshooting Techniques

Discussion

This section contains information concerning your system's hardware and how you can use software to directly interface with it. Here you will see how the BIOS and DOS functions you were introduced to in the previous section actually make use of your computer's memory. Armed with this knowledge, you will be able to develop programs that can take full advantage of programs built into your system.

Basic Idea

An **interrupt** is a special control signal that causes your computer to temporarily stop its main program. Once this happens, it will then go to another program called the **interrupt handler** (sometimes called the **interrupt service routine**). Your computer will take instructions from the interrupt handler. Once the requirements of this routine are completed, the computer will return control back to where it left off in the main program. This process is called *servicing the interrupt*. In the process of servicing the interrupt, the μP saves the contents of some of its internal registers into the stack. Doing this allows these internal registers to be used by the service routine. When completing the service routine, the μP restores these registers from the stack. This allows the μP to pick up where it left off when the interrupt occurred. This concept is illustrated in Figure 11.26.

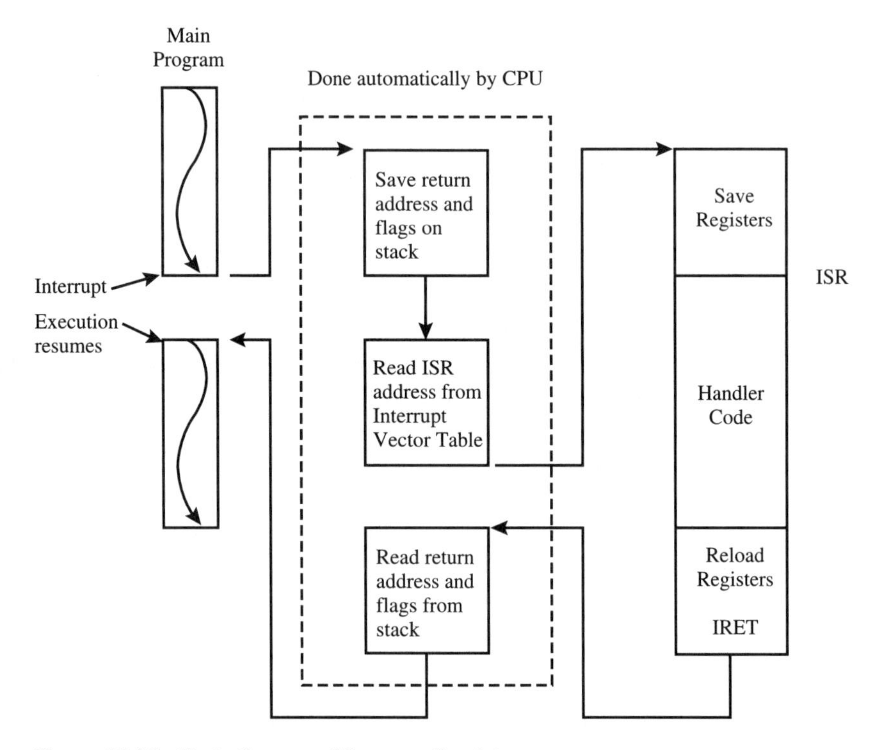

Figure 11.26 Basic Concept of Interrupt Servicing

What an Interrupt Does

An interrupt provides access to the ROM BIOS and DOS routines. For example, the BIOS and DOS interfacing functions presented in the previous section are really interrupts. When the 80x86 μP is used, it reserves the first 1024 bytes to store a series of 4-byte addresses known as **interrupt vectors**. These are really far pointers that contain the offset and the segment address of the interrupt handler. Some of the more common interrupts will be presented in this section. For a complete list of all the interrupt vectors and their corresponding handlers, refer to the *IBM Personal Computer Technical Reference Manual*.

Because the first 1024 bytes of memory hold a series of 4-byte addresses, this means that there are 1024/4 = 256 different interrupt vectors that can be stored in this section of memory. However, depending on how current your system is, only about 31 of these vector locations are actually used. The rest are available for your use (or use of your software, such as your C development system).

To see what is contained in part of this memory location, Program 11.9 uses the Turbo C `peek()` function to get the contents.

Program 11.9

```
#include <stdio.h>
#include <conio.h>

main()
{
        unsigned int offset;
        unsigned char value;
        int count;
        unsigned char address;

        clrscr();
        printf(" \n\n" );

        for(count = 0; count <= 240; count += 15)
        {
                address = count;
                printf(" \n 0:%2X => " ,address);

                for(offset = count; offset < count + 16; offset++)
                {
                        value = peek(0, offset);
                        printf(" %2X " , value);
}
```

The output of Program 11.9, along with an explanation, is shown in Figure 11.27.

Interpreting a Vector Table

A **vector table** is a sequence of memory locations that contains the memory locations (addresses) of other programs. Each vector contains 4 bytes. As an example, the Turbo C function for interfacing with the ROM BIOS that does the keyboard work is `bioskey()`. This function does little more than actually call interrupt number 16_{16}. An **interrupt num-**

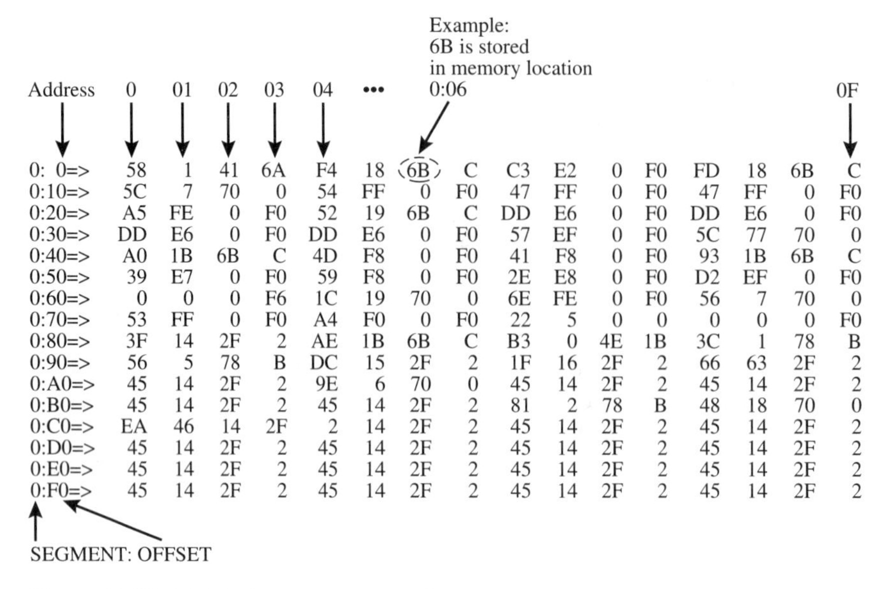

Address	0	01	02	03	04											0F
0: 0=>	58	1	41	6A	F4	18	6B	C	C3	E2	0	F0	FD	18	6B	C
0:10=>	5C	7	70	0	54	FF	0	F0	47	FF	0	F0	47	FF	0	F0
0:20=>	A5	FE	0	F0	52	19	6B	C	DD	E6	0	F0	DD	E6	0	F0
0:30=>	DD	E6	0	F0	DD	E6	0	F0	57	EF	0	F0	5C	77	70	0
0:40=>	A0	1B	6B	C	4D	F8	0	F0	41	F8	0	F0	93	1B	6B	C
0:50=>	39	E7	0	F0	59	F8	0	F0	2E	E8	0	F0	D2	EF	0	F0
0:60=>	0	0	0	F6	1C	19	70	0	6E	FE	0	F0	56	7	70	0
0:70=>	53	FF	0	F0	A4	F0	0	F0	22	5	0	0	0	0	0	F0
0:80=>	3F	14	2F	2	AE	1B	6B	C	B3	0	4E	1B	3C	1	78	B
0:90=>	56	5	78	B	DC	15	2F	2	1F	16	2F	2	66	63	2F	2
0:A0=>	45	14	2F	2	9E	6	70	0	45	14	2F	2	45	14	2F	2
0:B0=>	45	14	2F	2	45	14	2F	2	81	2	78	B	48	18	70	0
0:C0=>	EA	46	14	2F	2	14	2F	2	45	14	2F	2	45	14	2F	2
0:D0=>	45	14	2F	2	45	14	2F	2	45	14	2F	2	45	14	2F	2
0:E0=>	45	14	2F	2	45	14	2F	2	45	14	2F	2	45	14	2F	2
0:F0=>	45	14	2F	2	45	14	2F	2	45	14	2F	2	45	14	2F	2

Figure 11.27 IBM PC Memory Dump of Interrupt Vector Table

ber is an integer that indicates to the μP the memory location of the interrupt vector. This effectively makes it a pointer to a pointer. All interrupt vectors use 4 bytes of memory to store the segment and offset address of the interrupt handler. Thus, to find the address of this interrupt vector, multiply the value of the interrupt number by 4. (Be sure to first convert the interrupt number to decimal.) As an example, the address of the interrupt vector for the keyboard interrupt (interrupt 16_{16}) is obtained as follows:

$$16_{16} = 22_{10} \text{ X } 4 = 88_{10} = 58_{16}$$

This means that the address for this ROM BIOS service routine is contained in memory location 58_{16}. Looking at Figure 11.27, from the memory dump, you can see that the line containing this address is

```
0:50 => 39 E7 0 F0 59 F8 0 F0 2E E8 0 F0 D2 EF 0 F0
```

Counting over to get the 4-byte address for interrupt vector #16, you have

```
2E E8 0 F0
```

To interpret the meaning of this address, you must take into account two things about how the IBM PC stores an address in memory:

1. The least significant byte is stored first.
2. The offset part of an address is stored before the segment part of the address.

Figure 11.28 shows how to construct the actual address for the keyboard BIOS interrupt handler.

As shown in Figure 11.28, the absolute address of the keyboard service routine is at $FE82E_{16}$.

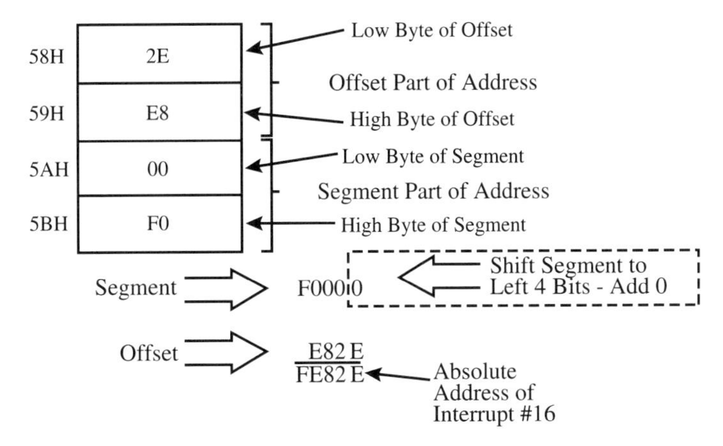

Figure 11.28 Determining the Absolute Address of an Interrupt Handler

Typical Interrupts

As you know, interrupts can be used to get ROM BIOS or DOS service routines. Table 11.12 lists some of the most commonly used BIOS and DOS interrupts.

Program 11.10

```
#include <stdio.h>
#include <dos.h>

#define int_number 0x12          /* The BIOS interrupt number. */

typedef struct                   /* Registers as 16-bit words. */
{
        unsigned int ax;
        unsigned int bx;
        unsigned int cx;
        unsigned int dx;
        unsigned int si;
        unsigned int di;
        unsigned int flags;
} registers_16;

typedef struct                   /* Registers as 8-bit words. */
{
        unsigned char al;
        unsigned char ah;
        unsigned char bl;
        unsigned char bh;
        unsigned char cl;
        unsigned char ch;
        unsigned char dl;
        unsigned char dh;
```

```
} registers_8;

typedef union                        /* Registers as 1 or 2 bytes. */
{
        registers_16 reg_16;
        registers_8  reg_8;
} registers_16_or_8;

main()
{
        registers_16_or_8 registers;
        unsigned int size;

        /* Call the interrupt. */

        int86(int_number, &registers, &registers);

        /* Get value from AX register. */
        size = registers.reg_16.ax;

        printf("\nThe memory size of your system is %d K." ,size);
}
```

Program 11.10 shows an application of a simple ROM BIOS interrupt application. The C program uses the Turbo C function `int86()`. This is a general 80x86 μP software interrupt used with `<dos.h>`. The format of the function is

```
int86(int number, &inregs, &outregs);
```

Where

`number` = The interrupt number.
`inregs` = Register values sent to service routine.
`outregs` = Register values returned from service routine.

The implementation of Program 11.10 requires two structures, each of which defines a set of internal registers for the 80x86 μP. The first structure defines this set as 16-bit words whereas the second defines them as 8-bit words. Memory is allocated for use by either structure through a C union. Note that each of the data types within the registers is declared as `unsigned int` (two bytes) in the first structure and as `unsigned char` (one byte) in the second structure. Program 11.10 uses interrupt number 12. As you can see in Table 11.12, this is a BIOS interrupt for checking the size of your system's memory.

Creating Interrupt Handlers

Many of the interrupt vectors in your IBM PC are unassigned. This means that you (or a software vendor) can use this space to write your own interrupts. You do this by placing a C far pointer into one of the unused interrupt vector locations. With this, whenever a predefined event takes place, such as a system clock time, keypress, or some other action, these handlers can then be used to pass certain instructions to the main program.

Table 11.12 Common Interrupt Numbers—IBM PC

BIOS	Function
5	PRINT SCREEN
10	VIDEO I/O
11	EQUIPMENT CHECK
12	MEMORY SIZE CHECK
13	DISKETTE I/O
14	COMMUNICATIONS PORT I/O
15	CASSETTE I/O
16	KEYBOARD I/O
17	PRINTER I/O
18	ROM BASIC
19	BOOTSTRAP START-UP
1A	TIME OF DAY
1B	KEYBOARD BREAK (user setable)
1C	TIMER TICK (user setable)
DOS	**Function**
20	PROGRAM TERMINATE
21	FUNCTION REQUEST
22	TERMINATE ADDRESS—This is the address to which control transfers when a program terminates.
23	CTRL-BREAK EXIT ADDRESS— Done by a Ctrl-Break.
24	CRITICAL ERROR HANDLER
25	ABSOLUTE DISK READ
26	ABSOLUTE DISK WRITE
27	TERMINATE BUT STAY RESIDENT—Leaves program in memory.

Program 11.11 shows how to develop an interrupt handler function, install it, and then test it. This program uses the Turbo C type `interrupt` for a function. It produces a short beep at the hardware output port of your PC's speaker. It's not terribly exciting, but is a pretty safe way of illustrating the development of an interrupt handler. New functions are explained after the program.

Program 11.11

```
#include <dos.h>

/************************   Function Prototypes:   ************************/

void interrupt handle(void);
/* This function is the actual interrupt handler. Note that it is of      */
```

```
/* type interrupt.                                                          */

void install(void interrupt (*faddr) (), int vector);
/* This is the function that actually installs the address of the          */
/* interrupt handler in the location given by vector.                       */

void testhandle(int bcount, int vector);
/* This function simply tests the installed routine.                        */

/***************************************************************************/

/*----------------- Start of Main Program ------------------------------*/

main()
{
        install(handle,0x90);
        testhandle(200,0x90);
}

/*-----------------------------------------------------------------------*/

void testhandle(int bcount, int vector)
{
        _CL = bcount;
        geninterrupt(vector);
}

/*-----------------------------------------------------------------------*/

void interrupt handle()
{
        int count_1, count_2;
        char originalbits, bits;

        int bcount = _CL;

        /* Get the current control port setting. */
        bits = originalbits = inport(0x61);
        for(count_1 = 0; count_1 <= bcount; count_1++)
{
                /* Turn off the speaker for a short time. */
                outport(0x61, bits & 0xFC);

                for(count_2 = 0; count_2 <= 7500; count_2++);

                /* Now turn it on for some time.  */
                outport(0x61, bits | 2);

                for(count_2 = 0; count_2 <= 7500; count_2++);
```

```
        } /* End of for */

        /* Restore the original control port setting. */
        outport(0x61, originalbits);
} /* End of handle() */

/*---------------------------------------------------------------------*/

void install(void interrupt (*faddr)(), int vector)
{
        setvect(vector, faddr);
} /* End of install() */
```

Program 11.11 uses several of Turbo C's special functions for developing interrupt handlers and doing I/O, as follows.

`geninterrupt(int interrupt_number);`

This function initializes a software trap for the interrupt.

`inport(int port_number);`

This function reads a word from a hardware port signified by `port_number`.

`outport(int port_number, int value);`

This function outputs a word to a hardware port signified by `port_number`.

`setvect(vector, faddr);`

This function sets the value of the interrupt vector.

Conclusion

This section has given you an introduction to interrupts. Here you saw what an interrupt is, how it is stored in your computer, and how it can be used. You also saw the development of a program that shows you how to develop an interrupt handler. Check your understanding of this section by trying the following section review.

11.8 Section Review

1. Briefly describe what is meant by an interrupt.
2. Explain the meaning of an interrupt vector.
3. State what is meant by an interrupt handler.
4. Explain how an address is stored in the IBM PC.
5. Where in the memory of the IBM PC are the interrupt vectors stored?

11.9 Case Study: Printer Controller

Discussion

The case study for this chapter demonstrates hardware interfacing between your computer and an external device. The device in this case study is an IBM printer. This case study really shows the development of a function more than a complete program. However, this is an important development to follow because it illustrates the steps involved in analyzing an actual output port from a hardware standpoint and then implementing control over this port using the C language.

The Problem

Create a demonstration program that contains a function that will allow the user to access the parallel printer port and control the printer. The program cannot use any built-in interrupts or other preprogrammed printer control routines. The purpose of this program is to illustrate the control of a hardware port on the IBM PC using only the `inport()` and `outport()` functions or their equivalents. Essentially, the user will be able to enter a line of text that will be echoed to the monitor screen and then outputted to the printer when the Return key is pressed. The program terminates when an `'x'` is entered.

First Step—Stating the Problem

Stating the problem in writing yields:

- Purpose of the Program: Provide a program that will allow the program user to enter a line of text, see it on the monitor, and have it sent to the printer when the Return key is pressed. The C program is not to use any built-in printer functions or interrupts; only the `inport()` and `outport()` functions or their equivalents are allowed.
- Required Input: A string of characters from the keyboard.
- Process on Input: Program is not to use built-in interrupts or preprogrammed printer routines. Only `inport()` and `outport()` (or equivalent) may be used.
- Required Output: The character string from the keyboard to the monitor and then, when the Return key is pressed, to the printer.

Developing the Algorithm

The algorithm for this program involves repetition of the following steps until the user enters Ctrl-C:

1. Initialize the printer.
2. Get character from user.
3. Echo character to screen.
4. Store character in printer buffer using `outport()`.
5. Check printer status.
6. If user enters Return, print the buffer.

Program Development

The development of this program requires a knowledge of the 25-pin connector at the rear of your IBM PC that interfaces this computer with an IBM printer (or compatible). Therefore, the first phase of the development for this program will be a presentation of the technical hardware material concerning this interface connection. Since the `outport()` and `inport()` commands require the use of a nonresident program, Turbo C was chosen as the language for this program.

The Parallel Printer Port

An IBM PC can be connected to a parallel printer by means of a 25-pin connector. Figure 11.29 illustrates the pin numbers and the purpose of the key pins used for sending and receiving signals between the computer and the printer.

As you can see in the figure, there are actually three ports associated with the computer printer interface. The data port is used to send the printable characters to the printer. The status port is used to get information from the printer so the computer will know when and if to send information to it. The last port, the control port, is used to send control information to the printer.

Defining Data

The first step in the development of this program was to write the `#defines`. These are based on the hardware requirements of the computer printer interface.

Note that the defines are grouped so that the three separate ports are evident. A descriptive identifier is used for each port (such as `DATA` for the data port located at 378_{16}).

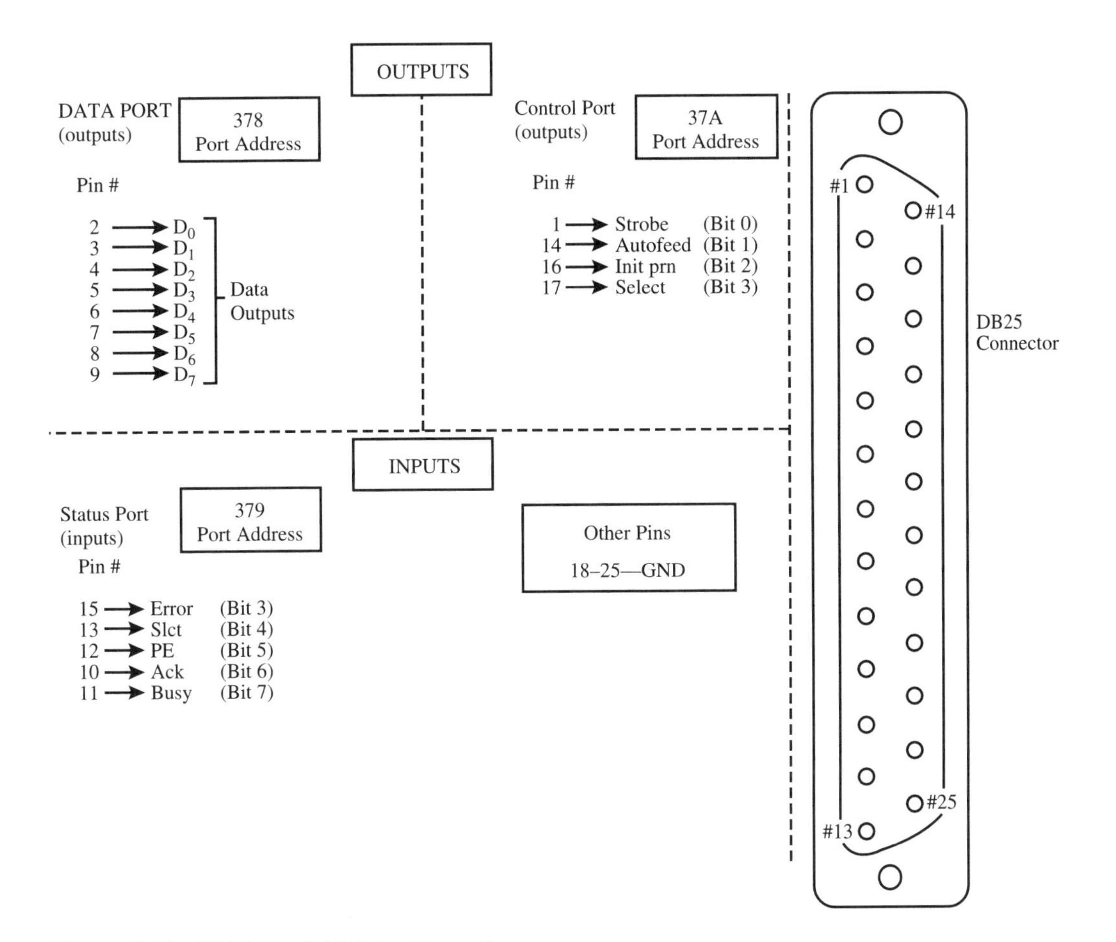

Figure 11.29 DB25 Parallel Printer Connections

Just below each port identification, any important values that will be needed for the program are defined. The values for all of the control characters are found in the printer manual. Thus, under the `DATA` definition, the line feed data character is 10, and `line_feed` is defined as that value. This is also true for the value of 13, which represents a carriage return. Note that the same approach is taken with the other two data ports that have corresponding `#define` statements for their associated signals for the use of the program.

The Completed Program

Program 11.12 is the completed program.

Program 11.12

```
#include <stdio.h>
#include <dos.h>
#include <conio.h>
#define TRUE 1
#define DATA 0x378                   /* Printer data port location.        */
                                     /* Use 0x3BC for IBM Printer Adapter  */

#define line_feed 10
#define carriage_return 13

#define CONTROL 0x37a                /* Printer control port location.     */
                                     /* Use 0x3BE for IBM Printer Adapter  */
#define strobe_hi 1                  /* Bit 0 of control port set high.    */
#define no_init_prn 4                /* Initialize and select printer.     */

#define STATUS 0x379                 /* Printer status port location.      */
                                     /* Use 0x3BD for IBM Printer Adapter. */
#define ack 0x40                     /* Bit 6. */

void strobe_and_wait_for_ack(void);

main()
{
        int character;               /* Character to be printed. */

        clrscr();

        /* Strobe and activate printer. */

        outport(CONTROL), strobe_hi);
        outport(CONTROL), strobe_hi | no_init_prn);
        do
        {
                do
                {
                        character = getchar();
                        outport(DATA, character);     /* Output a character. */
                        strobe_and_wait_for_ack();
```

```
            } while((character != '\n') && (character != 'x'));

            /* Output new line codes. */

            outport(DATA, carriage_return);
            strobe_and_wait_for_ack();

            outport(DATA, line_feed);
            strobe_and_wait_for_ack();
        } while(character != 'x');
} /* End of main() */

void strobe_and_wait_for_ack()      /* Strobe then wait for acknowledge. */
{
        int j;

        outport(CONTROL,no_init_prn);               /* Strobe is low.           */
        for(j = 0; j < 255; j++);                   /* Delay for pulse width.  */
        outport(CONTROL,strobe_hi | no_init_prn);   /* Strobe is high.         */
        if(inport(STATUS) & ack)                    /* Wait for acknowledge.   */
                while((inport(STATUS) & ack));
        while(!(inport(STATUS) & ack));
}
```

Program Analysis

The program contains two C do loops. The outer loop allows the user to keep outputting characters to the screen one line at a time, and then causes them to appear on the printer. The inner C do loop keeps getting characters from the keyboard, displaying them to the screen and placing them into the printer buffer. The printer buffer will not be printed until a carriage return is initiated by the program user. When this happens, the program goes into the outer loop, where the contents of the printer buffer are then printed.

Conclusion

This section demonstrated the use of C to interface an actual I/O port with an external device using `outport()` and `inport()` functions. You should not consider this to be a model printer control program. Rather consider it to be a foundation upon which you can build.

The Self-Test questions at the end of this chapter address specific questions about this program. For now, test your understanding of this material with the following section review.

11.9 Section Review

1. How are commands sent to the printer?
2. How is the pulse width of the strobe signal controlled?
3. What statements are used to "wait" for the acknowledge signal?

Interactive Exercises

DIRECTIONS

These exercises require that you have access to a computer and software that supports C. They are provided here to give you valuable experience and immediate feedback on what the concepts and commands introduced in this chapter will do.

Exercises

1. Program 11.13 illustrates an interaction with Turbo C pseudo variables. Try it.

Program 11.13

```
#include <stdio.h>

main()
{
        unsigned char lo;
        unsigned char hi;
        unsigned int hi_lo;

        _BL = 0xA;
        _BH = 0xB;
        lo = _BL;
        hi = _BH;
        hi_lo = _BX;

        printf(" Both together are %X" , hi_lo);
}
```

2. Program 11.14 presents an example of the Turbo C inline machine code. Try the program and see what it does. Did you experience any difficulty with your system?

Program 11.14

```
include <stdio.h>
#include <dos.h>

main()
{
        __emit__(0xB2,0x01);            /* mov dl,1 */
        __emit__(0xB4,0x02);            /* mov ah,2 */
        __emit__(0xCD,0x21);            /* int 21   */
        __emit__(0xCD,0x20);            /* int 20   */
}
```

3. Program 11.15 is a Turbo C program that returns an equipment code number. Can you interpret the meaning of the resultant number for your system?

Program 11.15

```
#include <stdio.h>
#include <bios.h>

main()
```

```
{
        printf(" Your equipment code number is %X" , biosequip());
}
```

4. You should know ahead of time what to expect from the output of Program 11.16. What did you get?

Program 11.16

```
#include <stdio.h>
#include <dos.h>

main()
{
        printf(" At location 0:50 is %X." , peek(0,0x50));
}
```

Self-Test

DIRECTIONS

Answer the following questions by referring to Program 11.12 in the case study section of this chapter.

Questions

1. State the meaning of `#define DATA 0x378` in the program.
2. What does DB25 mean in terms of your computer connections?
3. How many ports does the DB25 parallel printer connector have?
4. State the names and port numbers of the ports in the DB25 connector.
5. Give the purpose of each of the ports in the DB25 connector.
6. In the program, explain the meaning of the statement

   ```
   (inport(STATUS) & ack)
   ```

7. Explain how the acknowledge bit is tested in the program.
8. What does the program statement `outport(CONTROL, strobe_hi)` actually do?

End-of-Chapter Problems

General Concepts

Section 11.1

1. Name the general purpose registers contained in the 80x86 μP. State how they may be used.
2. What is a stack?
3. What is a LIFO stack?
4. Explain what is meant by a `push` and a `pop`.
5. Describe the process of segmentation.

Section 11.2

6. State some of the advantages and disadvantages of assembly language programs.
7. What is the short word used as a memory aid in assembly language programming called?

8. State how the stack grows as used by your C program and an IBM PC.
9. Where does C place locally declared variables on the stack? Function parameters?

Section 11.3

10. What is a memory model in C?
11. State the differences between the MEDIUM and the COMPACT memory models.
12. Give the advantage of confining your C code to just one 64K segment.
13. What is the difference between a far pointer and a near pointer?

Section 11.4

14. Is it possible to convert C source code to assembly language source code? Give an example.
15. Conventionally, what kind of file is given the extension `.ASM`?
16. State the difference between the assembly language instructions `[bp+4]` and `bp+4`.

Section 11.5

17. Explain what is meant by a pseudo variable. Name a C development system that uses pseudo variables.
18. Is there any problem with the statement `&_AH` in Turbo C? Explain.
19. What is it called when you include assembly language source code within your C source code?
20. Is a pseudo variable an accepted ANSI C standard?

Section 11.6

21. What is a utility as used in a C development system?
22. Which utility can help you keep your source, object and `.exe` files current?
23. Explain the purpose of a library utility.
24. In Turbo C, what is the purpose of the `GREP.EXE` utility?

Section 11.7

25. What does BIOS mean?
26. What does BIOS consist of?
27. State the advantage and disadvantage of using BIOS routines compared to using DOS routines.

Section 11.8

28. What is an interrupt?
29. State the difference between an interrupt vector and an interrupt handler.
30. What is the relationship between an interrupt handler and an interrupt service routine?
31. Explain how an address is stored by the 80x86 µP family.

Section 11.9

32. Why is `strobe_and_wait_for_ack()` so complex? What are all the `while` statements for?
33. How could a prompt string, such as "Ready," be output to the printer?

Program Design

For the following programs, use the structure that is assigned to you by your instructor. Otherwise, use a structure that you prefer. Remember, the program should be easy for anyone to read and understand (especially you if it ever needs modification). Note that for these programs, you will need a C system and an IBM PC or true compatible. Because the advanced programming topics in this chapter concern the details of hardware and software, the following questions include primarily the areas of electronics technology and computer science.

34. Figure 11.30 shows a hardware interface between the DB25 printer connector and an electrical circuit that operates relay drivers. You should refer to Figure 11.29 for information on the

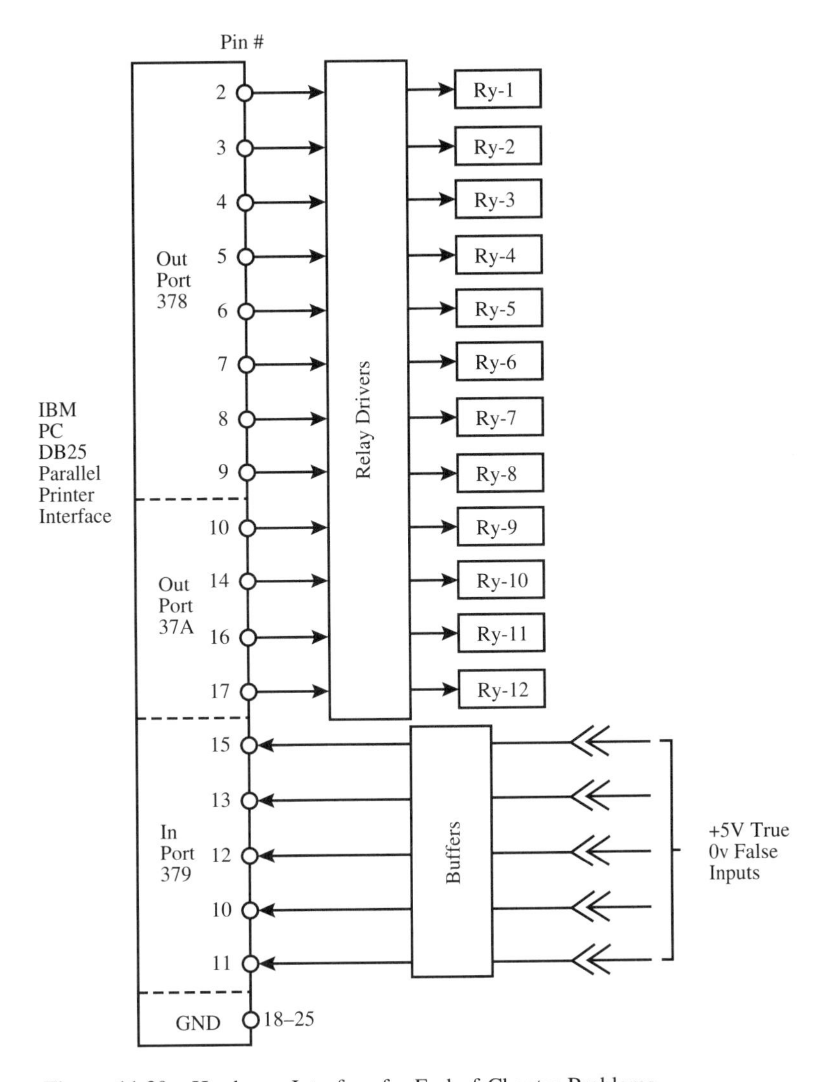

Figure 11.30 Hardware Interface for End-of-Chapter Problems

pins. Note from the figure that there are two output ports and one input port. Develop a C program that will allow the program user to activate any one of the output relays.

35. Modify the program in problem 34 to allow the program user to activate any one or more of the relays. Use computer graphics to actually show the condition of each of the relays.
36. Create a C program that will test the condition of each input for the hardware interface in Figure 11.30. The program should contain computer graphics that will actually show the condition of each of the switches.
37. Using the hardware interface circuit in Figure 11.30, develop a C program that will allow the user to have any one input pin determine the condition of one of the output pins.
38. Create a C program using computer graphics that will graphically show the condition of all the outputs and all the inputs for the hardware interface circuit in Figure 11.30. The user should be able to change the condition of any output.

39. Using your C system, develop a small assembly language program and link it to a C program.
40. Expand the requirements of problem 39 by having at least two values passed to the assembly language program. The assembly language program will then use these values to complete a simple process of your choice.
41. If your C system has inline assembly capabilities, redo the requirement of problem 40.
42. If your C system has inline machine language capabilities, develop a simple assembly language program using inline machine language.
43. Select an interrupt vector in your computer system. Then write a C program that will do a memory dump of the entire interrupt handler for the selected vector.
44. Develop a MAKE utility file that will assist you in the development of any of the programming assignments above. The programming assignment should have at least three separate files.
45. Using the library utility that comes with your C system, create a library of three of your most used C functions that are original.
46. Create a C program that uses the functions placed in your library from problem 45.

12 An Introduction to C++

Objectives

This chapter gives you the opportunity to learn:

1. About the differences between C and C++.
2. How to use `cin` for getting user input.
3. How `cout` operates.
4. What classes and objects are.
5. How to define your own classes.
6. What happens when an object is instantiated.
7. How to hide data and functions in a class.
8. How to share object data with friend functions.
9. The benefits of inheritance.
10. About virtual functions.
11. How to use file objects.

Key Terms

Class
Object
Overloading
Standard Output
I/O Manipulator
Argument
Escape Sequence
Insertor

Member
Member Function
Public Member
Instantiation
Encapsulation
Constructor
Destructor
Private Member

Friend Function
Inheritance
Base Class
Derived Class
Subclass
Protected Member
Virtual Function
Pure Virtual Function
Abstract Class
File Object
`ofstream`
`ifstream`
Character Stream
Extractor

Outline

Introduction

This chapter provides an introduction to the C++ programming language. Now that you know C, it should be an easy transition to begin using the new powerful features of C++. Aside from noticing some of the obvious differences between C and C++, there are several differences that are not so obvious. We will explore these differences and show you how to use C++ to create some basic C++ object-oriented programs.

In this chapter we will also examine the most powerful features of the C++ programming language: classes and objects. Simply put, a C++ **class** defines a structure that allows both variables and functions to be grouped together. The class definition is like a blueprint that shows how to build an object with the given structure. The **object** contains the variables and functions included in the class definition. One class may be used to create multiple copies of the same object or multiple objects that share some features but not others.

Classes allow data and functions to be hidden from other classes and even the main program. This adds a measure of security to a program by restricting access to data to those functions that use the data. Classes may also share data among themselves or keep their data private. Using classes eliminates the need to use global variables in a program. This is a good feature, since errors that result from the improper use of global variables are often hard to find. Finally, using classes in a C++ program helps to achieve software *reuse,* a method of using the same code to perform many different functions.

The power of classes and objects will become more apparent as you examine the chapter programs. Let's begin by reviewing some of the differences between C and C++.

12.1 C++ Fundamentals

Discussion

In this section we introduce some of the changes a C programmer must recognize and adopt to begin programming in C++. Although C++ can compile and execute C statements, C++ provides better ways of interacting with the program user and files in addition to the object-oriented capabilities.

C++ Comments

In C++, the old method of using /* and */ to indicate the beginning of a comment and the end of a comment, respectively, is still available. In addition to this method C++ programs can use the // to create a comment. For example, look at the following C++ statements:

```
//
// This is a C++ comment.  Everything after the // is ignored.
//
main()      // The comment can be placed anywhere on the line.
{
}
```

The old style of commenting is still useful when a multiline comment is written and using a combination of both is recommended. For example, single line comments can use the new // notation and comments longer than one line, such as a programmer's block, can be written using the /*   */ notation.

C++ Include Files

The C++ language uses #include files to define external function definitions just as you are used to doing when programming in C. The C++ programmer has to learn about several new #include files to perform the basic input and output operations. Just as #include <stdio.h> was necessary in almost every C program, the #include <iostream.h> is necessary in almost every C++ program. Unfortunately, some of the familiar #include files used in C are not available or have been moved to a new #include file in C++. They may even contain new function definitions. We will introduce several new #include files throughout this chapter.

Operator Overloading

One of the special features of the C++ language involves using the same symbols to perform different functions. For example, we will see that the << and >> symbols used to shift bits in C are also used to perform other functions in C++. When a single operator is used for more than one function, we say the operator is **overloaded.** It is possible to use ++, --, <<, >>, and others to perform many different functions. It is important to recognize operator overloading when you see it in a C++ program. It is also important to use operator overloading to your advantage when writing C++ programs.

Dynamic Memory Allocation

In a C++ program, the `new` operator is used to allocate dynamic storage at run time. If the requested memory is not available, `new` will return a NULL pointer. This is similar to using `malloc()` in a C program. It remains a very important task for the C++ programmer to manage memory carefully. Allocate only what you need and return it to the memory pool when you are done with it. In C++, memory is returned to the memory pool by using the `free()` operator.

Conclusion

This section introduced some of the obvious differences a C programmer sees when examining C++ programs. Check your understanding of this material with the following section review.

12.1 Section Review

1. How many different ways can a comment be entered into a C++ program? What are they?
2. Are the `#include` files you used in C defined in the same `#include` file in C++? If not, where are they located?
3. When does `new` return a NULL pointer? When does this situation occur?
4. What is an overloaded operator?

12.2 The cout Function

Discussion

This section introduces you to the powerful `cout` function used in C++. This is actually a separate C++ function (as `main()` is a function) and is contained in the input/output stream library that comes with your C++ system.

What cout Does

The `cout` function is used to write information to **standard output** (normally your monitor screen). A typical structure for this function is

```
cout << "character string" << variable << '.' ;
```

where the overloaded << symbols represent the **insertor** function. This function is used to insert characters or numbers into the standard output stream.

Characters are set off by single quotes (such as `'a' or '.'`), and strings are set off by double quotes (such as `"This is a string."`). An **I/O manipulator** instructs the `cout` function how to convert, format, and print its **arguments**. For now, think of an argument as the actual values that are within the parentheses of the function. As an example,

```
cout << "This is a C++ statement.";
```

Table 12.1 I/O Manipulators Used by `cout`

Manipulator Flag	Bit Value	Resulting Output
`ios::skipws`	0x0001	Skip white space
`ios::left`	0x0002	Left-align output
`ios::right`	0x0004	Right-align output
`ios::internal`	0x0008	Pad number with spaces
`ios::dec`	0x0010	Output decimal number
`ios::oct`	0x0020	Output octal number
`ios::hex`	0x0040	Output hexadecimal number
`ios::showbase`	0x0080	Show base indicator
`ios::showpoint`	0x0100	Show decimal point on float
`ios::uppercase`	0x0200	Use A..F in hexadecimal output
`ios::showpos`	0x0400	Output plus sign on integers
`ios::scientific`	0x0800	Output in scientific notation
`ios::fixed`	0x1000	Output with fixed decimal point
`ios::unitbuf`	0x2000	Flush streams after insertion
`ios::stdio`	0x4000	Flush `stdout`, `stderr` after insertion

when executed produces

```
This is a C++ statement.
```

With an I/O manipulator, the execution of

```
cout.setf(ios::hex);
cout << "The number 92 in hexadecimal is" << 92 << ".\n";
```

produces

```
The number 92 in hexadecimal is 5c.
```

I/O manipulators are used to control the way numbers are formatted and displayed by `cout` and the I/O operators. Here, we used the `ios::hex` manipulator to force `cout` to display a number in hexadecimal format. These manipulators are found in the `iomanip.h` file.

Table 12.1 lists the various I/O manipulators used by `cout`.

Program 12.1 illustrates the use of the different I/O manipulators.

Program 12.1

```
#include <iostream.h>
#include <iomanip.h>

main()
{
```

```
    cout << "The value 92 displayed as an integer is " << 92 << ".\n";
    cout.setf(ios::oct);
    cout << "The value 92 displayed in octal is " << 92 << ".\n";
    cout.setf(ios::hex);
    cout << "The value 92 displayed in hexadecimal is " << 92 << ".\n";
    cout.setf(ios::uppercase);
    cout << "The value 92 displayed in HEXADECIMAL is " << 92 << ".\n";
    cout << "The value 92.5 displayed as a float is " << 92.5 << ".\n";
    cout.setf(ios::scientific);
    cout << "The value 92.5 displayed in SCIENTIFIC notation is " << 92.5 << ".\n";
    cout.unsetf(ios::uppercase);
    cout << "The value 92.5 displayed in scientific notation is " << 92.5 << ".\n";
}
```

Execution of Program 12.1 results in the following output:

```
The value 92 displayed as an integer is 92.
The value 92 displayed in octal is 134.
The value 92 displayed in hexadecimal is 5c.
The value 92 displayed in HEXADECIMAL is 5C.
The value 92.5 displayed as a float is 92.5.
The value 92.5 displayed in SCIENTIFIC notation is 9.25E+01.
The value 92.5 displayed in scientific notation is 9.25e+01.
```

As output shows, the formats used to display the two numbers 92 and 92.5 are different in each line. The format used by `cout` can be modified through the use of the `cout.setf()` and `cout.unsetf()` functions. Each manipulator in Table 12.1 has a unique binary value that corresponds with a bit in a special area called `ios::flags`. So, when you use `cout.setf(ios::uppercase)`, you are actually setting a specific bit in `ios::flags` that is in turn used by `cout` to control formatting.

Notice how the change made to the `uppercase` flag affects the output of Program 12.1.

Escape Sequences

The `\n` is an example of an **escape sequence** that can be used by the `cout` function. The backslash symbol (\) is referred to as the escape character. You can think of an escape sequence used in the `cout` function as an escape from the normal interpretation of a string. This means that the next character used after the \ will have a special meaning, as listed in Table 12.2.

The Interactive Exercises for this chapter will give you some practice in using the `cout` escape sequences in your system.

I/O Manipulators

The `cout` function allows you to format your output. For example, the output of a type `float` might appear as: `16.000000` even though you do not need the six trailing zeros, they are still printed. The `cout` function provides I/O manipulators that allow you to control how printed values will appear on the monitor.

Table 12.2 Escape Sequences

Sequence	Meaning
`\n`	New line
`\t`	Tab
`\b`	Backspace
`\r`	Carriage return
`\f`	Form feed
`\'`	Single quote
`\"`	Double quote
`\\`	Backslash

Note: The double quote and the backslash can be printed by preceding them with the backslash.

We have already seen a few examples of how I/O manipulators work in Program 12.1. Two more useful manipulators are `setw()` and `setprecision()`. `setw()` is the set-width function, which controls how many spaces wide the field is for a number. For example, the three statements

```
cout << "The count is " << counter << "\n";
cout << "The count is " << setw(4) << counter << "\n";
cout << "The count is " << counter << "\n";
```

produce the following output when executed:

```
The count is 25
The count is   25
The count is 25
```

Notice that the output from the second `cout` statement contains two additional blank spaces before the number 25. This is the result of the `setw(4)` manipulator. Also, the output of the third `cout` statement is the same as the first. This is because the `setw()` function affects the way only the next number output with `cout` is displayed. It is necessary to use `setw()` more than once to format a group of numbers.

The `setprecision()` function controls the number of digits that follow the decimal point when displaying a `float`. Examine the following four statements:

```
cout << "The interest is " << 123.456 << "\n";
cout << "The interest is " << setw(10) << 123.456 << "\n";
cout << "The interest is " << setprecision(2) << 123.456 << "\n";
cout << "The interest is " << 123.456 << "\n";
```

and their respective outputs:

```
The interest is 123.456
The interest is    123.456
The interest is 123.46
The interest is 123.46
```

The second line of output clearly shows the effect of the `setw(10)` function, where three leading blanks are added to the seven-character number `123.456` (the decimal point counts as a character position). These additional blanks do not appear in the last two lines of output, since `setw()` is good only for the very next number displayed by `cout`.

The last two lines of output show the effect of `setprecision(2)`, where the `.456` portion is rounded up to `.46`. `setprecision()` remains in effect for all subsequent outputs.

Conclusion

This section presented some important details about the `cout` function. Here you saw the other escape sequences that could be used with the `cout` function as well as additional I/O manipulators. Test your understanding of this section by trying the following section review.

12.2 Section Review

1. Explain the use of an escape sequence in the `cout` function.
2. What is the backslash character (\) sometimes called when used in a `cout` function?
3. Give three escape sequences that are used in the `cout` function.
4. What is an I/O manipulator as used by the `cout` function?

12.3 Getting User Input with cin

Discussion

The real power of a technical C++ program is its ability to interact with the program user. This means that the program user gets to input values of variables. As you might guess, there is a built-in C++ function that lets you make this happen.

The cin Function

The `cin`function is a built-in C++ function that allows your program to get user input from the keyboard. You can think of it as doing the opposite of the `cout` function. Its use is illustrated in Program 12.2.

Program 12.2

```
#include <iostream.h>

// Getting user input.

main()
{
     float value;    // A number supplied by the program user.

     cout << "Input a number => ";
     cin >> value;
     cout << "The value is => " << value << "\n";
}
```

When Program 12.2 is executed, the output will appear as follows (assuming the program user inputs the value of 23.4):

```
Input a number => 23.4
The value is => 23.4
```

Note that the `cin` function has a similar format to that of the `cout` function. Instead of <<, we use the overloaded >> operator to direct information from the standard input to our variable. The >> operator is called an *extractor*. It is used to extract information from the input stream.

The type of variable used in the `cin` statement controls how the input stream is scanned and converted. In Program 12.2, since `value` is defined as a `float`, the `cin` statement knows what to look for in the input stream. A carriage return is used to terminate the input number.

The `cin` function can accept more than one input with just one statement, as follows:

```
cin >> number1 >> number2 >> character;
```

where the variable `number1` has been defined as a `float`, the variable `number2` a type `int`, and `character` a type `char`. In this case, the program user could type in three separate values separated by spaces. As an example:

```
52.7 18 t
```

Because it is easy for the program user to make errors entering data in this manner, multiple inputs with the `cin` function will be done one variable at a time.

An Application Program

You now know how to input data and display results. Program 12.3 solves for the voltage across a resistor when the values of the current and resistance are known. The mathematical relationship is:

$$\text{Voltage} = \text{Current} \times \text{Resistance}$$

Program 12.3

```
#include <iostream.h>

// Ohm's law

main()
{
     float voltage;          // Value of the voltage.
     float current;          // Value of the current.
     float resistance;       // Value of the resistance.

     cout << "Input the current in amps => ";
     cin >> current;

     cout << "Input the resistance in ohms => ";
```

```
    cin >> resistance;

    voltage = current * resistance;     // Compute the voltage.

    cout << "The value of the voltage is " << voltage << " volts.\n";

}
```

Assume that the program user will enter the values of 3 for the current and 4 for the resistance. Then execution of Program 12.3 would yield

```
Input the current in amps => 3
Input the resistance in ohms => 4
The value of the voltage is 12 volts.
```

What you just observed was a fundamental problem in technology solved by the C++ language. This was a very simple problem that you could have easily solved with your pocket calculator. But, for now, the point is to keep the problems simple so that they don't get in the way of understanding the C++ language.

Using the E Notation

Some mention should be made of using **E (exponential) notation** with C++. As pointed out earlier, C++ will accept E notation numbers for floating point types. You can do this for input as well as output (use `ios::scientific` to output a number in E notation). For example, in the last program, the program user could have entered the more practical values of 0.003 amps for the current and 2000 ohms for the resistance. This could have been done using E notation:

```
Input the current in amps => 3E-3
Input the resistance in ohms => 2E3
The value of the voltage is 6 volts.
```

It should be pointed out that C++ will accept either an uppercase E or a lowercase e for this kind of data representation.

Conclusion

This section brought you to the point where you can now see the basic operations used with `cin`. Check your understanding of this section by trying the following section review.

12.3 Section Review

1. State the purpose of the `cin` function.
2. How does the `cin` function know what variable type to use for inputting the user data?
3. Does use of the `cin` function produce a carriage return to a new line?
4. How can floating point values be entered by the program user in a C++ program?
5. State what you must do in order to have values outputted to the screen in E notation.

12.4 Classes and Objects

Discussion

In this section we will discuss some of the properties of C++ classes and how a class is used to create an object. This information will help you understand why classes are such a powerful method of handling many types of applications.

Let us define a class called `gas_pump` that contains the features common to all gas pumps, such as:

- Reset fuel charge to $0.00.
- Reset fuel pumped to 0.0 gallons.
- Display fuel charges.
- Display fuel pumped.
- Determine pump status: off, on, pumping.

In C++, we might use the following statements to define the `gas_pump` class

```
class gas_pump
{
     float charge;
     float gallons_pumped;
     float calculate_cost(float pumped);
     void display_totals(float price, float pumped);
     int pump_status(int pump_number);
};
```

But there are many different kinds of gas pumps, such as regular, unleaded, and diesel. Even among a single type of pump, such as unleaded, there are different unleaded gasses available based on their octane rating. The `gas_pump` class can be used to create a more specific class called `unleaded` that contains all of the common features, plus additional features required by the unleaded pump.

The `class` keyword is used to begin a class definition. The class definition format is similar to the format used by the `struct` keyword and is diagrammed in Figure 12.1.

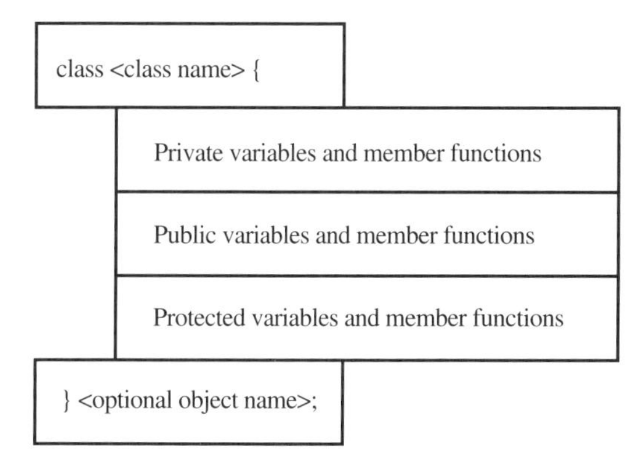

Figure 12.1 Basic Structure of a Class Definition

The difference here is that function names may now be used as well. Program 12.4 illustrates a simple class called sum.

Program 12.4

```
#include <iostream.h>

class sum {
public:
        int total;
        void add(int value);
} number;

void sum::add(int value)
{
        total += value;
}

main()
{
        number.total = 0;
        number.add(1);
        number.add(2);
        number.add(3);
        number.add(4);
        cout << "The sum is " << number.total << "\n";
}
```

The class definition

```
class sum {
public:
     int total;
     void add(int value);
} number;
```

contains two members. A **member** of a class is a variable or function defined within the class definition. The first member is the variable `total`. The second member is the function `add()` (also called a **member function**). Both of these members are **public members**—that is, they are visible to the entire program (this is indicated by the keyword `public` used in the definition). Members of a class may be public, private, or protected. We will examine private and protected members later in this chapter.

Instantiating an Object

One thing to remember is that the class definition by itself does not reserve memory for the `total` variable, nor does it generate the code for the `add` function. These tasks are performed when the class definition is used to instantiate an object. **Instantiation** refers to creating an instance of an object. In Program 12.4, the class `sum` instantiates the object `number`. Both `total` and `add()` are encapsulated in the object `number`. **Encapsulation** refers to the joining together of data (`total`) and operations on that data (`add()`).

Since total and add are public members, they may be accessed from the main program after number is instantiated. The statement

```
number.total = 0;
```

is used to set total to zero and is exactly like a statement we would use to access the total variable if it was part of a struct definition. The additional statements

```
number.add(1);
number.add(2);
number.add(3);
number.add(4);
```

use a similar format to call the member function add().

Taking instantiation a little further, suppose sum is defined like this:

```
class sum {
public:
     int total;
     void add(int value);
};
```

No object is instantiated here because no object name is specified after the closing brace. To instantiate number now, we must use the statement

```
sum number;
```

after the class definition and before any reference to a member of number. This is illustrated in Program 12.5.

Program 12.5

```
#include <iostream.h>

class sum {
public:
        int total;
        void add(int value);
};

void sum::add(int value)
{
        total += value;
}

        sum number;

main()
{
        number.total = 0;
        number.add(1);
        number.add(2);
        number.add(3);
        number.add(4);
        cout << "The sum is " << number.total << "\n";
}
```

There is no difference in execution between Programs 12.4 and 12.5. Both programs produce the following output:

```
The sum is 10
```

Program 12.6 shows a third way of instantiating `number` that produces the same results as Programs 12.4 and 12.5.

Program 12.6

```
#include <iostream.h>

class sum {
public:
        int total;
        void add(int value);
};

void sum::add(int value)
{
        total += value;
}

main()
{
        sum number;

        number.total = 0;
        number.add(1);
        number.add(2);
        number.add(3);
        number.add(4);
        cout << "The sum is " << number.total << "\n";
}
```

Later in the chapter we will see that the place in the program where an object is instantiated controls the accessibility of the object.

Defining Member Functions

Any member function included in a class definition requires its own function definition as well. The name of the class associated with the function must be included in the function declaration. This is the only difference between an ordinary function and a member function. This is illustrated in Figure 12.2.

Compare the structure shown in Figure 12.2 with this function definition:

```
void sum::add(int value)
{
     total += value;
}
```

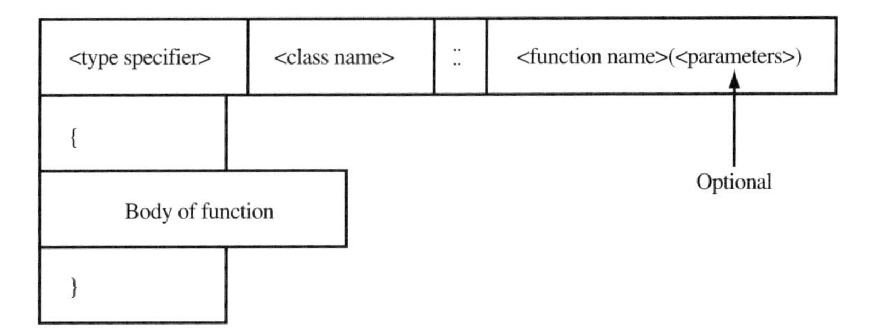

Figure 12.2 Structure of a Member Function Definition

Here, the `add` function for the class `sum` is defined. The `::` operator (called the *scope resolution operator*) is required by the C++ compiler to associate `add` with the class `sum`. This is necessary, since other class definitions may also include an `add` function.

Since `add` is a member function of `sum`, it automatically has access to the contents of `total`. The statement

```
total += value;
```

compiles because `total` is a member of the class `sum`. It is not necessary to include the class name in the variable reference. This brings up an important point: All variables and functions defined within a class are visible to each other.

Conclusion

This section illustrated how to define a class and instantiate objects of the class. Test your understanding of this section by trying the following section review.

12.4 Section Review

1. What is a class in C++? What does it do?
2. State the difference between a class and an object.
3. Explain what it means to instantiate an object.
4. What are class members? How are they used?

12.5 Constructors and Destructors

Discussion

In the last section you saw how to define and create a simple object. This section presents additional information about special functions you may define in the class that can be used to automatically prepare an object for use when it is instantiated or to help clean up when you are finished with an object.

When an object is instantiated, it may contain variables that require initialization. In Programs 12.4 through 12.6, the `total` variable is initialized with the statement

```
number.total = 0;
```

This is an acceptable method for variable initialization. There are, however, other ways to perform this operation. One alternative is to place the initialization code in a member function and call the member function to initialize the variable. This method is shown in Program 12.7.

Program 12.7

```
#include <iostream.h>

class sum {
public:
        int total;
        void initialize(void);
        void add(int value);
} number;

void sum::initialize()
{
        total = 0;
}

void sum::add(int value)
{
        total += value;
}

main()
{
        number.initialize();
        number.add(1);
        number.add(2);
        number.add(3);
        number.add(4);
        cout << "The sum is " << number.total << "\n";
}
```

Here the `initialize()` function is used to set `total` to zero when it is called from `main()` via

```
number.initialize();
```

Unfortunately, these methods require the programmer to remember to initialize `total`. You can imagine what might happen if the programmer forgets to do this. To avoid this situation, we take advantage of a class feature called a constructor. A **constructor** is a member function that is automatically executed when the class is used to instantiate an object. Program 12.8 shows how this is accomplished.

Program 12.8

```
#include <iostream.h>

class sum {
public:
```

```
        int total;
        sum(void);
        void add(int value);
} number;

sum::sum()
{
        total = 0;
}

void sum::add(int value)
{
        total += value;
}

main()
{
        number.add(1);
        number.add(2);
        number.add(3);
        number.add(4);
        cout << "The sum is " << number.total << "\n";
}
```

The name of the constructor is always the same as the class name, as you can see from the class definition:

```
class sum {
public:
     int total;
     sum(void);
     void add(int value);
} number;
```

Notice that there is no type specifier used with the constructor. This is because constructors do not return any data.

The function definition for the constructor, as with any member function, requires a specific format, as shown in Figure 12.3.

The constructor for the `sum` class looks like this:

```
sum::sum()
{
     total = 0;
}
```

Once again, no type specifier is used.

Constructor functions are executed during instantiation. In Program 12.8, `number` is instantiated at the end of the class definition. This is when the constructor is called to initialize `total`.

Program 12.9 defines a new class called `counter` containing the variable `count`, which must be initialized.

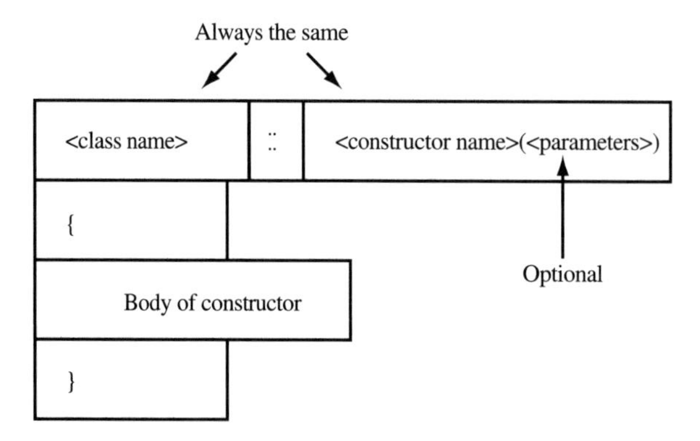

Figure 12.3 Structure of a Constructor

Program 12.9

```
#include <iostream.h>

class counter {
public:
        int count;
        counter(void);
        void adder(void);
};

counter::counter()
{
        count = 0;
        cout << "The initial count is " << count << "\n";
}

void counter::adder()
{
        count++;
        cout << "The count is now " << count << "\n";
}

main()
{
        counter doloop;

        cout << "Beginning the loop...\n";
        do
        {
                doloop.adder();
        } while (doloop.count < 5);
        cout << "Out of the loop.\n";
}
```

The execution of Program 12.9 is as follows:

```
The initial count is 0
Beginning the loop...
The count is now 1
The count is now 2
The count is now 3
The count is now 4
The count is now 5
Out of the loop.
```

Clearly, the output message

```
The initial count is 0
```

demonstrates that the constructor has been executed.

Consider Program 12.10.

Program 12.10

```
#include <iostream.h>

class counter {
public:
        int count;
        counter(void);
        void adder(void);
};

counter::counter()
{
        count = 0;
        cout << "The initial count is " << count << "\n";
}

void counter::adder()
{
        count++;
        cout << "The count is now " << count << "\n";
}

void loops(void);

main()
{
        cout << "In main()...\n";
        loops();
        cout << "Back to main().\n";
}

void loops()
{
        counter doloop;
```

```
        cout << "Beginning the loop...\n";
        do
        {
                doloop.adder();
        } while (doloop.count < 5);
        cout << "Out of the loop.\n";
}
```

Recall that constructors are executed when their associated objects are instantiated. In Program 12.10, the `doloop` object is not instantiated until the `loops()` function is executed by `main()`. The resulting output indicates this:

```
In main()...
The initial count is 0
Beginning the loop...
The count is now 1
The count is now 2
The count is now 3
The count is now 4
The count is now 5
Out of the loop.
Back to main().
```

Note that it is impossible for the `counter` constructor to have executed before `main()`, since all `main()` does is

```
cout << "In main()...\n";
loops();
cout << "Back to main().\n";
```

The object `doloop` is instantiated inside `loops()`, via

```
counter doloop;
```

The output of Program 12.10 supports the fact that constructors execute when their associated objects are instantiated.

Passing Parameters to Constructors

It may be necessary to pass a parameter to a constructor to control the way initialization is performed. For example, information about the amount of memory to allocate or the name of a file to open may need to be passed to a constructor. Parameter variables are included in the constructor definition in the same way as any other function definition. Program 12.11 demonstrates this principle.

Program 12.11

```
#include <iostream.h>

class sum {
public:
```

```
        int total;
        sum(int value);
        void add(int value);
} number(0);

sum::sum(int value)
{
        total = value;
}

void sum::add(int value)
{
        total += value;
}

main()
{
        number.add(1);
        number.add(2);
        number.add(3);
        number.add(4);
        cout << "The sum is " << number.total << "\n";
}
```

The `sum()` constructor now contains the integer parameter `value`. Passing an actual integer to `sum()` during instantiation is accomplished by following the class definition with the object name and constructor parameter, as in:

```
class sum {
public:
     int total;
     sum(int value);
     void add(int value);
} number(0);
```

If the object is instantiated elsewhere, use

```
sum number(0);
```

to pass a value to the constructor.

Destructors

There are times when a program (or a single function) must perform some housekeeping task before it completes. For example, the results of an analysis must be displayed, allocated memory must be returned to the operating system, or open data files must be closed. These housekeeping chores can be performed automatically by destructors. A **destructor** executes when the function that instantiated it terminates. Program 12.12 shows an example of how a destructor is used.

Program 12.12

```
#include <iostream.h>

class sum {
public:
        int total;
        sum(int value);
        ~sum(void);
        void add(int value);
} number(0);

sum::sum(int value)
{
        total = value;
}

sum::~sum()
{
        cout << "The sum is " << total << "\n";
}

void sum::add(int value)
{
        total += value;
}

main()
{
        number.add(1);
        number.add(2);
        number.add(3);
        number.add(4);
}
```

The destructor has the same name as the constructor and the class. The compiler recognizes the destructor definition when it sees the ~ symbol that precedes the destructor name. This is illustrated in the class definition:

```
class sum {
public:
     int total;
     sum(int value);
     ~sum(void);
     void add(int value);
} number(0);
```

Since the constructor and destructor functions always share the same name, the ~ symbol is required to tell them apart. It actually means "not a constructor."

The ~ symbol must also be used in the destructor function definition, as indicated in Figure 12.4.

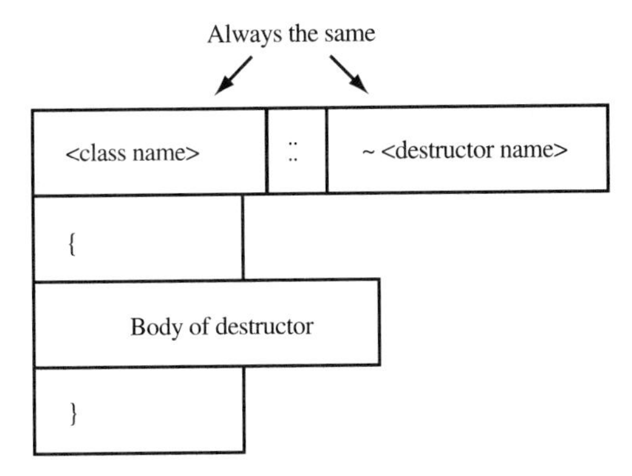

Figure 12.4 Structure of a Destructor

The destructor function from Program 12.12 looks like this:

```
sum::~sum()
{
     cout << "The sum is " << total << "\n";
}
```

When Program 12.12 executes, it displays the following message:

```
The sum is 10
```

Note that it is impossible for this message to have come from `main()`, since `main()` does not have a single `cout` statement.

Program 12.13 updates Program 12.9 so that a destructor is used to indicate the final value of the loop counter.

Program 12.13

```
#include <iostream.h>

class counter {
public:
        int count;
        counter(void);
        ~counter(void);
        void adder(void);
};

counter::counter()
{
        count = 0;
        cout << "The initial count is " << count << "\n";
}

void counter::adder()
{
        count++;
```

```
        cout << "The count is now " << count << "\n";
}

counter::~counter()
{
        cout << "The final count is " << count << "\n";
}

main()
{
        counter doloop;

        cout << "Beginning the loop...\n";
        do
        {
                doloop.adder();
        } while (doloop.count < 5);
        cout << "Out of the loop.\n";
}
```

Since the object `doloop` is instantiated in `main()`, the destructor executes when `main()` terminates. This is shown in Program 12.13's execution:

```
The initial count is 0
Beginning the loop...
The count is now 1
The count is now 2
The count is now 3
The count is now 4
The count is now 5
Out of the loop.
The final count is 5
```

As you can see, the output from the destructor occurs after the last line of output generated by `main()`.

Program 12.14 demonstrates the operation of a destructor that is executed when a function other than `main()` terminates.

Program 12.14

```
#include <iostream.h>

class counter {
public:
        int count;
        counter(void);
        ~counter(void);
        void adder(void);
};

counter::counter()
{
```

```
        count = 0;
        cout << "The initial count is " << count << "\n";
}

void counter::adder()
{
        count++;
        cout << "The count is now " << count << "\n";
}

counter::~counter()
{
        cout << "The final count is " << count << "\n";
}

void loops(void);

main()
{
        cout << "In main()...\n";
        loops();
        cout << "Back to main().\n";
}

void loops()
{
        counter doloop;

        cout << "Beginning the loop...\n";
        do
        {
                doloop.adder();
        } while (doloop.count < 5);
        cout << "Out of the loop.\n";
}
```

The `loops()` function is responsible for instantiating the object `doloop` in Program 12.14. This causes the destructor to execute when `loops()` terminates and returns to `main()`. One look at the output from Program 12.14 proves this point.

```
In main()...
The initial count is 0
Beginning the loop...
The count is now 1
The count is now 2
The count is now 3
The count is now 4
The count is now 5
Out of the loop.
The final count is 5
Back to main().
```

Clearly, the `loops()` function is responsible for all text displayed between the first and last lines of output.

In keeping with the concept of a class, where data and operations on the data are bundled together, constructors and destructors fit right in. A single class definition can now provide all that is necessary to support operations on an object in a program.

Conclusion

This section presented some important reasons why you may want to use constructor and destructor functions and when constructor and destructor functions execute. Test your understanding of this section by trying the following section review.

12.5 Section Review

1. What is a constructor? What does it do?
2. Can parameters be passed to a constructor? How?
3. Describe the purpose of a destructor.
4. When does a destructor execute?

12.6 Multiple Objects of the Same Class

Discussion

Now that you are familiar with the necessary tools to define and instantiate objects, let us examine how we may use many objects of the same class or arrays of objects.

Just as we may define multiple variables of the same type, via

```
int x,y,z;
char a,b,c;
```

we may also declare multiple objects of the same class. Each object will contain an identical set of variables and member functions. Program 12.15 demonstrates this principle.

Program 12.15

```
#include <iostream.h>

class counter {
public:
        int count;
        counter(int start);
        ~counter(void);
        void adder(void);
};

counter::counter(int start)
{
        count = start;
        cout << "The initial count is " << count << "\n";
}
```

```
void counter::adder()
{
        count++;
        cout << "The count is now " << count << "\n";
}

counter::~counter()
{
        cout << "The final count is " << count << "\n";
}

main()
{
        counter one(5), two(10), three(15);

        one.adder();
        two.adder();
        three.adder();
}
```

The `counter` class contains all of the features covered so far. The integer variable `count` is initialized by the constructor function through the `start` parameter. The `adder()` function increments `count` and the destructor displays the final value of `count`. Three separate objects are instantiated: `one`, `two`, and `three`. It is very important to realize that each of these objects has its own internal storage space for `count`, as illustrated in Figure 12.5.

The actual variables names are `one.count`, `two.count`, and `three.count`.

Unlike member variables, member functions are not duplicated when multiple objects are instantiated. One set of member functions is installed in memory, and their access is

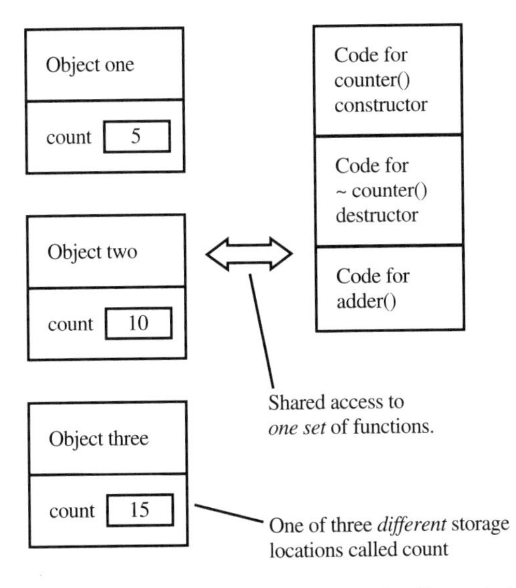

Figure 12.5 Memory Allocation after Instantiation of Three Objects

shared by all of the objects. This makes sense, because only one function at a time may be called when a C++ program is executing. For instance, it is not possible for the `one.adder()` and `two.adder()` function calls to take place at the same time. Instead, they take place in program order, with `one.adder()` completing before `two.adder()` is allowed to begin. This allows the C++ compiler to *reuse* the code for each member function.

The output of Program 12.15 deserves some examination.

```
The initial count is 5
The initial count is 10
The initial count is 15
The count is now 6
The count is now 11
The count is now 16
The final count is 16
The final count is 11
The final count is 6
```

The first three lines of output indicate that objects are instantiated in the order that they appear in the program. The second three lines are the result of the three member function calls in `main()`. The last three lines, which are output by the destructor functions, show us that multiple objects are destructed in *reverse* order of their instantiation. This implies that the C++ compiler uses an object stack to keep track of multiple objects.

Program 12.16 is a rewritten version of the vowel-counting program we were introduced to earlier. In this version, multiple objects are used to keep track of the individual vowel counts.

Program 12.16

```
#include <iostream.h>
#include <string.h>
#include <ctype.h>

class vcount
{
public:
        char vowel;
        int count;
        vcount(char letter);
        void testvowel(char letter);
};

vcount::vcount(char letter)
{
        count = 0;
        vowel = letter;
}

void vcount::testvowel(char letter)
{
        if (vowel == letter)
                count++;
```

```
}

main()
{
        char text[] = "The cat in the hat is so very, very fat.";
        char vstr[] = "AEIOU";
        static int vowels;
        vcount a('A'), e('E'), i('I'), o('O'), u('U');
        int j;

        cout << "The input string is => \"" << text << "\"";
        for(j = 0; j < strlen(text); j++)
        {
                text[j] = toupper(text[j]);
                if (strchr(vstr,text[j]) != '\0')
                        vowels++;
                a.testvowel(text[j]);
                e.testvowel(text[j]);
                i.testvowel(text[j]);
                o.testvowel(text[j]);
                u.testvowel(text[j]);
        }
        cout << "\n";
        cout << "The input string contains " << vowels << " vowels.\n";
        cout << "There are " << a.count << " A's, ";
        cout << e.count << " E's, " << i.count << " I's, ";
        cout << o.count << " O's, and " << u.count << " U's.\n";
}
```

The execution indicates that separate variables were instantiated:

```
The input string is => "The cat in the hat is so very, very fat."
The input string contains 10 vowels.
There are 3 A's, 4 E's, 2 I's, 1 O's, and 0 U's.
```

If the `count` variables were not separate, it would not be possible to see the five different vowel counts that appear in the output.

Arrays of Objects

Another way to work with multiple objects is to create an array of objects. This is done in the same way we declare `int` or `float` arrays. Program 12.17 demonstrates how to do this.

Program 12.17

```
#include <iostream.h>
#include <string.h>
#include <ctype.h>

class vcount
{
```

```
public:
        int count;
        vcount(void);
} vowelcounts[5];

vcount::vcount()
{
        count = 0;
}

main()
{
        char text[] = "OUR instructor SPEAKS clearly in EACH class.";
        char vstr[] = "AEIOU";
        static int vowels;
        int j;

        cout << "The input string is => \"" << text << "\"";
        for(j = 0; j < strlen(text); j++)
        {
                text[j] = toupper(text[j]);
                if (strchr(vstr,text[j]) != '\0')
                        vowels++;
                switch(text[j])
                {
                        case 'A' : vowelcounts[0].count++;
                                break;
                        case 'E' : vowelcounts[1].count++;
                                break;
                        case 'I' : vowelcounts[2].count++;
                                break;
                        case 'O' : vowelcounts[3].count++;
                                break;
                        case 'U' : vowelcounts[4].count++;
                }
        }
        cout << "\n";
        cout << "The input string contains " << vowels << " vowels.\n";
        cout << "There are " << vowelcounts[0].count << " A's, ";
        cout << vowelcounts[1].count << " E's, ";
        cout << vowelcounts[2].count << " I's, ";
        cout << vowelcounts[3].count << " O's, and ";
        cout << vowelcounts[4].count << " U's.\n";
}
```

The class definition for `vcount` contains a five-element array declaration. The individual elements of the `vowelcounts` object array are accessed directly in the `switch()` through their respective index values [0] through [4]. Multidimensional arrays may also be used.

Object Pointers

Whenever multiple objects are used in a program it may be useful to use pointers to access an object. Pointers to objects work in much the same way as pointers to structures, except that the -> operator may now also point to a member function. Program 12.18 shows an example of how a pointer to an object is used.

Program 12.18

```
#include <iostream.h>

class sum {
public:
        int total;
        void add(int value);
};

void sum::add(int value)
{
        total += value;
}

main()
{
        sum *sumptr, number;

        sumptr = &number;
        sumptr->total = 0;
        sumptr->add(1);
        sumptr->add(2);
        sumptr->add(3);
        sumptr->add(4);
        cout << "The sum is " << sumptr->total << "\n";
}
```

The pointer variable `sumptr` is declared when the `number` object is instantiated:

```
sum *sumptr, number;
```

Then the pointer is initialized with the address of the object:

```
sumptr = &number;
```

Now, instead of using the object name with dot notation (as in `number.total` and `number.add()`), we use the -> operator:

```
sumptr->total = 0;
sumptr->add(1);
sumptr->add(2);
sumptr->add(3);
sumptr->add(4);
cout << "The sum is " << sumptr->total << "\n";
```

In summary, anywhere a pointer to a structure is used, we may instead use a pointer to an object containing the same structure, with the addition of being able to point to member functions as well.

Conclusion

As you have seen, using multiple objects and arrays of objects provides a convenient way to standardize the variables and functions used to work with each object. Test your understanding of this section by trying the following section review.

12.6 Section Review

1. Why create multiple objects of the same class?
2. Are there any limitations on the number of object instantiations?
3. What is an object pointer? How is it used?
4. In what order are multiple objects destructed?

12.7 Private Members and Friend Functions

Discussion

This section introduces you to some important information about limiting the visibility of data and functions in a class. Normally, when the visibility of a member is restricted, it cannot be accessed from outside the class. C++ provides some convenient ways to allow access to the members on a restricted basis.

Program 12.19 is an exact copy of Program 12.4 except for one difference: the removal of the `public` keyword from the class definition.

Program 12.19

```
#include <iostream.h>

class sum {
        int total;
        void add(int value);
} number;

void sum::add(int value)
{
        total += value;
}

main()
{
        number.total = 0;
        number.add(1);
        number.add(2);
        number.add(3);
        number.add(4);
        cout << "The sum is " << number.total << "\n";
}
```

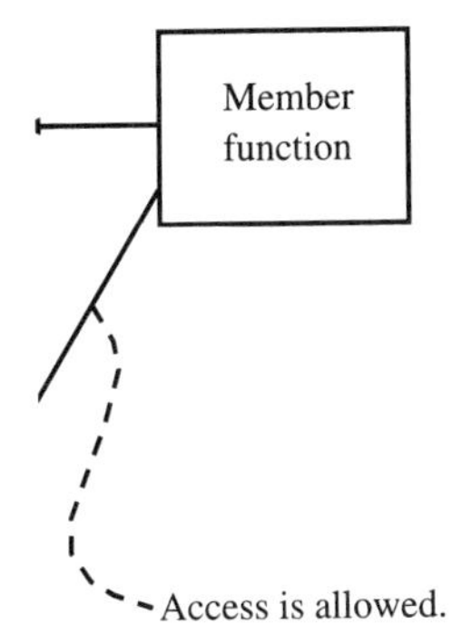

Figure 12.6 Public vs. Private Access

The removal of the `public` keyword results in a program that will not compile! Let us see why.

Up to this point in the chapter, all class definitions began with the `public` keyword. This was done to give all parts of the program access to member variables and functions, similar to the way all parts of a program have access to a global variable.

By default, member variables and functions of a class are considered private when declared. A **private member** is a member that may only be referenced by other members of the class. For example, in Program 12.19 the `total` variable may be accessed by the member function `add()`, but not by any statement in `main()`. The `add()` function itself is not accessible by `main()`, since it is also private. Figure 12.6 illustrates the differences between public and private access.

The whole idea behind classes is to limit the visibility of class members. This is often called *information hiding,* and is usually used to provide a measure of security to program design and execution. By using a class definition, the appropriate access for each member item can be controlled (limited) in such a way that only the functions that need access to the member variable may have it.

So, the class definition

```
class sum {
     int total;
     void add(int value);
} number;
```

declares two private members, `total` and `add()`. This is why the compiler generates errors when it encounters `number.total` and `number.add()`, since these members are private to the class and not visible to `main()`.

The position of the `public` keyword in a class definition controls which member items are visible to the rest of the program. Specifically, all members declared *after* the `public` keyword are public. Any members declared *before* the `public` keyword are, by default, private. This is illustrated in Program 12.20.

Program 12.20

```
#include <iostream.h>

class sum {
```

```
        int total;
public:
        void add(int value);
} number;

void sum::add(int value)
{
        total += value;
}

main()
{
        number.total = 0;
        number.add(1);
        number.add(2);
        number.add(3);
        number.add(4);
        cout << "The sum is " << number.total << "\n";
}
```

A slightly different class definition makes the `add()` function public (visible to `main()`) while keeping `total` private:

```
class sum {
     int total;
public:
     void add(int value);
} number;
```

Program 12.20 does not compile because of the two `number.total` references in `main()`, since `total` is still private.

Why make a member item private? Why restrict access to members at all? For small programs, this is usually not an issue. For large, complex programs, it is desirable to make the program as *modular* as possible—that is, each module performs a specific function and does not necessarily require access to all other parts of the program. For example, consider a word processing program made up of many modules. The modules perform different functions, such as file I/O, editing, spell checking, word counting, and page formatting. Each module contains variables and functions specific to its task. If it were possible for one module to change a variable used by another module, even by accident, the results could be disastrous. In simple terms, it is best only to share information that must be shared. The use of private members helps localize data and operations on that data, which in turn makes the job of debugging the program easier, since we know where to look when something does not work.

Friend Functions

When member data of a class is private, only member functions of the same class are allowed access to the data. This may pose a problem if it is necessary to access the private data from a function outside the class, since access is restricted. As we have already seen, if we attempt to access private data from outside the class, the program will not even compile. The problem can be eliminated by changing the access to public, but this opens up the data to modification from anywhere in the entire program. Also, suppose that a program contains a number of different class definitions. It may be necessary to access private data located in more than one of the classes. Should everything in the program be made public to solve this situation?

Fortunately, C++ allows us to reach a compromise between private and public data by using a friend function. A **friend function** has access to private data of a class, just as the member functions of the class do, but it is not itself a member function. Let us examine a situation where friend functions are required as shown in Program 12.21.

Program 12.21

```
#include <iostream.h>
#include <string.h>
#include <ctype.h>

class vcount
{
        int count;
public:
        vcount(void);
} vowelcounts[5];

vcount::vcount(void)
{
        count = 0;
}

main()
{
        char text[] = "OUR instructor SPEAKS clearly in EACH class.";
        char vstr[] = "AEIOU";
        static int vowels;
        int j;

        cout << "The input string is => \"" << text << "\"";
        for(j = 0; j < strlen(text); j++)
        {
                text[j] = toupper(text[j]);
                if (strchr(vstr,text[j]) != '\0')
                        vowels++;
                switch(text[j])
                {
                        case 'A' : vowelcounts[0].count++;
                                break;
                        case 'E' : vowelcounts[1].count++;
                                break;
                        case 'I' : vowelcounts[2].count++;
                                break;
                        case 'O' : vowelcounts[3].count++;
```

```
                                break;
                        case 'U' : vowelcounts[4].count++;
                }
        }
        cout << "\n";
        cout << "The input string contains " << vowels << " vowels.\n";
        cout << "There are " << vowelcounts[0].count << " A's, ";
        cout << vowelcounts[1].count << " E's, ";
        cout << vowelcounts[2].count << " I's, ";
        cout << vowelcounts[3].count << " O's, and ";
        cout << vowelcounts[4].count << " U's.\n";
}
```

Program 12.21 is a modified version of Program 12.17. The modifications were made to the class definition, changing the `count` variable from public to private. This single change prevents the program from compiling. The reason is that `vowelcounts[].count` is not visible from anywhere in `main`, since `count` is private to the class `vcount`.

Program 12.22 solves this problem by using two friend functions. These functions are `checkvowel()` and `showvowels()` and are identified as friend functions in the class definition.

Program 12.22

```
#include <iostream.h>
#include <string.h>
#include <ctype.h>

class vcount
{
        int count;
public:
        vcount(void);
        friend void checkvowel(char letter);
        friend void showvowels(void);
} vowelcounts[5];

vcount::vcount(void)
{
        count = 0;
}

main()
{
        char text[] = "OUR instructor SPEAKS clearly in EACH class.";
        char vstr[] = "AEIOU";
        static int vowels;
        int j;

        cout << "The input string is => \"" << text << "\"";
        for(j = 0; j < strlen(text); j++)
        {
```

```
            text[j] = toupper(text[j]);
            if (strchr(vstr,text[j]) != '\0')
                vowels++;
            checkvowel(text[j]);
    }
    cout << "\n";
    cout << "The input string contains " << vowels << " vowels.\n";
    showvowels();
}

void checkvowel(char letter)
{
    switch(letter)
    {
            case 'A' : vowelcounts[0].count++;
                       break;
            case 'E' : vowelcounts[1].count++;
                       break;
            case 'I' : vowelcounts[2].count++;
                       break;
            case 'O' : vowelcounts[3].count++;
                       break;
            case 'U' : vowelcounts[4].count++;
    }
}

void showvowels()
{
    cout << "There are " << vowelcounts[0].count << " A's, ";
    cout << vowelcounts[1].count << " E's, ";
    cout << vowelcounts[2].count << " I's, ";
    cout << vowelcounts[3].count << " O's, and ";
    cout << vowelcounts[4].count << " U's.\n";
}
```

As shown, all that is necessary to make a function a friend function is the addition of the `friend` keyword at the beginning of the function declaration in the class definition:

```
class vcount
{
     int count;
public:
     vcount(void);
     friend void checkvowel(char letter);
     friend void showvowels(void);
} vowelcounts[5];
```

Notice from the function definitions for `checkvowel()` and `showvowels()` that these functions are not members of the `vcount` class. However, as friend functions, they are allowed access to the `count` variable, which allows Program 12.22 to compile and execute correctly.

Conclusion

The information presented in this section allows you to control the visibility of members within an object. Armed with this knowledge, your programs can be written in such a way that, if you work within a larger group of programmers, your code and data may be protected. If you need to share functions or data, you allow access to them on an as-needed basis. Test your knowledge of this section by trying the following section review.

12.7 Section Review

1. What is a private member? How is a private member defined?
2. Why do we need friend functions? How are friend functions defined?
3. State the purpose of information hiding. How is this accomplished?
4. Describe what is meant by member visibility.

12.8 Inheritance, Virtual Functions, and Pure Virtual Functions

Discussion

This section introduces some powerful new concepts and techniques that you can use to create classes that contain some predefined members.

A powerful feature of C++ classes involves the use of **inheritance.** That is, a new class definition inherits some or all of the functionality from another class called the **base class.** A class that inherits the features of the base class is called a **derived class,** or **subclass.**

Let us consider a practical situation. A program must be written to aid a customs agent in computing the amount of tariff on a motor vehicle when it is imported to the United States. When we speak about a motor vehicle, several types of vehicles come to mind, such as cars, trucks, and buses. A `tariff()` function will be used to compute the amount of tariff on the vehicle based on the make, model, weight, and country of origin. For example, consider this class definition:

```
class motor_vehicle {
public:
      char make[20];
      char model[20];
      float weight;
      float tariff(char origin);
}
```

This definition is used to define features common to all motor vehicles. This class definition is the base class. Each type of motor vehicle shares this common information, but in addition can be further defined into smaller classes, such as cars, trucks, and buses. These subclasses include the common information defined in the base class and also information specific to the type of motor vehicle. For instance, the number of doors on a car, wheels on a truck, and people standing and sitting on a bus are all important *specific* information required by each subclass. These items, and others, are defined as follows:

```
class car : public motor_vehicle {
public:
      int numberofpassengers;
```

```
     int numberofdoors;
}

class truck : public motor_vehicle {
public:
     int numberofwheels;
     char typeload[20];
}

class bus : public motor_vehicle {
public:
     int numberstand;
     int numbersit;
}
```

The subclasses car, truck, and bus include the common features of the base class motor_vehicle plus their own distinguishing features. The public motor_vehicle portion of each subclass definition causes the public members of the base class to become public members in each subclass.

There are many ways to inherit members from a base class, as indicated in Table 12.3.

Note the new keyword protected shown in Table 12.3. A **protected member** has features similar to both public and private members. A protected member of a base class may be inherited, ending up as either private or protected in the new subclass. Compare this to a private member in the base class, which may never be inherited.

The reasons for using one type of inheritance over another are based on the overall design goal of the program. If security and information hiding are necessary, inheritance using private and/or protected members is required. To allow the most visibility possible, use public members and inheritance instead.

Virtual Functions

Let us reexamine the gas_pump class defined previously:

```
class gas_pump
{
     float charge;
     float gallons_pumped;
     float calculate_cost(float pumped);
     void display_totals(float price, float pumped);
     int pump_status(int pump_number);
};
```

Table 12.3 Rules of Inheritance

Member in Base Class	Inherited Member in Public Subclass	Inherited Member in Private Subclass	Inherited Member in Protected Subclass
Public	Public	Private	Protected
Private	Not inherited	Not inherited	Not inherited
Protected	Protected	Private	Protected

The code for the calculate_cost() function, which computes the cost of the gasoline pumped, is as follows:

```
float gas_pump::calculate_cost(float pumped);
{
     return(pumped * 1.35);
}
```

The $1.35 price is the cost of one gallon of low-octane unleaded fuel.

Now, suppose that the gas_pump class is used to create two new subclasses: premium and super_premium. The calculate_cost() function in these new subclasses will be exactly the same as it appears in the base class. Unfortunately, this leads to a problem, because the costs of premium and super-premium fuels are higher. One solution would be to pass the cost of one gallon of fuel to calculate_cost() as a parameter. A different solution involves the use of a **virtual function** to do the different price calculations. This requires the programmer to write new versions of the calculate_cost() function, one for each new subclass. Each version will have its own specific price formula.

In order to use virtual functions, the declaration in the base class must indicate that a function is virtual. This is done as follows:

```
class gas_pump
{
     float charge;
     float gallons_pumped;
     virtual float calculate_cost(float pumped);
     void display_totals(float price, float pumped);
     int pump_status(int pump_number);
};
```

Note that the declaration for calculate_cost() has been preceded by the virtual keyword. This informs the compiler that more than one version of calculate_cost() will be used. The original calculate_cost() function defined in the base class will be used with objects instantiated by the base class. Subclasses that do not supply a new version of calculate_cost() will, by default, use the function defined in the base class. However, in the premium and super_premium subclasses, the calculate_cost() function must be rewritten to use the appropriate fuel cost. To use a new calculate_cost() function in the premium subclass, the following statements are needed:

```
class premium : public gas_pump
{
     float calculate_cost(float pumped);
}
```

These statements tell the compiler to use everything defined in the base class gas_pump, except for the calculate_cost() function, which will be different (not inherited). The new calculate_cost() function must be defined in the new premium class. The same is true for the second subclass, super_premium. The actual code for each new calculate_cost() function looks like this:

```
float premium::calculate_cost(float pumped)
{
     return(pumped * 1.45);
```

```
}

float super_premium::calculate_cost(float pumped)
{
     return(pumped * 1.70);
}
```

Objects instantiated by the `premium` class will have a fuel cost of $1.45 and objects instantiated by the `super_premium` class will use $1.70 for the fuel cost.

Suppose that three objects are instantiated in the following way:

```
gas_pump pumpA;
premium pumpB;
super_premium pumpC;
```

The respective function calls for each object will be:

```
pumpA.charge = pumpA.calculate_cost(pumpA.gallons_pumped);
pumpB.charge = pumpB.calculate_cost(pumpB.gallons_pumped);
pumpC.charge = pumpC.calculate_cost(pumpC.gallons_pumped);
```

Even though the function name is the same in each statement, in actuality three different `calculate_cost()` functions are called.

The use of virtual functions greatly increases your ability to tailor a predefined base class to your needs. As the gas pump example shows, making the `calculate_cost()` function virtual allows us to create any type of gas pump we want. You are encouraged to develop a working gas pump program using the `gas_pump`, `premium`, and `super_premium` classes.

Program 12.23 shows how a virtual function may be used to report temperature in different ways.

Program 12.23

```
#include <iostream.h>

        int tempK;

class probe
{
public:
        int samples;
        probe(void);
        virtual void read(void);
};

probe::probe()
{
        samples = 0;
}

void probe::read()
{
        samples++;
        cout << "Reading " << samples << " is ";
```

```
        cout << tempK << " degrees Kelvin\n";
}

class fahrenheit_probe : public probe
{
public:
        void read(void);
};

void fahrenheit_probe::read()
{
        samples++;
        cout << "Reading " << samples << " is ";
        cout << ((tempK - 273) * 9.0/5.0 + 32) << " degrees
        Fahrenheit\n";
}

class celsius_probe : public probe
{
public:
        void read(void);
};

void celsius_probe::read()
{
        samples++;
        cout << "Reading " << samples << " is ";
        cout << (tempK - 273) << " degrees Celsius\n";
}

main()
{
        probe kelvin;
        celsius_probe celsius;
        fahrenheit_probe fahrenheit;

        kelvin.read();
        celsius.read();
        fahrenheit.read();
        cout << "\n";
        tempK = 273;
        kelvin.read();
        celsius.read();
        fahrenheit.read();
}
```

The base class `probe` defines the virtual function `read()` that reports the value of the global temperature variable `tempK`.

```
class probe
{
public:
```

```
    int samples;
    probe(void);
    virtual void read(void);
};
```

In probe, the read() function reports the temperature in degrees Kelvin. The two subclasses fahrenheit_probe and celsius_probe define their own read functions to report temperature in degrees Fahrenheit or degrees Celsius, respectively.

```
class fahrenheit_probe : public probe
{
public:
    void read(void);
};

class celsius_probe : public probe
{
public:
    void read(void);
};
```

The tempK variable has been made global so that all three instances of the temperature probe use the same Kelvin temperature as a starting point.

The execution of Program 12.23 is as follows:

```
Reading 1 is 0 degrees Kelvin
Reading 1 is -273 degrees Celsius
Reading 1 is -459.4 degrees Fahrenheit

Reading 2 is 273 degrees Kelvin
Reading 2 is 0 degrees Celsius
Reading 2 is 32 degrees Fahrenheit
```

It is clear that celsius.read() and fahrenheit.read() are performing their associated conversions, because their outputs are different. Also, since tempK is a global variable, its initial value is automatically set to zero by the compiler. This explains why the first Kelvin reading is zero.

Using static Variables in a Class Definition

Recall that storage space for member variables is reserved when an object is instantiated. The storage space may be initialized by a constructor and used during operations on the object. When the function that instantiated the object terminates, the storage space for the member variables is returned, and the contents are lost. This constitutes the lifetime of the member variable. Figure 12.7 illustrates this process graphically.

We use a static variable declaration to reserve memory that does not disappear when its associated function terminates. In other words, once a static variable is initialized, its value is available as long as the program continues execution. The same is true for static member variables defined by a class.

In Program 12.23, the global variable tempK was used to provide all instantiated temperature probes with the same temperature value. Sharing a variable in this way may also

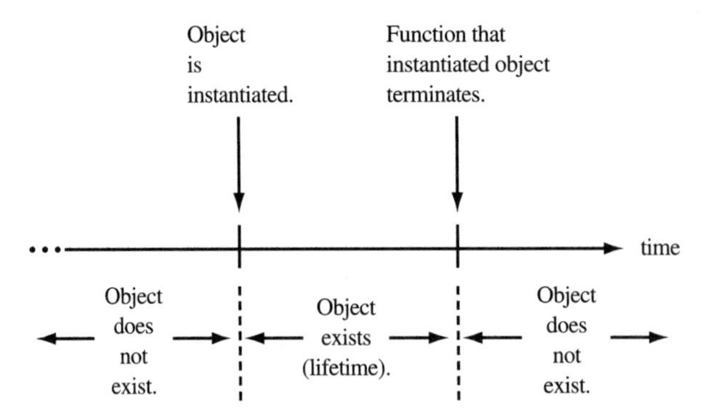

Figure 12.7 Lifetime of an Object

be done through the use of a `static` member variable. The `tempK` variable is declared `static` in the following way:

```
class probe
{
public:
     int samples;
     probe(void);
     virtual void read(void);
     static int tempK;
};
```

This declaration makes `tempK` a member of the class `probe`. To actually reserve memory for `tempK`, another declaration is needed in the global area of the program (preceding the `main()` block). This declaration does not use the `static` keyword and is different from an ordinary global variable declaration:

```
int probe::tempK;
```

The difference here is that `tempK` is only accessible by member functions of the class `probe` (or any derived classes). As usual, the `static` variable is initialized to zero automatically by the compiler.

Now, every `probe` object that is instantiated will have access to `tempK`, as in:

```
probe kelvin;

kelvin.tempK = -273;
```

Only one copy of `tempK` will exist, and its value will be retained as long as the program executes.

Program 12.24 demonstrates the use of `static` class members.

Program 12.24

```
#include <iostream.h>

class probe
{
```

```
public:
        int samples;
        probe(void);
        virtual void read(void);
        static int tempK;
};

probe::probe()
{
        samples = 0;
}

void probe::read()
{
        samples++;
        cout << "Reading " << samples << " is ";
        cout << tempK << " degrees Kelvin\n";
}

class fahrenheit_probe : public probe
{
public:
        void read(void);
};

void fahrenheit_probe::read()
{
        samples++;
        cout << "Reading " << samples << " is ";
        cout << ((tempK - 273) * 9.0/5.0 + 32) << " degrees Fahrenheit\n";
}

class celsius_probe : public probe
{
public:
        void read(void);
};

void celsius_probe::read()
{
        samples++;
        cout << "Reading " << samples << " is ";
        cout << (tempK - 273) << " degrees Celsius\n";
}

        int probe::tempK;

main()
{
        probe kelvin;
        celsius_probe celsius;
```

```
        fahrenheit_probe fahrenheit;

        kelvin.read();
        celsius.read();
        fahrenheit.read();
        cout << "\n";
        kelvin.tempK = 273;
        kelvin.read();
        celsius.read();
        fahrenheit.read();
}
```

Program 12.24's execution is identical to that of Program 12.23.

Pure Virtual Functions

When a base class definition contains a declaration for a virtual function, but not a definition for the virtual function, the function is referred to as a **pure virtual function.** For example, in Program 12.23 the `probe` class contained a virtual function called `read`:

```
class probe
{
public:
int samples;
     probe(void);
     virtual void read(void);
};
```

The definition of the `read` function in the base class contained code that reported the temperature in degrees Kelvin. Two subclasses, `fahrenheit_probe` and `celsius_probe`, supplied their own function definitions for `read` as well. Now, suppose that the `probe` class did not supply a definition for the `read` function. This would mean that the `read` function would be defined only in the subclasses `fahrenheit_probe` and `celsius_probe`. This is what it means to be a pure virtual function. Only subclasses of a base class define the function.

In order to tell the C++ compiler that a function is pure virtual, it is necessary to set the function equal to zero in its declaration, as follows:

```
class probe
{
public:
     int samples;
     probe(void);
     virtual void read(void) = 0;
};
```

When a class definition contains at least one pure virtual function, it is called an **abstract class.** With no definition for `read` in the `probe` class, it is not possible to instantiate objects of the abstract class `probe`. This means that only subclasses containing a function definition for `read` may instantiate objects, as `fahrenheit_probe` and `celsius_probe` do.

In Program 12.25, the `read` function is now a pure virtual function.

Program 12.25

```
#include <iostream.h>

        int tempK;

class probe
{
public:
        int samples;
        probe(void);
        virtual void read(void) = 0;
};

probe::probe()
{
        samples = 0;
}

class fahrenheit_probe : public probe
{
public:
        void read(void);
};

void fahrenheit_probe::read()
{
        samples++;
        cout << "Reading " << samples << " is ";
        cout << ((tempK - 273) * 9.0/5.0 + 32) << " degrees Fahrenheit\n";
}

class celsius_probe : public probe
{
public:
        void read(void);
};

void celsius_probe::read()
{
        samples++;
        cout << "Reading " << samples << " is ";
        cout << (tempK - 273) << " degrees Celsius\n";
}

main()
{
        celsius_probe celsius;
        fahrenheit_probe fahrenheit;
```

```
        celsius.read();
        fahrenheit.read();
        cout << "\n";
        tempK = 273;
        celsius.read();
        fahrenheit.read();
}
```

Note that the subclass definitions do not require any special keywords to indicate that `read` is a pure virtual function.

The execution of Program 12.25 looks like this:

```
Reading 1 is -273 degrees Celsius
Reading 1 is -459.4 degrees Fahrenheit

Reading 2 is 0 degrees Celsius
Reading 2 is 32 degrees Fahrenheit
```

Pure virtual functions are intended for use when we do not plan on instantiating any objects of a base class, but we do want to preserve and use inheritance to control visibility.

Conclusion

This section introduced you to some of the most important concepts and techniques available when you work with classes. Test your knowledge of this section by trying the following section review.

12.8 Section Review

1. What is a base class?
2. What is a subclass? How is it related to a base class?
3. State the rules for inheriting public members. How can inheritance be controlled?
4. Why use a static variable definition in class definitions?
5. Describe the difference between a virtual function and a pure virtual function.

12.9 File Objects

Discussion

In this section, you will discover how to save data to the disk and how to get it back. Doing this will allow you to create C++ programs that will let the program user store data entered into the program as well as get the data back when the program is used again.

A File Creation Program

Program 12.26 shows a C++ program that creates a disk file called `MYFILE.DAT` on the active drive. The program does not store any information in the file. However, when the program is executed, the new file `MYFILE.DAT` will exist on the disk in the active drive.

Program 12.26

```
#include <iostream.h>
#include <fstream.h>

main()
{
        ofstream new_file;      // This is the file object.

        // Create a file called MYFILE.DAT

        new_file.open("MYFILE.DAT");

        // Close the created file.

        new_file.close();
}
```

Program Analysis

Program 12.26 starts by creating a **file object** that is of a data type called **`ofstream`**:

```
ofstream new_file;      // This is the file object.
```

This is a class type defined in most versions of C++.

Next, the C++ `new_file.open()` function is called. In order to use this function, a legal DOS filename is required in the argument.

```
new_file.open("MYFILE.DAT");
```

This will actually create a file named `MYFILE.DAT` on the active drive and open it for writing in data.

The next statement is necessary when you are finished with the open files. If you don't use it, you may damage any data in your open files.

```
new_file.close()
```

This is the built-in C++ `close()` function.

Putting Data into the Created File

Program 12.26 created a disk file using a C++ program. However, nothing was put into the file. Program 12.26 illustrates the minimum requirements for creating a disk file, but the file is empty. Program 12.27 illustrates the creation and opening of the same file, but this time the character C is put into the file.

Program 12.27

```
#include <iostream.h>
#include <fstream.h>

main()
{
        ofstream new_file;      // This is the file object.

        // Create a file called MYFILE.DAT
```

```
        new_file.open("MYFILE.DAT");

        // Put a letter into the opened file:

        new_file << 'C';

        // Close the created file.

        new_file.close();
}
```

Program Analysis

Program 12.27 does the same thing as Program 12.26 with one exception: it puts the letter C into the opened file. This is done with the << operator as follows:

```
new_file << 'C';
```

This statement puts a single character into the opened file.

You now have a file called `MYFILE.DAT` that contains the character C.

Reading Data from a Disk File

Program 12.27 showed you how to put a single character into a file created by a C++ program. What you now need is a method of allowing the C++ program to get data from an existing disk file. Program 12.28 shows the basic requirements for doing just this.

Program 12.28

```
#include <iostream.h>
#include <fstream.h>

main()
{
        ifstream old_file;      // This is the file object.
        char file_character;    // Character to be read from the file.

        // Open the existing file called MYFILE.DAT

        old_file.open("MYFILE.DAT");

        // Get the first letter from the opened file.

        old_file >> file_character;

        // Display the character.

        cout << "The character is " << file_character << "\n";
```

```
        // Close the file.

        old_file.close();
}
```

For the file created in Program 12.27, Program 12.28, when executed, will display

```
The character is C
```

Program Analysis

Program 12.28, for reading data from an existing file, is very similar to Program 12.27, for creating and writing data into a file. This time an **`ifstream`** (input filestream) is instantiated into `old_file`.

```
ifstream old_file;      // This is the file object.
```

This is the same requirement as for creating and inputting data to a disk file. For this program, a variable to hold the character to be received from the file is declared:

```
char file_character;    // Character to be read from the file.
```

The file is opened using the same built-in C++ function `new_file.open()`. Again, its argument contains the file name to be opened.

```
old_file.open("MYFILE.DAT");
```

Next, the file character variable is read from the file with the >> operator.

```
old_file >> file_character;
```

Now the character read from the file is displayed on the monitor:

```
cout << "The character is " << file_character << "\n";
```

Last is the important step of closing the opened file when you are finished with it:

```
old_file.close();
```

Saving a Character String

The previous programs save and retrieve only a single character from a file on disk. However, they illustrate the fundamental requirements for a disk file with C++.

Program 12.29 illustrates a method of saving a character string to a disk file. In this case, the program user may input any string of characters from the keyboard and have them automatically saved to a file called `MYFILE.DAT`. The program will continue getting characters from the program user until a carriage return is entered.

Program 12.29

```
#include <iostream.h>
#include <fstream.h>
#include <conio.h>
```

```
main()
{
        ofstream new_file;        // This is the file object.
        char file_character;      // Character to be read from the file.

        // Create a file called MYFILE.DAT

        new_file.open("MYFILE.DAT");

        // Put a stream of data into the opened file.

        while((file_character = getche()) != '\r')
               new_file << file_character;

        // Close the created file.

        new_file.close();
}
```

Note that like the other file programs, Program 12.29 uses the built-in C++ `ofstream` class function which means that the file is being opened for writing. Again, when all file activity is completed, the file is closed with the built-in C++ `close()` function.

The main difference between this program and the others is the `while` loop.

A character is written into the file like this:

```
new_file << file_character;
```

Note that this statement is included in a C++ `while` loop:

```
while((file_character = getche()) != '\r')
        new_file << file_character;
```

Here the loop will continue as long as the program user does not press the -RETURN/ENTER- key.

Reading a Character Stream

A **character stream** is a sequence of bytes of data being sent from one place to another (such as from computer memory to the disk).

To read a character stream from a disk file, you use the same tactics as those for reading characters from the keyboard. The difference is that your program will continue to read a stream of characters until it encounters an end-of-file marker (EOF). This is automatically placed at the end of a disk file when the file is created. When this marker is read by the program, it indicates that the end of the file has been reached and there is nothing further to read. Program 12.30 illustrates.

Program 12.30

```
#include <iostream.h>
#include <fstream.h>

main()
```

```
{
        ifstream old_file;        // This is the file object.
        char file_character;      // Character to be read from the file.

        // Open an existing file called MYFILE.DAT

        old_file.open("MYFILE.DAT");

        // Get a stream of characters from the opened file
        // and display them on the screen.

        while(!old_file.eof())
        {
                old_file >> file_character;
                cout << file_character;
        }

        // Close the created file.

        old_file.close();
}
```

Program 12.30 is very similar to the other file programs. The C++ `open` function is used to open the file for reading. The main difference is in the C++ `while` loop:

```
while (!old_file.eof())
{
     old_file >> file_character;
     cout << file_character;
}
```

Here, the C++ `while` loop continues until an EOF marker is reached, indicating the end of the file. While the loop is active, the characters received from the disk file are directed to the monitor screen for display.

The Possibilities

Table 12.4 shows the possible modes you can encounter when working with disk data.

As you can see from Table 12.4, there are many possibilities when working with disk information. All of these possibilities will be presented in this section.

Besides the considerations presented in Table 12.4, in the C++ language there are four different ways of reading and writing data. These are listed in Table 12.5.

The next step in a disk I/O program is to open the file. As shown in Figure 12.8, the built-in C++ function `file.open("DOSNAME", file_mode, file_attribute);` is used. As you saw before, `"DOSNAME"` is any legal DOS file name, which may include an extension. (It should be noted that this may have also included an MS-DOS path name such as `\MYFILES\MYFILE.DAT`. The purpose of the file mode is to identify the type of file that will be opened. The file attribute may often be omitted since a plain file is usually sufficient.

Table 12.4 Disk File Modes

Condition	Meaning
`ios::out`	The disk file does not exist, and you want to create it on the disk and add some information.
`ios::in`	The disk file already exists, and you want to get information from it.
`ios::ate`	The disk file already exists, and you want to add more information to it while preserving the old information that was already there.
`ios::trunc`	The disk file already exists, and you want to get rid of all of the old information and add new information.
`ios::binary`	The file is opened in binary mode.
`ios::in | ios:out`	The file is opened for input and output.
`ios::nocreate`	Do not create file if it does not exist.

Table 12.5 Different Methods of Reading and Writing Data

Method	Comments
One character at a time	Inputs and outputs one character to the disk at a time.
Read and write data and strings	Inputs and outputs a string of characters to the disk.
Mixed mode	Used for I/O of characters, strings, floating points, and integers.
Structured	Used for I/O of array elements and structures.

```
#include <iostream.h>
#include <fstream.h>
  •
  •
  •
fstream file;
  •
  •
  •
file.open ("Filename", file_mode, file_type);
  •
  •
  •
// Interaction with file.
  •
  •
  •
file.close();
  •
  •
  •
```

Figure 12.8 Standard C++ Structure for Disk I/O

Table 12.6 C++ File Attributes

Value	Meaning
0	Plain file
1	Read-only file
2	Hidden file
4	System file
8	Archive file

Note in Table 12.6 how these file directives relate to the different ways you can treat a disk file.

Example File Program

Program 12.31 demonstrates several important points about C++ disk files. This program allows the user one of the following options for creating and reading from a simple message file:

1. Create a new file by a user-supplied file name.
2. Create a new file by a user-supplied file name and place a message in it.
3. Read a message from an existing file.
4. Append a message to an existing file.
5. The user is told if a given file for reading or appending does not exist.

Program 12.31

```
#include <string.h>
#include <stdlib.h>
#include <stdio.h>
#include <conio.h>
#include <iostream.h>
#include <fstream.h>

main()
{
        int file_mode;            // User file mode selection.
        char file_name[13];       // Name of the disk file.
        int user_choice;          // User choice for file activity.
        char file_character;      // File character to be saved.
        fstream iofile;           // Input or Output file object.

        // Display user options.

        cout << "Select one of the following by number:\n";
        cout << "1] Create a new file.    2] Write over an existing file.\n";
        cout << "3] Add new data to an existing file.\n";
        cout << "4] Get data from an existing file.\n";

        // Get and act on user input.
```

```
do
{
        cout << "Your selection => ";
        cin >> user_choice;

        switch(user_choice)
        {
                case 1  : break;  // Just create the file.
                case 2  : file_mode = ios::out;
                          // Write over existing data.
                          break;
                case 3  : file_mode = ios::app;
                          // Append data to existing file.
                          break;
                case 4  : file_mode = ios::in;
                          // Get data from existing file.
                          break;
                default : {
                                cout << "That was not one ";
                                cout << "of the choices.\n";
                                user_choice = 0;
                          }
        }
} while(user_choice == 0);

// Get the file from the user.

cout << "Enter the name of the file => ";
cin >> file_name;

// Open file.

iofile.open(file_name, file_mode);
if(iofile.bad())
{
        cout << "Cannot open file " << file_name << "\n";
        exit(-1);
}

// Write to or read from the file.

switch(user_choice)
{
        case 1 :
        case 2 :
        case 3 : {
                        cout << "Enter string to be saved: \n";
                        while((file_character = getche()) != '\r')
                              iofile << file_character;
                 }
                 break;
        case 4 : {
```

```
                                while(!iofile.eof())
                                {
                                        iofile >> file_character;
                                        cout << file_character;
                                }
                        }
                        break;
        }

        iofile.close();
}
```

Note: The function getche gets a character from the keyboard and echoes it to the screen. The function getch does the same thing except it does not echo the character to the screen.

Execution of Program 12.31 produces

```
Select one of the following by number:
1] Create a new file.   2] Write over an existing file.
3] Add new data to an existing file.
4] Get data from an existing file.
Your selection => 2
Enter the name of the file => MYFILE.01
Enter string to be saved: Saved by a C++ program.
```

The message above can be retrieved at a later date, added to, or written over. Note that the program will tell the user if the desired file does not exist. With this program, any new file must first be created by the user.

Program Analysis

Two file classes are instantiated, ifstream for the input file and ofstream for the output file.

```
ifstream infile;
ofstream outfile;
```

User options are then displayed using the cout function:

```
cout << "Select one of the following by number:\n";
cout << "1] Create a new file.   2] Write over an existing file.\n";
cout << "3] Add new data to an existing file.\n";
cout << "4] Get data from an existing file.\n";
```

The user prompt for input becomes part of a C++ do loop. The reason for this is to repeat the prompt if the user input is not valid:

```
do
{
        cout << "Your selection => ";
        cin >> user_choice;
```

Note that next a C++ cin function is used for receiving the user input.

A C++ `switch()` function is then used to select one of five possible conditions. Note that the first two conditions will both open files for writing:

```
switch(user_choice)
{
        case 1   : break;   // Just create the file.
        case 2   : file_mode = ios::out;
                   // Write over existing data.
                   break;
        case 3   : file_mode = ios::app;
                   // Append data to existing file.
                   break;
        case 4   : file_mode = ios::in;
                   // Get data from existing file.
                   break;
        default : {
                         cout << "That was not one ";
                         cout << "of the choices.\n";
                         user_choice = 0;
                   }
}
```

The C++ `default` condition will activate if one of the first four cases is not selected. Here the user is told that none of the possible selections was selected. Also note that in the `default` case, the variable `user_choice` is set to 0. This is done to cause a repeat of the C++ `do` loop and allow the user to try the selection again.

```
while(user_choice == 0);
```

Assuming the user makes a correct selection, the program will now ask the user for the name of the file. Note that again the C++ `cin` function is used:

```
cout << "Enter the name of the file => ";
cin >> file_name;
```

The next part of the program makes use of the fact that if, for any reason, a file cannot be opened, the program terminates.

```
iofile.open(file_name, file_mode);
if(iofile.bad())
{
     cout << "Cannot open file " << file_name << "\n";
     exit(-1);
}
```

In the statement above, the C++ `open` function is being used to open the file. The variable `file_name` contains the name of the file entered by the user, and the variable `file_mode` will contain one of the file mode directives `(ios::in, ios::out, ios::app)`. The file object `iofile` will be used to replace all references to the opened file. If `iofile.bad()` returns TRUE, the following compound statement will be executed:

```
{
        cout << "Cannot open file " << file_name << "\n";
        exit(-1);
}
```

Otherwise, these statements are skipped by the program.

Next, a C++ `switch` is used to determine what action is to be taken with the now opened file. Note that in `case 1` the user opts to simply create a new file, so there is nothing to input or output:

```
switch(user_choice)
{
        case 1 : break;
```

In `case 2` or `3`, the program user has opted to save a string to the file (by writing over an existing string or appending a new string). Here, the C++ `while` loop is used to input the string using the C++ `cin` function, and then write a character `file_character` to the file. The `while` loop stays active until the program user presses the -RETURN/ENTER- key, producing a carriage return (`\r`).

```
case 2 :
case 3 : {
                cout << "Enter string to be saved:\n";
                while((file_character = getche()) != '\r')
                        iofile << file_character;
         }
         break;
```

In the fourth case, the user chooses to read a string from a selected file. Here, `iofile >> file_character` is used to get a character from the file. This process is in a `while` loop that will continue until the EOF marker is reached. The `cout` function is used to direct the resulting characters to the monitor screen.

```
case 4 : {
                while(!iofile.eof())
                {
                        iofile >> file_character;
                        cout << file_character;
                }
         }
         break;
```

The program then terminates with the necessary file closing:

```
iofile.close();
```

Mixed File Data

The previous file programs were limited to working with strings. Program 12.32 illustrates a method of working with files that allows you to input numerical as well as string data. The program illustrates a process of entering the mixed data from the parts structure program of the previous chapters. This is accomplished by writing the mixed file data with a blank between each field. The `cin` function can be used to read and write all types except for a character string. `gets()` is used on strings.

Program 12.32

```
#include <iostream.h>
#include <fstream.h>
```

```
main()
{
        char part_name[15];        // Type of part.
        int quantity;              // Number of parts left.
        float cost_each;           // Cost of each part.
        ofstream outfile;          // File object.

        // Open a file for writing.

        outfile.open("PARTS.DAT");

        // Get data from program user.

        cout << "Enter part type, quantity, cost each, separated by spaces:\n";
        cout << "Press -RETURN- to terminate input.\n";

        cin >> part_name >> quantity >> cost_each;

        outfile << part_name;
        outfile << " ";
        outfile << quantity;
        outfile << " ";
        outfile << cost_each;

        // Close the opened file.

        outfile.close();
}
```

Execution of Program 12.32 produces:

```
Enter part type, quantity, cost each, separated by spaces:
Press -RETURN- to terminate input.
Resistor 12 0.05
```

Assuming that the program user carefully follows instructions, the `cin` function will receive the required data and it will be stored in the opened file by the statements:

```
outfile << part_name;
outfile << " ";
outfile << quantity;
outfile << " ";
outfile << cost_each;
```

Program Analysis

Program 12.32 has several weaknesses. It doesn't protect user input. It also doesn't let the user know if there was a problem opening the file. However, it does illustrate a simple C++ program that enters mixed data types into a file. The key to the program is the use of blank characters between the data fields, which allows the user to format the input data to the disk in many different ways—in this case as a string, then as an integer, and finally as a float.

Retrieving Mixed Data

Program 12.33 demonstrates how the mixed data stored on the disk may be retrieved and sent to the monitor screen.

Program 12.33

```
#include <iostream.h>
#include <fstream.h>

main()
{
        char part_name[15];      // Type of part.
        int quantity;            // Number of parts left.
        float cost_each;         // Cost of each part.
        ifstream infile;         // File object.

        // Open a file for reading.

        infile.open("PARTS.DAT");

        // Get data from disk file.

        while(!infile.eof())
        {
                infile >> part_name;
                infile >> quantity;
                infile >> cost_each;
                cout << part_name << " " << quantity << " ";
                cout << cost_each << "\n";
        }

        // Close the opened file.

        infile.close();
}
```

Assuming that the file contained the data entered by the previous program, execution of Program 12.33 would yield:

```
Resistor 12 0.05
```

Let's examine some better ways to use C++ files.

Binary Record Input

Program 12.34 shows an example of saving C++ data structures to a file. This is a powerful and important method of saving complex data to disk files. Observe that this program is similar to that of the mixed data type. However, the important difference is that it stores a data structure to the file. This is done using a new object function called `write()`. To use a binary file instead of text all we need to do is add the `ios::binary` after the file directive.

Program 12.34

```
#include <conio.h>
#include <iostream.h>
#include <fstream.h>
#include <stdio.h>

#define NUMBERS 15

class part_entry
{
public:
        char part_name[15];      // Type of part.
        int quantity;            // Number of parts left.
        float cost_each;         // Cost of each part.
} parts_data[NUMBERS];

main()
{
        int number;                          // Index to array.
        ofstream outfile;                    // File object.

        // Open a file for writing.

        outfile.open("PARTS.DAT", ios::binary);

        // Get data from program user.

        number = -1;
        do
        {
                ++number;
                cout << "\n";
                cout << "Name of part => ";
                gets(parts_data[number].part_name);
                cout << "Number of parts => ";
                cin >> parts_data[number].quantity;
                cout << "Cost per part => ";
                cin >> parts_data[number].cost_each;

                // Prompts user for more input.

                cout << "Add more parts (y/n)? => ";
        } while (getche() == 'y');

        // Write structure to opened file.

        outfile.write((char *)parts_data, sizeof parts_data);

        // Close the opened file.

        outfile.close();
}
```

Assuming that the user enters the given value, execution of Program 12.34 will yield

```
Name of part => Resistor
Number of parts => 12
Cost per part => 0.05
```

This data will now be saved to the disk as a block of information in a binary format under the file name of `PARTS.DAT` on the current drive.

Program Analysis

Program 12.34 first defines a class called `part_entry` as a class consisting of three members.

```
class part_entry
{
public:
        char part_name[15];          // Type of part.
        int quantity;                // Number of parts left.
        float cost_each;             // Cost of each part.
} parts_data[NUMBERS];
```

The program then declares the required `ofstream` class. For user input, a character string is also declared.

```
main()
{
        int number;          // Index to array.
        ofstream outfile     // File object.
```

A file on the current drive called `PARTS.DTA` is opened for writing in the binary mode (if the file does not exist, it will be created).

```
// Open a file for writing.

outfile.open("PARTS.DAT",ios::binary);
```

The program user is now prompted to enter data. Note the use of the member of operator (.). This is done inside a C++ `do` loop so that the user may enter more than one structure of data:

```
// Get data from program user.

number = -1;
do
{
        ++number;
        cout << "\n";
        cout << "Name of part => ";
        gets(parts_data[number].part_name);
        cout << "Number of parts => ";
        cin >> parts_data[number].quantity;
        cout << "Cost per part => ";
        cin >> parts_data[number].cost_each;
```

Data is then entered into the opened file using the C++ `write` function. This function has two arguments:

```
outfile.write(buffer, size)
```

The `buffer` contains the address of the data that is to be written into the file. The `size` is the size of the buffer in bytes. The actual `write` operation in Program 12.34 is as follows:

```
outfile.write((char *)parts_data, sizeof parts_data);
```

Note that the C++ `sizeof` function is used to automatically determine the size, in bytes, of the data to go to the file.

The user is prompted for more data. The input loop will end if the character `y` is not entered:

```
        cout << "Add more parts (y/n)? => ";
} while(getche() == 'y');
```

The file is then closed.

```
// Close the opened file.

outfile.close();
```

Record Output

Program 12.35 illustrates the retrieval of the block of data entered by Program 12.34. Note that this program uses the same type definition of the parts structure as did Program 12.34.

Program 12.35

```
#include <conio.h>
#include <iostream.h>
#include <fstream.h>

#define NUMBERS 15

class part_entry
{
public:
        char part_name[15];      // Type of part.
        int quantity;            // Number of parts left.
        float cost_each;         // Cost of each part.
} parts_data[NUMBERS];

main()
{
        int number;                      // Index variable.
        ifstream infile;                 // File object.

        // Open a file for reading.
```

```
        infile.open("PARTS.DAT", ios::binary, 0);

        // Get data from file and display.

        number = -1;

        infile.read((char *)parts_data, sizeof parts_data);

        do
        {
                ++number;
                cout << "\n";
                cout << "Part name => " << parts_data[number].part_name;
                cout << "\n";
                cout << "Number of parts => " << parts_data[number].quantity;
                cout << "\n";
                cout << "Cost of part => " << parts_data[number].cost_each;
                cout << "\n";
                infile.read((char *)parts_data, sizeof parts_data);
        }while(infile);

        // Close the opened file.

        infile.close();
}
```

When executed, Program 12.35 will look for the file `PARTS.DAT` on the current drive and open it in the binary mode for reading (the file must already exist). The data is read using the C++ `read()` function, with the following statement:

```
infile.read((char *)parts_data, sizeof parts_data);
```

This is very similar to the `write()` function.

Conclusion

This section demonstrated some key points concerning disk file I/O. Here you saw the method of storing and retrieving different kinds of data to and from disk files. You also saw how to do this with blocks of structured data. Check your understanding of this section by trying the following section review.

12.9 Section Review

1. List the four disk file I/O conditions that are possible.
2. State the different methods of reading and writing data with the C++ language.
3. Give the three basic C++ file type commands.
4. What is the purpose of the C++ `ofstream` class?
5. State the purpose of a block read or write function.

12.10 Example Programs

Discussion

In this section we will examine four new programming applications. You are encouraged to study them to gain a full understanding of how they work. Try to determine why pure virtual functions were used in some cases instead of ordinary virtual functions or no virtual functions at all.

Reversing a String with a Stack

Program 12.36 shows how a stack object is created and used to reverse a character string. We have already seen the stack used for this purpose. What is new here is the ability to create multiple stack objects and use the push and pop member functions to access both stacks.

Program 12.36

```
#include <iostream.h>
#include <stdio.h>
#include <string.h>

#define STACKSIZE 32

class stack
{
        char stackbuff[STACKSIZE];
        int stackptr;
public:
        stack(void);
        int push(char item);
        int pop(char *item);
} astack;

stack::stack()
{
        stackptr = -1;
}

int stack::push(char item)
{
        if (stackptr == STACKSIZE)
                return 0;
        else
        {
                stackptr++;
                stackbuff[stackptr] = item;
                return 1;
        }
}

int stack::pop(char *item)
```

```
{
        if (stackptr == -1)
                return 0;
        else
        {
                *item = stackbuff[stackptr];
                stackptr--;
                return 1;
        }
}

main()
{
        char text[STACKSIZE], letter;
        int k;

        cout << "Enter a string: ";
        gets(text);
        for (k = 0; k < strlen(text); k++)
                astack.push(text[k]);
        cout << "\n";
        cout << "The reversed string is: ";
        while (astack.pop(&letter))
        {
                cout << letter;
        }
        cout << "\n";
}
```

A sample execution is as follows:

```
Enter a string: Multiple stack objects!

The reversed string is: !stcejbo kcats elpitluM
```

Displaying the Date

Program 12.37 shows how a virtual function may be used to display a person's birthdate in one of two different formats:

American: Month-Day-Year
European: Day-Month-Year

Two versions of the virtual function `showdate()` are used for this purpose.

Program 12.37

```
#include <iostream.h>

class calendar {
public:
        int month, day, year;
```

```
        calendar(void);
        virtual void showdate(void);
};

calendar::calendar()
{
        month = 12;
        day = 8;
        year = 1970;
}

void calendar::showdate()
{
        cout << month << " - " << day << " - " << year;
}

class european_calendar : public calendar
{
public:
        void showdate(void);
};

void european_calendar::showdate()
{
        cout << day << " - " << month << " - " << year;
}

main()
{
        calendar usdate;
        european_calendar edate;

        cout << "You were born on ";
        usdate.showdate();
        cout << "\n";
        cout << "You were born on ";
        edate.showdate();
        cout << "\n";
}
```

The execution of Program 12.37 looks like this:

```
You were born on 12 - 8 - 1970
You were born on 8 - 12 - 1970
```

Your Weight on Different Planets

For a little excitement, it might be fun to see what your Earth weight becomes if you go to a number of different planets. Program 12.38 uses the pure virtual function `weight()` to convert Earth weight (in pounds) into Mercury, Mars, and Jupiter weights.

Program 12.38

```
#include <iostream.h>

class planets
{
public:
        virtual void weight(float Earthweight) = 0;
};

class first : public planets
{
public:
        void weight(float Earthweight);
};

void first::weight(float Earthweight)
{
        cout << "You weigh " << (0.25 * Earthweight);
        cout << " pounds on Mercury.\n";
}

class fourth : public planets
{
public:
        void weight(float Earthweight);
};

void fourth::weight(float Earthweight)
{
        cout << "You weigh " << (0.36 * Earthweight);
        cout << " pounds on Mars.\n";
}

class fifth : public planets
{
public:
        void weight(float Earthweight);
};

void fifth::weight(float Earthweight)
{
        cout << "You weigh " << (2.64 * Earthweight);
        cout << " pounds on Jupiter.\n";
}

main()
{
        first Mercury;
        fourth Mars;
        fifth Jupiter;
        float Earthweight;
```

```
        cout << "Enter your weight on Earth (in pounds): ";
        cin >> Earthweight;
        Mercury.weight(Earthweight);
        Mars.weight(Earthweight);
        Jupiter.weight(Earthweight);
}
```

The execution of Program 12.38 for a person weighing 170 pounds on Earth is as follows:

```
Enter your weight on Earth (in pounds): 170
You weigh 42.5 pounds on Mercury.
You weigh 61.2 pounds on Mars.
You weigh 448.8 pounds on Jupiter.
```

Clearly, this person should not move to Jupiter anytime soon.

Conclusion

In this section we examined some useful applications that demonstrate some of the power and functionality of objects. Check your understanding of this material using the following section review.

12.10 Section Review

1. How many stack objects can be created in Program 12.36?
2. Why is it legal for `stackbuff` to be defined as `private` in Program 12.36?
3. Why is `showdate()` in Program 12.37 defined as virtual?
4. Is it necessay to initialize the pure virtual function `weight()` to 0 in Program 12.38? What happens if it is omitted?

12.11 Troubleshooting Techniques

Discussion

In this section we examine some of the problems a C programmer might experience while writing programs in C++. The C++ language provides a significant amount of new functions and features. Everything that used to work in C will probably work in C++; however, most functions specific to C++ cannot be written using C. The topics in this section will help you successfully make the transition from being a C programmer to a C++ programmer.

Syntax Errors

One of the biggest challenges a new C++ programmer faces is just getting his or her programs to compile. It is common for many errors to be attributed to the definition of classes and the rules of inheritance. C++ programs will not compile correctly if access is not allowed to members of a class. These errors can be caused by missing scope resolution operators or mistyping the name of a class or member.

Logic Errors

It is also possible to have statements compile that are not logically correct (as you are probably already aware). In C++ this is even more likely to happen, and, therefore, it may become even more difficult to locate program bugs. New C++ features such as operator overloading can account for some of these problems since a stray plus sign, minus sign, or any operator which may be overloaded can accidently be appended to a variable. These statements will compile correctly but produce unexpected results.

Objects

It is wise to remember an object need not be used to solve every simple type of application. Classes and objects deal with abstract ideas, some of which can become quite complex. When objects are required, the class must be designed carefully to meet the program requirements. It is necessary to spend ample time at the beginning of the design process when designing objects instead of trying to recover from a bad design.

Conclusion

The C++ language offers many new features to an experienced C programmer. Objects are a powerful tool to be used wisely. Test your understanding of this material by trying the following section review.

12.11 Section Review

1. Why is it more difficult to get C++ programs to compile?
2. What are the rules of inheritance? Why are they important?
3. When and why should objects be used in a C++ program?
4. Why is it more difficult to locate errors in C++ source code?

12.12 Case Study: Card Casino

Discussion

In this section we will combine many of the topics discussed in this chapter. The card casino program we will develop will use the programming techniques applicable to classes and objects. The goal of the design is to end up with a program that plays a hand of blackjack or acey-deucey.

The Problem

Any card game requires the programmer to devise a structure that represents the deck of cards. There must also be a way to pick a random card from the deck. Two ways this may be done are as follows: First, initially shuffle the deck, then pick cards from the top. Second, pick cards randomly from the deck. The second method will be used in the card casino program.

Once a card has been picked, its face value and suit are usually displayed. For example, "King of Hearts," or "7 of Diamonds," or "Jack of Spades" might be output. Finally, the cards in a player's hand must be evaluated to determine what to do next. Should the dealer take a hit? Did the player win?

The problem for the programmer is to determine what type of structures will be used to implement these requirements.

Developing the Algorithm

The following structures and definitions will be used:

- A global integer array called `deck` that contains 52 elements. Each element is initially set to zero.
- A function called `pickcard()` that picks a card by accessing a random element of `deck`. Once a card is picked, its element value is set to 1 to prevent it from being picked again.
- A function called `showcard()` that displays the face value and suit of the card. The card is represented by an index value from 0 to 51 (chosen by `pickcard()`). Cards 0 through 12 are clubs, cards 13 through 25 are diamonds, then come hearts and spades. A numeric suit value is easily found by dividing the card index by 13.
- Face values within a suit (card index mod 13), assigned as shown in Table 12.7.
- A member array called `hand` that stores up to five card indexes.
- A pure virtual function called `evaluate()` that determines the value of the `hand` array in blackjack, or tests the inside/outside range of the third card in the `hand` array in acey-deucey.

The Overall Process

The card casino program works in the following way: Initially, the player chooses a game, blackjack or acey-deucey. If the choice is blackjack, cards are dealt as follows:

```
One to player.
One to dealer (computer).
Second to player.
Second to dealer (not displayed).
```

Table 12.7 Face Value Assignments

Index	Card Face	Value
0	2	2
1	3	3
2	4	4
3	5	5
4	6	6
5	7	7
6	8	8
7	9	9
8	10	10
9	Jack	10
10	Queen	10
11	King	10
12	Ace	11

The player's hand is then evaluated. The player may choose to take hits until the value of the hand exceeds 21 or the player stands. The dealer takes a hit on anything under 17. The player wins on a tie.

If the choice is acey-deucey, two cards are dealt. The player is asked to choose if the next card is inside or outside the range of the two cards. One more card is dealt to test the player's guess.

Program 12.39 shows how the card casino games are implemented. Note the use of default function parameters in `pickcard()`.

Program 12.39

```
#include <iostream.h>
#include <stdlib.h>
#include <ctype.h>

        int deck[52];

class cardgame
{
public:
        int cardindex;
        cardgame(void);
        int hand[5];
        void showcard(int card);
        void pickcard(char say);
        virtual int evaluate(void) = 0;
};

cardgame::cardgame()
{
        cardindex = 0;
}

void cardgame::pickcard(char say = 'N')
{
        int pick;

        do
        {
                pick = rand() % 52;
        } while (deck[pick]);

        deck[pick] = 1;
        hand[cardindex] = pick;
        cardindex++;
        if (say == 'Y')
                showcard(pick);
}

void cardgame::showcard(int card)
{
        switch(card % 13)
```

```
        {
                case 9 : cout << "Jack";
                         break;
                case 10: cout << "Queen";
                         break;
                case 11: cout << "King";
                         break;
                case 12: cout << "Ace";
                         break;
                default: cout << (2 + card % 13);
        }
        cout << " of ";
        switch(card / 13)
        {
                case 0 : cout << "Clubs";
                         break;
                case 1 : cout << "Diamonds";
                         break;
                case 2 : cout << "Hearts";
                         break;
                case 3 : cout << "Spades";
        }
        cout << "\n";
}

class blackjack : public cardgame
{
public:
        int evaluate(void);
};

int blackjack::evaluate()
{
        int values[] = {2, 3, 4, 5, 6, 7, 8, 9, 10, 10, 10, 10, 11};
        int k, points, aces;

        points = 0;
        aces = 0;
        for (k = 0; k < cardindex; k++)
        {
                points += values[hand[k] % 13];
                if (12 == hand[k] % 13)
                        aces++;
        }
        while ((aces > 0) && (points > 21))
        {
                points -= 10;
                aces--;
        }
        return points;
}
```

```
class aceydeucey : public cardgame
{
public:
        int evaluate(void);
};

int aceydeucey::evaluate()
{
        int c1, c2, c3;

        c1 = hand[0] % 13;
        c2 = hand[1] % 13;
        c3 = hand[2] % 13;
        if ((c1 < c3) && (c3 < c2))
                return 1;
        else
                return 2;
}

void playbj(void);
void playad(void);

main()
{
        int choice;

        srand( (unsigned) time(NULL) );
        cout << "Choose a card game:\n";
        cout << "1] Blackjack\n";
        cout << "2] Acey-Deucey\n";
        cout << "Choice ? ";
        cin >> choice;
        switch(choice)
        {
                case 1 :  playbj();
                          break;
                case 2 :  playad();
                          break;
                default : cout << "OK, we will play later.\n";
        }
}

void playbj()
{
        blackjack player;
        blackjack dealer;
        char hit;

        cout << "Your card is the ";
        player.pickcard('Y');
        cout << "Dealer's card is the ";
```

```
        dealer.pickcard('Y');
        cout << "Your next card is the ";
        player.pickcard('Y');
        cout << "Dealer takes second card.\n\n";
        dealer.pickcard();
        cout << "You have " << player.evaluate() << "\n";
        do
        {
                cout << "Do you want a hit? ";
                cin >> hit;
                hit = toupper(hit);
                if (hit == 'Y')
                {
                        cout << "You are dealt the ";
                        player.pickcard('Y');
                        cout << "You have ";
                        cout << player.evaluate() << "\n";
                }
        } while ((hit == 'Y') && (player.evaluate() < 22));
        cout << "\n";
        if (player.evaluate() < 22)
        {
                cout << "Dealer has " << dealer.evaluate() << "\n";
                while (dealer.evaluate() < 17)
                {
                        cout << "Dealer takes a hit...\n";
                        cout << "The card is the ";
                        dealer.pickcard('Y');
                        cout << "Dealer has " << dealer.evaluate() << "\n";
                }
                if (dealer.evaluate() < 22)
                {
                        if (player.evaluate() >= dealer.evaluate())
                                cout << "You win!\n";
                        else
                                cout << "You lose.\n";
                }
                else
                        cout << "You win!\n";
        }
        else
                cout << "You lose.\n";
}

void playad()
{
        aceydeucey player;
        int response;

        cout << "The cards are:\n";
        player.pickcard('Y');
```

```
    player.pickcard('Y');
    cout << "\n";
    cout << "Is the next card:\n";
    cout << "1] In\n";
    cout << "2] Out\n";
    cout << "Choice ? ";
    cin >> response;
    cout << "The next card is the ";
    player.pickcard('Y');
    if (response == player.evaluate())
            cout << "You win!\n";
    else
            cout << "You lose.\n";
}
```

A sample execution of Program 12.39 for blackjack is as follows:

```
Choose a card game:
1] Blackjack
2] Acey-Deucey
Choice ? 1
Your card is the 5 of Clubs
Dealer's card is the 10 of Diamonds
Your next card is the King of Hearts
Dealer takes second card.

You have 15
Do you want a hit? y
You are dealt the 5 of Hearts
You have 20
Do you want a hit? n

Dealer has 20
You win!
```

A sample execution of Program 12.39 for acey-deucy is as follows:

```
Choose a card game:
1] Blackjack
2] Acey-Deucey
Choice ? 2
The cards are:
4 of Clubs
King of Spades

Is the next card:
1] In
2] Out
Choice ? 1
The next card is the 9 of Hearts
You win!
```

Conclusion

Program 12.39 is a good illustration of the use of pure virtual functions (the `evaluate()` function), multiple objects of the same class (the `player` and `dealer` objects of the `blackjack` class), and default function parameters (the `say` parameter of `pickcard()`). You are encouraged to add your own card game to Program 12.39, or to change the way the deck is represented and used.

12.12 Section Review

1. Why is the `evaluate()` function pure virtual?
2. Describe how the `hand` member is used in both card games.
3. Can objects of the base class `cardgame` be instantiated? Why or why not?
4. What method is used to simulate card shuffling? Is it random? Explain.

Interactive Exercises

DIRECTIONS

These exercises require you to have access to a computer with software that supports C++. They are provided here to give you valuable experience and immediate feedback on what the concepts and commands introduced in this chapter will do.

Exercises

1. Predict what the output of Program 12.40 will be, and then try it.

Program 12.40

```
#include <iostream.h>
#include <iomanip.h>

main()
{
     cout << "The number is " << 15 << "\n";
     cout.setf(ios::dec);
     cout << "The number in decimal is " << 15 << "\n";
     cout.setf(ios::oct);
     cout << "The number in octal is " << 15 << "\n";
     cout.setf(iso::hex);
     cout << "The number in hex is " << 15 << "\n";
}
```

2. What is different about the outputs in Program 12.41?

Program 12.41

```
#include <iostream.h>
#include <iomanip.h>

main()
```

```
{
    cout << "The number is " << 15.5 << "\n";
    cout.setf(ios::scientific);
    cout << "The number in scientific notation is " << 15.5 << ".\n";
}
```

3. Program 12.42 is a fun one. See if you can predict what the output will be before you try it.

Program 12.42

```
#include <iostream.h>

main()
{
    cout << "The number is " << 10 / 4 << "\n";
    cout << "The number is " << 10.0 / 4.0 << "\n";
}
```

4. Figure out Program 12.43 with pencil and paper. Then try it. Do your results agree with the computer's?

Program 12.43

```
#include <iostream.h>

main()
{
    float result = 10;

    result = 2 * (3 + 11)/8 - 3;

    cout << "The result is " << result <<"\n";

}
```

5. Program 12.44 is a good test question. Make sure you try it!

Program 12.44

```
#include <iostream.h>

main()
{

    cout << "What is this ==> \\\n";
    cout << "and this ==> \"";

}
```

6. For Program 12.45, see if you can predict what the output will be in each case. Then enter and execute the program. Be sure to write what your system gave you in your notes.

Program 12.45

```
#include <iostream.h>
#include <iomanip.h>

     float number_1 = 125.738;

main()
{

     cout << setprecision(2);
     cout << "In float notation 125.738 = " << number_1 << "\n";
     cout.setf(ios::scientific);
     cout << "In E notation 125.738 =  " << number_1 << "\n";

}
```

7. Will Program 12.46 compile? If not, why not?

Program 12.46

```
#include <iostream.h>

class abc
{
        void letter(void);
} abcobj;

void abc::letter()
{
        cout << "A";
}

main()
{
        abcobj.letter();
}
```

8. Will Program 12.47 compile? If not, why not?

Program 12.47

```
#include <iostream.h>

class abc
{
protected:
        void letter(void);
} abcobj;

void abc::letter()
{
        cout << "A";
}
```

```
main()
{
        abcobj.letter();
}
```

9. Predict the output of Program 12.48 and then run it.

Program 12.48

```
#include <iostream.h>

class abc
{
public:
        abc(void);
        void letter(void);
        ~abc(void);
} abcobj;

abc::abc()
{
        cout << "B";
}

void abc::letter()
{
        cout << "A";
}

abc::~abc()
{
        cout << "C";
}

main()
{
        abcobj.letter();
        cout << "D";
}
```

10. Predict the output of Program 12.49 and then run it.

Program 12.49

```
#include <iostream.h>

class velocity
{
public:
        virtual void rate(float speed);
};

void velocity::rate(float speed)
```

```
{
        cout << "The speed is " << speed << " miles/hour.\n";
}

class new_velocity : public velocity
{
public:
        void rate(float speed);
};

void new_velocity::rate(float speed)
{
        float fps;

        fps = speed * 5280.0 / 3600.0;
        cout << "The speed is " << fps << " feet/sec.\n";
}

main()
{
        velocity car1;
        new_velocity car2;

        car1.rate(60.0);
        car2.rate(60.0);
}
```

11. Program 12.50 will place a new file on your active disk. After you try the program, look at your disk files. What new file do you see?

Program 12.50

```
#include <iostream.h>
#include <fstream.h>

main()
{
        ofstream outfile;

        outfile.open("NEWFILE.DAT");

        outfile.close();
}
```

12. Program 12.51 puts something in the new file created on your disk by Program 12.50. Make sure to execute Program 12.50 before trying Program 12.51.

Program 12.51

```
#include <iostream.h>
#include <fstream.h>

main()
```

```
{
        ifstream infile;

        infile.open("NEWFILE.DAT");

        infile.put('C');

        infile.close();
}
```

13. Program 12.52 gets from the file what Program 12.51 put into it. Try it and see if you get what you expected.

Program 12.52

```
#include <iostream.h>
#include <fstream.h>

main()
{
        ifstream infile;
        char ch;

        infile.open("NEWFILE.DAT");

        infile.get(ch);
        cout << "The character is " << ch << "\n";

        infile.close();
}
```

14. What changes are made to `NEWFILE.DAT` when Program 12.53 is executed?

Program 12.53

```
#include <iostream.h>
#include <fstream.h>

main()
{
        ofstream outfile;

        outfile.open("NEWFILE.DAT",ios::app);
        outfile.put('!');
        outfile.close();
}
```

15. What does Program 12.54 do? Are any errors generated? Is it necessary for MORENEW.DAT to exist before Program 12.54 is run?

Program 12.54

```
#include <iostream.h>
#include <fstream.h>

main()
```

```
{
        ofstream outfile;

        outfile.open("MORENEW.DAT",ios::app);
        outfile.put('!');
        outfile.close();
}
```

Self-Test

DIRECTIONS

Answer the following questions by referring to Program 12.39.

Questions

1. Why is the integer `deck` declared as a global variable?
2. Which function uses a default parameter?
3. What card (face and suit) is associated with `deck[30]`?
4. What does this statement do?

```
if (12 == hand[k] % 13)
        aces++;
```

5. What game is the statement of question 4 used in?
6. How might an enumerated list be used to represent faces and suits?
7. Suppose that the `hand` array contains three cards whose numbers are 6, 36, and 46. What does `blackjack::evaluate()` return?
8. What does `aceydeucey::evaluate()` return for the same cards as in question 7?
9. How would a third card game, such as poker, be added to Program 12.39?
10. Which member items of the class `cardgame` can be made private without affecting execution?

End-of-Chapter Problems

General Concepts

Section 12.1

1. What must all C++ programs start with?
2. State the purpose of the `//` used in C++.
3. What indicates the beginning and the end of program instructions in the C++ language?

Section 12.2

4. What is the main difference between how strings and characters are represented in a `cout` function?
5. State what in a `cout` function specifies how it is to convert, print, and format its arguments.
6. What is the name given to the actual values within the parentheses of a function?
7. How many arguments must a `cout` function contain?

Section 12.3

8. What in a `cout` function causes a vast departure from the normal interpretation of a string?
9. Name the character in the `cout` function that is referred to as the escape character.
10. State what determines the number of digits to the right of the decimal point in a displayed value when the `cout` function is used.

Section 12.4

11. How are member functions defined?
12. How can global variables be eliminated by using classes?
13. Where is the `public` keyword used?
14. What is encapsulation? How is it done?

Section 12.5

15. How is a parameter passed to a constructor?
16. Can a constructor function return a data value?
17. How is a destructor defined?
18. Explain what is meant by software reuse.

Section 12.6

19. How is an array of objects defined?
20. How is an array of object pointers defined?
21. What affects the visibility of class members?
22. How does a pointer to an object differ from a pointer to a structure?

Section 12.7

23. How do friend functions access private members of an object?
24. How does the use of friend functions affect access to public members?
25. What happens when a friend function tries to access private members without permission?
26. Why not make all class members public? Is there a disadvantage to doing this?
27. Why not make all class members private?

Section 12.8

28. How is a virtual function defined? How is it different from a pure virtual function?
29. How many steps are involved in defining a static variable?
30. How is inheritance used to control the visibility of a member?
31. What does the `protected` keyword do? How does it affect visibility?

Section 12.9

32. How is a file class declared in C++? Show by example.
33. When disk file I/O is being performed with C++, what must always be done before the program is terminated?
34. State the different methods of reading and writing data with the C++ language.
35. Give the three basic C++ file command strings.
36. If the command for writing to a disk file in C++ is given, and the file does not exist, what will happen?
37. State the two kinds of data streams used by C++.

Section 12.10

38. How is inheritance used in Program 12.37?
39. How can the sixth planet be added to Program 12.38?
40. Why is `weight()` defined as a pure virtual function in Program 12.38?
41. Is it useful to sort dates if they are stored in the American or European format? What is the ideal format for a date to be stored?

Section 12.11

42. Why is it important to keep track of objects? What can happen if you do not?
43. Is it more difficult to get a C++ program to compile compared to a C program? Why?
44. Can a protected member be inherited?
45. Can most C++ programs be rewritten using C?

Section 12.12

46. Explain how face values and suits are represented in the card deck.
47. How many hand indices are used in the blackjack game? In acey-deucey?
48. Who wins on a draw in blackjack? In acey-deucey?
49. How are multiple objects of the same class used in the card casino program? What are the objects?

Program Design

When writing the following C++ programs, utilize the new class features discussed in this chapter.

50. Complete the design of the gas pump program. Add diesel fuel to the types of gas that may be pumped. Charge $2.15 a gallon for the diesel fuel. Allow a maximum of 100 gallons of any fuel to be pumped. Also, fuel may not be pumped from a pump that is off. Update the pump status as necessary.
51. Develop a C++ program that generates mathematical questions such as:

```
What is the sum of -6 and 12?
What is the difference of 23 and 176?
What is the product of 55 and -8?
```

    Use a pure virtual function to handle the different math operations. Generate 10 questions and keep track of the number of correct answers in each section.
52. Write a C++ program that simulates the operation of a checking/savings account. The initial balance in both accounts is $100. A fee of $0.50 is charged for withdrawals from the checking account. No fee is charged for withdrawals from savings. A maximum of $50 may be withdrawn at any time from either account. There is no limit to the amount of money that may be deposited in either account.
53. Create a C++ program that uses a constructor to initialize the contents of an array that contains the first 10 prime numbers. Use a destructor to display the contents of the array.
54. Develop a C++ program that uses an array of objects to keep track of the score in a bowling game. Each frame in the game should be represented by its own object. A friend function should be used to compute the final score of the game.
55. A C++ program is needed to keep track of hours worked for 10 employees. There are three grades of employees: G-1, G-2, and G-3. G-1 employees are paid $7.50 per hour and do not work overtime. G-2 employees are paid $8.10 per hour for each hour up to 40 hours and $12.15 per hour for each hour worked over 40 hours. G-3 employees must work at least 36 hours per week and make a flat rate of $600 per week plus $20 for each year of service. The program must compute the weekly pay for each of the 10 employees.
56. A library needs a C++ program to keep track of the number and types of publications it owns. There are four types of publications: magazines, other periodicals, hardcover, and oversize. Among hardcover books there are three different categories: educational, fiction, and nonfiction. Magazines and other periodicals both have two subcategories: English and foreign. The program should display the number and type of each book, including subcategories, and the overall number of books for the entire library.
57. A small C++ database is needed to keep track of telephone numbers. There are three kinds of phone numbers: extensions, local numbers, and long distance numbers. Extensions are four-digit numbers, local numbers contain seven digits, and long distance numbers require ten

digits. The program should display the entire telephone number database and also allow numbers to be added, changed, and deleted.

58. Write a program that demonstrates the operation of three different kinds of stacks: integer, floating point, and character. The associated `push()` and `pop()` functions should be pure virtual functions.
59. Develop a C++ program that keeps track of integrated circuit part numbers. The integrated circuits come in five different package sizes: 8-pin, 14-pin, 16-pin, 24-pin, and 40-pin. Each package has its own part number, cost, and manufacturer. The program must display the total number of integrated circuits and the inventory of integrated circuits for each manufacturer.
60. Develop a C++ program that will allow the program user to use the computer as a simple word processor. The program must allow the user to save and retrieve text files, name each file, and delete old files.
61. Write a C++ program that searches the input file for every occurrence of a string. The input file and search string must be specified on the command line, as in:

```
SEARCH <filename> <string>
```

62. Write a C++ program that makes a list of all words found in an input file. Write the sorted list to an output file, along with the number of times each word appears in the input file.

Answers

Answers to Section Reviews—Chapter 1

Section 1.1

1. The purpose of an editor is to allow you to enter source code. It is needed to achieve this purpose.
2. The reason for using a compiler is so the computer will have access to a program that it understands.
3. A compiler converts the source code into a code that the computer understands.

Section 1.2

1. Three advantages of the C programming language are: portability, computer control, and flexibility. Other advantages are listed in Table 1.1.
2. Portability means that a program you write on one computer will operate on another computer system with few if any changes.
3. All C programs must start with the keyword `main()`.
4. An example of a C comment is

```
/* This is a comment. */
```

5. Braces indicate the beginning and the end of the program instructions.

Section 1.3

1. Program structure refers to the appearance of the source code. Good program structure will make the source code easy to read, modify, and debug.
2. It isn't necessary to give a structure to a C program for it to compile without errors. Good program structure makes things easier for the programmer, not the computer.
3. Block structure makes the program easier to read and understand. An example would be the structure used for a business letter.
4. The reason for structured programming is to make the program easy to understand, modify, and debug.
5. The programmer's block is used to explain all of the important points about the program. It contains:
   a. Program name
   b. Developer

c. Description of the program
d. Explanation of variables
e. Explanation of constants

Section 1.4

1. A set of characters is used to write a C program. These characters consist of the uppercase and lowercase letters of the English alphabet, the ten decimal digits of the Arabic number system, and the underscore (_). Whitespace characters are used to separate the items in a C program.
2. In C, a token is the most basic element recognized by the compiler.
3. The major data types used by C are numbers, characters, and strings.
4. In C, a keyword is a predefined token that has a special meaning to the C compiler.
5. The data type that handles the largest number is the `double`.

Section 1.5

1. The purpose of the `printf()` function is to write information to standard output.
2. Characters have single quotes, and strings have double quotes.
3. The format specifier in a `printf()` function specifies how it is to convert, print, and format its arguments.
4. An argument is the actual values within the parentheses of a function.
5. You must have as many arguments as format specifiers. Extra arguments are ignored.

Section 1.6

1. A C function is an independent collection of declarations and statements.
2. An identifier is the name you give to key parts of your program.
3. All identifiers must start with a letter of the alphabet (uppercase or lowercase) or the underscore (_). The remainder of the identifier may use any arrangement of letters (uppercase or lowercase), digits (0 through 9), and the underscore—and that's it—no other characters are allowed (this means spaces are not allowed in identifiers). Do not use reserved words for your own identifiers.
4. The first 32 characters of an identifier are recognized by C.

Section 1.7

1. You can think of a variable as a specific memory location set aside for a specific kind of data and given a name for easy reference.
2. In C, you must declare all variables before using them. To declare a variable, you must declare the type and identifier of the variable.
3. Three fundamental C type specifiers are: `char`, `int`, and `float`.
4. Initializing a variable means to combine its declaration with an assignment operator.
5. To prevent a new variable from coming into your program as a result of a typing error.

Section 1.8

1. The common arithmetic operators used in C are: + => (addition), – => (subtraction), * => (multiplication), / => (division), % => (remainder).
2. In integer division, C will truncate the remainder. The significance of this is that a division such as 3/5 will be evaluated to 0.
3. In C, `3 - 2 = result;` is not allowed. You cannot have an assignment to an expression.
4. Precedence of operation means the order in which arithmetic operations are performed.
5. An example of a compound assignment is: `value -= 5;`.

Section 1.9

1. The escape sequence in the `printf()` function causes an escape from the normal interpretation of a string.

2. When used in a `printf()` function, the backslash character is called the escape character.
3. Three escape sequences used in the `printf()` function are: `\n` => (newline), `\t` => (tab), and `\b` => (backspace).
4. A field width specifier as used by the `printf()` function determines the minimum number of spaces to the left of the decimal point and the maximum number of spaces to the right of the decimal point.

Section 1.10

1. The `scanf()` function allows your program to get user information from the keyboard.
2. The `scanf()` function is told what variable identifier to use by including the identifier in its argument preceded by an ampersand (`&`) with no space (such as `&variable`).
3. In most systems, the `scanf()` function produces a carriage return to a new line automatically.
4. Floating point values may be entered as whole numbers or by use of E notation.
5. To have values outputted to the screen in E notation, simply change the format specifier in the `printf()` function from `%f` to `%e`.

Section 1.11

1. There are three error messages. They are: fatal error messages, compilation error messages, and warning messages.
2. A fatal error message will terminate the compilation process.
3. Case sensitivity means that the C compiler makes a distinction between uppercase and lowercase letters. This means that if you use identifiers with capital letters and declare them as such, you must consistently use the exact same capital letter configuration for the remainder of your program.
4. It is good practice to look for semicolons when you do not understand the reason for the error message, because a missing semicolon usually distorts the meaning of the program to the point where the compiler cannot conclude that the error was a missing semicolon. This results in another kind of error message depending on what follows the missing semicolon.
5. A nested comment is one comment placed inside another. It is not legal, and will prevent the program from executing. Some compilers will allow you to select this option, but it is not recommended.

Section 1.12

1. The formula to convert from centimeters to inches is:

```
inches = centimeters / 2.54;
```

2. Add a new resistor R3 and a second equivalent resistance REQ2, both as floats. Put R3 in parallel with REQ to get the new REQ2 as follows:

```
REQ2 = (R3 * REQ) / (R3 + REQ);
```

3. `float` data types are used because the input and output values are real numbers. Using integers would result in a loss of accuracy due to rounding.

Section 1.13

1. The first step in the development of a program is to state the requirements in writing.
2. The items that should be included in the program problem statement are the purpose of the program, required input (source), the process on the input, and the required output (destination).
3. The first step in the actual coding of the program is to outline it using nothing more than comments.
4. The process used to develop the final program consists of coding each section of the program separately, then compiling and executing it. Any program bugs are thus removed from the program a section at a time.

Answers to Self-Test—Chapter 1

1. This program may not compile and execute on your system because an `#include <stdio.h>` directive may be required. If this is the case, be sure to use standard format and place it before anything else in the source code, starting on the far left part of the screen.
2. The program solves for the circuit current. The program user must input the values of the circuit voltage and the circuit resistance. The program will then display the value of the resulting current. This was determined by reading the comments at the beginning of the program.
3. There are three variables in the program. They are all of type `float`. This was determined by reading the `/*Declaration block.*/`.
4. The `puts()` function was used to explain the program to the user because it provides an automatic return to a new line. No other formatting commands were needed other than to output a series of strings to the monitor screen.
5. The program user may input the values of the voltage and resistance as whole numbers, as numbers with decimal fractions, or by E notation. This was determined by observing the `scanf()` function. It uses the `%f` (float) for input.
6. The output will be displayed using E notation. This was determined by observing the `printf()` function. It uses the `%e` (E notation) for presenting output values from the program variables.

Answers to Odd End-of-Chapter Problems—Chapter 1

Section 1.1

1. The program used to enter C source code is called an *editor.*

Section 1.2

3. All C programs must start with `main()`.
5. The opening and closing braces (`{` and `}`) indicate the beginning and end of program instructions in C.

Section 1.3

7. The purpose of a programmer's block is to present all of the important information about the program.
9. No, it isn't necessary to have a programmer's block for a C program to compile.

Section 1.4

11. The underscore (_).

Section 1.5

13. Characters are represented by single quotes, and strings by double quotes.
15. The name given to the actual values within the parentheses of a function is *arguments.*

Section 1.6

17. The name given to an independent collection of declarations and statements in C is a *function.*
19. The first 32 characters of an identifier are recognized by C.

Section 1.7

21. The three fundamental C type specifiers are `char`, `int`, and `float`.
23. The requirement that all variables must be declared prevents a new variable from coming into your program as a result of a typing error.

Section 1.8

25. The unique characteristic of integer division in C is that the answer will be *truncated.*
27. The C statement `result *= 5;` means `result = result * 5;`

Section 1.9

29. The *escape sequence* in a `printf()` function causes a vast departure from the normal interpretation of a string.
31. The *field width specifier* determines the number of digits to the right of the decimal point.

Section 1.10

33. As described in this chapter, the purpose of the `scanf()` function is to get input from the program user.
35. The program user may enter values in whole numbers (no decimal part), in numbers with decimal fractions, and in E notation.

Section 1.11

37. A fatal error message immediately terminates the compiling process.
39. No, compiler error messages do not always identify the problem in your program. It depends on the type of problem. A misplaced or omitted semicolon can really throw a compiler off.

Section 1.12

41. If `int` data types are used in Programs 1.21-1.24 the results are not always accurate. For example, in program 1.21 the ratio of feet per mile (5280) over seconds per hour (3600) would change from 1.46 to 1. The resulting speed in feet per second would not be accurate.

Section 1.13

43. The first step in the coding of a program is to enter comments that divide the program into major sections.

Answers to Section Reviews—Chapter 2

Section 2.1

1. The C compiler does not require that a C program be structured. Structuring is done to make it easier for the programmer and others who will be responsible for it to read, debug, and modify it.
2. Block structure means the program will be constructed so that there are several groups of instructions rather than one continuous listing of instructions.
3. Each program block should begin with a remark that explains what the block will do.
4. Program blocks may be separated by spaces or comments that form lines (`/*------*/`) across the program code.
5. The body of a program block is highlighted by indenting the program lines from the left margin.
6. The three types of blocks are sequential, branch, and loop.

Section 2.2

1. A C function is a specific part of a C program that is designed to return a value.
2. A type `void` does not return any value. An example would be a function that does not return a value.
3. A function prototype declares the function type, name, and parameters at the beginning of the C program.

4. The purpose of using a function prototype is to let the compiler know what to expect in your program. Doing this ensures that the proper amount and type of memory will be allocated to each function before it is actually used.
5. Calling a function means invoking it into action. This is done by using the function identifier in the calling function (as was done in `main()` in this section).
6. The `exit()` function is used as the last function call in `main()` in order to ensure that all of the necessary computer "house-cleaning" has been done. This allows you to go and use another program after your C program has terminated. The value used in `exit()` is 0. By convention, this means a successful termination.

Section 2.3

1. The parameter of a function identifies the type of variable that will be passed to the function.
2. A formal parameter is the identifier that is used to identify the argument type. An actual parameter is the identifier that contains the value to be passed. These must be the same by being of the same data type. They may be different in that the formal and actual arguments do not have to have the same identifiers.
3. Passing values between functions means having a value obtained in one function influence the operation or value of a called function.
4. The only difference between the coding of a function prototype and the head of a function declaration is that the prototype must end with a semicolon whereas the declaration must not have the ending semicolon.
5. One method of passing a value back to the calling function is by using the C command `return()` where the value to be returned is placed in the function argument.

Section 2.4

1. Yes, a function may pass more than one value to a called function. The number of values to be passed must all be identified by formal parameters.
2. Yes, a function may call more than one function. The requirement here is that the function has been defined.
3. No, it makes no difference in what order called functions are defined within the body of the program.
4. The meaning of a called function calling another function is that any function may call another function regardless of how that function was activated. This means that a function may call itself as well.
5. Recursion means a function calling itself.

Section 2.5

1. A preprocessor directive is a special instruction to the preprocessor that causes an action before the program is compiled and executed.
2. The `#include` directive instructs the preprocessor to substitute one set of tokens for another set of tokens.
3. A macro is a preprocessor instruction.
4. Constants are usually defined in C by using preprocessor commands. Conventionally, constants defined in this manner are all uppercase.
5. Yes, a parameter may be used with `#define`. An example would be `#define cube(x) x*x*x`.

Section 2.6

1. The advantage of creating your own header files is that you can build a library of data specific to your technology area. This will relieve you of having to replicate the same code over and over again.
2. As presented in this section, the information contained in your header files consists of a series of `#define` statements.

3. The extension given to a C header file is `.h`.
4. You call your header files into your C program by the statement `#include "file.h"`, with `file` being the legal DOS name of your header file.

Section 2.7

1. The `#ifdef` statement is a conditional compilation directive that compiles statements according to the presence/non-presence of a defined/undefined variable.
2. A `#define` statement creates the variable used in the `#ifdef` statement. If the programmer leaves out the `#define` statement, the `#ifdef` will not find the variable and therefore compile accordingly.
3. The `#endif` statement is used to indicate the end of a conditional compilation block.
4. The `printf()` statement is compiled only if both `VARX` and `VARY` are `#defined`.

Section 2.8

1. The main goal in the development of the case study for this section was to increase readability and understanding of what the program will do while still preserving the fundamental characteristics of the C language.
2. The following information should be included in the programmer's block:

   I. Program Information
      A. Name of program
      B. Name of programmer
   II. Program Explanation
      A. What the program will do
      B. What is required for input
      C. What process will be performed
      D. What the results will be
   III. Describe All Functions
      A. Use function prototypes
      B. Explain purpose of each prototype
      C. Define all variables and constants

3. Using `#define` statements can result in the saving of much program code. More importantly, these statements allow for the eventual development of header files, whereby these same statements may be used in future programs.
4. The last thing that is usually asked of the program user in a typical technology program is whether the program is to be repeated.

Answers to Self-Test—Chapter 2

1. Four functions are defined in the program: `main()`, `explain_program()`, `get_values()`, and `calculate_and_display()`.
2. The total number of functions used in the program is seven. The extra three are `exit()`, `printf()`, and `scanf()`.
3. The identifiers used for formal parameters in the program are `f`, `l`, `r`, and `v`.
4. The identifiers used for actual parameters in the program are `resistor`, `inductor`, `frequency`, and `voltage_s`.
5. The `get_values()` function passes values to the `calculate_and_display()` function.

6. Values are passed from one function to another by using the identifiers of the calling function as actual arguments of the same type and number. An example from the program is:

```
calculate_and_display(frequency, inductor, resistor, voltage_s);
```

7. There are eight variable identifiers used in the program: `f`, `l`, `r`, `v`, `resistor`, `inductor`, `frequency`, and `voltage_s`.
8. The minimum change required to dispaly the value of the inductive reactance would be in the function `calculate_and_display()`. Just use the value returned by `X_L(f,l)` in a `printf()` statement.
9. Yes, the program will accept E notation because the input variables are of type `float`.

Answers to Odd End-of-Chapter Problems—Chapter 2

Section 2.1

1. Block structure is a method of breaking the program into distinct groups of code so the program is easier to read and understand.
3. The loop block has the ability to go back and repeat a part of the program.

Section 2.2

5. The part of a C program that gives the compiler specific information about the functions that will be defined in the program is called the *function prototype*.
7. A function is called by using the function identifier within another function along with any actual arguments.

Section 2.3

9. The type assigned to a function when no value is to be returned by it is type `void`.
11. The C statement `return()` is used to return a value from the called function to the calling function.

Section 2.4

13. Yes, a function may call more than one other function.
15. Yes, a called function may call another function.

Section 2.5

17. The preprocessor directive presented in this chapter is `#define`.
19. An example of using parameters wih the `#define` is

```
#define square(x) x*x
```

Section 2.6

21. You invoke your own header file by using

```
#include "myfile.h"
```

assuming your header file is named `myfile.h`.

Section 2.7

23. The conditional compiler directive that begins a conditional block of code is `#ifdef`.

Answers to Section Reviews—Chapter 3

Section 3.1

1. Relational operators are symbols that indicate the relationship between two quantities.
2. The relational operators used in C are

   > Greater than
   >= Greater than or equal to
   < Less than
   <= Less than or equal to
   == Equal to
   != Not equal to

3. The two conditions allowed for relational operators are TRUE and FALSE.
4. The statement that a relational operator in C returns a value means that a value of 1 is returned if the relation is TRUE and a value of 0 is returned if the relation is FALSE.
5. The difference between the C operation symbols = and == is as follows. The single equals sign (=) is an assignment operator. It assigns the value on the right side of the operator to the memory location of the variable on the left side. The double equals sign (==) tests to see if the value of the data on the right is equal to the value of the data on the left. No assignment or transfer of information takes place.

Section 3.2

1. An open branch offers an option to the program that may or may not be executed depending on a given condition. In either case, program execution will always go forward to the next statement.
2. The `if` statement has the form

   ```
   if (expression) statement
   ```

   This means that if `expression` is TRUE, `statement` will be executed, and if `expression` is FALSE, `statement` will not be executed.
3. A compound statement consists of more than one statement and is enclosed by braces `{}`.
4. Yes, an `if` statement may call a function. This is no different from the use of a compound statement as part of the `if` statement.

Section 3.3

1. A closed branch is a branch that forces the program to take one of two alternatives.
2. The difference between the `if` and the `if...else` is that the `if` represents an open branch whereas the `if...else` represents a closed branch.
3. Yes, compound statements may be used with the `if...else`. The requirement is that the statements must be enclosed in brackets `{}`.
4. Function calls may be used with the `if...else`. This is done by placing the function call for each statement in the `if...else`.
5. The `if...else if...else` statement in C will essentially give an option among three choices. The first two choices will depend on the condition of the first two expressions. If both expressions are FALSE, the third condition (the one following the last `else`) will be executed.

Section 3.4

1. A bitwise complement in C means that the binary equivalent of the value to be complemented will have each of its 1s converted to a 0 and each of its 0s converted to a 1. The resulting binary number will then be converted back to the base of the original value.

2. The meaning of the bitwise AND operation is that the binary equivalents of the two values to be bitwise ANDed have each of their bit pairs ANDed. The resulting binary value is then converted back to the base of the original value.
3. The meaning of the bitwise OR operation is that the binary equivalents of the two values to be bitwise ORed have each of their bit pairs ORed. The resulting binary value is then converted back to the base of the original value.
4. In a bitwise XOR operation, the binary equivalents of the two values to be bitwise XORed have each of their bit pairs XORed. The resulting binary value is then converted back to the base of the original value.
5. A bitwise shift in C means converting the value to be shifted to its binary equivalent and then shifting the resulting binary number left or right the required number of bits. The new binary value is then converted back to the base of the original value.

Section 3.5

1. The logical AND operator compares the TRUE/FALSE conditions of two expressions. If both are TRUE, the final result is TRUE; otherwise, the result is FALSE. The `&&` symbol is used for the logical AND in C.
2. The logical OR operator compares the TRUE/FALSE conditions of two expressions. If both are FALSE, the final result is FALSE; otherwise, the result is TRUE. The || symbol is used for the logical OR in C.
3. The operation of the logical NOT causes the opposite to take place. Thus, NOT TRUE is FALSE, and NOT FALSE is TRUE.
4. An example of combining a relational and a logical operation is `(3 == 3)&&(5 < 10)`.
5. C evaluates the AND operation by first evaluating the expression on the left. If this is FALSE, the expression on the right is not evaluated. C evaluates the OR by first evaluating the expression on the left as before. However, this time, if the expression is TRUE, the expression on the right is not evaluated.

Section 3.6

1. Mixing data types means performing operations on data of different types.
2. The type `int` is upgraded to `float`, and the resulting addition is a type `float`.
3. The rule for mixed data types in C is that the data types are upgraded according to the highest-ranking type in the expression.
4. A cast in C forces a change in data type.
5. An lvalue expression is an expression that refers to a memory location.

Section 3.7

1. The purpose of the `switch` statement in C is to allow a selection from several different options.
2. The other keywords that must be used with the `switch` statement are `case` and `break`. An optional keyword `default` may also be used.
3. The purpose of the keyword `default` in the `switch` statement is to activate a statement if there was no match in any of the `switch` selectors.
4. Function calls may be made from a `switch` statement. All that is necessary is that the function definition be available to the program.

Section 3.8

1. Omission of the `break` in a C `switch` causes execution of the statement following the selected `case` label.
2. To make sure that the `default` is always executed, omit all `break` statements in the C `switch`.

3. The conditional operator consists of three expressions. If the first expression is TRUE, evaluation of the second expression takes place; otherwise, evaluation of the third expression takes place.
4. Any value other than zero makes the conditional operator TRUE. A value of zero makes the conditional operator FALSE.

Section 3.9

1. Negative 1 is interpreted as true (since it is not zero), so `eval(0,-1)` is the same as `eval(0,1)`.
2. To convert numbers as large as 65535 in Program 3.26, `n` and `m` must each be defined as an `unsigned int`.
3. A function call is recursive if it calls itself or calls another function that calls itself.
4.
```
if (xyz & 1)
      printf("even\n");
else
      printf("odd\n");
```

Section 3.10

1. The statements will compile and execute. When executed, the `a == 5` statement is evaluated and the true or false result thrown away because the programmer used `==` instead of `=`. The `printf()` statement displays an unknown quantity for variable `a`, which has not been initialized.
2. Yes, neither expression ever results in zero.
3. A `break` statement is used to prevent more than one case statement from responding to a `switch`. It is important to use `break` so that the intended operation of the `switch` is performed.
4. If you do not use braces, you can only use one statement for the true action and one statement for the false action.

Section 3.11

1. The first step in the development of any program is to state the purpose of the program in writing.
2. The purpose of a troubleshooting flow chart is to aid the technician in servicing a particular system.
3. A program stub is a part of the program that is intentionally left incomplete with just enough coding to ensure that the program flow will activate it at the correct time.
4. For the program developed in the case study, the user input represents the actual measurements and observations of the hypothetical robot.

Answers to Self-Test—Chapter 3

1. There are six functions in the program: `main()`, `explain_program()`, `arm()`, `power_unit()`, `light_check()`, and `arm_drive_disconnect()`.
2. No, there are no open branches.
3. Yes, a closed branch is contained in the function `arm_drive_disconnect()`.
4. The function that uses the C `switch` is `light_check()`.
5. The meaning of

```
measurement = (measurement>30)?30 : measurement;
```

   is that if the value of measurement is greater than 30, it will be set equal to 30. Otherwise, its value will not be changed.
6. There are two variable identifiers: `measurement` and `light_status`.
7. There is only one function that returns a value: it is `arm()`.

8. The value returned is the voltage measurement inputted by the program user.
9. The variable `light_status` is not of type `float` because a C `switch` cannot use a type `float`.

Answers to Odd End-of-Chapter Problems—Chapter 3

Section 3.1

1. The six relational operators presented in this chapter, and their meanings, are:

   `>` Greater than
   `>=` Greater than or equal to
   `<` Less than
   `<=` Less than or equal to
   `==` Equal to
   `!=` Not equal to

3. The = sign in C means assignment; it does not mean equal to.

Section 3.2

5. The concept of an open branch is that the program always goes forward to new information and that an option that may or may not be taken exists.
7. A compound statement in C consists of two or more statements enclosed by opening and closing braces `{}`.

Section 3.3

9. In a closed branch, program flow always goes forward, and there are two options, one of which must be taken.
11. A compound `if...else` in C uses the `if...else if...else` statement and makes a selection from more than two alternatives.

Section 3.4

13. A. 3_{16} B. 1_{16} C. AF_{16}
15. A. 7_{16} B. F_{16} C. F_{16}

Section 3.5

17. A logical operation is any operation that will produce one of two values, usually designated by TRUE and FALSE.
19. The logical OR operation in C compares two statements. If both are FALSE, then the result of the OR operation will be FALSE. For any other combination of the two statements, the result will be TRUE. The symbol for the logical OR operation in C is ||.

Section 3.6

21. Yes, it is legal in C to add a type `int` to a type `char`. This is called *mixing data types*.
23. A cast in C forces a change in data type.

Section 3.7

25. The C `switch` is used when there is a selection from several different options.
27. The keyword `break` in a C `switch` is used to terminate the selected statement.

Section 3.8

29. The purpose of the keyword `default` in a C `switch` is to identify the part of the C `switch` that will be executed if no match is made with any of the `case` statements.

Section 3.9

31. Yes, recursion can be used to generate a truth table. The recursive function must adjust the Boolean variables at each level of recursion.
33. Yes, recursion can be used to display all the factors of a number. At each level of recursion, a counter may be advanced and used as a divisor.

Section 3.10

35. Consider this example:

```
100 &  001 = 0
100 && 001 = 1
```

The same input patterns produce two different outputs. Thus, these operations are different.

Section 3.11

37. The property of the C language that allows the information in a troubleshooting flow chart to be developed into an interactive computer program is the decision-making statements of the language.

Answers to Section Reviews—Chapter 4

Section 4.1

1. The four major parts of the C `for` loop are
   - The value at which the loop starts.
   - The condition under which the loop is to continue.
   - The changes that are to take place for each loop.
   - The loop instructions.
2. Yes, you can have more than one statement in a C `for` loop. You must enclose the statements between opening and closing braces `{}`.
3. The meaning of `++Y` is that *Y* is to be incremented before an operation on the variable.
4. The comma operator allows you to have two sequential C statements.

Section 4.2

1. The construction of the C `while` loop is

```
while(expression)
    statement
```

2. A `while` loop will be repeated as long as `expression` is TRUE (not zero).
3. The loop condition is tested first, before statement execution.
4. A good use of a `while` loop is in programming conditions when you do not know how many times the loop is to be repeated.

Section 4.3

1. The construction of the C `do while` loop is

```
do
    statement
while(expression);
```

2. The `do while` loop will be repeated as long as `expression` is TRUE.

3. The loop condition is tested in a C `do while` loop after `statement` is executed.
4. Generally, the `while` loop is preferred over the `do while` because it is considered good programming practice to test the condition before execution rather than after.

Section 4.4

1. A nested loop is one program loop inside another.
2. The structure used with nested loops should make it clear where each loop begins and where it ends. This can be done by indenting the body of each loop and using comments to make it clear where each loop ends.
3. All three of the C loop types may be nested.
4. The C `do` loop will always be activated at least once. This may cause problems if it is used as a part of a nested loop, because every time its outer loop is activated, the C `do` will also become active at least once, no matter what the condition of its loop counter.

Section 4.5

1. A recursive function is a function that calls itself.
2. Recursive functions are implemented through the use of a run time stack that contains parameter areas for each new invocation of the recursive function.
3. One of the functions must return to break the recursive call chain.
4. If recursion is not stopped, eventually all available memory will be filled with copies of the recursive functions data and a run time error will occur.
5. The following `for()` statement performs the same job as `fact()`:

```
for (i = 1; i <= N, i++)
     fact *= i;
```

   The `fact` variable must be initialized to 1 prior to the `for()` loop.

Section 4.6

1. Yes, Fibonacci numbers can be generated without a loop by using recursion.
2. The program must initialize the `state` variable to 1 to guarantee that the user enters a number for the first input sequence. Also, the `switch` statement assumes that the system is in a known state and thus requires the `state` variable to be initialized so that it will work correctly the first time.
3. Group 1 contains a nested pair of `for()` loops, both of which repeat N times. So, N repetitions of N gives N^2. Groups 2 and 3 perform in the following way:

N	1	2	3	4	5	6	7	...
Loops	1	3	6	10	15	21	28	...

   This is a well-known mathematical progression of numbers. Each number has a value equal to $N + (N - 1) + (N - 2) +$ etc. A general equation for this progression is $N*(N + 1)/2$. Group 4 contains three nested `for()` loops, resulting in N repetitions of N repetitions of N, or N^3.
4. There are 24, as follows:

```
abc   abd   acb   acd   adb   adc
bac   bad   bca   bcd   bda   bdc
cab   cad   cba   cbd   cda   cdb
dab   dac   dba   dbc   dca   dcb
```

5. Let each of the three letters be generated by recursive calls. Provide a letter counter to stop recursion at the third letter.
6. Since each guess eliminates half the remaining numbers, it takes $\log_2 1024$, or 10 guesses maximum to reduce the range to a single number.

Section 4.7

1. A run time error is an error that will take place during program execution.
2. No, a compiler does not catch run time errors. The reason is that, by definition, a run time error is not an error in the programming syntax; it is an error in program design.
3. A debug block usually contains a visual display of some data and the ability to step through the function.
4. An auto debug function is a convenient way of activating or deactivating the debug feature.

Section 4.8

1. The first step in the design of the case study program was the same as for the design of any program: to state the problem in writing.
2. A C `for()` loop was selected in this program because the loop needed a counting loop with a definite beginning and ending and increment value.
3. The beginning and ending values of the program loop were determined by calculations done on user inputs during program execution.
4. The C `return()` had its required calculations done within its argument.

Answers to Self-Test—Chapter 4

1. There are five function prototypes used in the program.
2. The type of loop used in the program is a C `for` loop.
3. The initial loop condition is `counter=below_fr`.
4. The final loop condition is `counter<=above_fr`.
5. The change each time through the loop is `counter+=freq_change`.

Answers to Odd End-of-Chapter Problems—Chapter 4

Section 4.1

1. A compound statement is a C statement that consists of more than one statement. It is set off by being enclosed between braces `{}`.
3. The comma operator allows you to have two sequential C statements.

Section 4.2

5. The construction of the C `while` loop is

```
while (expression)
  statement
```

7. A C `while()` loop will be repeated as long as `expression` is TRUE (not zero).

Section 4.3

9. The construction of the C `do while()` loop is

```
do
   statement
while(expression);
```

11. A C `do while()` loop will be repeated as long as `expression` is TRUE (not zero).

Section 4.4

13. A nested loop is one loop contained inside another.
15. Yes, there is a problem with nesting a C `do` loop. The C `do` will always be executed at least once by the outer loop.

Section 4.5

17. The difference between direct and indirect recursion is determined by whether a function calls itself (direct) or not (indirect).
19. Recursion simulates the operation of a loop by causing the statements to be re-executed at each level of recursion.

Section 4.6

21. Program 4.16 changes states by changing the value of the `state` variable. This is done whenever the user enters the correct information.
23. Coin reservoirs could be added to Program 4.19 by adding variables to hold the number of each type of coin. The function calls must then be modified to keep track of the change activity. For example, the coin bucket for quarters would be incremented when quarters are deposited and decremented when a quarter is given as change.

Section 4.7

25. A debug function usually contains a visual display of some data and the ability to step through the function.

Answers to Section Reviews—Chapter 5

Section 5.1

1. A computer's memory may be visualized as a stack or pile of memory locations, each identified by a unique number.
2. An instruction causes some computer action. Data is the values that are acted upon.
3. An address is a number that represents a specific memory location.
4. Immediate addressing is when the data immediately follows the instruction.
5. Direct addressing is when the instruction directs the CPU where in memory the data is located.

Section 5.2

1. A computer word is a group of bits treated as a unit.
2. The word size of a C `char` type is 1 byte (8 bits).
3. The computer represents signed numbers by using the twos complement notation system.
4. The difference between a signed and an unsigned data type in C is that the MSB is not treated as a sign bit. Thus the range of values that may be represented is the same, but the magnitude of the unsigned is larger.
5. The C data type that uses the smallest amount of memory is the `char`. The largest amount of memory is used by the `long double` (which may be the same as a `double` depending on your system).

Section 5.3

1. A pointer is a data type that represents the address of another memory location.
2. A pointer is called a pointer because it can be thought of as pointing to another memory location.

3. A pointer gets the address of another memory location by having the address assigned to it with the `&` operator. As an example, to store the address of a variable `x` in a pointer `p`, you do the assignment `p = &x`.
4. To pass a value to a variable using a pointer, you must make sure that the pointer contains the address of the variable (see answer 3 above). Once this is done, a value can be passed by using the `*` immediately preceding the pointer variable: `*p = 12`. This passes the value of 12 to the variable `x`, provided the variable `p` contains the address of `x`.
5. A pointer is declared by placing a space followed by the `*` sign immediately preceding the pointer variable name.

Section 5.4

1. The two ways of passing a value from a called function to the calling function are to use the C `return()` and to use a pointer as the function formal argument.
2. The limitation of using the C `return()` to pass a value back to the calling function is that only one value can be returned.
3. More than one value can be returned to the calling function from the called function through the use of pointers in the formal argument of the called function.
4. The mechanism for passing values to the called function that uses pointers in its formal argument is the use of the addresses (`&`) of the actual variables that will receive the values from the called function.
5. Separate functions must be used in order to facilitate good program design.

Section 5.5

1. A local variable is a variable declared within a function. It is known only to that function.
2. The scope of a variable applies to those part(s) of a program in which the variable is known.
3. A global variable is declared at the beginning of the program before the function `main()`. A local variable is declared within the function that will use it.
4. No, it is not considered good programming practice to use global variables because the value of such a variable may be changed by any function within the program.
5. Values may be passed between functions by using function arguments.

Section 5.6

1. The effect of the C keyword `const` is to cause the value assigned to the data type not to be changed (intentionally or otherwise) within the program. If the program user attempts to do this, a warning message will be given at compile time.
2. An automatic variable is a variable that is local to the function in which it was declared and whose life is equal to that of the declaring function.
3. A static variable is a variable that is local to the function in which it was declared, but whose life is equal to that of the program.
4. A register variable is a variable that requests that the compiler keep this variable in one of the internal registers of the system microprocessor rather than in system memory.

Section 5.7

1. The `scanf()` function requires an address because it places its scanned data into memory. This allows the `scanf()` function to be able to write lots of different things into memory.
2. No, the value of the `pattern` variable is not changed in the `tobin()` function.
3. Yes, a pointer is necessary to access the data stored in `random`. Without a pointer, the `random` variable would display the same constant value.

Section 5.8

1. The address operator in C is the ampersand sign `&`. It returns the value of the address of the variable with which it is used.

2. The indirection operator in C is the asterisk `*`. It treats the value stored within it as a memory location for data.
3. If you forget to use the address operator, the location read or written by a pointer will not be correct.

Section 5.9

1. Global variables are used to avoid having to pass a lot of variables back and forth between functions. The alternative would be to include all variables as parameters in the function definitions.
2. `getkey()` sends a character back using `return()` and under special conditions also changes the value of the global variable `dir`.
3. No.
4. `xymove()` enforces edge boundaries by comparing the `r` and `c` values against their allowable minimum and maximum settings.
5. To use a 25 by 40 display, all of the following must be changed:

   - The initial `c` value
   - The `#define` for `vbuff`
   - The `clear()` function
   - The boundary checking statements in `xymove()`

Answers to Self-Test—Chapter 5

1. The output of each of the `printf()` functions in `main()` will be

```
the constant is 57532.
The value of memory_location_1 is 375
The contents of this_value are => b
The result is F.
```

2. There are two data values that are global for the entire program:

```
const unsigned int number_1 = 57532;
int *look_at;
```

3. There is only one data value that is global for a part of the program:

```
extern char new_value;
```

   Its scope is all of the functions following its declaration.

4. The value of 375 for the variable `memory_location_1` is received by the statement in `function_1`:

```
*look_at = 375;
```

5. The statement in `main()` that causes `memory_location_1` to get the value of 375 is

```
function_1();
```

6. The output of the last `printf` function in `main()` is *F* because the binary values to be ANDed are

$$\begin{array}{l} 1111_2 \Leftarrow 15_{10} \\ \underline{1111_2 \Leftarrow 15_{10}} \\ 1111_2 \Leftarrow F_{16} \end{array}$$

7. `function_1()` does not require pointers in its argument because it uses a pointer within its definition. `function_2()` requires the use of pointers within its parameter list because it is returning more than one value to the calling function as parameters.
8. No, `function_1()` does not need to be of type `double`. The reason is because the function itself does not return a value to the calling function.
9. `function_1()` "knows" the pointer `*look_at` because it is a global variable pointer.
10. `function_1()` causes the value of the variable `memory_location_1` to change because it is initialized to the address of `memory_location_1`:

```
look_at = &memory_location_1;
```

Answers to Odd End-of-Chapter Problems—Chapter 5

Section 5.1

1. A way of visualizing a computer's memory as suggested in this chapter is as a pile of storage locations.
3. The process by which the CPU gets an instruction from memory and then executes the instruction is called a *fetch/execute cycle.*

Section 5.2

5. Some of the most common word sizes used by computers are 8-bit, 16-bit, 32-bit, and 64-bit.
7. A. 0110_2 B. 1000_2
9. A. 1111_2 B. 0001_2 C. 0110_2
11. A. –6 B. –8 C. –100

Section 5.3

13. The indirection operator in C is the asterisk (`*`). When it precedes a data type, it will represent the data whose address is contained in the data type preceded by the indirection operator.
15. A. The value of `pointer` is the address of `data`.
    B. The value of `data` is 5.
    C. The value of `*pointer` is 5.

Section 5.4

17. Separate functions are used in a C program in order to facilitate good program design.
19. The `&` operator is used in the arguments of a called function in order to return values back to the calling function.

Section 5.5

21. A variable declared within a function whose life lasts only as long as the function is active is called a *local* or *automatic variable.*
23. A variable declared at the beginning of the program before `main()` is called a *global variable.* Its scope is the entire active program.

Section 5.6

25. The C keyword used to ensure that an assigned value cannot be changed during program execution is `const`.
27. The name of the variable class that requests the compiler to keep the variable in one of the internal registers of the microprocessor is `register`.

Section 5.7

29. A recursive function that converts a number into binary using a pointer looks like this:

```
void tobin(unsigned char *number)
{
      unsigned char rem;

      if (*number == 0)
            return;
      rem = *number & 1;
      *number /= 2;
      tobin(number);
      printf("%d ",rem);
}
```

Section 5.8

31. The address of `data` may be obtained by using the address operator with the variable: `&data`.
33. The C symbol used for equality is ==. The symbol used for assignment is =.

Section 5.9

35. `vbuff(10,50,0)` produces the address B800:06A4. The 06A4 portion is found by converting (10 * 160) + (50 * 2) = 1700 into hex.
37. A command to turn the screen upside down would begin by reading characters from the first and last row of the display and swapping them. Then characters from the 2nd row and next-to-last row are read and swapped. This process continues until rows 12 and 13 are swapped.

Answers to Section Reviews—Chapter 6

Section 6.1

1. A string is an arrangement of characters.
2. You indicate a `char` string in C by the square brackets `[]`.
3. For a string consisting of five characters, six array elements are required. The last array element will contain the NULL terminator which is required in C so that the computer knows where in memory the string ends.
4. The element number of the first character in a C string array is 0.
5. The relationship between pointers and string array elements is that a pointer may be used to access an individual character of the string array just as an individual string element may be used to access the same element.

Section 6.2

1. Twenty-seven characters are placed into memory. Twenty-six are for the alphabetic characters and one is for the `'\0'` character.
2. Only the character string `"Oh,"` would be scanned in.
3. `"\0"` represents the string of characters `'\'`, `'0'`, and `'\0'`, which has a length of 2. `'\0'` represents the single character NULL and has a length of zero.
4. Using `char string[80] = "Hello"` allows the text of the string to grow up to 80 characters, whereas `char string[] = "Hello"` provides storage only for the five characters between the quotes (plus a sixth for the NULL character).

Section 6.3

1. Using `gets()` allows the entire user input string to be displayed. Replacing `gets()` with `scanf()` would cause the program to scan only the user input string up to the first blank. If the user entered his or her entire name, only the first name would be outputted.
2. When a string is passed to a function, only the address of the first character in the string is passed. The string can be of any length, and it must be terminated with a `'\0'` character.
3. If a string did not contain a NULL character, any function accessing the string would search memory until it found one. This would artificially increase the length of the string and lead to unpredictable execution.

Section 6.4

1. The function `getchar()` waits for the user to press a key on the keyboard. The ASCII code of the key pressed is returned.
2. The C character classifications are

   alphanumeric, alphabetic, control, digit, printable, lowercase, punctuation, space, uppercase, and hexadecimal.

   Examples of several of these types are as follows:

Character(s)	*Classification(s)*
'a' to 'z'	Alphanumeric, alphabetic, lowercase, printable
'A' to 'Z'	Alphanumeric, alphabetic, uppercase, printable
'0' to '9'	Alphanumeric, digit, hexadecimal
'.'	Punctuation

3. The program will output all characters that are not punctuation or digit characters.
4. Yes, if the `isalpha()` function is used.
5. On a PC, the integer 40000 requires 16 bits of storage, with the most significant bit high. This high level is interpreted by the `printf()` function as the signed integer –25536.

Section 6.5

1. The statement `#include <string.h>` is needed to make string functions available to a C program.
2. `Strcat`, `strlen`, `strcpy`, `strstr`, `strcmp`.
3. The `strlen()` function will search memory until it finds the NULL character. If the NULL character is not in the correct location, the string length will be invalid.
4. To perform a string substitution, the input string `"abcde"` must be searched (with `strstr()`) for a substring equal to the search string `"bc"`. If found, the replacement string `"howdy"` must take the place of the search string `"bc"`. This involves clever use of the `strcpy()` function.
5. The length of the first string argument must be large enough to contain the additional characters of the string being catenated.
6. The function `strchr()` is case sensitive because it considers lowercase and uppercase characters to be different. Thus, `'a'` and `'A'` are not the same character.
7. `Strchr`, `strstr`, `strcmp`.

Section 6.6

1. A character array defined as `[7][10]` requires 70 memory locations for character storage.
2. Define a new integer variable to keep track of the number of comparisons. Increment the variable inside the second `for()` loop. Experiment with different orderings of the input strings.

3. The array is as follows after each outer-loop execution:

kristen	kenny	turner	kimberly	victoria
kenny	kristen	kimberly	turner	victoria
kenny	kimberly	kristen	turner	victoria
kenny	kimberly	kristen	turner	victoria (no changes!)

4. The bubble sort contains a nested pair of `for()` loops. Since a comparison is made during each pass of the inner loop, we get approximately N^2 comparisons.
5. The strength of rectangular arrays is that strings can be easily catenated up to a fixed length. The weakness of rectangular arrays is that some locations are wasted when strings are of unequal length. The strength of ragged arrays is efficient storage of all string elements. The weakness of ragged arrays is the inability to use some built-in string functions (such as `strcat`) without danger of overwriting other string space within the array.
6. The `""` in `names[3]` resulted from the sequence of `strcpy()` statements used to swap `"turner"` with `"kenny"` in the ragged array. Since each string in the array had a predefined length, the compiler created a pointer to the beginning of each substring in the array. These pointers were not updated during the `strcpy()` function, and thus pointed to the wrong string positions within the array. Work it out character by character on paper.

Section 6.7

1. The value 0x30 is the ASCII bias associated with the digit '0'. Subtracting this value from each ASCII digit gives the actual numerical digit value.
2. The postorder expression is: 20 2 6 4 - / 3 1 7 + * + 4 / -
3. The postfix stack is a string. Adding or removing items from the stack is accomplished through the use of the `pfsidx` index variable.
4. If a lowercase ASCII code is ANDed with the pattern 0xdf, the equivalent uppercase ASCII code is generated.
5. The `lchar` and `rchar` pointers begin at opposite ends of the input string. As long as the characters pointed to by `lchar` and `rchar` match, each pointer is advanced one position towards the center of the input string. When the pointers pass each other checking is complete.
6. A tokenizer breaks an input expression down into its basic components. For example, the statement

   ```
   int count = 0;
   ```

   contains five tokens: `int`, `count`, `=`, `0`, and `;`.
7. Using transposition encoding on the input string `"ken is here"` gives the following result:

   ```
   k e n
   i s h    =>    "kieesrnhe"
   e r e
   ```

Section 6.8

1. The number of locations reserved for a string are important because they determine how much the string can "grow."
2. `strlen()` may give an incorrect result for the length of an uninitialized string because it will have to keep looking in memory until it finds a NULL terminator. The string reported by `strlen()` could be much larger than it should be.
3. Yes, it is possible to correctly swap two elements in a ragged array. Simply pad each element with trailing blanks in the definition so that each element is the same length, as in:

   ```
   char *rag[3] = { {"start   "},
                    {"stop    "},
                    {"buckets"} };
   ```

Section 6.9

1. Text formatting is desired for a number of reasons: it gives the formatted output a professional look, it assists in readability, and it allows the user to enter text in any nonformatted way.
2. The length of the next word is determined by searching the input string for the next space character (or return character).
3. No formatting is needed if the output buffer is completely filled with text.
4. Changing the `WIDTH` value results in a formatted output whose width matches the value of `WIDTH`. This may result in additional lines of output when the value of `WIDTH` is lowered and fewer lines of output when `WIDTH` is increased in value.
5. If `WIDTH` is changed to 10, some words in the input text will exceed this value, causing unpredictable results.
6. Yes, `WIDTH` should be restricted to at least the length of the longest word in the input text. The program may need to scan the entire input text to determine what this value is, and not proceed with formatting if `WIDTH` is too small.
7. Blanks are inserted in `expand_line()` by adding one blank at a time between every two words in the output buffer.
8. A ragged array may help with expansion if each word in the output buffer is stored in its own position within the ragged array. The array might then be used to create the properly formatted output buffer by using multiple `strcat()` operations.

Answers to Self-Test—Chapter 6

1. The *X* is used in an ISBN code when the last remainder value is 10. The last `if` statement generates the *X* when this occurs.
2. Priorities are determined by examining the value of the stack pointer (`pfsidx`) and the data saved on top of the stack. If the stack is empty, no priorities are needed. Otherwise, priorities are set according to this order (highest to lowest): (, * or /, + or -.
3. When ')' is encountered, the stack is popped until a '(' is seen. All items popped off the stack are written to the output.
4. The `vowels` and `vcount` variables are declared as `static ints` so that they are automatically initialized to zero.
5. A palindrome has been completely checked when `lchar` is greater than `rchar`.
6. Checking for a token in the `doubles` array requires that we use one-character lookahead when reading tokens. This requires careful manipulation of the input character stream.
7. Once a double quote is seen, a string is extracted by reading all input characters until the second double quote is seen.
8. The encoder matrix is filled one column at a time.

Answers to Odd End-of-Chapter Problems—Chapter 6

Section 6.1

1. The index of the first character in a C string is `[0]`.
3. A C character string contains any number of ASCII character codes terminated by a NULL character.

Section 6.2

5. Three initialization techniques for the string `"Data"` are as follows:
   1. `char string[] = "Data";`
   2. `char string[4] = "Data";`
   3. `char string[] = {'D', 'a', 't', 'a'};`
7. A blank in the input string causes `scanf()` to terminate without scanning the rest of the string.

Section 6.3

9. The storage space for a string must be reserved within the calling function.

Section 6.4

11. `Getchar()` does not echo characters to the display.

Section 6.5

13. `strlen()` searches memory until it finds a NULL character, resulting in an incorrect string length.
15. Harris comes before Harrison in the phone book. If two strings of unequal length are identical in every position of the shorter string, the shorter string comes first in an alphabetical ordering.
17. `strexe()` (for string execute) might be useful. `strexe()` will execute the mathematical expression represented by the input string. For example, `strexe("2 + 3 * 4")` will result in the integer value 14. `strdel()` (for string delete) might also be useful. In this function, the search string is deleted everywhere it is found in the input string. So, `strdel("unhappy", "un")` results in the string `"happy"`.

Section 6.6

19. Braces `{}` are used to separate individual strings.
21. A ragged array is a two-dimensional array of strings of unequal length.

Section 6.7

23. `static ints` are used to initialize all vowel counts to zero.

Section 6.8

25. When using strings, remember to reserve the right amount of storage, initialize the string, and work properly with the strings' dimensions.

Section 6.9

27. A very long character string is defined by placing double quoted substrings on each successive line of your C source file. The `;` goes after the last substring.
29. When the last word has been read, the output buffer may not be empty. The buffer is output without expansion, since none is needed on the last line of output.

Answers to Section Reviews—Chapter 7

Section 7.1

1. You can let C know how many elements an array will have by placing a number equal to the number of array elements inside the array brackets: `[N]`, where N = the number of elements.
2. The index of the first element of a C array is always 0.
3. If `int value[]` is declared, then `value`, `&value`, and `&value[0]` are equal, and all contain the starting address of the array.

4. A global array is initialized; a local array is not.
5. A `char` array requires 1 byte per element, and an `int` array requires 2 bytes per element (on a PC).

Section 7.2

1. The basic idea behind array applications is to manipulate the value of the array index.
2. The programming method employed in order to get arrayed values from the program user is to use a C `for` loop and increment the array index.
3. The method used to cause a series of entered values to be displayed in the opposite order from which they were entered is to use a C `for` loop that increments the array index on the input and a C `for` loop that decrements the array index on the output display.
4. The method used to extract a minimum value from a list of entered values is to compare these values with each other by placing the first value in a variable. The variable is then compared with the value of each array element using a C `for` loop. If the array element is smaller, a switch is made.
5. The entire list of values is searched for the highest value by initializing the high value to the first element of the array and then comparing the remaining elements with the high value. Each time the current high value is lower than the new array element, the high value is changed to the element value.

Section 7.3

1. Merge sort breaks the input numbers into individual groups. Comparisons are made to order the numbers in each group. Then additional comparisons are made to merge the ordered groups together. The total number of comparisons is less than the number required by bubble sort because each number in the input list does not have to be compared against every other number.
2. Any sorting algorithm must make at least one pass through the numbers. If the numbers are already in the correct order, the program may exit after the single pass.
3. The type of comparison used to control the `switch` of elements determines the sort order.
4. (a) Bubble sort:

```
7  3  1  8   First pass
3  1  7  8   Second pass
1  3  7  8   Third pass
1  3  7  8   Fourth pass
```

   (b) Bucket sort: one pass required to fill the buckets.
   (c) Merge sort:

```
8  7        3  1   First level of recursion
7  8        1  3   Second level of recursion
1  3  7  8         Merge at first level
```

5. The bucket sort is efficient because it requires only a single pass through the numbers to sort them. It is limited by the fact that the highest number being sorted must be known in advance.
6. Three levels of recursion are required:

```
Level 1:  0  3  8  1  9              2  6  4  7  5
Level 2:  0  3     8  1  9           2  6     4  7  5
Level 3:  0  3     8     1  9        2  6     4   7  5
```

7. The two techniques use recursion in different ways to obtain a sorted list. Merge sort always divides the list into equal parts. Quick sort divides the list into two halves, one containing elements smaller than the pivot and the other containing elements larger than the pivot. The pivot value may create many additional levels of recursion when it does not divide the list into equal parts.
8. There is no difference between sorting numbers and sorting character strings, because both are represented by binary values within memory.

Section 7.4

1. A way of intializing an array is by using the braces `{}`. All elements of the array, separated by commas, are placed within the braces.
2. To declare an array with more than one dimension, use a set of square brackets `[]` for each additional dimension desired.
3. A formal array parameter that uses more than one dimension in C must have the sizes of its other dimensions stated above one dimension.
4. The method used in the programs of this section that allows arrayed variables to be added is to use a variable array index (`array[index]`) and then the += operator.
5. You must ensure that each element of the array is set to 0 when using the += operator with arrays.
6. (a) [1][3] (b) [3][1] (c) [2][4] (d) illegal

Section 7.5

1. The `rand()` function is divided by 10 using remainder arithmetic to generate a sequence of random numbers from 0 to 9.
2. Hash tables are useful because they have a very fast search time.
3. The order in which the elements are accessed in Program 7.28 is [0][1], [2][0], [1][2], [0][1] blocked, [2][2], [1][1], [0][0], [2][2] blocked, [1][0], [0][2], [2][1].

Section 7.6

1. Bounds checking involves making sure that an array index does not exceed the dimension of the array.
2. Unpredictable results occur if an illegal index is used during run time.
3. The `inbounds` definition for the `points` array is:

```
#define inbounds(r,c) ((r >= 0) && (r < 4) && (c >= 0) && (c < 7))
```

Section 7.7

1. The board is automatically initialized to 0 because it is a global variable.
2. Recursion is stopped in `solve_row()` when the row equals 8 and `solve_row()` returns without making another call. This breaks the recursive loop.
3. Backtracking is a problem-solving technique where the program has the ability to back up to a previous decision and change it.
4. `inbounds()` returns a TRUE when the specified array index is legal.
5. When `legal(3,5)` is called, the board positions in row 3 and column 5, and both diagonals from position (3,5), are examined.

Answers to Self-Test—Chapter 7

1. The switch-detection statement causes the program to terminate when no switches have been performed in a single pass. This results in fewer comparisons when the list is in partial order to begin with.
2. If `temp_value` is not used, one of the two elements being swapped will be overwritten.
3. Once the bucket array has been loaded, one pass through the array will print out the input numbers in sorted order by displaying the index number of each bucket element that has been set.
4. It is important for the initialized bucket array to contain a value that does not occur anywhere in the list of input numbers. To use positive and negative numbers in the bucket sort, two bucket arrays can be used to hold the groups of positive and negative numbers. The negative-

number bucket array is scanned from its highest index to its lowest index to output sorted negative numbers.

5. Bubble sort requires a single N-element storage block, since the numbers are sorted by exchanging them. In merge sort, each new level of recursion requires storage space for the entire array.
6. The midpoint of the array is chosen by adding the index values of each end together and dividing by 2.
7. The pivot determines how the array is divided into two sub-arrays.
8. The median element of an array is determined by sorting the array and taking the element in the middle.
9. The comparison function is defined in such a way that elements of any type can be used as inputs to `qsort()`. The type of the element is fixed at compile time. `qsort()` is then capable of sorting many different types of numbers.
10. Merge sort is more efficient. Merge sort will break the input array into sub-arrays of equal size with a minimum number of recursive calls. With `qsort()`, because the array is sorted into descending order, the pivot value does not partition the array into equal sub-arrays at any level of recursion. Thus, N levels of recursion with N – 1 comparisons at each level are required.

Answers to Odd End-of-Chapter Problems—Chapter 7

Section 7.1

1. A numeric array is a group of numbers accessed through an array variable.
3. The index of the first element in a ten-element array is `[0]`. The index of the last element is `[9]`.
5. A static array declaration initializes all array elements to zero.

Section 7.2

7. Use a loop to examine each element of the array. For each element larger than the current maximum value, replace the maximum value.

Section 7.3

9. The differences are as follows:

    Bubble sort moves the highest element to the end of the input array after each pass. It usually requires multiple passes.

    Bucket sort sets flags in a bucket array for each number in the input array. It requires a single pass.

    Merge sort breaks the input array into multiple sub-arrays of equal size. Comparisons are made only on sub-arrays of two elements. Sub-arrays are then merged with additional comparisons. Multiple levels of recursion are usually required.

    Quick sort partitions the input array into sub-arrays through use of a pivot. Sub-arrays may not be of equal size. Pivot elements end up in their correct positions in the sorted array. Multiple levels of recursion are usually required.

11. Two ordered arrays are merged by comparing their elements one by one. The lower array element gets written to the output array. The pointer to the lower array element is advanced. When the pointer moves past the end of either array, the remaining elements of the other array are copied to the output.

13. The choice of a pivot determines how the array will be partitioned at each level of recursion. To minimize the number of levels, the pivot at each level should be the median value of each sub-array.

Section 7.4

15. A pointer to the matrix is passed.

Section 7.5

17. Dynamic linking uses a FAT to store allocated sectors. Each sector entry in a chain points to the next (or last) sector.
19. An array is needed to store the numbers because the differences cannot be computed until after the average is known.

Section 7.6

21. The compiler cannot perform bounds checking when the value of the index is unknown at compile time. This is usually the case when the index value is supplied by the user.

Section 7.7

23. To add a counter variable to `legal()`, declare a `static int` within `legal()` and increment it upon entry.

Answers to Section Reviews—Chapter 8

Section 8.1

1. In C, the keyword `enum` means *enumerated.*
2. An `enum` data type is used to describe a discrete set of integer values.
3. For the code `enum numbers {first, second, third}`, the integer value of `first` is `0`.
4. The main purpose of using a C `enum` is to make your program more readable.
5. Yes, a C `enum` constant can be set to a given integer value. An example is `enum numbers {first = 15}`.

Section 8.2

1. The C `typedef` allows you to create your own name for an existing data type.
2. The resulting new name of the data type from the code `typedef char name[20]` is `name`.
3. The main purpose of using the C `typedef` is to create synonyms for existing data types.

Section 8.3

1. The information on a check from a checking account can be treated as a structure because it contains information that is a collection of different types of data that are logically related.
2. A C structure consists of a collection of C data elements of different types that are logically related.
3. The structure of a C structure as presented in this section is the keyword `struct` followed by a beginning brace `{` and ending with a closing brace followed by a semicolon `};`. Between the braces are the structure members:

```
type variable_identifier;.
```

4. A structure member variable is programmed for getting data and outputting data by using the member of operator (`.`).
5. An example of problem 4 would be

```
structure_name.member_name
```

Section 8.4

1. A structure tag is an identifier that names the structure type defined in the member declaration list of the structure.
2. An example of a structure tag would be

```
struct tag
     {
        member-declaration-list;
     }
```

3. A structure can be of a variable type. This can be done by using the C `typedef` and the structure variable identifier as the name of the type.
4. A structure member is pointed to by the use of the -> operator.
5. The three operations that are allowed with structures are
   - Assign one structure to another with the assignment (=) operator.
   - Access one member of a structure (`.` or `->`).
   - Get the structure address (using the `&` operator).

Section 8.5

1. The purpose of a C `union` is to enable one of several different data types to be stored at the same starting memory address.
2. A C `union` is declared in the same way a C structure is declared. The difference is that the keyword `union` is used in place of `struct`.
3. A C structure may be initialized by the program. However, it must be global or static.
4. An array of structures may be declared in the same manner as any array. The difference is that the array type has been declared as a structure with defined members.
5. An individual member of an array of structures is accessed by `structure_variable[N].member_variable`, where `N` is the number of the array.

Section 8.6

1. A structure can be an array. In the sense of arrays, a structure may be treated as a data type.
2. A structure can contain another structure. A defined C structure may be used as a member type within another structure.
3. Yes, it is possible for an array to be a structure type that contains an array as one of its members. Since you can have a structure as an array and a structure may contain an array as one of its members, then the combination is also possible.

Section 8.7

1. A node usually consists of two parts: a data field and a pointer field.
2. A linked-list of characters can be spread out over many different memory locations. A character string is stored in a single contiguous block of memory locations. The character string will have a fixed number of locations assigned to it (including one for `'\0'`) and thus cannot be easily extended. A linked-list of characters may be easily extended simply by adding more nodes.
3. Any type of data can be stored in a node.
4. The NULL pointer indicates the end of a linked-list (or sometimes an initialized node).
5. Run time memory allocation is accomplished through the use of the `malloc()` function.
6. Basic operations on each advanced data structure are as follows:

Linked-list:	Insert, Delete, Search
Stack:	Push, Pop
Queue:	En-queue, De-queue
Binary tree:	Add Node (tree), Delete Node (tree), Search

7. FIFO stands for First In First Out and describes the basic operation of a queue. LIFO stands for Last In First Out and describes the basic operation of a stack.

Section 8.8

1. `number_of_nodes` is incremented at the beginning of every recursive call to `count_nodes()`.
2. A stack is used to reverse a string by pushing elements from the string onto the stack and then popping them back off. Since a stack is a LIFO structure, the string comes out backwards.
3. The recursive calls are made in this order (using a depth-first search): first, the root node is examined. 7 is greater than 6, so the left child node is examined next. This node's value (5) is less than 6, so the right child of this node is examined next. This third node does equal 6, so only three recursive calls are needed.
4. Using a breadth-first search, the nodes are accessed in this order: 7, 5, 13, 3, 6.

Section 8.9

1. Pointers initialized by `malloc()` should be checked before being used because the allocation may have failed and the pointer may be set to NULL.
2. Even if only a small number of nodes are used in a program, they may still be repeatedly allocated and discarded if the program gets into a loop. Eventually the system will run out of memory and crash.
3. If the list pointer is NULL the list is empty and there is nothing to delete.
4. A 1000-element binary tree would be better suited because of its quick search time (10 or fewer comparisons), compared to a worst case of 1000 comparisons with the linked list.

Section 8.10

1. Multiple program counters are maintained when subroutines are used by the mechanism that supports recursion.
2. To add new instructions to MiniMicro, put their names in the `iset enum` and add case statements to execute them. To add new data types, the fetch statements would have to be modified to all memory reads of different sizes.
3. The program prints the average of the four numbers (23).

Answers to Self-Test—Chapter 8

1. The instruction set is defined through the use of an `enum` statement.
2. No, there is no significance. The instructions may be defined in any order.
3. The accumulator must be defined as a global variable because it must be available to recursive calls to `exec()`.
4. `exec(50)` causes MiniMicro to fetch the instruction from address 50 and execute it.
5. The `iop` variable is ignored when PRINT, RET, or STOP is executed.
6. AND, OR, and XOR must be added to `iset` and each requires its own case statement. For example, AND requires

   ```
   case AND : acc &= iop; break;
   ```

7. Use `acca` and `accb` for the two accumulator names. Add new instructions that specify accumulator A or B, as in ADDA and ADDB.
8. The JNZ instruction is executed by the statement:

   ```
   case JNZ : ip = (acc != 0) ? iop : ip; break;
   ```

   If the accumulator is zero, `ip` is unchanged. Otherwise, `ip = iop` and we get the effect of a jump to address `iop`.
9. The program displays `10 9 8 7 6 5 4 3 2 1 0` and then halts.

Answers to Odd End-of-Chapter Problems—Chapter 8

Section 8.1

1. The `enum` (enumerated) data type is used in C to describe a discrete set of integer values.
3. No, the C `enum` does not create a new data type. Its purpose is to make your source code more readable.

Section 8.2

5. No, the C `typedef` does not create a new data type. It allows you to create your own name for an existing type.
7. The resulting new name of the data type for the given code is `structure`.

Section 8.3

9. An example of a common everyday system that utilizes the concept of a C structure is a personal checkbook.
11. The C member of operator identifies the structure to which the member belongs. The symbol is the period (.).

Section 8.4

13. An example of the use of a structure tag is

```
struct tag
       {
           member-declaration-list;
       }
```

15. The three operations that are allowed with structures are assigning one structure to another, accessing one member of the structure, and getting the structure address with the `&` operator.

Section 8.5

17. A C structure and a C union are both declared in the same way, the difference being that the keyword `union` is used in place of the keyword `struct`.
19. The code represents a member (identified as `number`) of the third element (2) of an array of structures called `variable`.

Section 8.6

21. Yes, a C array of structures can contain an array as one of its members. Since you can have a structure as an array and a structure may contain an array as one of its members, then the combination is also possible.

Section 8.7

23. The sizes of the data and pointer fields typically determine the size of a node, with the data field size being the most subject to change for different node definitions.
25. Insertion at head or tail: A new node must be obtained and its data field initialized. For insertion at the head of the list, the new node's pointer field is loaded with the current pointer to the head of the list. The new node's address becomes the new pointer to the list. For insertion at the tail of the list, the pointer field of the new node is loaded with NULL and the pointer field of the last node in the list is loaded with the address of the new node.

    Deletion at head: Replace pointer to list with the address in the pointer field of the first node in the list. Deletion at tail: Search the list to find the second-to-last node. Replace its pointer field with NULL. In both cases, the deleted node should be returned to the storage pool (via the `free()` function).

27. Recursion is used to access a node in a binary tree by passing the pointer to the left child or right child to a new invocation of the function accessing the binary tree. Recursion terminates when the pointer becomes NULL.

Section 8.8

29. A stack can be used to check for palindromes by pushing the string onto the stack one character at a time and then popping and comparing the stack data with the input string. Since the string data comes off the stack in reverse order, the character coming off the stack should match the character at the current position in the input string (if the input string truly is a palindrome).

Section 8.9

31. Memory allocated by `malloc()` but not returned by `free()` is unusable. Eventually the system could run out of free memory.

Section 8.10

33. To add a STORE instruction to MiniMicro, add STORE to the `iset enum`. Then add the following case statement:

```
case STORE : mem[iop] = acc; break;
```

35. A bank of eight general purpose registers could be declared like this:

```
int regs[8];
```

A three-bit field must be assigned to each instruction that specifies a register.

Answers to Section Reviews—Chapter 9

Section 9.1

1. A legal DOS file name has the form

```
[Drive:]FileName[.EXT]
```

Where

`Drive:`	=	The optional name of the drive that contains the disk that you want to access.
`FileName`	=	The name of the file (up to eight characters).
`.EXT`	=	An optional extension to the file name (up to three characters).

Both Drive: and .EXT are optional. If Drive: is not specified, then the active drive will be used. The colon (:) following the drive letter is necessary.

2. A file pointer in C is a pointer of type `FILE`. An example would be `FILE *file_pointer;`
3. The built-in C function for opening a disk file is `fopen()`.
4. The file pointer is assigned to the DOS file name by

```
file_pointer = fopen("FILENAME.EXT", "r");
```

5. The difference between the C file opening commands for writing to and reading from a file is the character directive used as an argument in the file command: `"w"` means write, and `"r"` means read. Note that the double quotes are used.
6. When finished with an open file, you must always close it. This is accomplished with the built-in C function `fclose(file_pointer)`.

Section 9.2

1. The four possible disk file I/O conditions are

   - The disk file does not exist and you want to create it on the disk and add some information.
   - The disk file already exists and you want to get information from it.
   - The disk file already exists and you want to add more information to it while preserving the old information that is already there.
   - The disk file already exists and you want to get rid of all of the old information and add new information.

2. The different methods of reading and writing data with the C language are

   - One character at a time
   - Character strings
   - Mixed mode
   - Block (or structure)

3. The three basic C file type commands are

   - `"a"` for append to existing file
   - `"r"` for read an existing file
   - `"w"` for write to an existing file (this will create the file if it does not exist).

4. The C `fscanf()` function allows you to read a file in mixed mode.
5. The purpose of a block read or write function is to store or retrieve a block of data at a time.

Section 9.3

1. The two kinds of data streams used in I/O operations are text and binary.
2. A buffer is a temporary storage place in memory.
3. The three standard files used by a C program are standard input, standard output, and standard error.
4. The redirection operators used by DOS are the `<` and the `>`. The command `A < B` means that file A will get its data from file B rather than from standard input (the keyboard), and `A > B` means that the data from file A will be sent to file B rather than to the monitor screen.
5. A command line argument allows you to invoke commands from the DOS prompt with an executable C program.

Section 9.4

1. If a file is opened properly its file pointer will not equal NULL. This is what is looked for after an `fopen()`.
2. The `More...` message is erased by outputting backspace characters and writing blanks to the same character positions.
3. The `(argc > 1)` test is needed to ensure that a file name was supplied on the command line.
4. When 13 is added to the letter value, letters from N to Z will have values greater than 26. To perform wraparound, 26 is subtracted to get back to a 1 to 13 range.

Section 9.5

1. `fopen()` returns a NULL if it cannot open the file and `fseek()` returns 0 if successful and a non-zero error code when unsuccessful.
2. One difference between a text file and a binary file is how the carriage return and line feed characters are treated.

Section 9.6

1. The user's authorization is checked by comparing strings from `argv` with `"Parts Access 123"`.
2. If the data file `PARTS.DAT` cannot be opened, the program prints an error message and then exits.
3. The data file `PARTS.DAT` is opened and closed `in read_data()` and `enter_data()`.

Answers to Self-Test—Chapter 9

1. There are four structure members. They are `part[20]`, `quantity`, `price`, and `record_number`.
2. The function `read_menu()` is of type `char`. It returns a character back to the calling function.
3. No global variables are used in the program.
4. The purpose of the arguments in function `main()` is to get command line input from the user. This will allow authorization for putting data into the disk files. These arguments are called *command line arguments.*
5. In the C `switch` in function `main()`, the `case'E'` was placed just before the `default` so that if `authorization` were FALSE, then the message in the `default` section would be displayed.
6. No `scanf()` functions are used in the program because of the potential problems inherent with this function.
7. The program will respond to either uppercase or lowercase input from the program user in response to the menu. This is because of the `toupper()` function used on the input. This converts any character input to uppercase.
8. The member `price` must be of type `long` because the `atof()` function converts the user input to a type `long`.
9. The information in the disk file is in binary mode. This is because the `fopen()` function for entering and receiving data uses the `"b"` extension as part of its file directives.

Answers to Odd End-of-Chapter Problems—Chapter 9

Section 9.1

1. Data can be saved and reused when stored in a disk file. The user does not have to reenter the dtaa each time.
3. The file residing in the current directory of drive B: is `FILER.02`.
5. All open files must be closed (with `fclose()`) before exiting any C program.

Section 9.2

7. The three basic file command strings are

`fopen(FILENAME, "a")`	to append to `FILENAME`
`fopen(FILENAME, "r")`	to read from `FILENAME`
`fopen(FILENAME, "w")`	to write to `FILENAME`

Section 9.3

9. The two kinds of data streams used by C are the text stream and the binary stream.

11. The physical representations of each standard C file are as follows:

Standard input:	Keyboard
Standard output:	Display
Standard error:	Display

Section 9.4

13. File extensions are generated by using the `strcat()` function to add `".TXT"` or `".SCR"` to the end of the filename string.

Section 9.5

15. In a binary file `0x0d` and `0x0a` are just bytes of data. In a text file they are used to indicate the end of a line.

Section 9.6

17. The old data from a `PARTS.DAT` is lost.

Answers to Section Reviews—Chapter 10

Section 10.1

1. In order to obtain color, a color graphics card must be installed and a color monitor attached.
2. In order to change the text from 80 columns to 40 columns, an installed graphics card is required.
3. Text can have 16 different colors. All these different colors can be displayed at the same time on the screen.
4. Eight different background colors are available. The `clrscr()` command is used to bring the full screen to the desired color.
5. There are basically two choices available to indicate the desired color, the color number and the color name. For example, `textcolor(0)` and `textcolor(BLACK)` both produce the same effect.

Section 10.2

1. The two different kinds of screens available on your monitor are text and graphics.
2. Text mode has all predefined images, whereas graphics mode allows you to define your own.
3. The one piece of hardware is an installed graphics adapter. The two pieces of software are `graphics.h` and the appropriate graphics driver.
4. The number of different colors available with color graphics depends on the installed color graphics hardware in your system and its mode of operation.
5. A pixel is the smallest element possible on the graphics screen.

Section 10.3

1. It's important to know about your system's modes and colors so you don't mistakenly try to look for lines that are drawn in the same color as the graphics screen (you won't see them).
2. A color palette means the set of colors available for the current active mode.
3. There are four color palettes available for the IBM CGA color graphics card. Three colors per palette are available.
4. Yes, there are modes where color is not available. This information can be found in Table 10.3.
5. You can find out the number of modes for your system by referring to Table 10.3 or by using a program that contains the built-in Turbo C function `getgraphmode()`.

Section 10.4

1. The effect of changing a graphics mode could be the changing of the color palette and/or the number of pixels available on the graphics screen.
2. There are four line styles in Turbo C plus the ability to have two thicknesses of these four styles. They are a solid, dashed, dotted, or center line style available in either of two thicknesses.
3. The built-in Turbo C shapes presented in this section were the rectangle and the circle. The circle's line style is limited to two line thicknesses.
4. A flood fill will fill an enclosed area or the surrounding area with a given color. The difference is determined by the screen coordinates used in the flood fill function depending on whether they are within the area to be filled or outside of the area (not to be filled).

Section 10.5

1. Turbo C has a built-in function for creating bars because bar graphs are commonly used to present various kinds of technical information.
2. Using Turbo C built-in functions, a flat or three-dimensional bar may be displayed.
3. The options available for displaying a three-dimensional bar using Turbo C built-in functions are placing or omitting a top on the bar. This feature is available in case you wish to stack three-dimensional bars.
4. Using built-in Turbo C functions, you can distinguish one bar from another by using the available area fill command and filling the bars with different patterns and/or different colors.
5. Some of the options available when using text in Turbo C graphics are the text style (font), text size, and text direction (horizontal or vertical).

Section 10.6

1. Scaling is a process that uses numerical methods to ensure that all of the required data appears on the graphics screen and utilizes the full pixel capability.
2. Coordinate transformation is the process of using numerical methods to cause the origin of the coordinate system to appear at any desired place on the graphics screen.
3. The values of the scaling factors are determined by the number of horizontal and vertical pixels and the minimum and maximum values of X and Y.
4. The coordinate transformation values are determined by the horizontal and vertical scaling factors, the minimum value of X, and the maximum value of Y.

Section 10.7

1. A first-person game is a game that looks through the player's eyes when generating a virtual game world.
2. The ray casting technique begins by casting a ray in a player's field of view. Horizontal and vertical intersections with walls are compared. A strip from the closest wall is scaled (based on distance to wall) and drawn on the screen.
3. Casting every ray as fast as possible is important to the real-time aspect of the game. Too much casting time will cause flicker or even worse results.
4. The game world is represented by a rectangular array of characters. Each character corresponds to a game cell that represents a 64 by 64 section of the game world. Different characters are used to represent the walls, floor, and player.
5. Walls are represented by a 4096 byte two-dimensional array that corresponds to a 64 by 64 pixel pattern.
6. Texture mapping is the technique of transferring a scaled pattern to the display screen.
7. The trig tables are built at the beginning of the game to save time later during ray casting.

Section 10.8

1. `printf()` statements are useful when troubleshooting a graphical application because they can display the results of the operations being performed (such as coordinate calculations).

2. Output redirection is a DOS feature that allows the output of a program that would normally go to the display to be sent to an output file instead.

Answers to Self-Test—Chapter 10

1. There are two structures in the program. Their identifiers are `wave_values` and `user_input`.
2. There are no global variables in the program.
3. There are five function prototypes in the program: `explain_program()`, `press_return()`, `auto_initialization()`, `sine_wave_display()`, and `program_repeat()`.
4. The local variable is `user_input in_values`. This variable contains four members.
5. The function `program_repeat()` is of type `int` because it returns a TRUE or FALSE condition to the calling function.
6. The reason why each of the strings in function `explain_program()` is terminated with `\n\r` is because the `cprintf()` function is used. This function, unlike the `printf()` function, does not give an automatic carriage return for the `\n`.
7. The `cprintf()` function is used in the function `explain_program()` in order to display the text in color.
8. The user input variables are passed to `sine_wave_display()` by passing the name of a structure through its argument.
9. The purpose of the C `while` in the function `program_repeat()` is to ensure that the only acceptable input response by the user is a Y or N (upper or lowercase).

Answers to Odd End-of-Chapter Problems—Chapter 10

Section 10.1

1. In order to produce text in color using Turbo C, your system must contain a color monitor and an installed color graphics system and be an IBM PC or true compatible.
3. The maximum number of colors text may have is 16. Yes, they can all be displayed at the same time.

Section 10.2

5. A text screen may only display text and use text commands. A graphics screen may have individual pixels manipulated by the program user.
7. A pixel is the smallest element possible on the graphics screen.

Section 10.3

9. The built-in Turbo C function for determining the modes available on your system is

```
getmoderange(GraphDriver,LoMode,HiMode)
```

11. The purpose of the built-in Turbo C functions `getmaxx` and `getmaxy` is to get the maximum size of the current graphics screen.

Section 10.4

13. The palette on the graphics screen can be changed by using the built-in Turbo C function `setgraphmode(mode)`.
15. Two ways of creating a rectangle using Turbo C are by using the built-in Turbo C `line(X1,Y1,X2,Y2)` function to create the four sides of the rectangle or the `rectangle(X1,Y1,X2,Y2)` function.

17. There are only two line styles for creating circles in Turbo C. They are normal width and thick width and are set by the last `setlinestyle()` function.

Section 10.5

19. A Turbo C fill-style represents different methods of filling in the area of a bar in order to help distinguish one bar from another.
21. There are 12 different fill-styles available for bars in Turbo C.

Section 10.6

23. Scaling is a process that uses numerical methods to ensure that all of the required data appears on the graphics screen and utilizes the full pixel capability.
25. Coordinate transformation is used so that the full range of the resulting graph generated by the mathematical function may be viewed on the graphics screen.
27. The coordinate transformation values are determined by the horizontal and vertical scaling factors, the minimum value of X, and the maximum value of Y.

Section 10.7

29. Virtual Maze is a static game world because it has no movable objects or objects that change their appearance.
31. A 30-pixel high strip leaves (200 – 30) / 2 = 85 pixels each for floor and ceiling.
33. Each cell of the game map contains a character that describes the object in the cell's position (player, wall, or floor).

Section 10.8

35. Output redirection is helpful in a graphics application because it preserves information that may get lost on the display screen.

Section 10.9

37. Converting from degrees to radians is necessary because `sine_wave_display()` uses the `sin()` function, which requires its input angle to have units of radians.

Answers to Section Reviews—Chapter 11

Section 11.1

1. The four major groupings of the registers in the 80x86 μP are general purpose registers, pointer and index registers, segment registers, and flag register.
2. The difference between a register inside the μP and a memory location in the computer's memory is that a register can modify the bit pattern stored within it by shifting it or performing an arithmetic or logic operation on it. A memory location can only store the bit pattern, but never change it.
3. The microprocessor interacts with the computer's memory in two fundamental ways: (1) It selects a memory location and copies a bit pattern from one of its internal registers to that memory location. This is called a *write operation.* (2) It selects a memory location and copies a bit pattern from that memory location into one of its internal registers. This is called a *read operation.*
4. The registers inside the μP that can be treated as 8-bit or 16-bit registers are the general purpose registers.

5. A 16-bit μP can access up to 1Mb of memory by using the process of segmentation. This is where the address of a segment register is shifted left 4 bits and added to the address of the register containing the offset. The result is a 20-bit binary number capable of addressing a full 1Mb of memory.

Section 11.2

1. Three advantages of assembly language programs are that they are more compact, run faster, and allow more control over your computer than higher-level languages.
2. Some of the disadvantages of assembly language programs are lack of portability and ease of making programming errors.
3. A mnemonic is a short word used as a memory aid to indicate a process on a specific microprocessor.
4. When a C function is called, the first internal register placed on the stack is the instruction pointer (IP).
5. C stores locally declared variables at the top of the stack and function parameters at the bottom of the stack. (Keep in mind that the stack grows down, so the bottom has the highest memory address.)

Section 11.3

1. The advantage of confining all of your C code to a 64K segment is that you can use it on systems in which memory is limited to this amount.
2. A memory model is the method used by the C system of allocating segmented memory to an executed C program.
3. In the MEDIUM memory model, far pointers are used for code but not for data. This means that code may occupy more than one 64K segment whereas data may not. Just the opposite is true in the COMPACT memory model.
4. A far pointer requires 4 bytes and may cross 64K segments. A near pointer requires 2 bytes and is confined to a single 64K segment.

Section 11.4

1. To convert C source code to assembly language source code, you need a C compiler on your active disk along with the C file to be converted.
2. The extension normally used for an assembly language source file is `.ASM`.
3. The assembly language instruction `push bp` means to copy the contents of the base pointer register to the top of the stack.
4. `bp+4` is an address whereas `[bp+4]` is the value contained at that address.

Section 11.5

1. In Turbo C, a pseudo variable is an identifier that corresponds to one of the 80x86 family internal registers.
2. An example of the use of a pseudo variable is `_AH = 5;`
3. No, the address operator (`&`) cannot be used with a pseudo variable, because such a variable represents an internal register of the μP and not a memory location.
4. Inline assembly allows you to include assembly language programming within your C source code.
5. No, pseudo variables and inline assembly are not accepted ANSI C standards. They are available as part of the Turbo C language development system.

Section 11.6

1. The purpose of a MAKE utility is to help keep your source, object, and `.EXE` files current.
2. A MAKE utility compares the date and time stamps on each of your related files to help keep them updated.

3. The use of the Turbo C TOUCH utility is to change the date and time stamp for one or more files.
4. A library utility is a separate program that allows you to create your own libraries that can then be accessed by your C programs.
5. The name of the Turbo C utility that will search files for a given string is `GREP.EXE`.

Section 11.7

1. BIOS consists of programs that are permanently resident in the computer's permanent memory, whereas DOS consists of programs that are externally loaded into the computer's read/write memory.
2. The advantages of using BIOS routines are that they are fast and offer direct access to the computer's I/O. The disadvantage is that they are machine specific.
3. The advantages of using DOS routines are that they can make adjustments for hardware differences between manufacturers, and that C programs using them are more portable between differently manufactured computer systems. The disadvantage is that they are neither as fast nor as direct as BIOS routines.
4. A BIOS function is `biosequip()`, which returns an integer value that indicates the hardware and peripherals used by the computer system.
5. A DOS function would be `_dos_getdate(date)`, which returns the current system date.

Section 11.8

1. An interrupt is a special control signal that causes your computer to temporarily stop its main program, perform a special task (called *servicing the interrupt*), and then return back to the main program.
2. An interrupt vector contains the offset address and the segment address of the interrupt handler.
3. An interrupt handler is the program that services the interrupt.
4. An address is stored in the IBM PC as follows. The least significant byte is stored first. The offset part of an address is stored before the segment part of the address.
5. The interrupt vectors are stored starting at memory location 0, and for the next 1024 bytes.

Section 11.9

1. Commands are sent to the printer by outputting command codes to the control port.
2. The pulse with of the strobe signal is controlled by the number of iterations performed by the `for()` loop.
3. The last two statements in `strobe_and_wait_for_ack()` are used to wait for the acknowledge signal.

Answers to Self-Test—Chapter 11

1. The meaning of `#define DATA 0x378` in the program is to create a descriptive identifier for the printer data port number 378_{16}.
2. DB25 is the identifying name of a special 25-pin connector that interfaces your IBM PC to its printer.
3. The DB25 parallel printer connector has three ports.
4. The names and port numbers of the ports in the DB25 connector are DATA port #378_{16}, CONTROL port #$37A_{16}$, and STATUS port #379_{16}.
5. The DATA port transmits the printable data to the printer, the CONTROL port controls various printer functions, and the STATUS port reads the necessary conditions of the printer.
6. The meaning of the statement `(inport(STATUS) & ack)` is to get data from the port number represented by STATUS, and then do a bitwise ANDing with the number represented by `ack`. Since `ack` is defined as 0×40, this represents testing the condition of bit 6.

7. The acknowledge bit is tested in the program by the statement

```
(inport(STATUS) & ack)
```

Here, the identifier `ack` is defined as the value 0×40 which means that bit 6 is being tested.

8. The statement `outport(CONTROL, strobe_hi)` sets bit 0 of the control port high or TRUE.

Answers to Odd End-of-Chapter Problems—Chapter 11

Section 11.1

1. The general purpose registers contained in the 80x86 μP are the accumulator, base register, count register, and data register. They may be used as eight separate 8-bit registers or four separate 16-bit registers.
3. A LIFO stack refers to memory where the last bit of data entered into the stack will be the first bit of data read from the stack.
5. The process of segmentation allows 16-bit internal registers to access memory that contains more locations than could be addressed by only 16 bits. This is achieved by using the values in two separate registers where the contents of one register are shifted left 4 bits, then added to the contents of a second register. The resulting 20-bit binary number is capable of addressing up to 1Mb of internal memory.

Section 11.2

7. The short word used in assembly language programming as a memory aid is called a *mnemonic*.
9. C places locally declared variables at the top of the stack and function parameters at the bottom.

Section 11.3

11. The differences between the MEDIUM and the COMPACT memory models is that the MEDIUM memory model uses far pointers for code but not for data whereas in the COMPACT memory model, just the opposite is the case.
13. A far pointer uses 4 bytes and may cross 64K segments. A near pointer requires 2 bytes and is confined to a single 64K segment.

Section 11.4

15. Traditionally, the files given the extension `.ASM` are assembly language source code.

Section 11.5

17. A pseudo variable is an identifier that corresponds to one of the 80x86 family internal registers. A C development system that uses pseudo variables is Turbo C.
19. Including assembly language source code within your C source code is called *inline assembly*.

Section 11.6

21. A utility, as used in a C development system, is a separate program that is designed to assist you in the development and management of your C programs and associated files.
23. A library utility helps you create your own libraries that can then be accessed by your C programs.

Section 11.7

25. The term **BIOS** stands for **B**asic **I**nput **O**utput **S**ystem.
27. The advantage of using DOS routines is that they can make adjustments for hardware differences between manufacturers. The disadvantage is that they are not as fast or direct as BIOS routines.

Section 11.8

29. An interrupt vector contains the offset address and the segment address of the interrupt handler. An interrupt handler is the actual program that services the interrupt.
31. An address is stored by the 80x86 μP family as follows. The least significant byte is stored first. The offset part of an address is stored before the segment part of the address.

Section 11.9

33. The characters of the "Ready" string can be sent to the printer one by one, waiting for the acknowledge signal inbetween.

Answers to Section Reviews—Chapter 12

Section 12.1

1. A comment can be entered two ways into a C++ program: (1) in the `/*  */` form and (2) beginning with the `//`.
2. Some of the `#include` files used in C are not available or have been moved to a new `#include` file in C++.
3. `new` returns a NULL pointer when it cannot allocate the requested memory. This situation occurs when the operating system runs out of memory.
4. An overloaded operator is an operator that performs more than one function. The << and >> operators are examples of overloaded operators.

Section 12.2

1. The escape sequence in the `cout` function causes an escape from the normal interpretation of a string.
2. When used in a `cout` function, the backslash character is called the escape character.
3. Three escape sequences used in the `cout` function are: `\n` => (newline), `\t` => (tab), and `\b` => (backspace).
4. An I/O manipulator as used by the `cout` function determines the minimum number of spaces to the left of the decimal point and the maximum number of spaces to the right of the decimal point.

Section 12.3

1. The `cin` function allows your program to get user information from the keyboard.
2. The `cin` function is told what variable type to use by the way a variable is declared.
3. In most systems, the `cin` function converts a carriage return to a new line automatically.
4. Floating point values may be entered as whole numbers or by use of E notation.
5. To have values outputted to the screen in E notation, simply change the I/O manipulator in the `cout` function.

Section 12.4

1. A class in C++ is the blueprint used to create objects.
2. The difference between a class and an object is that the class is the blueprint and the instance of a class is the object itself.
3. To instantiate an object means to turn a class definition into an object.
4. Class members are the individual functions or variables defined within a class. Class members' uses vary between being totally public and totally private.

Section 12.5

1. A constructor is a function that executes during instantiation of an object. It can be used to allocate memory, initialize parameters, or perform many other functions.

2. Yes, parameters can be passed to a constructor. Parameters are passed during object instantiation.
3. A destructor is a function that if defined for an object automatically executes when an object goes out of existence.
4. A destructor executes when an object goes out of existence.

Section 12.6

1. It is useful to create multiple objects of the same class when dealing with any area in which information is categorized. All areas can be included in this context.
2. The number of object instantiations is limited only by the memory available.
3. An object pointer is the address of an object. It is useful to use pointers when objects are passed to functions.
4. Multiple objects are destructed in the reverse order of the instantiation.

Section 12.7

1. A private member is the default when members are defined in a class definition. No special actions or keywords are necessary to define it.
2. A friend function is used to allow a function outside of the class full access to all class members. Friend functions are defined by using the friend keyword inside of a class definition.
3. The purpose of information hiding is to keep object members as private as possible. This method of coding allows objects to be stand-alone programs or program functions.
4. A member's visibility is defined by the access keyword in a class definition. Private and protected members are the least visible, while public members are globally visible. All combinations are allowed.

Section 12.8

1. A base class is the class definition used to create derived classes.
2. A subclass is a class defined to include members of a base class although not defined within the subclass. It is related to a base class in a parent-type relationship.
3. All subclasses inherit all public members of the base class. Inheritance is controlled using private (default) and protected keywords.
4. A static variable definition in class definitions allows a type of global access to a variable within a class. A static memory definition is accomplished in two steps, within the class definition as static and before the class definition using scope resolution.
5. The difference between a virtual function and a pure virtual function is that a virtual function defines a base class definition for the function, while a pure virtual function does not define a base class definition for the function; special notation to the compiler is required for all pure virtual functions.

Section 12.9

1. The four possible disk file I/O conditions are

   - The disk file does not exist and you want to create it on the disk and add some information.
   - The disk file already exists and you want to get information from it.
   - The disk file already exists and you want to add more information to it while preserving the old information that is already there.
   - The disk file already exists and you want to get rid of all of the old information and add new information.

2. The different methods of reading and writing data with the C++ language are

   - One character at a time
   - Character strings
   - Mixed mode
   - Block (or structure)

3. The three basic C++ file type commands are
    - append to existing file
    - read an existing file
    - writing to an existing file (this will create the file if it does not exist)
4. The `ofstream` class is the object used by C++ to work with output files.
5. The purpose of a block read or write function is to store or retrieve a block of data at a time, such as a C++ class.

Section 12.10

1. You may create as many stack objects as you want or until the operating system runs out of memory.
2. `stackbuff` can be defined as a private member because it is only accessed by other members of the same class.
3. `showdate()` is defined as virtual because the base class contains a definition for `calendar::showdate()`. The derived class `european_calendar::showdate()` overrides the base class definition.
4. Yes, it is necessary to initialize the pure virtual function to 0. This syntax is required by the compiler. If a pure virtual function is not initialized the program will not compute.

Section 12.11

1. It is more difficult to get C++ programs to compile because there are many more ways to use incorrect syntax.
2. The rules of inheritance are shown in Table 12.3 in Section 12.8. They are important because they will prevent your program from compiling if they are applied improperly.
3. Objects should be used in a C++ program when their benefits are required (such as inheritance and the ability to instantiate multiple copies).
4. Two reasons it is more difficult to locate errors in C++ source code are the improper use of overloading and the cryptic syntax required in class definitions.

Section 12.12

1. The `evaluate()` function is pure virtual because there is no definition for `evaluate()` in the base class. Each game is determined by a totally different function.
2. The `hand` member is used in both card games to store the cards defined in the object defined for each player's hand.
3. Objects of the base class `cardgame` cannot be instantiated. The pure virtual function `evaluate()` prevents it.
4. The method used to simulate card shuffling is controlled by a built-in C++ function used to return random numbers. It is not completely random, since the programmer must check for duplicate numbers and store numbers already selected.

Answers to Self-Test—Chapter 12

1. The integer `deck` is declared as a global variable to allow access to the cards during execution of either game.
2. The `sayit()` function uses a default parameter to determine whether or not to print the suit and face value of a card while dealing.
3. `deck[30]` is the 9 of hearts.

4. The statement:

```
if (12 == hand[k] % 13)
     aces++;
```

   determines if `hand[k]` is an ace of any suite.
5. The statement of question 4 is used in `blackjack`.
6. An enumerated list could be used to explicitly represent faces and suits by declaring each card using `enum`.
7. The three cards whose numbers are 6, 36, and 46 are evaluated in `blackjack::evaluate()` to return a value of 27 for the blackjack hand.
8. An `aceydeucey::evaluate()` of the same cards as in question 7 would be 1, indicating a winning hand.
9. A third card game, such as poker, could be added to Program 12.39 by defining all functions necessary to deal and evaluate the new game.
10. No member items of the class `cardgame` can be made private without affecting execution.

Answers to Odd End-of-Chapter Problems—Chapter 12

Section 12.1

1. All C++ programs must start with `main()`.
3. The opening and closing braces (`{` and `}`) indicate the beginning and end of program instructions in C++.

Section 12.2

5. AN I/O manipulator is used to convert, format, and print the arguments in a `cout` function.
7. A `cout` function must contain at least one argument.

Section 12.3

9. The (\) character is referred to as the escape character.

Section 12.4

11. Member functions defined inside of a class definition.
13. The `public` keyword is used in a class definition to allow public access to members.

Section 12.5

15. A parameter is passed to a constructor during instantiation of an object.
17. A destructor is defined by including a destructor function in the class definition. The destructor function has the same name as the constructor with a ~ preceding the name.

Section 12.6

19. An array of objects is defined the same as any other array. The number of occurrences of the array is determined at instantiation.
21. The visibility of class members is determined by the keyword entry in a class definition and by inheritance.

Section 12.7

23. Friend functions are granted access to private members of an object by declaring friend functions within the class definition.

25. A friend function that tries to access private members without permission is not really a friend. The program will not compile.
27. All class members can be declared private. The use of friend functions is required.

Section 12.8

29. Two steps are involved in defining a static variable. First it is defined in the class definition itself. Second, it is defined using the scope resolution operator, outside and before a class definition where the variable is used.
31. The `protected` keyword defined in a base class allows for all subclasses to inherit all protected members. It is visible to the derived classes.

Section 12.9

33. All open files must be closed (with `close()`) before exiting any C++ program.
35. The three basic C++ file command strings are

`infile.open("FILENAME",ios::ate);`	to append to `FILENAME`
`infile.open("FILENAME");`	to read from `FILENAME`
`outfile.open("FILENAME");`	to write to `FILENAME`

37. The two kinds of data streams used by C++ are the text stream and the binary stream.

Section 12.10

39. The sixth planet can be added by creating a new class called `sixth` that defines a function `weight()` with the appropriate weight conversion factor. New statements are necessary in `main()` to instantiate from class `sixth` and call the `weight()` function.
41. It is not very useful to sort dates unless they are stored in the form year/month/day.

Section 12.11

43. Yes, it is more difficult to get a C++ program to compile than a C program. The syntax of C++ is much more complex.
45. Many C++ programs can be rewritten in C. However, the C++ program will lose the object functionality and other benefits that C++ provides.

Section 12.12

47. The blackjack game uses all occurrences. Acey-deucey uses three.
49. In `blackjack` multiple objects of the same class are used to create the player and dealer hand. In acey-deucey, one object is defined.

Appendix A: C Reference

C Keywords:

auto	do	goto	signed	unsigned
break	double	if	sizeof	void
case	else	int	static	volatile
char	enum	long	struct	while
const	extern	register	switch	
continue	float	return	typedef	
default	for	short	union	

C Statements

In C, a statement consist of keywords, expressions, and other statements. Statements are used to control the flow of program execution. These include statements to execute loops, transfer control (decision making), and select other statements. What follows is a summary of C statements, listed in alphabetical order.

Assignment =

Assigns value of expression on right to variable on left.

Example:

resistance = 12;

Compound

A statement inside the body of another statement. The example of the `break`, below, illustrates a compound statement.

`break`

Causes the innermost `do`, `for`, `switch`, or `while` statements to end.

Example:

```
while(count > 0)
  {
       if(count > 5) break;
  }      /* Loop terminates when count > 5. */
```

`continue`

Used with the `do`, `for`, and `while` statements. Causes control to pass to the next iteration of these statements.

Example:

```
while(count > 0)
  {
       if(count == 5) continue;
  }      /* Loop skipped for count equal to 5. */
```

`do statement while(expression)`

Body of the `do` statement is executed one or more times until the `while` expression is FALSE (zero), and then control passes to the next statement.

Example:

```
do
 {
      value = value + 1;
 }
while (value < 5);
/* Loop continues as long as value < 5. */
```

`for` loop

```
for(expression1 ; expression2 ; expression3)
  <statements>
```

Process is as follows:

`expression`$_1$ is first evaluated (done only once).

`expression`$_2$ is evaluated and if TRUE, `statements` are executed.

`expression`$_3$ is evaluated each pass through the loop.

Example:

```
for(count = 1 count < 5; count++)
  {
      printf("The count is %d.\n",count);
  }
   /* Loop continues while count < 5. */
```

`if` statement

```
if(expression)
  statement1
else
  statement2
```

If `expression` is TRUE (nonzero), then `statement`$_1$ is evaluated; otherwise, `statement`$_2$ is evaluated.

Example:

```
if(value != 5)
   printf("Value is not equal to 5.");
else
   printf("Value is equal to 5.");
/* Executes first printf() if value is not equal to 5,
   otherwise second printf() function is executed. */
```

`return` statement

Causes termination of the execution of the function in which it is activated. Program flow returns to the calling function.

Example:

```
return(value);
/* Returns the value of value to the calling function */
```

`switch` statement

Causes transfer of control to a statement within the body of the `switch`.

Example:

```
switch (value)
 {
   case 1 : {
              resistor = 10;
              printf("Resistor is 10 ohms.")
              }
              break;

   case 2 : printf("No value assigned.")
             break;

   default  : printf("Invalid selection.");
   }
   /* Value of value determines one of the three selections. */
```

`while` loop

```
while(expression)
     statement
```

Body of `statement` is executed zero or more times until `expression` is FALSE (zero).

Example:

```
while(value < 5)
  {
     value++;
     printf("Value is %d.\n",value);
  }
  /* Loop continues while value <5. */
```

Appendix B: ANSI C Standard Math Functions

This appendix lists the ANSI C library routines for mathematical functions. They are listed alphabetically in the following table.

Function	Include File and Description	Example

`abs()`	Returns the absolute value of the integer argument n.
	`#include <stdlib.h>`
	`int abs(int n);`
		v = abs(–3);
		returns a value of
		v = 3

`acos()`	Computes the arc cosine of the argument whose value is between 1 and –1.
	`#include <math.h>`
	`double acos (double x);`
		angle = acos(0.5);
		returns a value of
		angle = pi/3

`asin()`	Computes the arc sine of the argument whose value is between 1 and –1.
	`#include <math.h>`
	`double asin(double x);`
		angle = asin(0.707);
		returns a value which is close to:
		angle = pi/4

`atan()`	Computes the arc tangent of the argument whose value is between $-\pi/2$ and $\pi/2$ radians.
	`#include <math.h>`
	`double atan(double x);`
		angle = atan(1.0);
		returns a value of
		angle = pi/4

`atan2()`	Computes the arc tangent of the ratio of two arguments. This function can use the sign of the results to determine in which quadrant of a Cartesian coordinate system the angle is contained.
	`#include <math.h>`
	`double atan2(double x, double y);`
		angle = atan2(1, 2);
		returns the arc tangent whose angle is 1/2 in radians.

`ceil()`	Returns the ceiling of a double argument. This is the smallest integer value that is equal to or just exceeds the value of the argument. Useful in rounding a double value up to the next integer value.
	`#include <math.h>`
	`double ceil(double x);`
		ceiling = ceil(3.2);
		returns a value of
		ceiling = 4.0

`cos()`	Returns the cosine of the argument in radians.
	`#include <math.h>`
	`double cos(double x);`
		ang_cos = cos(0);
		returns a value of
		ang_cos = 1

`cosh()`	Returns the hyperbolic cosine of the argument. If the value is larger than the system can handle, a range error will occur.
	`#include <math.h>`
	`double cosh(double x);`
		value = cosh(x);
		returns the value of the hyperbolic cosine of x.

(Continued)

Function	Include File and Description	Example

div()	Returns the result of integer division in the form of a structure which has a resulting integer quotient and remainder. This is defined in the header file as:
	typedef struct
	{
	   int quot; /* Quotient */
	   int rem; /* Remainder */
	} div_t
	#include <stdlib.h>
	div_t div(int number, int denom);
		value = div(14, 3);
		returns a value of
		value.quote = 4
		value.rem = 2

exp()	Returns the exponential e^x (base e, where e = 2.7182818) of the argument. #include <math.h> double exp(double x);	value = exp(1); returns a value of value = 2.7182818. . .
fabs()	Returns a type double that is the absolute value of the argument. #include <math.h> double fabs(double x);	value = fabs(–3.94); returns a value of value = 3.94
floor()	Returns the largest integer value that is less than or equal to the argument. #include <math.h> double floor(double x);	value = floor(8.25); returns a value of value = 8.0
fmod()	Returns the value of the floating point remainder resulting from the value of the quotient of the arguments while ensuring that the value returned is the largest possible integer value. #include <math.h> double fmod(double x, double y);	value = fmod(x, y); returns a value: n = floor(x/y) value = n – n*y;
frexp()	Returns the mantissa m and the integer exponent n of the floating point argument. #include <math.h> double frexp(double x, int *expptr);	man = frexp(x,&exp); returns a value of man = mantissa exp = exponent.
labs()	Returns the absolute value of a long integer argument. #include <stdlib.h> long labs(long n);	value = labs(–63490L); returns a value of value = 63490
ldexp()	Returns the floating point value of a double value argument with an exponent of the base 2. #include <math.h> double ldexp(double x, int exp);	value = ldexp(x,exp); returns a value of value = x * 2^{exp}

ldiv()	Returns the result of long integer division in the form of a structure that has a resulting integer quotient and remainder. This is defined in the header file as:
	typedef struct
	{
	   long quot; /* Quotient */
	   long rem; /* Remainder */
	} ldiv_t
	#include <stdlib.h>
	ldiv_t div(long number, long demon);
		value = ldiv(63463L,63460L);
		returns a value of
		value.quot = 1
		value.rem = 3

(Continued)

Function	Include File and Description	Example

`log()`	Returns the natural logarithm of the double valued argument.
	`#include <math.h>`
	`double log(double x);`
		value = log(3);
		returns a value of
		value = 1.098612

`log10()`	Returns the logarithm base 10 of the double valued argument.
	`#include <math.h>`
	`double log10(double x);`
		value = log1(3);
		returns a value of
		value = 0.47712

`modf()`	Returns the fractional and integer parts of a floating point argument.
	`#include <math.h>`
	`double modf(double x, double *intptr);`
		value = modf(x,&int_pt);
		returns floating point value in location given by int_pt.

`pow()`	Returns the value of the arguments x and y where the relationship is x^y.
	`#include <math.h>`
	`double pow(double x, double y);`
		value = pow(2,3);
		returns: value = 8

`rand()`	Returns a pseudorandom integer between 0 and RAND_MAX as defined in the header file.
	`#include <stdlib.h>`
	`int rand(void);`
		value = rand();
		returns a random integer to value.

`sin()`	Returns the sine of a double radian valued argument.
	`#include <math.h>`
	`double sin(double x);`
		value = sin(x);
		returns the sine of x.

`sinh()`	Returns the hyperbolic sine of a double valued argument.
	`#include <math.h>`
	`double sinh(double x);`
		value = sinh(x);
		returns the hyperbolic sine of x

`sqrt()`	Returns the square root of a positive valued integer argument.
	`#include <math.h>`
	`double sqrt(double x);`
		value = sqrt(9.0);
		returns a value of
		value = 3

`srand()`	Sets the seed for the random number generator function determined by the value of the unsigned argument.
	`#include <stdlib.h>`
	`void srand(unsigned n);`
		srand(n);

`tan()`	Returns the tangent of the argument expressed in radians.
	`#include <math.h>`
	`double tan(double x);`
		value = tan(x);
		returns the tangent of x.

`tanh()`	Returns the hyperbolic tangent of a double argument.
	`#include <math.h>`
	`double tanh(double x);`
		value = tanh(x);
		returns the hyperbolic tangent of x.

Appendix C: ASCII Character Set

Character*	Code	Character	Code	Character	Code	Character	Code
NUL	0	blank	32	@	64	‘	96
SOH	1	!	33	A	65	a	97
STX	2	“	34	B	66	b	98
ETX	3	#	35	C	67	c	99
EOT	4	$	36	D	68	d	100
ENQ	5	%	37	E	69	e	101
ACK	6	&	38	F	70	f	102
BEL	7	‘	39	G	71	g	103
BS	8	(	40	H	72	h	104
HT	9	)	41	I	73	i	105
LF	10	*	42	J	74	j	106
VT	11	+	43	K	75	k	107
FF	12	,	44	L	76	l	108
CR	13	–	45	M	77	m	109
SO	14	.	46	N	78	n	110
SI	15	/	47	O	79	o	111
DLE	16	0	48	P	80	p	112
DC1	17	1	49	Q	81	q	113
DC2	18	2	50	R	82	r	114
DC3	19	3	51	S	83	s	115
DC4	20	4	52	T	84	t	116
NAK	21	5	53	U	85	u	117
SYN	22	6	54	V	86	v	118
ETB	23	7	55	W	87	w	119
CAN	24	8	56	X	88	x	120
EM	25	9	57	Y	89	y	121
SUB	26	:	58	Z	90	z	122
ESC	27	;	59	[	91	{	123
FS	28	<	60	\	92	|	124
GS	29	=	61	]	93	}	125
RS	30	>	62	↑	94	-	126
US	31	?	63	_	95	DEL	127

*These 32 characters (code numbers 0 through 31) are known as **control characters.**

Index